# PACIFIC FORTRESS

## A HISTORY OF THE SEACOAST DEFENSES OF HAWAII

## GLEN M. WILLIFORD

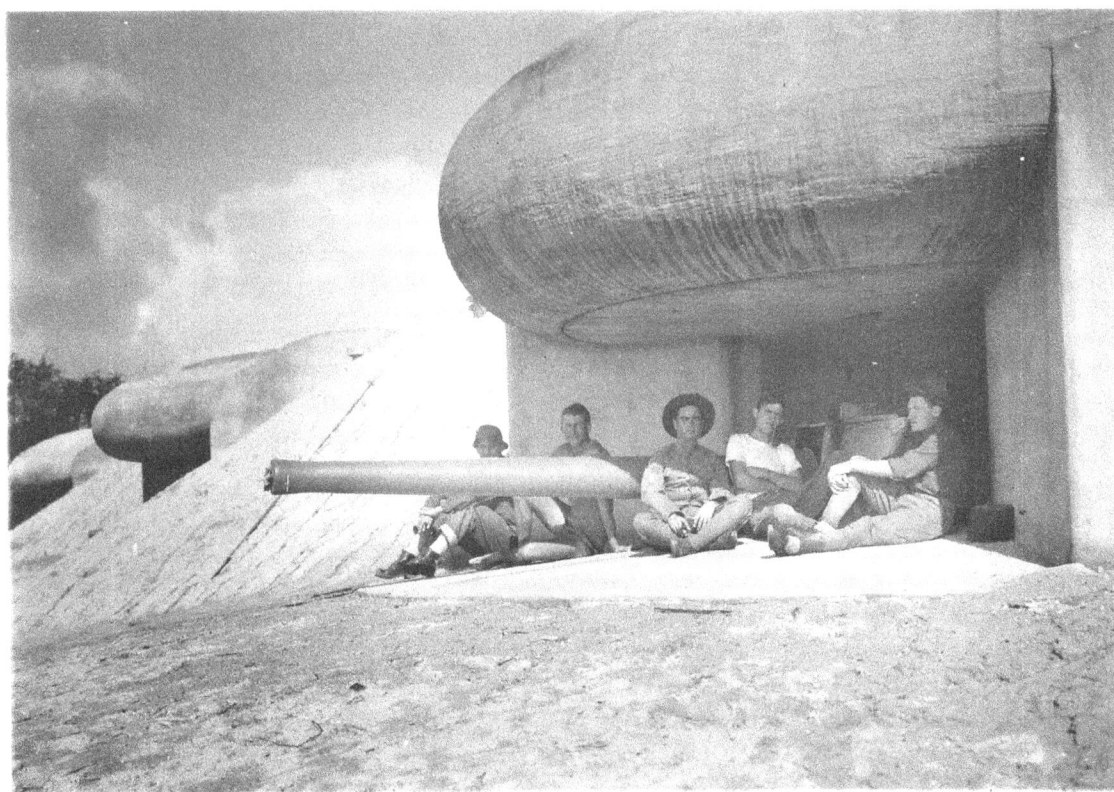

REDOUBT PRESS - MCLEAN, VIRGINIA — 2023

Library of Congress Cataloging-in-Publication Data
Library of Congress Catalog Number: 2023938850
Williford, Glen M.
Pacific Fortress: A History of the Seacoast Defenses of Hawaii
— 1st Edition
p.  cm.
includes index

ISBN 978-1-7323916-5-9
1. American seacoast defenses—history—20th century. 2. Oahu Hawaii military—history
3. World War II Hawaii attack and defense
I—title

Front Cover Images: Upper—Coat of Arms for the Coast Defenses of Honolulu (left) and Pearl Harbor (right)
Lower—One of the 8-Inch turrets being moved to its new seacoast emplacement site. *(Berhow Collection)*
Rear Cover Images: Clockwise from upper left—A view of Battery Boyd on Ford Island *(NARA)*; firing one of the
mortars at Battery Hasbrouck, Fort Kamehamea *(Schmidt Collection)*; One of the 16-inch guns of Battery Williston
at Fort Weaver *(USAMH)*; Station "U" on the Puu O Hulu ridge *(NARA)*; manning crew for Battery Jackson, Fort
Kamehameha. *(Schmidt Collection)*; insignia for the Coast Defenses of Pearl Harbor.
Frontspiece Page image: 1916 photograph of one of the 4.7-inch guns of Battery Barri. *(Schmidt Collection.)*
Symbols on Spine: Unit insignia for the 15th (upper) and 16th (lower) Coast Artillery Regiments

Cover design by Mark Berhow

**Redoubt Press** is a division of McGovern Publishing, which along with Three Sisters Press, publishes books on subjects of historical interest. Under the Redoubt Press label, we have published *The American Defences of the Panama Canal* by Terrance McGovern, *The Concrete Battleship - Fort Drum, El Fraile Island, Manila Bay,* by Francis J. Allen, *A Legacy in Brick and Stone,* by John R. Weaver II, *Pacific Rampart, A History of Corregidor and the Harbor Defenses of Manila and Subic Bays* by Glen M. Williford, and *Pacific Fortress, a History of the Seacoast Defenses of Hawaii* by Glen M. Williford. Under the Three Sisters Press label, we have published *The Chesapeake Bay at War! – The Coastal Defenses of Chesapeake Bay During World War Two* by Terrance McGovern, *Seacoast Cannon Coloring Book* by Brian B. Chin, and *The Delaware Bay at War! – The Coastal Defenses of Delaware Bay During World War Two* by Terrance McGovern. McGovern Publishing is interested in new titles, especially those dealing with fortifications, please contact Terry McGovern at 703/538-5403 or at tcmcgovern@att.net if you have a title that you are seeking to have published.

Visit www.mcgovernpublishing.com

**Redoubt Press**
A Division of McGovern Publishing
1700 Oak Lane, McLean, Virginia 22101 USA
tcmcgovern@att.net

Dedicated to:

Alex M. Holder, Jr.

A contributor, advisor, companion, and dear friend in the pursuit of discovering
the legacy of American Coast Defense.

# PREFACE

*Pacific Fortress, A History of the Seacoast Defenses of Hawaii* is an apt descriptor for the American defenses on Oahu, one of the Hawaiian Islands. For forty years, the seacoast defenses of the advanced island base of Pearl Harbor and adjacent areas were the strongest of any Pacific nation. It is a suitable title for this companion volume to *Pacific Rampart, A History of Corregidor and the Harbor Defenses of Manila and Subic Bays.* Like that previous work this book also describes the construction and service of harbor defenses—in this case those of Pearl Harbor, Honolulu, and Kaneohe Bay from their start in 1907 to the ending of the use of coast defense by the United States in 1949.

The American seacoast harbor defenses of Hawaii received little historical attention during their service or for the forty years following their demise after the Second World War. However, starting in the 1990s, several articles, many published in the Coast Defense Study Group's *Coast Defense Journal,* began to emerge that described various aspects of these defenses. Most were location-specific, describing a given fort or military reservation often with a view to explaining the remains that could be seen by a visitor. While generally accurate in their facts, they often did not describe how these sites fit into the overall defensive or historical aspects of the overall system. This work is an attempt to describe these defenses in one comprehensive volume.

This book is heavily dependent on the work of four groundbreaking, published historians. Foremost is the work of William C. Gaines. A good friend and collaborator of many years, Bill left two large unpublished manuscripts when he passed away in 2018. These plus several fort-specific articles in the *Coast Defense Journal,* were heavily used as the source for many of the details included in *Pacific Fortress.* Certain portions of his "Red-Book" were edited for size and content and inserted into this work as was appropriate. Bill had a real affection for the history and assignment accounting of coast artillery individual units; while not helpful for readability, his research and knowledge of the change in units needed to be preserved in an accessible resource.

A second valuable resource was the collection of articles by Oahu resident John D. Bennett. John also extensively published in the *Coast Defense Journal* in the 1990s-2010s. Many of his articles were on individual sites or types of emplacements on Oahu and, in addition to the usual historical background, included information and photographs of surviving remains. While the latter is not a subject covered in this historical treatment, readers are referred to John's works about what remains today.

A third historian was William H. Dorrance. Bill was also a Hawaiian resident and interested in both the military and railroad history of Oahu. He published a book on Fort Kamehameha in 1993 and several other coast defense articles in the *Coast Defense Journal.* These, and some of the research notes he shared, were used for this work.

Finally, a very specific piece of research needs to be mentioned. E. R. (Ray) Lewis and David P. Kirchner's article on the Oahu Turrets, published in full form in *Warship International* in 1993, is an excellent piece on the generation of ex-naval turrets emplaced in Oahu during the Second World War. The article and its referenced sources were important resources used in this book.

A work like this can be organized in several ways. The history of the Hawaiian defenses could be approached by time (chronologically), location, or military unit. I have chosen a hybrid approach, following mostly a chronological development of the systems of defenses deployed. Over time the Oahu seacoast defenses can be described as a series of initiatives or generations of works of a particular type, layered onto the island every few years. For many readers the details of just what geographical locations were built or used will be of little interest. Lists of these sites with brief informational descriptions are included in referenced appendices, rather than interrupting the main body of the text.

Some comments about what is not covered here. This is a book about the seacoast defenses on the island of Oahu. That includes both the harbor defense and anti-aircraft branches of the Coast Artillery Corps. It is the story of the American period for which the Coast Artillery Corps provided defense, not the preceding Hawaiian period or postwar period antiaircraft and missile defenses. The full defense of Oahu also called for beach defense, an active garrison field army, and, later, for military aircraft. Those subjects are not covered in any detail here and could justifiably be covered in their own books. The Coast Artillery in Oahu represented only a small portion of the army's assets on the island. The attack on Pearl Harbor in 1941 has had a large amount of historical scrutiny, written about probably as much as any single American battle. This is not a new analysis or even a complete coverage of that event, though it does present the limited impact on and by the Coast Artillery during the raid.

Attention should be made of the terminology referring to these types of defenses. Most of the various command functions of these defenses over time refer to them as "Hawaiian", even though almost entirely concentrated on the single island of Oahu. Also in general terms, "seacoast artillery", "coast artillery", and "harbor defenses" are used interchangeably. Do note however that when used as titles for specific military organizations, American coast artillery in its later stages often included both seacoast and anti-aircraft based branches, and "harbor defense" was an often used term for defenses specific to a geographical harbor or region.

Comments should be made about the spelling treatment of Hawaiian place names. Today efforts are being made to use the proper native spelling of Hawaiian words, including special writing marks like macrons and glottal stops. Thus, the big island of Hawaii is considered correctly spelled as Hawai'i. Many locations are preferably expressed as several separate words rather than the previous consolidated or run-together phrases. However, this spelling treatment was not used in the period of this historical study and is not used in the quoted or cited works of the time. With recognition and respect for the preservation of the indigenous language treatment, to present a seamless description of what are at time unfamiliar names, the older name versions will be used here.

The undertaking of a project of this magnitude would be almost impracticable without the assistance afforded me for photographic collections by my fellow members of the Coast Defense Study Group: Bolling Smith, Karl Schmidt, Terry McGovern and John Bennett. The U.S. Army Museum of Hawaii made available their collection of photographs and library resources freely and should be thanked.

A work of this size and complexity about a relatively obscure branch of military technology means only a small, select audience will appreciate it. Only the proactive assistance of Redoubt Press managed by Terry McGovern and Mark Berhow have made this publication possible; they are as much responsible for its ultimate appearance as anyone.

Glen Williford
Zionsville, Indiana
2023

# CONTENTS

# APPENDICES

# ENDMATTER

# ILLUSTRATION SOURCES

| | |
|---|---|
| Bennett Collection | John D. Bennett personal photograph collection |
| Berhow Collection | Mark A. Berhow personal photograph collection |
| CDJ | The Coast Defense Journal, published by the Coast Defense Study Group |
| Gaines Collection | William Gaines personal photograph collection |
| Guidry | Site sketches made by Lee Guidry |
| Historic Hawaii | Historic Hawaii Foundation, Honolulu |
| Library of Congress | Library of Congress, Prints & Photographs Division |
| McCarthy Collection, CSA | William M McCarthy Collection, California State Archives |
| McGovern Collection | Terrence C, McGovern personal photograph collection |
| NARA | National Archives and Records Administration, Archives II, Still Picture Branch. |
| NHC | Naval History and Heritage Command, Washington D.C. |
| NPS | National Park Service, Gateway National Recreation Area, NJ |
| Olbrych Collection | Soldier photograph collection owned by Glen Williford |
| *Pacific Ocean Engineers* | Book by Erwin N. Thompson. Honolulu, no date. |
| Rowbottom | Ordnance diagrams made by Dan Rowbottom |
| Schmidt Collection | Karl Schmidt personal postcard and photograph collection |
| Scrapbook | Hawaiian Antiaircraft Artillery Command, Scrapbook, Fort Shafter, HI |
| Smith Collection | B.W. Smith personal photograph collection |
| Army Technical Manual | U.S. Army Technical Manuals (TM) |
| USAMH | U.S. Army Museum of Hawaii, Fort DeRussy, Oahu |
| USAMHI | United States Army Military History Institute, Carlisle Barracks, PA |
| Welch Collection | Colonel Shawn Welch personal photograph collection |
| Williford Collection | Glen M. Williford personal photograph and map collection |

Kahuku Pt.

Pupukea

Laie

Kawailoa

Haleiwa

Kaena Pt.　Kawaihapai

Mokuleia

Punaluu
Kahana

Waialua

Kaaawa

Makua

Pvt. Property

SCHOFIELD　BARRACKS

Kualoa
Ka Lae O Kaoio Pt.

Keaau
Makaha

WAIANAE BAY

Kahaluu

KUWAAHOE

Mokapu Pt.

WAIANAE BAY
(BEACH SITE)

Kaneohe

Kailua

LUALUALEI

Puu O Hulu
Maili

IKEAAHALA

NANAKULI

Waipahu

WAIMANALO

FORD'S IS

AIEA

Honouliuli

SALT LAKE RED HILL

Ewa

FORT SHAFTER

Puuloa

PUNCH BOWL
SUGAR LOAF

Kahe Pt.

FORT WEAVER

ROUND TOP

L H

L.H. Oneula

FORT
KAMEHAMEHA

L H

Makapuu Pt.

Barber's Pt.

HONOLULU

FORT RUGER

ENTRANCE CHANNEL
TO PEARL HARBOR
NAVAL STATION

FORT DERUSSY

Koko Head

FORT ARMSTRONG RESERVATION
FOR SUBMARINE DEFENSE ON
KAAKAUKUKUI REEF

L H

Black Pt.

Diamond Head

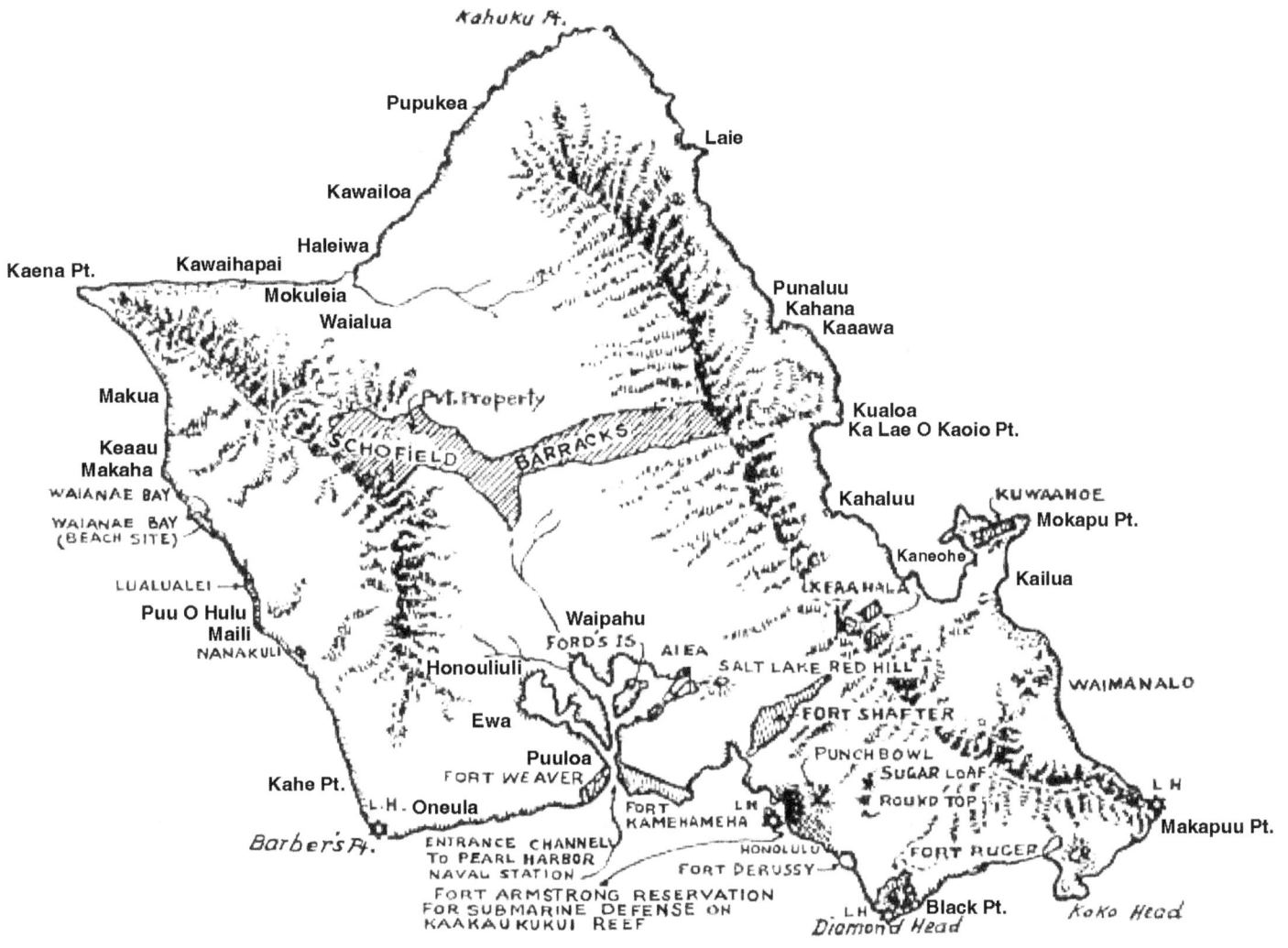

TERRITORY OF HAWAII

EDITION OF JUNE,7,1921.
REVISIONS JUNE 9,1922

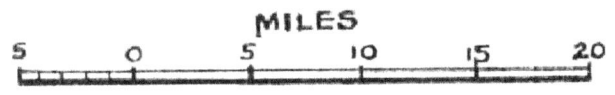

OAHU

MILES

5　0　5　10　15　20

Revised June 9, 1922 map of Oahu with key army military sites and additional location names denoted (revision by authors, 2023).

# CHAPTER 1
## THE HAWAIIAN DEFENSIVE HISTORY

The general geography for the Hawaiian Island chain is widely known, but it is probably useful to review some of the factors that complicated the defensive plans. The chain itself (stretching from the big island of Hawaii to the Midway atoll) is extensive—some 1700 miles in length—but isolated. The islands were formed from volcanic activity over millions of years in what otherwise is one of the deeper regions of the Pacific Ocean. Surrounding waters often cover an ocean depth reaching 16,000-feet. Distance from the major commercial island, Oahu, is more than 2000 miles from the California coast. There are no other islands or harbors between the American continent and these isles.

Oahu itself was formed from two massive, adjacent shield volcanoes. Over time the outer edges of the volcanic craters have eroded away, leaving two roughly parallel mountain chains with an intervening flat valley. Oahu is roughly diamond shaped, measuring 44 miles north to south, and 30 miles west to east. Parallel to the eastern shore is the Koolau mountain range, rising to 3100-feet heights. This is the windward side of the island, with more rainfall and lusher vegetation. In the west is the Waianae range, taller at 4000-feet and somewhat shorter in length. This defines the leeward side, which is drier. These are imposing ranges of rough volcanic terrain. Few roads, even today, cross the ranges west to east. Few passes existed a century ago, the only major roads were the Pali and Likelike Highways that ran from Honolulu to Kaneohe Bay. The western, northern, and eastern coasts were often rocky or with bluffs, just a few small beaches and almost no harbors. Kaneohe Bay on the southern windward side was the remnant of the center cinder cone of the eastern volcano but was shallow and filled with debris and coral.

The southern coast of Oahu is more interesting from a commercial and military standpoint. It embraces several smaller volcanic craters (Punchbowl, Diamond Head, Koko Head). There are low-lying beaches or extended flat land behind much of the central and western coastline, including the area what would eventually evolve into the vibrant city and port of Honolulu. To the northwest of Honolulu was an extensive inlet of several bays that defined Pearl Harbor. Initially these bays were shallow with a substantial coral bar that hindered entrance. This bar would eventually be dredged out and removed to allow the passage of the larger naval vessels. Between the mountain ranges sat the central plain. With its rich volcanic soil, it was the site of most of the island's important agricultural production, and a major conduit for north-south transportation.

The Islands of Hawaii were relatively neglected during the early rush to acquire and colonize habitable islands in the Pacific. At least early on they were not actively secured by one of the European political entities. A rather confused, and non-official effort by the Russian American Company to colonize the islands in 1815 ended in failure. But a partially built leftover Russian storehouse and fort, authorized for modification by King Kamehameha I in a location now near Fort and Queen Street in Honolulu, became the most sophisticated pre-American fortification in the islands. It was developed by the locals into a typical earthwork of ramparts around an enclosed enceinte with smoothbore weapons around the walls. It was named Fort Kekuanohu, though also mentioned names are Fort Kamehameha I, Fort Ka'ahumanu, or Kapapu Barracks.

This new fort (papu in Hawaiian) had a seaward face built of blocks of cut coral and faced or covered with local adobe-type clay. It had a substantial trace, estimated to have perimeter walls of about 300 to 340-feet length, with a height of 12-feet and thickness of 20-feet. Inside were a guardhouse with prison cells, powder magazines, garrison barracks and at least for a while a government or governor's residence. Not unexpectedly in a nation without a metal industry, the guns were an assortment of many sizes from many sources. Reports in 1830 mention 40 guns, in 1838 there were 52 guns. In 1846 63 guns were noted with a garrison of 286 men and a reported maximum of 70 guns reported in 1849. The fort was used and looted in a dispute with French forces in 1849.

The fort was certainly built and armed with a view of projecting defense over entry to the Honolulu harbor, but like many other locations it became a source of government power and authority unto itself. It had served as a courthouse location for several years, then as a jail and police headquarters. It did not last long into the latter part of the century—the armament was sold off in 1853, and the walls and other structures demolished in 1857. The major reason for its discontinuance could well have been the completion of a new prison at Iwilei. Some of the coral blocks found future employment as part of the harbor breakwater. About the only thing it really left behind is being the source of the name "Fort Street" that runs immediately adjacent to the original site. (1)

Only one major pre-American fort existed on Oahu. Fort Kekuanohu was begun by Russian traders and completed by local Hawaiians about 1815. *Historic Hawaii.*

A secondary work was erected on the King's property at Puowaina (today's Punchbowl) also in the early 1800s. *Gaines Collection.*

A secondary work was constructed at what is now called the Punchbowl Crater. This volcanic tuff cone rises behind downtown Honolulu and was originally called Puowaina. During the Hawaiian monarchy the south rim of the crater was the position of a small gun battery. This location seems always to have been part of the king's or crown property. Apparently, it was just a level platform for guns with an earthen wall protecting at least the southern or front approach. Armament was probably no more than two to six guns. In 1826 it was reported to have been armed with two 32-pdr guns, and by 1887 it is mentioned as having room for six guns, but not currently being manned or used. Apparently just used as a saluting battery in the late 1840s, it was disarmed about 1853. All the crown lands, amounting to a little under 160 acres, became U.S. federal property after the island's acquisition. The fort site stayed in federal hands and became part of the Punchbowl Military Reservation in General Orders No. 21 on January 27, 1906. (2)

The United States formally recognized Hawaiian independence in a treaty signed on December 23, 1826. Earlier in September 1820, John Coffin Jones, Jr. received an appointment as Agent for Commerce and Seamen on behalf of the United States to the Kingdom of Hawaii in the port of Honolulu. This post was converted to a consul position in July 1844. Throughout the 1820s American missionaries descended on the Hawaiian Islands. Dominant were those from Protestant sects out of New England, though followed by a later wave of Roman Catholic missionaries. Commercially American whalers frequented the waters and ports. The big commercial driver, however, was sugar cane. The soil and weather conditions were ideal for cane production, and the high labor demands could be met by Asian workers, especially from Japan. American demand for the sugar was strong on the West Coast, and much of the invested capital and management personnel developing the industry was American. By the 1870s a flourishing and profitable trade on sugar was operating between the two countries. A large and wealthy American business class began to agitate for annexation or at least closer political ties with the U.S. David L. Gregg presented his credentials as U.S. Commissioner to the Kingdom of Hawaii on December 20, 1853. Once accepted, he operated as a legation to the island nation. Subsequently officers operated as the U.S. Representative to the Kingdom and the first U.S. Envoy.

A series of treaties from 1826-84 began to cement the two political entities together. In 1826 a Treaty of Friendship, Commerce, and Navigation was signed. This was the first treaty that the Kingdom of Hawaii had signed with any foreign country. It was never ratified by Congress, although both countries acted in accordance with its articles. An updated variation of this treaty was signed in 1849. With the growth of the sugar business a new issue arose. The planters in Hawaii vociferously requested a treaty to remove the import duties on Hawaiian sugar (actually, all agricultural products). In return they would allow import of all products duty-free into Hawaii from the U.S. After some negotiation it was signed on January 1875 and renewed in 1884. (3)

American military interest in possessing or at least having access to the Hawaiian Islands, nominally at least an independent nation, was high through much of the late 19th century. Two army officers of note, Major General John M. Schofield (himself a Secretary of War 1868-1869) and Lieutenant Colonel Barton S. Alexander of the Engineers were sent by specific instruction of Secretary of War William M. Belknap to the Sandwich Islands (as Hawaii was sometimes known). Sailing on navy steamer USS *California* (ex-*Minnetonka*) in 1873 they were to scout and report on the possible location and defensibility of a projected naval port in the islands. This sounds strange today, dispatching a team to locate fortifications in a friendly but still independent nation. However, the pair were received in an open and friendly fashion, and they had frank discussions with the King and family and local leaders. In their reported opinion, the government of Hawaii might well consider exchanging rights to acquiring and building a navy base and seacoast defenses for the duty-free status on sugar they were anxious for. While that did not subsequently occur, it was the background of their report.

A full technical survey was not possible, but these two officers were professional experts, particularly the engineer Alexander. The small harbor of Honolulu was the only harbor of any current viability in the island chain. Furthermore, it was "a small harbor laying seaward from the land, and only protected from the sea by outlying coral reefs". It could not be adequately defended from the shore—being right on the edge of the sea. It would be of no use to a navy as a harbor of refuge in a war with another powerful maritime nation.

Major General John Schofield under orders of the Secretary of War conducted in 1873 a survey of the military potential and required defenses of Oahu. *Gaines Collection.*

The only site possible to meet the needs for a naval harbor was the large estuary of the Pearl River. Located about seven miles west of Honolulu it was a protected body of relatively deep water extending inland about six miles. The shores on either side were suitable for constructing shore batteries—which could be advanced sufficiently to offer defensive positions unlike positions around Honolulu. Also known as Pearl Harbor, their enthusiasm for the location was tempered only by the presence of a blocking coral and sand bar which greatly reduced the water depth for entry. They figured that the reef was 250-300 yards in width with about 12-18 feet of wat covering it at low tide. They thought the removal or cutting open of the coral reef was possible with an expense of about $250,000—as it turned out a major underestimation. Still, once opened the lochs or bays of the harbor were deep enough to accommodate any fleet, the shores amenable for necessary docks, wharves, shops of a stations, and the shoreline open for defensive batteries. Additionally, the land around the harbor was privately owned but not developed and probably available for ready (inexpensive) acquisition.

Schofield was tactful about the political nature of the acquisition. He used phrases like: "In case it should become the policy of the Government of the United States to obtain the possession of this harbor for naval purposes…it should be ceded by the Hawaiian Government…from what we could learn of the feelings of the Hawaiian Government on this subject…the cession …would probably be freely given by the Government of these Islands as a quid pro quo for a reciprocity treaty. Indeed, the sugar planters are so anxious for a reciprocity treaty…with the United States that many of them openly proclaim themselves in favor of annexation of these Island to the United States. Neither the Government nor the Native people of the Islands are, it is believed, prepared to consider the question of annexation, at the present time, even if the United States desired to propose it, but the cession of Pearl River Harbor for free trade is freely discussed and favorably considered by the government and the people." (4)

It appears that Schofield and Alexander, 30 years in advance, identified the proper location of a major Hawaiian naval base and the defenses it would require. They can be understandably excused about their naivety in how the United States might come to acquire sovereignty over the islands.

From the 1870s to the 1890s a major political struggle occurred in Hawaii between the native people with their constitutional monarchy and the mostly American businessmen and leaders in the island. The largely foreign farm labor population were mostly ignored in this struggle. In 1887 the Hawaiian League, representing the American sugar industry, forced a new constitution on the monarchy, stripping the monarch of most of his power and installing a non-native cabinet. A bloodless coup in 1893 overthrew even this government. It was supported by a U.S. Navy and Marine presence, and the U. S. counsel proclaimed Hawaii an American protectorate. This action was not endorsed by the U.S. Government in Washington. Consequently, the American businessmen proclaimed an independent Republic of Hawaii.

Continued efforts were made by the pro-American faction to achieve annexation, opposed by the native Hawaiians and their (former) Queen Liliuokalani. In June 1897 Congress began to pass legislation for annexation, but it was thwarted, or at least postponed, by the emergence of a popular referendum in the islands and presented to congress as a petition in opposition, The final deciding act, however, was the outbreak of the Spanish-American War. With the new national sense of threat from foreign powers and a desire to assert itself on the international stage, The 55th Congress approved a Joint Resolution for annexation of the Hawaiian Islands on July 7, 1898, that was signed by President William McKinley. (5)

Following annexation, the first American troops were dispatched to the islands. Just four days after finalization of the annexation, the 1st New York Volunteer Infantry and the 3rd Battalion of the 2nd US Volunteer Engineers arrived. They set up a tent camp at what was to be called Camp McKinley in Kapiolani Park, just outside Honolulu. With this beginning, Oahu was to be continuously garrisoned by the U.S. military. The engineer unit devoted itself to conducting extensive island surveying. The infantry regiment soon left, but was replaced on April 15, 1899, by four batteries of the regular 6th Artillery Regiment, commanded by Colonel Samuel Mills. Eventually two of its batteries (I and K) would become the 66th and 67th Coast Artillery Companies, the start of the coast artillery unit presence in the islands. On June 14, 1900, by congressional act, Hawaii became a United States Territory. (6)

Coast artillery units would come and go over the next ten years, in some cases it seems Oahu was used as a steppingstone for units going to the more turbulent requirements of the American garrison in the Philippines. The two original coast artillery companies moved to Manila in 1904 but were replaced by the 28th and 92nd from the U.S. mainland moving into Camp McKinley. The first permanent infantry regiment arrived in 1905.

During the early years of the U.S. acquisition of Hawaii, the small local navy dock and station was in Honolulu Harbor. *NARA*

In 1909 a cavalry regiment came, and the 1ˢᵗ Field Artillery arrived in 1910. The new post at Schofield Barracks became the home to an increasing number of army units, the eventual home for the Oahu U.S. mobile army. (7)

The primary rationale for building major defenses in Oahu was to protect the projected naval base. The Hawaiian Islands had a fully functional, if somewhat small, commercial port in Honolulu. This became the site of the first American naval station. Early on the attractiveness of Pearl Harbor as fleet anchorage and base was apparent, but it was shallow and would need extensive (and expensive) dredging before being useful. The property around the harbor and its lochs was purchased by the Navy for military purposes, but disputes over fishing rights stubbornly persisted for many more years. In these early years of the 20ᵗʰ century the small American battlefleet was kept concentrated in the Atlantic, which was always considered the ocean of the greater national interest. Even with the acquisition of Hawaii, Guam and the Philippines there was little interest for investing in naval facilities for which there was no immediate need.

Thinking began to change about 1907-1908. Japan was rapidly emerging as a major naval competitor, particularly after their success in defeating Russia in the Russo-Japanese War. A particularly nasty diplomatic dispute with the Japanese about immigration to the U.S. occurred in February 1907; there were even rumors of war. The voyage of the Great White Fleet (December 1907-March 1909) demonstrated the practicalities of moving the battlefleet from one ocean to another, but also revealed how woefully inadequate the naval facilities were on the American West Coast. In 1908 the Navy had just a single battleship drydock on the entire Pacific coast. The Joint Army-Navy Board finally recommended against building a major naval base in the Philippines, and rather depending on the facilities at Puget Sound, San Francisco and Hawaii to support any battlefleet sent to the Pacific with coal, docks, and stores. Oahu and specifically Pearl Harbor became the focus of expanded facilities for naval support in the Pacific starting in 1908 and reinforced by other reports in 1911. (8)

Dredging to allow naval warships to enter Pearl Harbor began in 1908 after major funds were appropriated by the U.S. Congress. *Gaines Collection.*

The initial significant Congressional appropriation for major harbor improvements in the Hawaiian Islands was made in an act on March 5, 1908. The Navy believed it needed a large, sufficiently deep, protected anchorage (for numerous large-draft battleships), a major coaling station, and at least one large dry dock with shops for maintenance and repairs. Over the next ten years those facilities were all developed. In December 1911 this new base began service. USS *California*, a large new armored cruiser crossed a yellow ribbon to enter Pearl Harbor through the newly deepened channel and anchored within. In 1916 the local facilities became the 14th Naval District. In August 1919 Drydock No. 1 was finally completed (after a prolonged and problem-ridden construction period), and the largest ships could be accommodated in Pearl Harbor. In 1921 the Naval Operating Base, Pearl Harbor became the formal title of the grouped navy activities. The final major development of the naval base infrastructure was the relocation of munitions storage to a new reservation at Lualualei, fifteen miles to the northwest of Pearl Harbor connected by a new road and railway to the piers of the navy base. It should also be noted that as developed Pearl Harbor was a "closed" harbor; no commercial or foreign ship access was possible except by special permission—it was reserved solely for the use of the U.S. Navy. (9)

Part of the considerable infrastructure required for a Pacific Base in Hawaii was an immense naval coaling station at Pearl Harbor, built from 1911-1918. *NARA.*

While a base with piers, docks, fuel, ammunition, and shops existed in the early years, there was not usually a large fleet or numerous ships stationed here. Temporary fleet exercises were conducted during the interwar years (major ones in 1925 and 1927). Squadrons of destroyers and submarines for local defense and escort were normally assigned to the base. The first significant fleet posting, though, was the Hawaiian Detachment of an aircraft carrier, cruisers, and destroyers in October 1939. Following the fleet maneuvers in April 1940 a political decision was made to retain indefinitely the major portion of the U.S. Fleet (shortly to be divided into the "Pacific Fleet" and the "Atlantic Fleet") with the Pacific Fleet based at the advanced position at Pearl Harbor.

In 1905 a new national assessment was made for the defense of American ports and Insular possessions. The last previous coastal defense initiative was known as the "Endicott Program", named for Secretary of War William C. Endicott. This was a ground-breaking effort that resulted in a massive expenditure over twenty years to provide forts, emplacements, weapons, and a professional Coast Artillery Corps between 1886 and 1906. When it started there were no overseas possessions of the U.S. that had enough naval or commercial value to justify fixed harbor defenses. That changed in 1898 with the almost simultaneous acquisition of Hawaii and the prizes of the Spanish American War—Puerto Rico, Guantanamo, Guam, and the Philippines. A new National Coast Defense Board (aka the Taft Board for its war secretary) was created in 1905 to update new continental needs and address defenses for new bases in what was collectively called the Insular Possessions. The report in 1906 included recommendations for Hawaii which proved to be the impetus for the congressional authorizations funding these defenses.

As published the Taft Board report held few specifics for individual defenses. The technology was updated with newer model guns, carriages, and emplacement designs. It still depended heavily on the proven disappearing carriages for heavy guns, light pedestal mounts for rapid-fire guns and groups of mortars. Submarine mines were still in use, but more attention was paid to electrification of systems and wider use of searchlights and networks of fire control stations. The report estimated $22.7 million dollars for equipping the insular possessions with defenses--$3,254,000 earmarked for Oahu. The defenses of Pearl Harbor and Honolulu were thought to require six 12-inch disappearing guns, four 6-inch guns, and sixteen 12-inch mortars, along with a full mine project and other supporting facilities. (10)

By the time of construction of the Taft generation of defenses, the processes of authorization, design and construction were well developed by the Army. Emplacement construction was supervised by the army's professional Corps of Engineers. Officers were well trained and had access to standardized emplacement plans on which to base their designs. However, they were also urged to adopt these plans to local terrain and tactical considerations. Once notified that funds were available, the local engineer would be authorized to draw up a plan for an already-approved element (battery or batteries) of the defense. This engineer would make the ultimate decision as to the precise placement and various design features, though his submission would have to be approved by Washington. In a similar fashion the army's Ordnance Corp designed and provided the weapons and the Signal Corp the more complex communications and fire control instruments.

Batteries were intended to be complete structures holding and protecting gun or mortar armament, its ammunition, and its power, maintenance and communication equipment. It was assumed that they would receive return fire from the same size gun they mounted, the guns and their ammunition magazines were massively protected by a recommended thickness of concrete and earth. In the 1900s the primary heavy guns were mounted on "disappearing carriages". This was a mechanically complicated device that raised the gun preparatory to firing and then used the recoil to lower it back behind the protective parapet for cleaning and reloading. In the early 1900s there was no need for protection from above, and the trajectory of ship guns meant that the major danger was from direct frontal impact, not plunging fire. Each emplacement had heavily protected magazines, separate ones for powder and shell, and often a set for each individual gun. Ammunition service from lower-level magazines was by electro-mechanical chain hoists. Gun crews were not expected to live in their batteries, rather they were quartered nearby in barracks and expected to man the emplacements only during training and combat. Still, they had minimal support features—such as latrines, storerooms, an emergency diesel generator for power if combat cut the underground cables from the central system, and a command and plotting room where firing solutions were calculated.

Heavy batteries (10, 12, and 14-inch guns) were expected to engage enemy capital ships. Mortar batteries were intended to fire salvos of plunging 12-inch rounds onto the same targets. Because of the arching fire by mortars, they were not required to have an actual line-of-sight view of their opponent, and they were usually emplaced in groups within a pit rather than individually mounted in their own, protected platform. Light guns (3- and 6-inch) were generally placed to protect minefields and searchlights. They would engage light enemy craft trying to sweep the mines or putting out the lights. Also, at several locations they protected possible landing beaches and could be used to help repel boats attempting to bring parties ashore. The 3-inch guns were emplaced in the lightest emplacements, on pedestals that left the guns exposed but also allowed rapid traverse to follow fast targets close offshore.

The defense of the Hawaiian Islands was always a clear mandate for the Army. This was unlike the situation with the Philippines or the Panama Canal. Those were defenses built for other strategic purposes in a further foreign location and not primarily defending a U.S. navy base. Hawaii strategically was on solid ground—it was an American territory with a large and important naval base. The Army's Coast Artillery was throughout the first half of the 20<sup>th</sup> century specifically charged with the primary defense of naval bases. The mission of the Army in Hawaii in 1919 was "to defend the Naval Base at Pearl Harbor against":

1) Damage from naval or aerial bombardment or by enemy sympathizers

2) Attack by enemy expeditionary force or forces supported or unsupported by an enemy fleet of fleets." (11)

Since its inception, the defense planning for Hawaii focused on the Japanese as the most likely antagonist. It possessed the only other significant naval fleet in the Pacific, and its government demonstrated an expansive and at time aggressive stance to territorial rights. With the adoption in the early post-WW1 period of the Joint Board's "Color" war plans, Japan was often referred to as enemy "Orange".

The pre-war planning documents for the defense of Hawaii seem to be realistic in terms of their assessment of Japanese capabilities. Except, perhaps, in one area. The perception that a real danger existed from possible sabotage by the local population proved false. There was no local organization with the desire or organization to act in this manner. While cultural ties remained strong and a certain sympathy to Japanese war aims was shared by many local Japanese-Americans, there was no corresponding dislike of America and their personal situation in Hawaii. Nor did the Japanese military attempt to organize, train, or equip the resident population to participate in sabotage activities. Legation personnel did make and report observations of military movements, but that happened around the world with most consular personnel during these times.

A report entitled Defense of the Naval Base – Oahu was made for the Chief of Staff by Brigadier General William Lassiter. It was completed on January 12, 1931 and distributed internally in both Hawaii and Washington. It provided a good oversight to the situation anticipated for a war with Japan and would not change much over the next ten years prior to the Pearl Harbor attack. The key passages about surprise element and forms of attack are instructive: (12)

SURPRISE ELEMENT

If our relations with Orange become very strained, it is to be anticipated that Orange would hold her commercial vessels in port and would clamp down a strict censorship on news. As a result we would pray to rumors of all kinds. Not knowing what was going to happen we would have to be ready for every conceivable form of attack.

It is considered that a light raiding force made up of aircraft carriers, light cruisers and destroyers might appear off our shores without notice, coincident with the declaration of war; but that the Orange grand fleet, escorting a large landing force of 100,000 men or more with their supplies and equipment could hardly arrive until we at least two weeks after we had recognized war as inevitable or after war had been declared.

FORM OF ATTACK

The forms of attack to be especially considered are:

a. Sabotage by alien sympathizers already on the Islands, involving destruction of telephone telegraph, cable and radio means of communication; bombing of military personnel in their barracks and quarters; destruction of oil tanks, ammunition and other stores; damage to drydocks, coast fortifications, etc.

b. A raiding air force, designed in conjunction with a. to destroy or greatly limit the value of Oahu as a base for our fleet. Obstructing the narrow entrance to Pearl Harbor or Honolulu Harbor, air bombing of the drydocks, oil tanks, and military and naval stores, would probably be the especial objectives of such an expedition; though men might be rushed ashore in Pearl Harbor and Honolulu Harbor from swiftly moving boats to seek definite demolition of the drydock and oil tanks.

c. Employment of the grand fleet and a large landing force to take possession of Oahu and, from it as a base, to dominate the waters of the Eastern Pacific.

Mission of the Separate Coast Artillery Brigade

a. To defend the Pearl Harbor Naval Base against air attack by the use of antiaircraft artillery

b. To prevent hostile vessels from entering the channels to Pearl Harbor and to Honolulu Harbor

c. To support the Hawaiian Division in defeating an enemy attempting to land and take Oahu

d. To keep hostile vessels outside of the range of the shore guns

e. To guard the vicinity of their fortifications and battery positions against close assault

In formulating his plans, the Brigade Commander will take the following into consideration:

a. The forms of attack most likely to be made are efforts to damage or destroy the Naval Base by air bombing, by gunfire from ships, by sinking obstructions in the channels, by run-by into the harbors, or else by landing in forms to capture Oahu. The Naval Base and the approaches thereto by air, sea, and land are the critical areas

b. Initial dispositions must be made on the basis that the attack will come at the outset of war, by surprise, under cover of darkness, and without reinforcement except such as may be obtained from local levies

c. Primary assignments of organizations to duties will have to be made so as to utilize available strength to the very best advantage to meeting what ought to be anticipated as the most probable first phase of an attack in force. However, each unit must be prepared.

Assessments had not significantly changed by 1940. A similar statement of perceived threats to the defense seem to continue to downplay the possibility of outright attack and still have the risk from sabotage as substantial: (13)

Category of Coastal Frontier Defense: Assumes the possibility, but not the probability, of a major attack. Defense measures specified envision the employment of seacoast, air, and antiaircraft elements, and the use of a general reserve.

Forms of Hostile Attacks:

1) Possible enemy attacks against the Oahu area in the order of probability are:

a. Submarine – torpedo and mine

b. Sabotage

c. Disguised merchant ship attack by blocking channels, by mines, or by air or surface craft

d. Air raids, carrier based

e. Surface ship raids

f. Major combined attack in the absence of the U.S. Fleet

2) Sabotage and Internal Dissension. The Orange population will have within it a division that is Loyal to Orange (to the extent of sabotage and other subversive actions). This group will probably be small, although formidable. Sabotage may include one or more of the following acts:

a. Destruction of electric light plants, gas works, and water supply reservoirs

b. Destruction of food supplies

c. Destruction of means of transportation, roads, and railroads

d. Arousing inhabitants to insurrection

The seacoast defenses of Oahu were an important element for the military defense of Hawaii, but it was not the only force with this assignment. From almost the beginning of the American annexation there was a mobile army present. Following the First World War, in 1921, the Hawaiian Division with a "square" four-regiment design was formed on Oahu and stationed at Schofield Barracks. Before the Second World War it absorbed the two island national guard regiments (the 198th and 199th) and was split into two triangular infantry divisions; the 24th and 25th. The mobile forces provided beach defense, manned some artillery batteries for the manpower-stretched coast artillery, and furnished garrisons for the other islands of the Hawaiian chain, and would have been primary combat force in the case of a successful invasion landing.

The army's air service (initially the Army Air Corps) provided the interceptor fighter aircraft for the islands. They worked closely with the anti-aircraft units of the Harbor Defense Command and were also the home for the air-search radar facilities deployed shortly before the war started. The Navy provided their own patrol aircraft to perform sector searches for approaching vessels, as well as conducting anti-submarine and surface craft patrols with destroyers and other light forces of the inshore patrol.

1853 Painting by Paul Emmert of the Honolulu fort. *Gaines Collection*

# CHAPTER 2
## THE TAFT FORTIFICATIONS

The first substantial appropriation for Hawaiian defenses was passed on June 26, 1906. Over the next several years annual appropriations added enough to effectively finance the construction of the first generation of defenses. Both Fort Kamehameha at Queen Emma Point and Fort Ruger at Diamond Head directly benefited from these first efforts. Work was supervised as usual by the Army's Corps of Engineers, often using local contracted labor or construction firms for much of the actual work. By this time the army's process for designing and building tactical structures was well standardized. The local army engineer would be told when funding was available for a certain project. He would then personally complete the plans and estimates for cost and submit to the Chief of Engineer's Office in Washington D.C.

The Taft Generation Forts: The Pearl Harbor and Honolulu Defenses 1908-1915. The first seacoast defense fort begun by the Americans was at Queen Emma Point, just to the east of the entry to Pearl Harbor. *Williford Collection.*

The very first military surveys done with a view of fortification for Oahu identified the benefits of placing heavy batteries at Queen Emma Point. The land in question was low-lying (much being less than ten feet above sea level), with scattered shallow ponds and covered with either luxuriant growth of rank grass or thick patches of algaroba and lantana brush. It was unoccupied and not too difficult or expensive to acquire. Much of the acreage which would later encompass the main part of Fort Kamehameha was acquired in the earliest years of the 20th century. This process was generally completed by 1907, when the final tracts were acquired, and the property named the Military Reservation at Queen Emma Point. In 1909, the fort was renamed in honor of Major General Emory Upton, who had served with distinction during the Civil War and had been a major proponent of reforms in the U.S. Army. Responding to a petition from local resident Archibald Cleghorn, the name was soon changed. It was renamed in honor of King Kamehameha I, the conqueror of Hawaii and first monarch of the Kingdom of Hawaii. (1)

In addition to the main part of the post, there were two other adjacent sections: one to the west at Bishop Point on the Pearl Harbor Channel above the garrison area, and another at Ahua Point at the east end of the reservation. The Ahua Point tract was not part of the original fort acquisition. In 1910 when army engineers laid out the fire control system for the fort's guns, they encountered a problem. A vertical base system was impractical because of the low elevation of Queen Emma Point, a horizontal base system was the only option. However, there was room only for a 1600-yard east to west baseline, inadequate for the maximum range foreseen for the 12-inch guns and mortars. There were three options: adopt two baselines, one running southeast to northwest, and the other the opposite diagonal. But crossing lines was potentially confusing and difficult

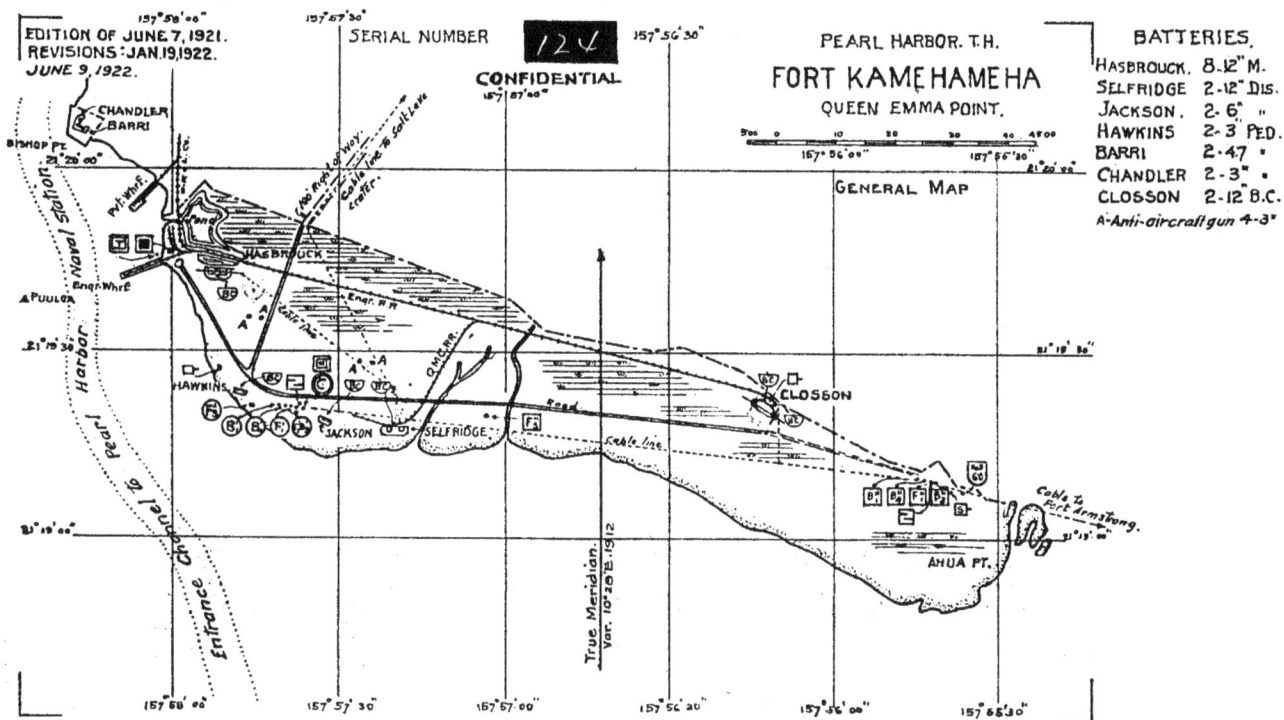

Engineer map of the Fort Kamehameha reservation in December 1919. *NARA.*

Statue of King Kamehameha I, namesake for the first U.S. Army coast artillery post on Oahu. *Gaines Collection.*

to implement. More land could be acquired across the channel on the west side of the Pearl Harbor entry. But that required running communication cables underwater that periodically had to be dredged. Finally, more land could be bought adjacent to the east, the Ahua parcels. It would cost $40/acre to purchase the 400 acres, but with a new wagon road and cable a baseline of 3500-yards was possible. In 1911 Ahua Point was added to the Fort Kamehameha reservation. Fort Kamehameha and its various dependencies were the locations for many seacoast and land defense gun batteries, as well as an important headquarters and coast artillery garrison post. (2)

From the earliest plans, the Queen Emma site was to be home for one, two-gun heavy artillery battery. To stay in range of potential offshore adversaries, it would need to be placed relatively close to the shoreline. It also seems that a decision was reached for these guns to be of a 12-inch bore size relatively early. Detailed planning came after authorizing funding in the summer of 1907.

One early decision had to be made on the counterweights. Disappearing batteries relied on a very heavy set of circular-shaped weights suspended below the mount in a deep pit to physically lift the gun on its upper carriage to the firing elevation. Most previous gun designs had stipulated these weights be made of lead. However, it was thought that in overseas possessions (mainly Philippines and Hawaii) the lead might prove too tempting to insurrectionist factions and be stolen and used to cast bullets. Strange as this logic might sound, all the Philippine batteries thus were required to be designed for cast iron weights. While cast iron was cheaper ($0.03/pound vs. $0.07/pound for lead), it was not as dense, and the counterweight well had to be much deeper to contain a specific weight. The designers of this new 12-inch battery were concerned. On a site so low already, to avoid the water level, the entire emplacement would need to be erected on an elevated position. They estimated this would add $38,770 to the cost. After discussion with Washington, economy prevailed and the go-ahead for lead counterweights was authorized. (3)

Local engineer Capt. C. W. Otwell submitted his proposed design for the emplacement on December 30, 1907. Otwell described it as basically following the standardized suggested plan (as described in published Engineer Mimeograph No. 66). However, it had an increased floor height for drainage, had no protection

Captain Curtiss Otwell. In his role as chief engineer for the U.S. Army in the islands, he personally designed the early gun batteries at Fort Kamehameha. *Williford Collection.*

on the battery flanks, and would not need separated or detached internal magazine rooms as he pointed out that there was no expected condensation in the weather experienced in Hawaii. The projected cost was quite high—at this point expected to be $308,846.77. That was explained to be a result of higher labor and material costs in the island, the remoteness of the site relative to railway lines, and the lack of water on site. The design was approved with minor changes, and $218,000 allocated from the Acts of 1906 and 1907 to start. As 1907 also saw a new crisis in Japanese American relations, it was requested that the work concentrate, if possible, on a first position so as to be ready for arming as quickly as possible if an emergency so required. (4)

The design otherwise closely followed the prescribed engineer monograph for batteries of this type. As with other Taft generation batteries of this size, the two guns were sited 240-feet apart in a common reinforced concrete emplacement. Two projectile magazines and two powder magazines were protected on the level below the central traverse. They fed powered hoists lifting the shells to a gallery on the upper level adjacent to the two loading platforms. Storerooms, latrines, and office, and a reserve power room were adjacent to the heavy gun blocks or on the flanks. A battery commander's station sat exactly between the guns to the rear and above the top of the traverse. To secure a wider field of fire, the direction of fire for each gun was slightly canted outward.

This armament consisted of two 12-inch M1895M1 guns on M1901 disappearing carriages. This gun was an older 35-caliber type, without the range of the 40-caliber Model 1901, but as the latter was found to have a short barrel life due to its higher velocity, only the Model 1895 was being used in new batteries. A suggestion was made, but apparently nothing came from it, to perhaps have a 40-caliber gun on hand to mount if longer range was required in a given situation. The battery was quickly begun, but work was not swiftly accomplished. At least the emplacement, now named Battery Selfridge, wasn't turned over to the garrison until August 4, 1913, having cost $440,000. By the time the 12-inch guns were being emplaced, the age of the improved 'Dreadnoughts' had arrived with battleships mounting 14-inch and 15-inch guns that outranged standard American 12-inch disappearing guns. The 12-inch guns were limited by their disappearing carriages to just a 13,200-yard range. Soon measures were taken to modify the 12-inch disappearing carriages in American coastal defenses to attain a greater range. Attaching a new elevation band and longer elevating arms increased the elevation to 15° and the range to 17,300-yards. This work was done to Battery Selfridge by the end of 1917. The 12-inch fired either a 1070 or 975-lb. projectile. (5)

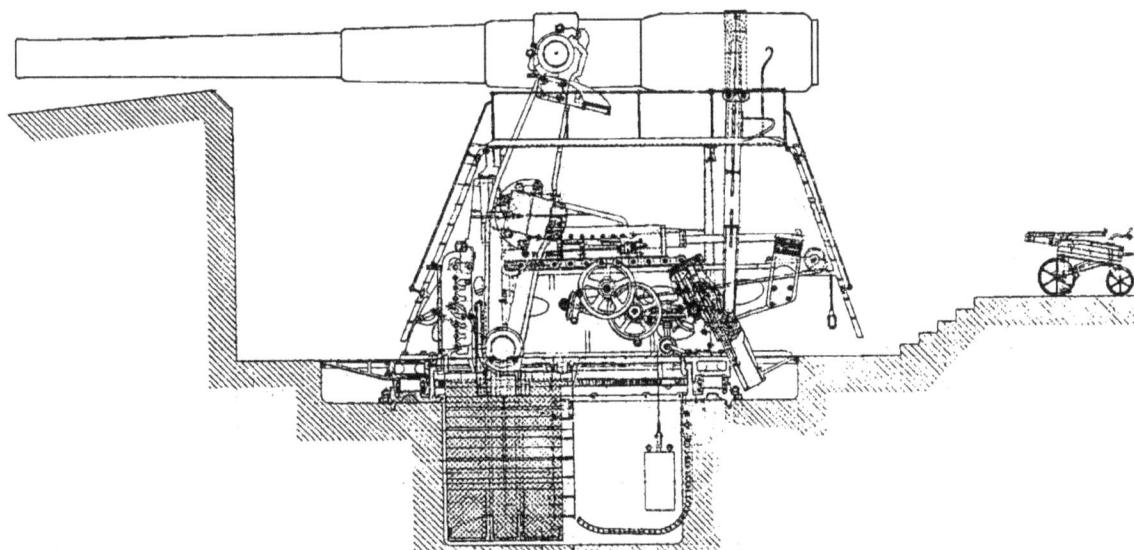

Ordnance Department side view of the Seacoast 12-inch Gun mounted on Model 1901 Disappearing Carriage, like was mounted at Battery Selfridge, Fort Kamehameha. *NARA.*

1919 Plan for Battery Selfridge. *NARA.*

The initial coast artillery garrison (the service branch which would actually man the defensive batteries) for the post, the 68th and 75th Co. CAC, arrived on January 14, 1913. In its early years the peacetime garrison grew from an initial two companies to seven companies by the First World War. Battery Selfridge was manned by the 75th Company. The 12-inch battery functioned as the primary gun battery in the Harbor Defenses of Pearl Harbor through World War I.

When the coast defense guns did fire, they often bothered the nearby residents. Disruption of routine, noise, and at time vibration and shock that could break windows and shake cabinets accompanied the firing. In some locations, like Fort DeRussy close to the heavily developed Waikiki commercial district, firings were delayed or conducted only with subcaliber guns. Even prior to installing the first gun at Fort DeRussy, Governor Samuel Dole submitted a request to the Secretary of War that the firing of coast artillery guns in the vicinity of the residential district of Honolulu would not be desirous and thus should not be allowed. His suggestion was not followed and the guns were duly emplaced at DeRussy as planned. (6)

Cutaway, side view through the gun pit for Battery Selfridge. Note thick areas of concrete and earth to provide frontal protection from enemy gunfire. *NARA.*

A couple of soldiers lounging on the steps leading to the upper (loading platform) level of Battery Selfridge. *Schmidt Collection.*

1922 aerial photograph of Battery Selfridge. Note the wide spacing between gun pits and the battery Commander's station located midway between the two pits. *NARA.*

However firings also could be the cause of a sort of holiday event. With the military families there was also a sense of pride and patriotism. Here are Anne Winslow's (wife of then resident army engineer, Major Eben Winslow) comments about witnessing the firing of Battery Selfridge's 12-inch disappearing guns at Fort Kamehameha on March 25, 1911: (7)

> Yesterday they fired the big guns down at "twelve inch" (Did I ever tell you the habit Eveleth and Bammy have of speaking of the various forts by the caliber of their guns?) This battery is located at Pearl Harbor and is quite a journey from here but we made the trip, the chickies and I, along with a whole train of men, and saw the sight and heard the sound, after which the Japanese can declare ware whenever they feel like it. They have no further terrors for me.

The coast artillery service used "subcaliber" guns to assist in training regimens. These were usually modifications of common service pieces like the field 75 mm guns and the infantry 37 mm guns. Mounted by a special carriage attachment they were generally placed on top of the large caliber tube itself. For some guns, like the 12-inch mortars, there was a special mounting to allow the subcaliber gun to be emplaced inside the larger gun's bore itself. Large guns literally wear out with repeated firing, some guns only having a service life of around a hundred rounds. To save this wear (and damage to the breech area by repeatedly ramming very heavy shells into them) firing the small subcaliber gun when it was time to pull the trigger during training made sense. All other aspects of collecting firing data, pointing and elevating the gun, etc. remained the same. Routine practice at Oahu's big 12, 14, and 16-inch guns all used such devices. They are very recognizable in photographs of the period.

Each of the two Oahu harbor defenses were to get a battery of eight mortars. These were to provide heavy fire at closer ranges in case an enemy fleet pressed itself in for a landing or close bombardment. At Queen Emma Point the battery was to be sited to the west of Selfridge, and to the north not too far from the engineer's wharf. Design and construction were begun not too long after work started on the 12-inch gun battery. It was found that this section of the land tract had more mud coverage, but not too deep underneath was plenty of coral rock which could support such a structure. The design selected carefully follow Mimeograph No. 27,

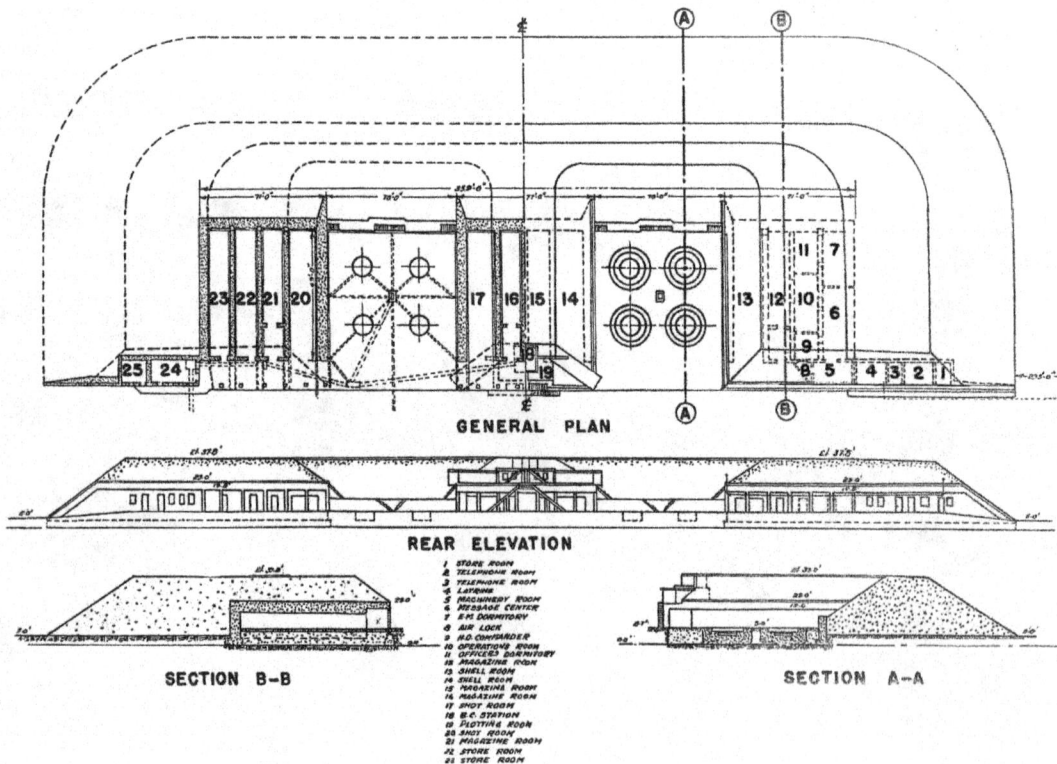

Plan, elevation, and section of Battery Hasbrouck in 1935. *NARA.*

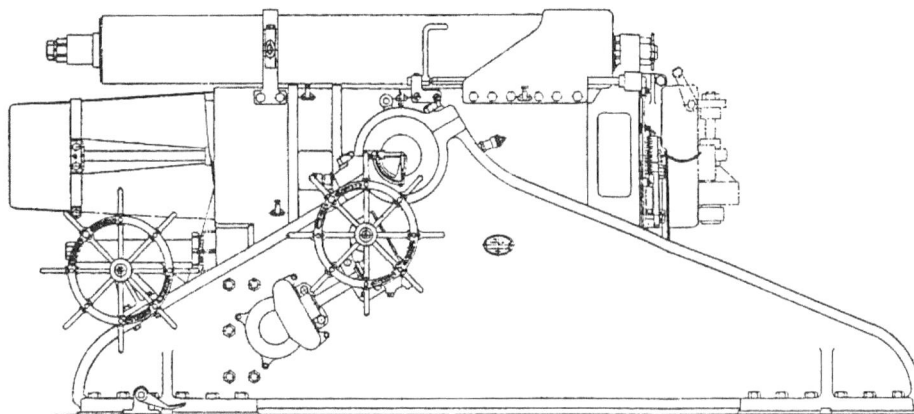

Ordnance Department side view of the 12-inch Model 1908 Mortar and Carriage, the type mounted at Battery Hasbrouck. *NARA.*

1922 Aerial view of Battery Hasbrouck, note the two pits, each holding four mortars. *NARA.*

Battery Hasbrouck and mortars about 1915. *Schmidt Collection.*

Mortar gun crew for one of the guns at Battery Hasbrouck, note shell on cart and ramrod, circa 1920. *USAMH.*

having a conventional plan of two four-mortar pits with intervening traverse. The original design was submitted on August 7, 1909. A significant change was expanded shell rooms to provide for the new long-point projectile that was just coming into coast artillery service. First funding (just $4793.26) was made available on March 11, 1912. Most work was done in 1912-1913. Transfer to troops was made on November 10, 1914, at a total construction cost of $274,160.65. (8)

Closeup of the data booth display of Battery Hasbrouck, where settings were displayed to the mortar crews. *USAMH.*

The structure was equipped with two 25 kW generators, both in the left flank in a special room for this purpose. Also, the battery was armed with the most modern type of mortar in the army's inventory currently available—the Model 1908 type mortar and carriage. While this gun was still the short, 10-caliber length of limited range, the new carriage and recoil system was a considerable improvement over the older M1896 type. After firing the barrel could be lowered for reloading while the recoil mechanism stayed in its elevated position. This mortar could fire either the 700 or 1046-lb. projectile up to 15,291-yards. Only 25 of this type of mortars were purchased by the army; just this battery, two batteries in the Philippines, and one battery pit in Boston received this type of gun. Even the mortars at nearby Fort Ruger were the older model. Apparently, there was some confusion during construction about the right type of mortar base ring to use, as the concrete had to be chiseled out to allow a change during building. Soon named Battery Hasbrouck, it was also placed in service by the 68th Company, CAC. The lack of a new post headquarters building forced the occupancy of one of Hasbrouck's magazines for this purpose for several early years. In 1924 a new battery commander's station was built at the battery. This structure was an observation station of corrugated iron construction on top of a ten-foot structural steel tower. This tower was erected at the center of the battery above the central traverse. The observation room of the station was equipped with three Warner-Swazey Model 1910 Azimuth Instruments. (9)

The defenses of both Pearl Harbor and Honolulu were rather deficient in supply of secondary armament. The lack of back channels on flank positions obviously did not require as much in the way of rapid-fire emplacements. Hawaii was too far from any possible enemy base to be threatened with torpedo boat attacks, still there was always the possibility of small boats attempting to land a raiding force. The Macomb Board of September 1912 acknowledged that the major batteries were almost complete, yet nothing had been done yet to provide any smaller emplacements. But then things moved quickly. Local engineers were requested to proceed with a design when informed that $95,000 was available on September 28th. Some discussion pursued about the relative directional facing the battery should have. Some factions wanted it to provide more coverage west over the channel and possible enemy field guns that might be landed and operated on the Puuloa side. Once again, the engineers in Washington prevailed and reminded all that the primary purpose of this (and all) coast defense batteries was to bear on the presumed position of a bombarding fleet. The position was, however, withdrawn a distance to the rear to avoid interfering with Battery Selfridge's fire. (10)

While Oahu was supplied the most modern plans for 6-inch disappearing guns that were being used in Panama, local engineers preferred the more conventional design like was being used at Battery Dudley at Fort DeRussy. The only significant change made to the design of Battery Jackson was to extend the length of the left flank wall to protect from flank fire, and to build the battery commander's station to the expanded Panama dimensions. This battery was completed and turned over to the Fort Kamehameha garrison on June 17, 1914. At that time the cost of the battery's construction had reached $86,067.25. Armament consisted of two 6-inch M1908 guns on M1905M2 Disappearing Carriages. This gun was the latest iteration of the standard seacoast intermediate gun. It fired a 108-lb. projectile at 15° elevation to a range of 14,400-yards. The disappearing carriage allowed good protection to the manning crew behind the parapet, and when well-trained could produce a high rate of fire. Battery Jackson was manned by the 91st Company, CAC, from the time of its arrival at Fort Kamehameha on January 15, 1915, through the end of World War I.

Continuing the gun line west from Battery Selfridge, in descending bore size, the final unit was to be a two-gun, rapid fire 3-inch battery. It was authorized for planning along with Battery Jackson in 1912. Final Plans were submitted on April 12, 1913. Major W.P. Wooten, CE, recommended a standard plan design, like that already approved for Battery Tiernon at Fort Armstrong. Using hired labor, he suggested it would cost $18,500. The same discussion about Jackson's orientation and coverage of the western shore was made with Battery Hawkins in mind. This battery consisted of two 3-inch M1903 Rapid Fire Guns on M1903 Pedestal Mounts, with gun center distance of 68-feet. These were the standard type of light, "rapid-fire" guns in use for many years. Hand-loaded and operated they were intended to defend the flanks of forts and engage light

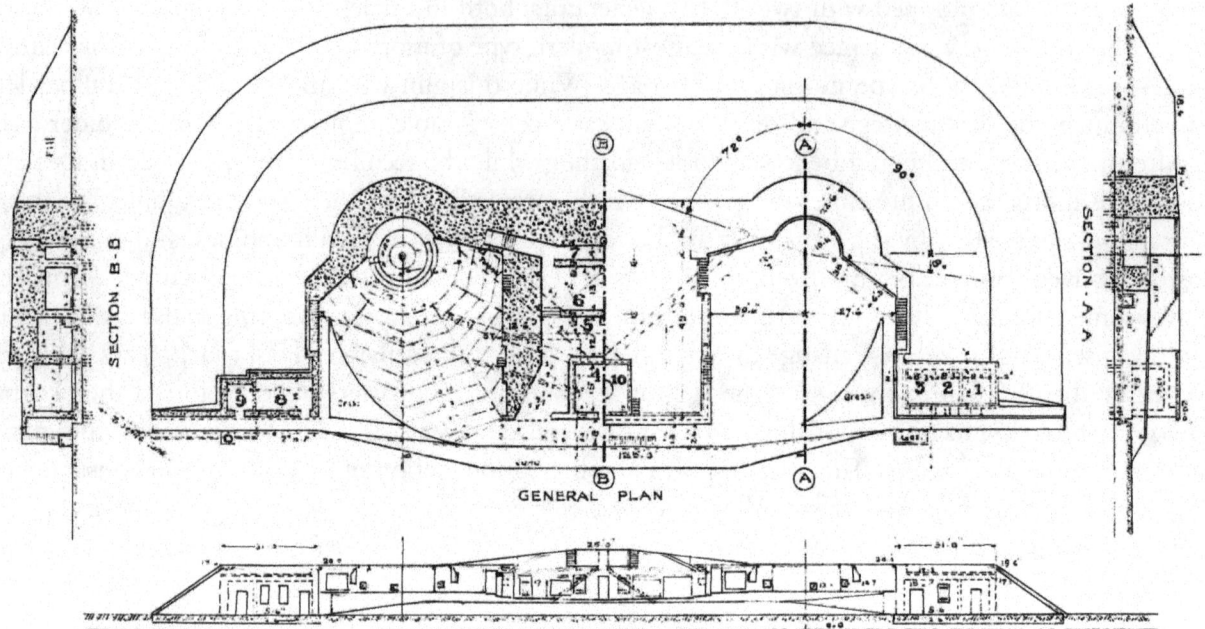

REAR    ELEVATION

1  Store Room
2  Radiator Fan Room
3  Engine Room
4  Plotting Room
5  Shell Room
6  Powder Magazine Room
7  Corridor
8  Enlisted Men's Latrine
9  Officers Latrine
10  B.C.Station

Plan, elevation, and section of Battery Jackson in 1919. *NARA.*

1932 Aerial photograph of the gun line at Fort Kamehameha. Note location of Battery Selfridge (lower emplacement) and Battery Jackson beyond. *NARA.*

6-inch disappearing gun Battery Jackson as completed. *USAMH.*

Closeup of gun and manning crew for Battery Jackson, Fort Kamehameha. *Schmidt Collection.*

landing boats and minesweepers of a hostile enemy. The gun fired a 12.9-lb. projectile at a muzzle velocity of 2800-f.p.s. to a range of 11,100-yards. It was mounted on a simple mechanical carriage allowing rapid traverse and had a protective steel shield. One design change was made, the reserve 25 kW generating plant for Battery Jackson was installed in Battery Hawkins' left flank rather than at the 6-inch battery itself. Work was begun in December 1912 and completed in March 1914. It was transferred to troops on March 6, 1914, for a construction cost of $22,200. (11)

Controlled mines were an integral element of American coast defense. The Coast Artillery operated their own types of mines and had the infrastructure to lay and control them. In times of peace the unloaded mines (usually in this era spherical cases) were stockpiled ashore in "torpedo storehouses". Torpedo was the original term used for underwater explosives, and only later became associated with just the automotive torpedo type.

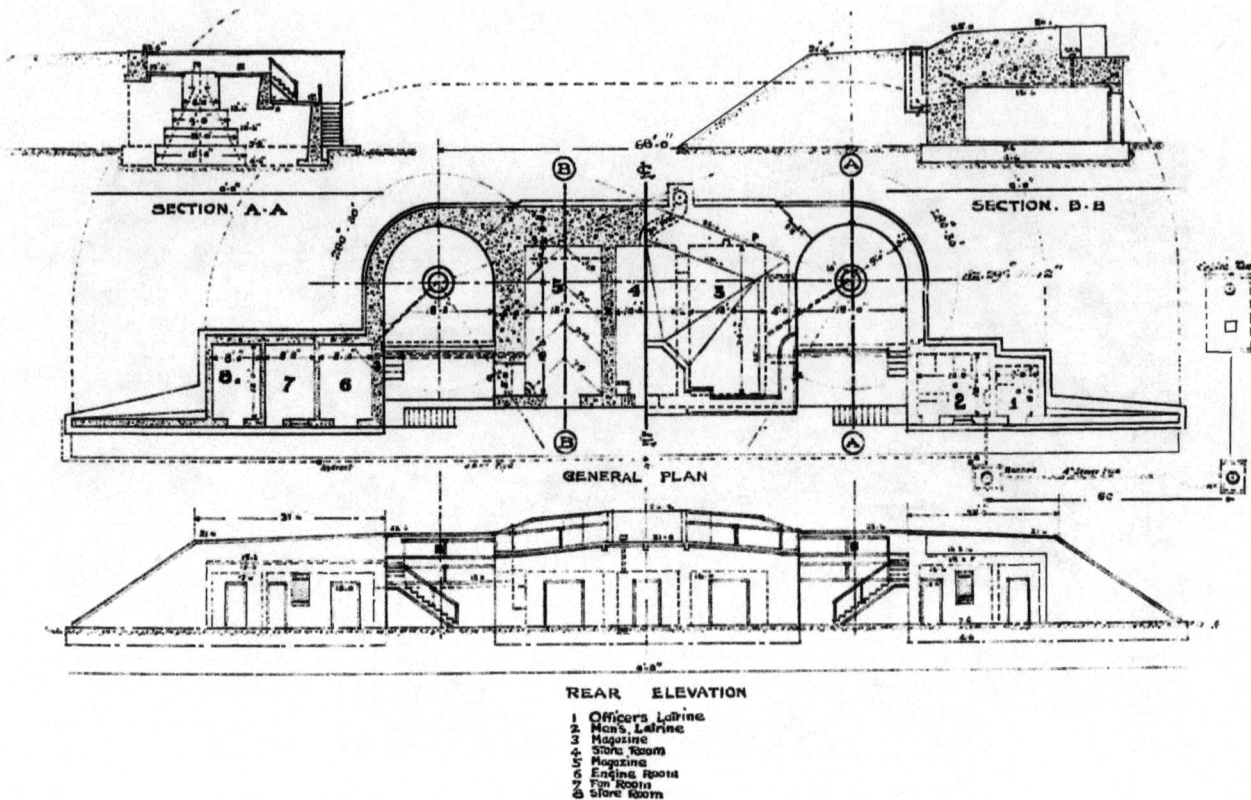

SECTION. A·A

SECTION. B·B

GENERAL PLAN

REAR   ELEVATION

1  Officers Latrine
2  Men's Latrine
3  Magazine
4  Store Room
5  Magazine
6  Engine Room
7  Fan Room
8  Store Room

Plan, elevation, and section of Battery Hawkins in 1919. *NARA.*

One of the 3-inch Guns and Pedestal Mounts for Battery Hawkins in 1936. *USAMH.*

Aerial photograph of Battery Hawkins in August 1922. *NARA.*

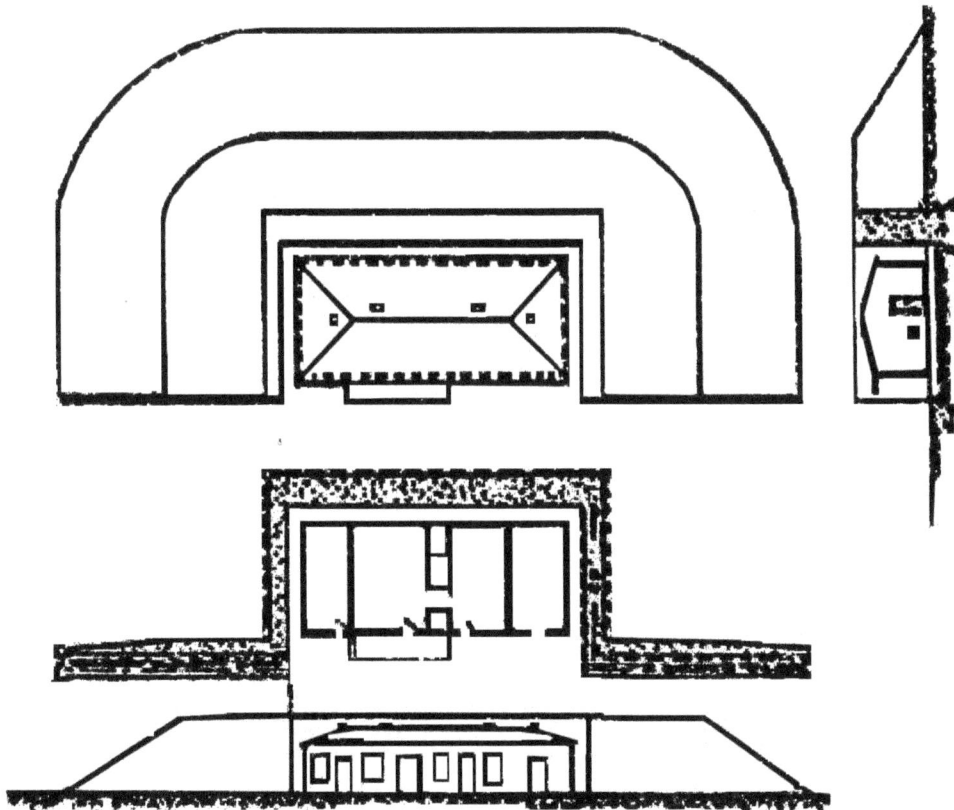

Elevation plan for the Submarine Mining Casemate at Fort Kamehameha. *NARA,*

These storehouses were lightly constructed structures providing mainly protection from just the weather. Explosive charges were kept in separate "storage dynamite rooms". Prior to placing, the mines were moved and loaded with explosives at the "mine loading room". Then, at this point usually by a railway-like rail track, they were moved to a special mine wharf where they were loaded onto dedicated army ships (army mine planters or smaller mine yawls) and taken out to pre-selected locations to be laid in groups. They could be either placed on the seafloor (if shallow enough) or anchored at a pre-determined depth to give maximum effect. Mines were connected to the shore by electrical cables and detonated on demand by a signal from a protected "mine casemate". Unlike the familiar navy mines with contact "horns", usually coast defense mines were manually controlled, though they could be set to a contact response if needed.

Calculations for range and firing of mines was facilitated by a network of concrete fire control stations. These were observation stations equipped with optical instruments to identify azimuth and distance to targets. They were tied into a system of mechanical plotting boards and used to calculate range and precise mines to fire. There were two separate minefields proposed for Oahu, both Pearl Harbor and Honolulu harbors were to be equipped with their own fields and supporting structures. The Pearl Harbor field was to be controlled from Fort Kamehameha, Honolulu from Fort Armstrong. Each defense was to be equipped to place and operate three groups of mines of 19 mines each. Funding to begin on both systems was forthcoming through the Congressional Act of May 27, 1908, which provided $129,000 for this purpose. (12)

However, it was not until April 1911 that the project moved forward with submission of definite plans for the Fort Kamehameha's mining casemate. It was to be located some 300-feet from the ocean shore, near the Engineer Wharf that extended into the harbor channel proper. Along with the proposal for the casemate were plans to build the cable hut, the two "M" observing stations, and to add a necessary tide station. From the beginning it was recognized that the closeness of the two Oahu mining projects allowed for some practical economy. Only at the Fort Armstrong reservation at Kaakaukakui Reef would there be stored the mines, cables, and loading room. Also, its wharf would host any resident army mine planter. Fort Kamehameha would have its own controlling casemate and observation rooms, and a small, covered boat house to keep a light yawl or boats to tender cable connections.

Photo probably taken in the First World War showing Battery Jackson (to the left) and the row of fire control stations just to the west.
*USAMH.*

The structure consisted of a reinforced concrete building concealed behind a sand parapet around three sides near the beach southeast of Battery Hasbrouck. There were questions concerning the expected shock from mortar firing to the sensitive mine control switchboard, but Washington assured local planners that no problem was anticipated. The structure contained several rooms. One served as the mine operating room, a smaller room contained the electric wet cell batteries, and a third served as a power generator room that contained a 25 kW generating set. It was turned over to the garrison on May 28, 1913, for a total cost of $12,811.33. (13)

The primary station for the mine command was in a thick grove of algaroba trees about 400-feet southeast of Battery Hawkins. The station, built in 1913 and transferred to the garrison on December 29, 1913, was a single-story building with two rooms built of reinforced concrete. It was equipped with two M1910 azimuth instruments, and a 14-foot by 16-foot plotting room behind the observation room. The secondary station for mine command 2 was housed in a small concrete building with a corrugated iron roof located some 400-feet from the shore and about 1,200-feet east of Battery Selfridge. The building was partially obscured from view to seaward by a parapet that extended up to the level of the observation slot of the station. This station was also transferred to the garrison on December 29, 1913. It was also equipped with a Warner Swazey M1910 Azimuth Instrument. (14)

Command and control of the gun and mortar batteries at Fort Kamehameha employed a fire control system that used a baseline between primary stations located near Batteries Jackson and Hawkins near the Pearl Harbor Channel, and a secondary group of stations at Ahua Point some 2,000-feet east of the primary group. The battle commander's station was located at the primary group. These stations along with others located at the gun and mortar batteries formed the Second Fort Command in 1915 composed batteries grouped into tactical fire commands. This system of fire control was continued during and after World War I, though later changes in designations were made.

A combined Battle Commander's Station, Meteorological Station, and Signal Station at Fort Kamehameha was built in 1912 and 1913 amid an algaroba thicket about 350-feet northwest of Battery Jackson. The structure was transferred to the coast artillery garrison on August 4, 1913, having cost $6990.18 to that date. It served as the command post or "C" Station for the Second Battle Command of the defenses. It was redesignated the C-1 Station for the Kamehameha Group of the HD of Pearl Harbor when the Oahu harbor defense command was separated into two harbor defense commands in 1921. (15)

The station was a two-story structure of concrete construction, with a third story of structural steel framework initially enclosed by walls of asbestos-covered corrugated iron. The lower floor contained a private room for the battle commander, a lavatory, a small storeroom, and seven telephone booths. The second floor of the building was fitted out as the Post Meteorological Station. The station's third story served as an observation station fitted out with an observing telescope. This observation level was open to the air except for an asbestos covered sun shield of corrugated iron.

Four fire control stations were constructed in 1913, and 1914, about 200-feet west of the "C" Station in a growth of trees that partially concealed them. They consisted of a row of four 25-foot tall structural steel towers. The first of these towers served as the Primary station for Fire Command 4 (F4'). It was completed by 1914 and transferred to the garrison on April 8, 1914, having cost $5163.20. The other three stations were identical to each other in size and construction consisting of structural steel towers provided with a single observing room. The walls and roofs of these observing levels were constructed of asbestos covered corrugated iron sheathing on a steel frame. Each station was equipped with a pedestal for a single observing instrument. One of the stations served as the secondary observing station for the Third Fire Command (F3") that was composed of Battery Hasbrouck's eight mortars. Another of the three fire control towers served as the primary station (B') for Battery Jackson, the final tower was assigned to Battery Selfridge as its primary station. All of these were built and transferred in 1913. These stations remained in service through most of World War II, although their functions changed in some cases as new batteries were assigned and others were deleted from the harbor defense plans. (16)

A searchlight position of the western side of the Queen Emma reservation was composed initially of a fixed 36-inch searchlight. Searchlight No. 9 was on a movable truck riding a 4-foot gauge rail. It used the electric plant of Battery Selfridge as power, but the light was stored in its own small shelter. It had been transferred on April 28, 1916. This light was removed following World War I but was replaced for a few years by two 60-inch General Electric searchlights that were temporally mounted on wooden towers. Then these lights were moved to Fort Weaver in 1922 and 1924 where they were permanently installed.

The first garrison troops arriving in 1913 found almost a total lack of habitat facilities. Tents and quarters in the city were found for some; the commanding officer's and second lieutenant's wives had to initially quarter in the completed but yet unequipped mine casemate. Over time an extensive complex for the post cantonment developed starting about 1913 in the area immediately across the street (Worchester Street) to the north of the main battery positions between Battery Selfridge in the east and Battery Hasbrouck in the west. By 1916 an extensive housing area for assigned troops was completed. There initially were nine 104-man barracks built on either side of the road. In the tropical weather of Oahu, they were rather simple in comparison to coast artillery barracks elsewhere. Each building was designed for a single company of troops, with bays for barracks, a sergeant room, dining hall and kitchen, latrines, and a recreation room. They were single story building of wood. The wings surrounded an enclosed central courtyard. Each building was approximately 140 by 160-feet. Later additional barracks were built, including a larger 168-man building in 1921. All of these served through the coast artillery period until after the Second World War.

There were 16 sets of married NCO quarters, again with additional units (ten more) added up until 1940. More impressive were 27 sets of officers' quarters. Of these 23 along the "bend" north of Worchester Avenue were for married company officers. Four were built on a side road near the parade ground for field grade officer families. They were described as built in the "Craftsman Style" and built of wood on lava stone and concrete pier foundations. As behooves the tropical location, barrack and housing was unheated and extensive use made of screen windows. A large officer's BOQ was also built. The usual post infrastructure buildings were constructed. Most were with the initial 1912-1920 building period.

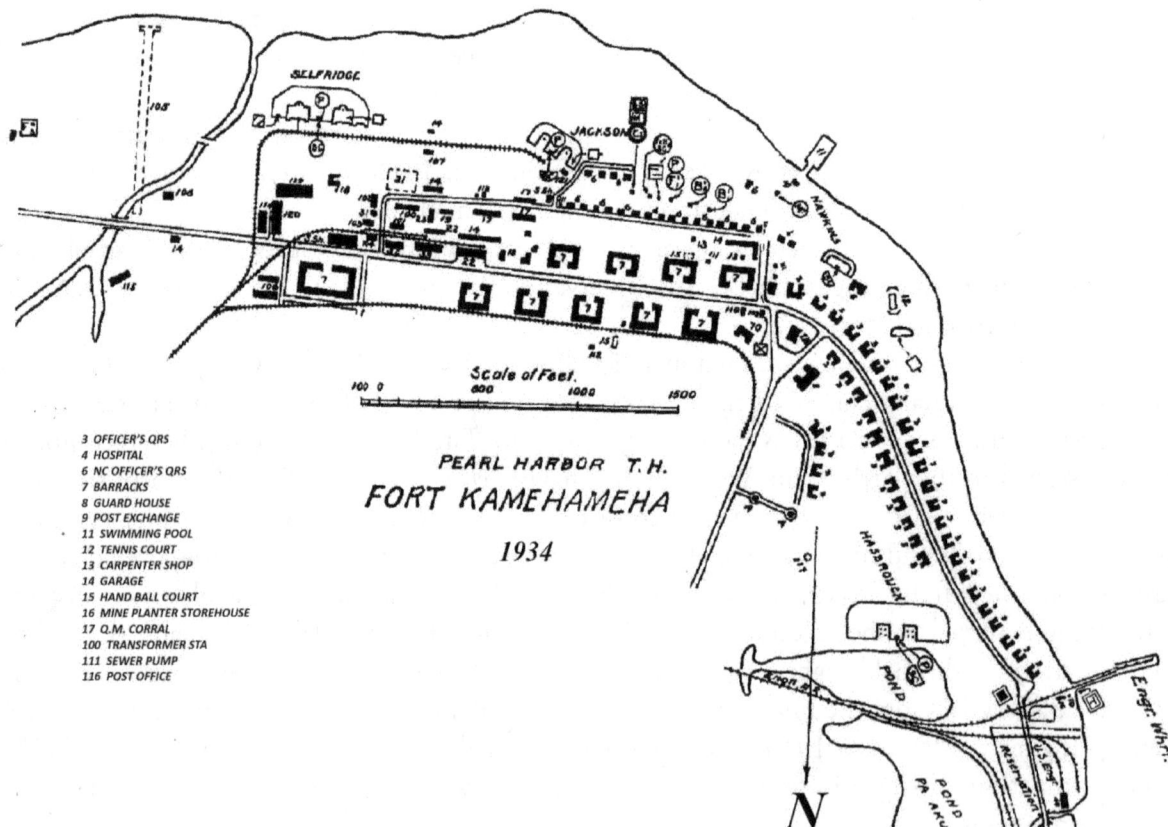

3 OFFICER'S QRS
4 HOSPITAL
6 NC OFFICER'S QRS
7 BARRACKS
8 GUARD HOUSE
9 POST EXCHANGE
11 SWIMMING POOL
12 TENNIS COURT
13 CARPENTER SHOP
14 GARAGE
15 HAND BALL COURT
16 MINE PLANTER STOREHOUSE
17 Q.M. CORRAL
100 TRANSFORMER STA
111 SEWER PUMP
116 POST OFFICE

Scale of Feet.
100  0       600          1000            1500

PEARL HARBOR T.H.
FORT KAMEHAMEHA
1934

N

1934 Map of Fort Kamehameha showing the main cantonment area of the post. Note the prominent barrack buildings and long rows of family quarters. *NARA*.

Fort Kamehameha Officers' Club about 1940. *Olbrych Collection.*

Headquarters Building in the late 1930s. *Olbrych Collection.*

Typical storehouse about 1934 at Fort Kamehameha. *Olbrych Collection.*

In 1913 the U.S. Army sent a detachment from its fledgling air service to operate a flying school in the Hawaiian Islands. As the early craft were all seaplanes, Fort Kamehameha was thought to have the required space, flat terrain, and beachfront to host this school. Army Transport *Logan* brought Lt. Harold Geiger and a small detachment of two Curtiss seaplanes to the fort that summer. The first flight was made on August 8, 1913. The flights met with mixed success, takeoff could only be done at high tide and the coastal winds proved problematic. Even hangars were built on the beach in front of Battery Selfridge. Later that year, after only a short period of a couple of months in evaluation, Geiger and the detachment was withdrawn. In March 1917 another unit, this time the 6th Aero Unit of the Army Signal Corps was sent with its three seaplanes to again operate from Fort Kamehameha. It was under the supervision of Captain John Brooks. This unit, now under the supervision of the Bureau of Military Aeronautics, was relocated to Ford Island in 1918. That location was in 1919 named Luke Field. (17)

Then in 1920 the Coast Artillery Corps investigated the use of captive balloons as observers for seacoast gun fire control. In surprising short time, the service pulled together recommendations for equipment and organized new units. The 21st Balloon Company, Army Air Service, we assigned to Fort Kamehameha on October 2, 1920. Besides new quarters for the company personnel, equipment sheds, a gas generating plant (the balloons used hydrogen gas generated chemically on site), a prominent and quite large hangar of wood and iron sheeting was built about 300-yards northeast of Battery Selfridge. All this ultimately proved premature. The results, when the rather expensive and manpower intensive system worked, were not marginally better than existing types of fire control. In not even two years, on July 27, 1922, the use of balloons for determining ranges was discontinued. The units were disbanded and absorbed into other formations; the structures repurposed where possible. (18)

The Ahua Point Section was at the eastern edge of the fort hosted the secondary fire control stations. The secondary stations (B") for Batteries Selfridge, Hasbrouck, Jackson and Fire Commands Three and Four were the first of the installations at Ahua Point. They were built in 1912 and 1913. The (F') for Fire Command Number 3 was turned over to the Fort Kamehameha garrison on August 22, 1913, while the other three were transferred on December 29, 1913. Each of these stations consisted of four identical single story observation rooms measuring ten-feet, six-inches square atop a 25-foot tall structural steel tower. The walls and roof of the observing rooms were constructed of asbestos covered corrugated iron attached to a steel frame. Each of the stations was equipped with a Warner Swayze M1910 azimuth instrument. The four stations were arranged in a line near the reservation's northeast boundary. A grove of algaroba trees ranging from 15 to 20-feet in height to the southwest of these stations partially screened the lower portions of these stations from visibility to seaward. (19)

The initial searchlight emplaced on Ahua Point was a 60-inch General Electric projector obtained in 1911. It was installed prior to August 1913 on a Scherzer Rolling Lift Tower that when elevated to its operating position placed the searchlight 70-feet above sea level. The light could be controlled either electrically or manually. It was powered by the Ahua Point power plant located in a concrete powerhouse located about halfway between the secondary fire control towers and the tower for Searchlight No. 6. This power plant was protected on three sides by an earthen revetment. The power plant consisted of a General Electric gasoline operated four-cycle engine that was directly connected to a 25 kW General Electric generator that was purchased in 1912. The cost of the power plant was $6300 while the cost of the searchlight and its tower was $12,500. This light designated as Searchlight No. 6 was transferred to the garrison on August 13, 1913. Searchlight Number 6 was redesignated as Searchlight No. 1 for the Harbor Defenses of Pearl Harbor in 1924 and renumbered again during the period between the two world wars, as Searchlight No. 9. (20)

From 1906 through World War I, five harbor defense installations composed what eventually became the Harbor Defenses of Honolulu. The first three installations were established by 1908: the Kaakaukukui Military Reservation at the entrance to Honolulu Harbor on December 30, 1899 (In 1910 that reservation was renamed Fort Armstrong); Fort Ruger at Diamond Head in 1906; and Fort DeRussy at Waikiki in 1908. Prior to World War I, two additional reservations were added; the Punchbowl Military Reservation in 1911 as a sub post of Fort DeRussy; and the Sand Island Military Reservation in 1916 as a sub post of Fort Armstrong. These five posts functioned as the Artillery District of Honolulu after 1917. These installations became a part of the wider Coast Defenses of Oahu in August 1917, an organizational structure that remained in place until 1921. During World War I, the Honolulu area coast artillery posts were garrisoned by a total of seven companies of coast artillery manning eight seacoast gun and mortar batteries and the sub-marine mine defenses of Honolulu Harbor.

On July 20, 1899 an Executive Order was issued claiming the former Crown Lands in Hawaii for military purposes. On November 10th of that year the US Navy claimed some 76 acres of Kaakaukukui Reef on the east side of the entrance to Honolulu Harbor. Of this acreage, only 42,050 square feet in the vicinity of Queen Emma Wharf was dry land. The claimed land was designated the Kaakaukukui Military Reservation on December 30, 1899. As much of this acreage was tidal it was unusable until it was filled and raised. A five-year lease of Queen Emma Wharf to the City of Honolulu's Department of Public Works was granted August 17, 1903.

A memorandum for the defense of Honolulu and its harbor prepared in April 1901 by Captain Daniel W. Ketcham of the Artillery Corps recommended planting of an extensive controlled minefield at the harbor entrance. The obvious location for controlling of the mines was at Kaakaukukui Reef. It was not until 1905 that the requisite dredging of the Fort Armstrong Channel, Honolulu's main channel commenced. The channel was dredged to a depth of 35-feet and widened to a width of 400-feet, the spoil being deposited on the reef and on Kaholaloa Reef on which Sand Island stood across the channel, raising both to an elevation of some seven-feet above sea level. The Army exchanged some of its claimed land for the Navy land at the reef. (21)

On August 2, 1909 the small army coast artillery post at the entrance to Honolulu Harbor was named Fort Armstrong in honor of Brevet Brigadier General Samuel Chapman Armstrong, born on the island of Maui. He was a veteran of the Civil War and former colonel of the 8th Regiment, U.S. Colored Troops during that war. General Armstrong had died on May 11, 1892. Dredging of the harbor and its channel created most of the fort's land area. Eventually, an area of 96 acres was converted from coral reef and shoals into a small coast artillery submarine mine base. In 1913, the 104th Company, CAC, arrived from Fort Washington, Maryland. The duties of the 104th Company of serving as the mine company for Honolulu Harbor.

Fort Armstrong served as the base for the controlled minefields and the U.S. Army mine planters (US-AMP) assigned to Honolulu Harbor. A similar-sized mine project was being built at Fort Kamehameha for the Defenses of Pearl Harbor. Each harbor was to be equipped with facilities to operate three groups of 19 mines each. The mine facility at Fort Armstrong served both defenses with a torpedo store house for the mine cases, magazines for the explosives, a loading room, the cable tanks and mine planter wharf.

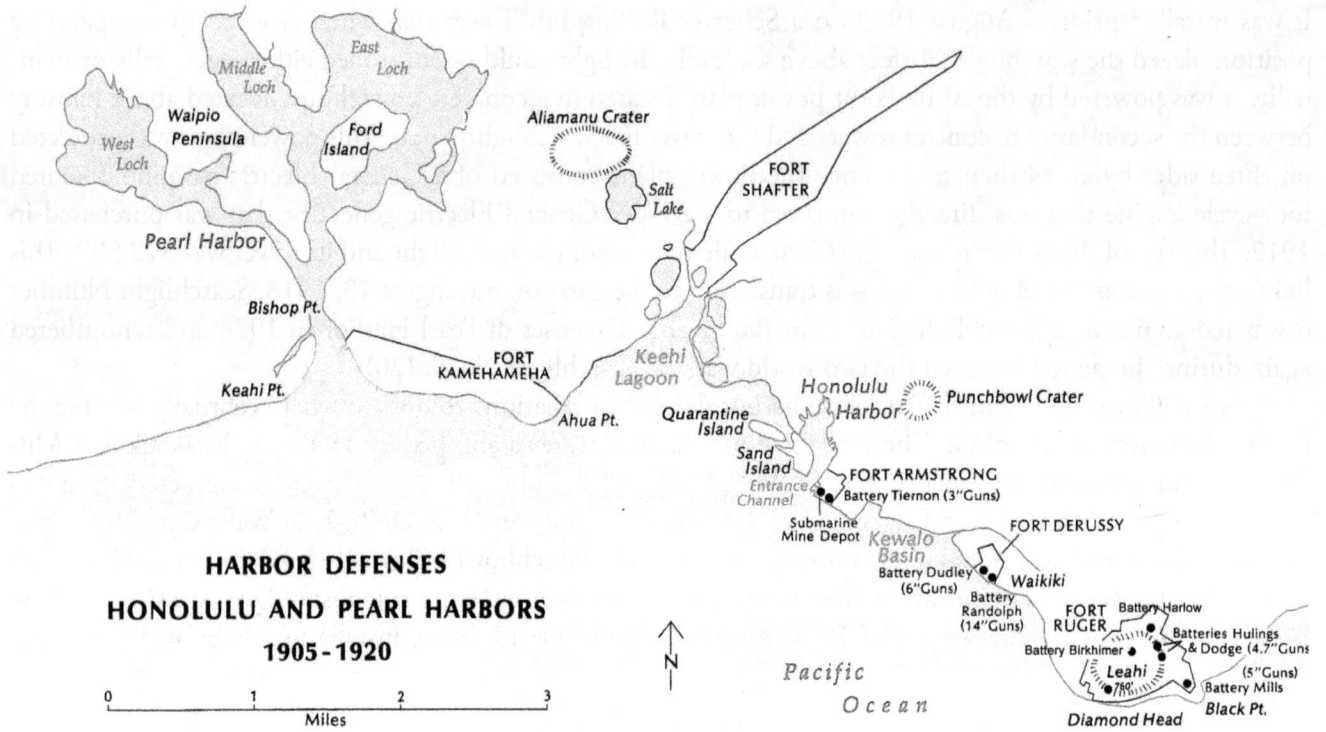

Map showing the major forts and batteries commissioned between 1905 and 1920 for the Harbor Defenses of Honolulu.
*Williford Collection from NARA map.*

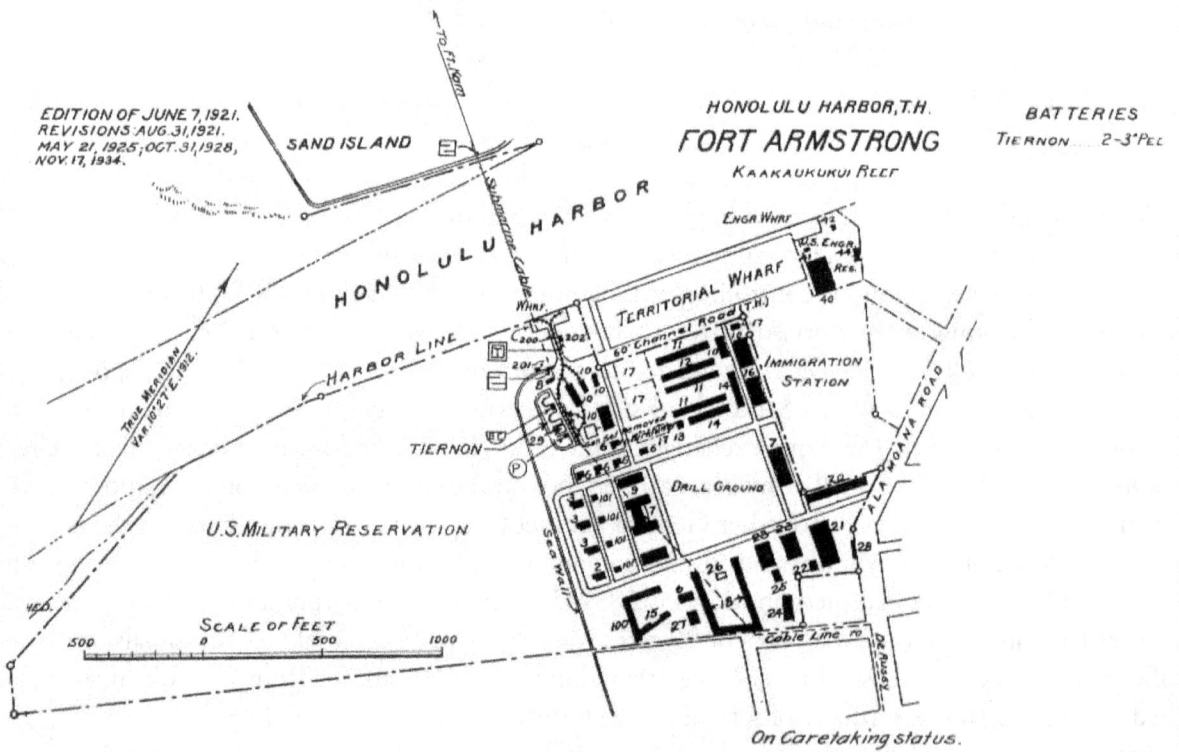

1934 Engineer plat showing main structures at Fort Armstrong. *NARA*

General Samuel Chapman Armstrong, namesake for the coast artillery post of Fort Armstrong. He was born in Hawaii and served with distinction during the American Civil War. *Gaines Collection.*

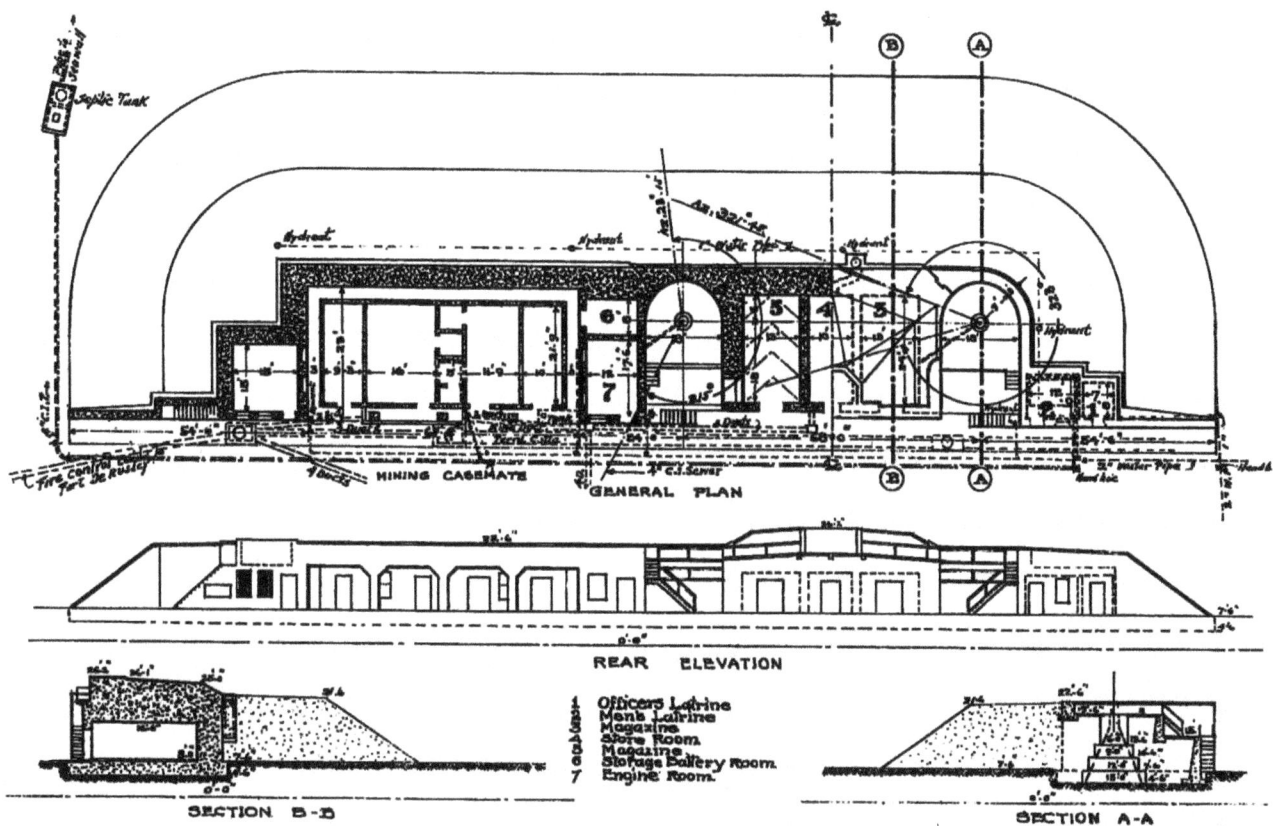

REAR ELEVATION

1 Officers Latrine
2 Mens Latrine
3 Magazine
4 Store Room
5 Magazine
6 Storage Battery Room
7 Engine Room

SECTION B-B

SECTION A-A

1919 Plan, elevation, and section for 3-inch Battery Tiernon and adjacent mining casemate at Fort Armstrong. *NARA,*

In 1912, the Corps of Engineers built a "T" shaped reinforced concrete pile mine wharf at Fort Armstrong. The wharf proper was a structure 55-feet wide and 175-feet long and connected with the shore by a concrete pile approach 152-feet long and 25-feet wide. The surface of the structure was of concrete slabs nine-inches thick and surmounted with macadam to the level of the rails of car tracks that connected with the post railway. It was completed and transferred to the garrison on June 15, 1911 for $31,453.26 in engineering cost. This wharf was used by the Distribution Box Boat Numbers 5 and 6 that worked with the lighthouse tender Columbine as an ad hoc mine planter in the pre-World War I years. (22)

Soon after the arrival of the 104th Company CAC in 1913, the USAMP *Major Samuel Ringgold* arrived to participate in practice mine planting. While in Hawaii the mine planter also towed targets for the gun batteries' target practice for a few months prior to returning to her homeport in at Puget Sound, Washington. In 1916, with the mine wharf completed, the Army exchanged Queen Emma's Wharf for property on Sand Island. Shortly after World War I, the newly commissioned 172-foot USAMP *Colonel Garland N. Whistler* and the Quartermaster Corps vessel *Gildart* were assigned to the Coast Defenses of Oahu.

Fort Armstrong's sole seacoast battery was for a pair of 3-inch M1903 Rapid Fire guns on M1903 pedestal mounts. These were the same type of guns also used at Battery Hawkins at Fort Kamehameha. Major E.E. Winslow submitted the initial design for the battery on April 13, 1909, after he had been notified that funds were allocated on March 8. He sited the battery in the center of the new reservation set aside for submarine mine purposes. He found good foundation which would not require piles if the foundation walls were just spread a bit. They used a conventional two-gun design as suggested in Engineer Mimeograph No. 30. In accordance with latest designs, a full battery commander's station was located on top of the traverse between the guns. The guns pointed to the southwest. It was begun in 1910, completed in 1911, and transferred to the Fort Armstrong garrison on June 15, 1911, having cost $20,000 to build. (23)

Battery Tiernon's initial manning detachment was also the 104th Company, CAC, then serving as the mine company for Honolulu Harbor. It was redesignated in 1916 as the 1st Company, Fort Armstrong, and, in August 1917 again redesignated as the 8th Company, CD of Oahu. During World War I Fort Armstrong was regularly garrisoned by the 8th Company and one or two other companies from Forts Kamehameha, DeRussy or Ruger on an alternating basis. In the first six months of 1919 the 8th Company and the 4th and 6th Companies, from Fort Kamehameha, manned Battery Tiernon.

Fort Armstrong was always a small post—in assignment, size, and garrison. It did not ever get much in the way of support structures or post buildings. After July 1919, the fort again became a one-company post. In March 1921, the 8th Company CD of Oahu was redesignated 1st Company, CD of Honolulu. In 1922, it was

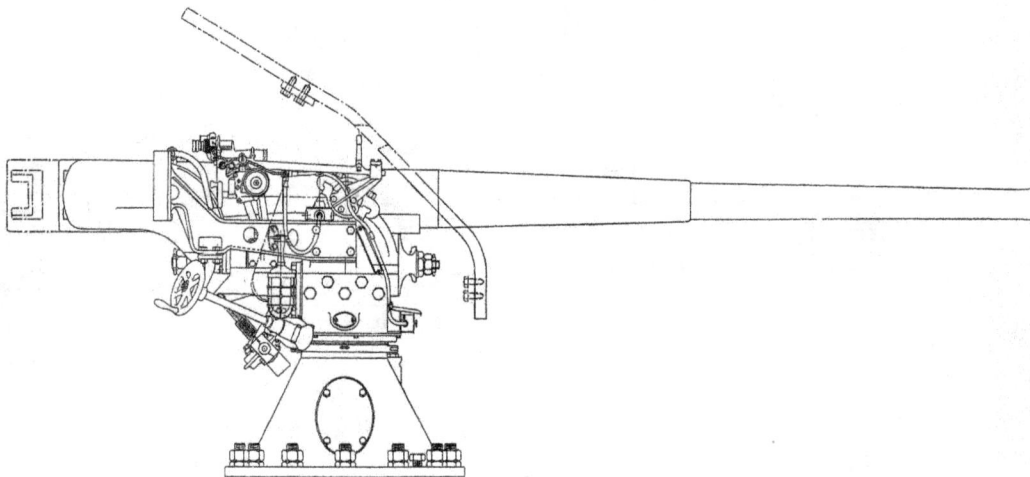

Army Ordnance Department sketch of the 3-inch Model 1903 and Pedestal of the type mounted in the Oahu 3-inch rapid-fire batteries. *NARA.*

Battery Tiernon approaching completion in September of 1910. *NARA.*

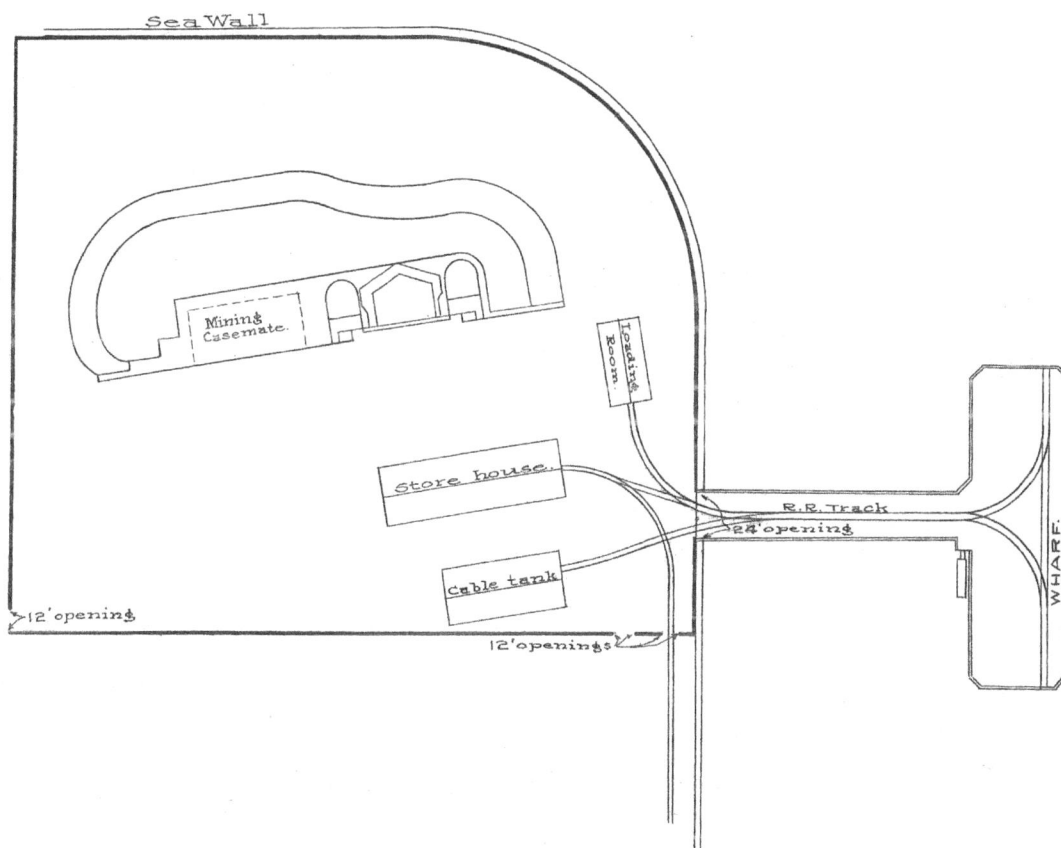

Sketch from 1916 of the major submarine mining buildings at Fort Armstrong. *NARA.*

again redesignated; this time, with its pre-1916 designation, becoming the 104th Company, CAC. The 104th Company was transferred to Fort DeRussy in late 1922, and Battery Tiernon was placed in caretaking status.

The mining casemate for Honolulu Harbor was located on the left flank of Battery Tiernon. Major Winslow submitted a plan for the structure on July 14, 1909. Initially this was a lightly built building of concrete coated metal lathe on wood. Protection to the front and sides was by separated concrete walls with earth embankment, high enough to protect the otherwise unprotected roof from plunging fire. It was also transferred to the Commanding Officer, Fort Armstrong, on June 15, 1911, for an initial cost of $7066.05. (24)

After construction second thoughts emerged. The local troops receiving the casemate were not convinced the protection was going to be adequate. Consequently, on October 17, 1914, the Chief of Engineers directed that new plans be prepared to adequately protect the mining casemate. Engineer Lt. Colonel Bromwell submitted a plan to rebuild it with full concrete outer walls. Being so close to Battery Tiernon, the two components would essentially become one building, though with separate entries and purposes. Unfortunately, funds were not immediately available for this, and work had to wait until the Fiscal Year 1917 budget, finally be released on February 17, 1917. An additional $12,500 was approved for construction and work finally done. (25)

The other tactical buildings for handling mines were also built. Winslow had submitted plans in April 1910 for a torpedo (mine) storehouse. Measuring 100 by 30-feet with an integral 10-ton traveling crane it could contain the mines for both the Pearl Harbor and Honolulu mine fields. It was built of galvanized iron nailed to a wooden frame. It was transferred to the CAC on June 19, 1913, at a cost of $6748.92. Likewise, the 45 by 21-foot concrete cable tank for the mine cables would be covered by a galvanized iron roof held up by pipes. It did, however, also require its own crane to lift the spools of cable. It was also transferred on June 19, 1913, for $8289.30. A separate lightly built loading room and dynamite storage shed were also built at this time. Searchlight Number 5 was a 36-inch seacoast searchlight mounted on a 4-foot gauge rail truck. It was built for $2077.90. When not in use it rolled on its rails and stored in the Torpedo Storehouse. (26)

Electric power for the mining station was provided by a power plant located in the left flank traverse of Battery Tiernon. This plant consisted of a General Electric 4-cycle gasoline engine that powered a 25 kW General Electric generator. The plant was transferred to the Fort Armstrong garrison on June 17, 1911. This power plant proved insufficient for the power demand for the post, and a reserve plant of the same type and capacity as the initial plant was purchased in 1913 and installed at a cost of $5800 in a separate 12 by 16-foot concrete power plant building in the rear of Battery Tiernon in 1914. This plant furnished electric service to Battery

The mine wharf at Armstrong, although by this late date it appears to be used more as a diving platform than for any mining activities. *USAMH.*

U.S. Army mineplanter *Ringgold,* photo taken when she was stationed in Puget Sound. *NHC.*

Aerial view of Fort Armstrong in July 1938. Fort, with its "T"-shaped mining wharf is in the lower left corner. *NARA.*

Officers' quarters at the Fort Armstrong reservation. *NARA*.

Tiernon, the mine casemate, and other torpedo structures, as well as to power Searchlight Number 5 in the event of accident to the main power plant. It was transferred on June 22, 1914. In 1916 a small shed was built to store a half-dozen wheeled 1-pdr Vickers-Maxim automatic guns as part of the Land Defense Program (27)

The primary observing station and plotting room for the Honolulu Harbor minefields was located atop the left flank of Battery Tiernon. The structure was constructed of reinforced concrete and consisted of an observing room measuring 12 by 14-feet that had a corrugated iron roof covered with asbestos for protection from the sun. The mine command plotting room measured 14-feet by 16-feet and was located behind and below the observing room and was provided with a roof of reinforced concrete. The station built in conjunction with Battery Tiernon and the mine casemate was transferred to the garrison of Fort Armstrong on June 17, 1911. (28)

Further to the east, the lovely beaches of Waikiki became the home for the next Taft defensive fort. The shoreline of today's Fort DeRussy was originally part of a wide sandy beach berm or sandbar that extended eastward from the mouth of a stream along the south shore of Oahu's Waikiki section. It was along this part of Waikiki that some of the earliest people of Hawaii settled from the late 14th or early 15th centuries. according to recent archaeological studies. The low-lying taro fields of Kalia extended for over a mile occupying the land between the numerous local streams. The streams carrying rainfall from the interior regularly flooded the taro patches and with those waters also came the freshwater fish. Over time the taro patches evolved into fishponds thus establishing an aquaculture. The fishpond aquaculture continued with only short interruptions through the 19th century. (29)

As early as 1901 consideration was given to the placement of seacoast guns in the vicinity of Waikiki. Initial proposals included two 10-inch guns on barbette carriages to supplement the fire of the gun and mortar batteries at Diamond Head as well as those projected for at the entrance to Pearl Harbor thus affording protection to both harbors. (30)

The section of Waikiki selected by the army for the battery was, in the early years of the 20th century, an area of quiet and generally undeveloped sandy beaches, low lying tidal marshes about three miles east of downtown Honolulu. Many of the island's influential and prominent families maintained beachfront cottages here. When the army selected this bucolic location on Oahu's south shore in 1904, the prospect of the army establishing artillery batteries there did not sit well with the residents who had no intention of willingly selling their property. Territorial Governor Sanford B. Dole went so far as to write to the federal government opposing the location selected by the army on the grounds that the firing of seacoast guns from the site would be undesirable. The governor's objection was ignored, and the army began the land acquisition process. Acquisition of the various tracts was not made easy by these long-time residents of Waikiki and implementing of eminent domain was necessary in most cases. The initial acreage at Waikiki was acquired on December 31, 1904, when 2.893 acres were acquired from Francis Elizabeth Hobron, the widow of Thomas H. Hobron.

As the late Hawaii historian William Dorrance noted: (31)

> The engineers were dealing with families who cherished their beachfront cottages and homes; several of them were not about to surrender their property without a fight. . . the lots eventually sold or surrendered, starting from the western or town end. were owned by members of the Coit Hobron, Waterhouse, Thomas Hobron, F.A. Schaefer, Afong and Pratt families. The army was dealing with prominent members of the territory's establishment. Coit and Thomas Hobron were sea captains who settled in Hawaii during the latter 1880s. Thomas was the more entrepreneurial of the two. He first settled in Makawao, Maui, where he began his island career managing Haliimaile Plantation. He moved to Honolulu and constructed the substantial (now-gone) Hobron Building in downtown Honolulu. John Thomas Waterhouse (1845-1904) arrived in Hawaii with his father in 1851, from Tasmania. The industrious father founded several enterprises, setting an example for his son who established the Waterhouse Trust Company. The son was a staunch annexationist. [The Waterhouse tract consisted of 2.812 acres were acquired from the estate and heirs of Henry Waterhouse, by decree of the United States District Court of the Territory of Hawaii in July 1905.] F.A. Schaefer (1836-1920) had arrived in Hawaii from Bremen, Germany, in 1857. After clerking for ten years for importer Melchers & Co., in 1867 he acquired control and reestablished the firm as F.A. Schaefer & Co. That company started Honokaa Sugar Company on the Big Island of Hawaii., one of the first Hawaiian plantations to plant macadamia nut trees as cash crops. Chung Afong (1825-1906) arrived from China in 1849 and went to work in his uncle's store. By 1889 he had struck out for himself and made his fortune through ship operations, importing, merchandising, and farming sugarcane and coffee on the Big Island. Afong had married a high-born Hawaiian chiefess and fathered a large and attractive family. In 1889 he left his family and his vast holdings in the hands of his good friend Samuel S. Damon (1845-1924), returning to China for the rest of his life. The army dealt with Damon in acquiring Afong's beachfront lot. The last lot was owned by the Pratt family. This prominent Kamaaina (born in Hawaii) family is descended from missionary Gerrit P. Judd (1803-1871) who arrived in Hawaii with the third company of missionaries. Then there were many close neighbors who commanded respect of their own. Included were the Waikiki residences of former Queen Lydia Lilioukalani (1839-1917), last monarch of the Kingdom of Hawaii, and Archibald R. Cleghorn (1835-1910) widower of heir to the throne Miriam Likelike (1851-1887), and father to the heir of the throne Princess Victoria Kaiulani (1875-1889). The army was not fencing with establishment pygmies in the battle to win acceptance of Waikiki fortifications.

The additional acreage acquired from F.A. Schafer, Chung Afong, and the heirs of Dr. Gerrit P. Judd; providing a total additional acreage of almost 11 acres. Much of the preliminary work in obtaining the requisite land tracts was undertaken by Captain John R. Slattery of the Corps of Engineers. Additional land along the Waikiki beachfront was acquired through 1906 until about 68 acres were included in the reservation, some of it being composed of sandy beaches, much of it being low lying marshlands.

Slattery was succeeded as District Engineer by Captain Curtis W. Otwell, Corps of Engineers in 1906. Otwell would oversee the final arrangements at Waikiki and the newly acquired property that would initially be named the Kalia Military Reservation in 1908. In November of that year Otwell was relieved as District Engineer by Major Eben Eveleth Winslow, Corps of Engineers, commanding officer of the 1st Battalion of Engineers. Winslow brought Company A of that battalion with him to undertake the task of creating a modern coast artillery post out of the marshy tracts and beachfront of Waikiki. (32)

As engineer in charge of the Honolulu District, Winslow had responsibility for the construction of the defenses from Diamond Head to the entrance to Pearl Harbor. Over the years the plans proposed for Waikiki's fortifications had been altered considerably. The 10-inch guns initially proposed had given way to a recommendation for a pair of 12-inch guns. In 1908 a Joint Army and Navy Board recommended that the Waikiki site be armed instead with a pair of 14-inch guns on disappearing carriages as well as a pair of 6-inch guns also on disappearing carriages. Winslow undertook the improvement of the small tract at Waikiki preparing it as the site for two modern seacoast gun batteries.

Upon their arrival, the engineer troops established a tent encampment on the newly acquired reservation and set about their tasks. Winslow and his family initially took up quarters in the Moana Hotel that stood about halfway between the Waikiki reservation and Diamond Head. Eventually the major acquired a private dwelling near the construction site. The army personnel also occupied 18 cottages and other buildings, most of them of somewhat flimsy frame construction that had been vacated by their civilian owners when the Army took possession of their property. Five of these structures were placed in service as officers' quarters while seven more were used as NCO quarters. The remaining civilian structures on the reservation were utilized as a kitchen for the engineer company, a lavatory, the temporary hospital, and a post exchange. Some of these fragile structures would be replaced by sturdy army buildings that were constructed during a $244,000 project undertaken between 1912 and World War I. In December 1914, a handful of these former civilian cottages were rebuilt as "new" officers' quarters using material from other old structures that were torn down. (33)

But the more immediate task of Winslow and his engineers was the construction of the massive concrete batteries for the 14-inch and 6-inch guns. Between 1908 and 1913, the Waikiki post was occupied by succession of engineer companies all of which were engaged in the construction of fortifications and the mapping of Oahu. To facilitate the construction of the gun batteries Major Winslow ordered a channel cut in the offshore reef. Through this channel large quantities of dredged sand and coral were deposited at the low-lying site of the new battery between November 1908 and the end of 1910.

This small coast artillery post was named in honor of Colonel and Brevet Brigadier General Rene Edward DeRussy of the Corps of Engineers in 1909. From 1812 until 1865 DeRussy had served as one of the nation's leading military engineers.

As soon as the engineers had settled into makeshift quarters at Waikiki, and the location for Fort DeRussy's two batteries had been staked out, the collection of building materials commenced and excavations for the battery foundations were begun. Plans for the two-gun 14-inch/34-caliber battery were submitted by Winslow on November 18, 1908. It closely followed prescribed characteristics contained in new Engineer Mimeograph No. 109. This was, however, the very first dual 14-inch emplacement built. The design was essentially an scaled-up dual 12-inch disappearing battery, like Battery Selfridge at Fort Kamehameha. The two gun pits were separated by 274-feet rather than 240-feet. The large central traverse between the guns covered two powder magazines and two projectile rooms. The latter were now being built with shot tables inside, the projectiles being stored at a level facilitating rolling them onto shot carts for transfer to hoists. On each side an additional shot room was located inside and below the gun loading platform. The battery commander's station was located on the top level directly between the guns, and the plotting room was just below it. The usual selection of latrines, storerooms, and reserve power plant completed the design. It was oriented to fire to the south and southwest. Winslow's plan submission had estimated a cost of approximately $440,000, once again a high amount compared to what might have been expected domestically. (34)

Actual construction of the emplacement finally got underway in December 1908. Eight months later work was started on the adjacent battery for two 6-inch guns also to be mounted on disappearing carriages. Winslow lost little time in getting a hydraulic dredge in place to begin pumping coral and sand from the waters off the beach onto the land to raise the elevation of the site and fill in the marshlands. The dredging also served another purpose: that of creating a basin capable of accommodating the barges that would be bringing the 14-inch guns to the shore at the battery site. Once the reclaimed land had been allowed to settle, construction of the enlisted barracks, officers' quarters, and the various other post structures could be undertaken. The coral used to cover much of the initial reclamation of the swampland was to be given a thick covering of loam so that the entire area could eventually be grassed over. Construction of the massive battery for the proposed 14-inch guns was well advanced and by 1910 the process of throwing up the earthen parapet on its seaward front was underway.

General René De Russy, Army Corps of Engineers and namesake for Fort De Russy on Waikiki Beach. *Gaines Collection.*

Construction was begun in December 1909, and its two emplacements were ready for their armament in September 1911. The completed battery held two 14-inch M1907M1 guns on M1907 Disappearing Carriages. The 14-inch gun was a powerful and elaborate weapon. It was deemed as much more capable than its smaller 12-inch sister at Fort Kamehameha. The gun with a muzzle velocity of 2150-f.p.s. at maximum 15° elevation could launch is 1660-lb. projectile to 16,900-yards. Not long after installation, during the latter part of the First World War, the range was extended. An Ordnance Department program had developed a modification of the carriage to replace the elevating disc, arms, and elevating band to allow the gun to be raised to a 20° elevation. Slight changes even had to be made in the emplacement to accommodate the new recoil that developed. With these changes the gun could now fire its projectile to a range of 20,000-yards.

There was a considerable delay waiting for parts and instruction for finishing the electric wiring, and in waiting for delivery of the armament. Battery Randolph was transferred to the garrison on October 31, 1913, after only a portion of its armament had arrived in Oahu. When the first gun, arrived aboard the Matson Navigation Company's steamship SS *Lurline II* in August 1913, the off-loading facilities in Honolulu Harbor were inadequate for handling the 69-ton gun and its carriage. Also the gun was considered too heavy for

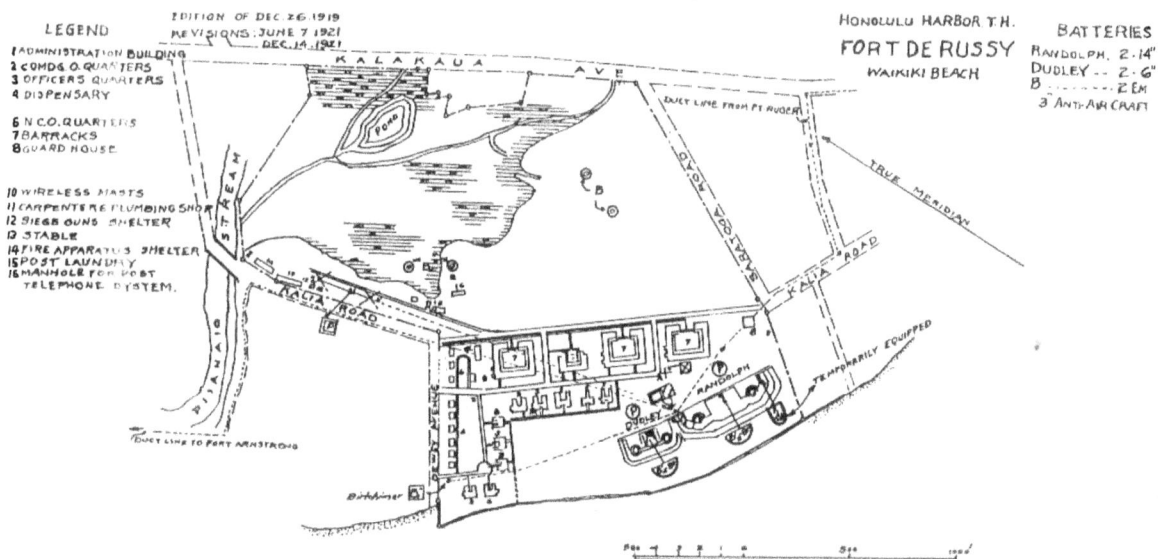

1921 Engineering plat for the layout of Fort DeRussy, Oahu. *NARA.*

Aerial photograph of Fort DeRussy from November 1932. Major gun batteries are located left, central. *NARA.*

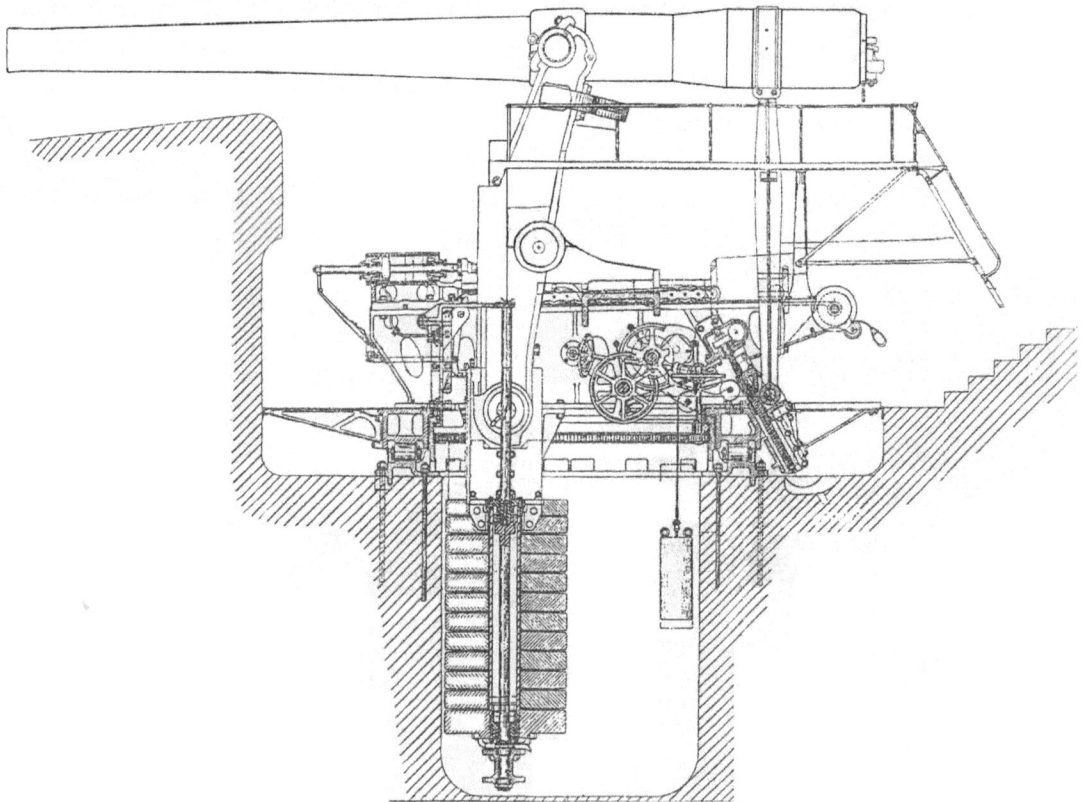

Ordnance Department side view drawing of the 14-inch Model 1907 gun and Disappearing Carriage as was mounted at Battery Randolph.
*NARA.*

transportation from Honolulu Harbor to the fort on the light rail system of the Honolulu Rapid Transit and Land Company. Consequently, it was necessary to take the steamship to Pearl Harbor where a floating crane with the capacity to lift the heavy gun, off-loaded the ordnance onto a barge and transported to the channel in front of Fort DeRussy. From there the gun could be winched up onto the emplacement and it's disappearing carriage.

The gun delivered to Battery Randolph was the first (and only) Model 1907 14-inch gun. Intended for test firing the new generation of guns, it was to be retained for proving ground usage. Probably due to the lengthy delays in finishing standard guns and carriages that were to follow, it was hurried off to Oahu in 1913. The next

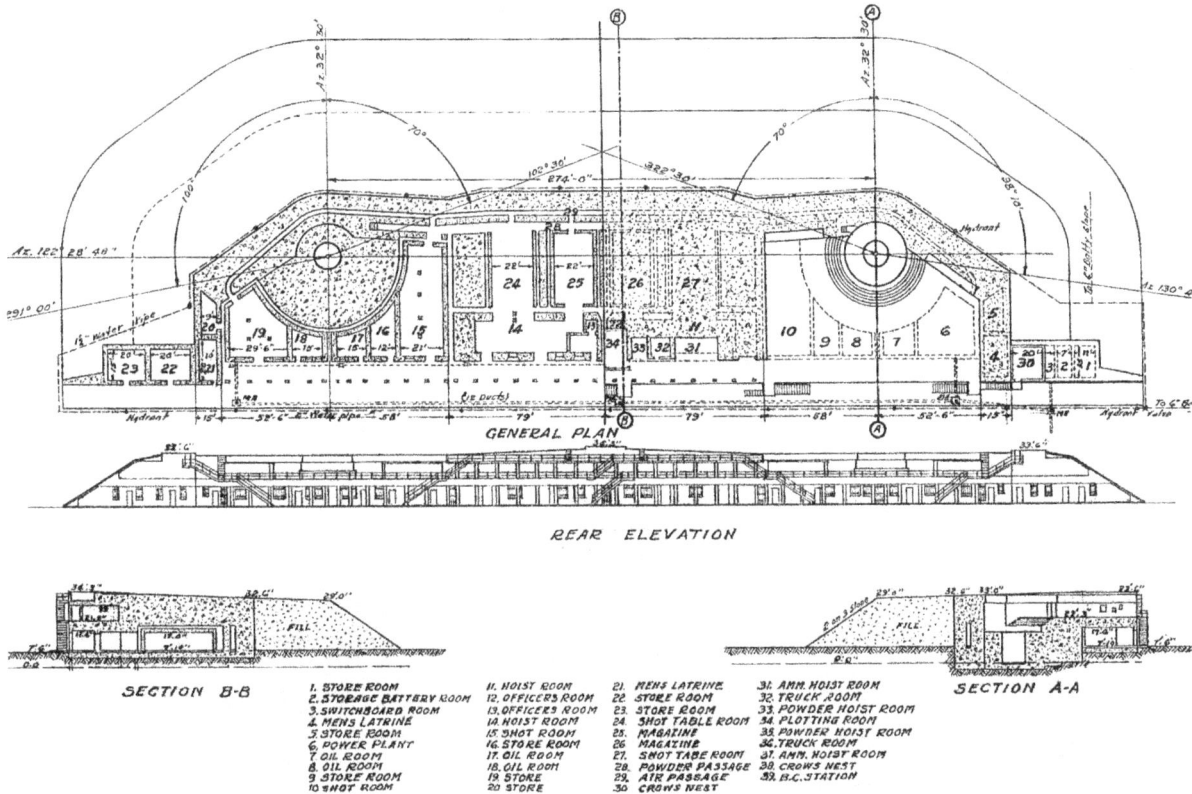

1929 plan, section, and elevation for 14-inch gun Battery Randolph at Fort DeRussy.
Note similarity to the plan for the slightly smaller Battery Selfridge at Fort Kamehameha. *NARA.*

Closeup of crew posing with their shell cart for loading Battery Randolph, Fort DeRussy. *USAMH.*

Photograph of the 14-inch gun at Battery Randolph.
Taken about 1918, the gun carriage has yet to be modified for increased elevation firing. *Schmidt Collection.*

A 1931 view of one of Battery Randolph's 14-inch gun with crew posing on top of it. *USAMH.*

four guns made were Model 1907M1 considerably different than that of the M1907. Gun No. 3 didn't get to the other pit of the DeRussy emplacement until two years later. Eventually the M1907 guns (serial No. 1) was returned to the U.S. for relining in 1923 and replaced with Model 1907M1 No. 1 (note the coincidence of serial numbers) in that year. Proof firing of the first gun was carried out on November 25, 1914. The second 14-inch gun and disappearing carriage for the battery were received and mounted in 1915 and was finally test fired on August 15, 1916. (35)

The power for the battery was supplied by commercial line—not a problem in central Honolulu. But as a safeguard a backup set and even a reserve generator for electric power (needed for the lighting of both Batteries Randolph and Dudley and for operations of the shot and powder hoists) was provided by a pair of gasoline operated four-cylinder General Electric engines that were direct connected to two 25 kW generators that were purchased in 1910 and 1911. The cost of the power plant was approximately $8000 and was turned over to the garrison on October 31, 1913. The emplacement was remodeled for long point projectiles in September 1916. The battery's primary fire control station was located at the Battery Commander's Station atop the central traverse of the battery. The battery's secondary station (B") was at the Point Leahi Fire Control Center at Diamond Head.

Upon its transfer to the coast artillery on October 31, 1913, the 10th Company, CAC, which arrived at Fort DeRussy on January 14, 1913, manned Battery Randolph. It continued to man the two 14-inch guns until 1924 when it was transferred, less personnel, to the HD of Cristobal in the Panama Canal Zone.

Battery Dudley for 6-inch guns on disappearing carriages was constructed at Fort DeRussy between September 1909, and July 1913. It was also the first of the batteries to be armed at this fort. Its armament consisted of two 6-inch M1908M1 guns on M1905M1 Disappearing Carriages. Major E. E. Winslow also drew up the plans for this battery, submitting them for approval on March 27, 1909. It was sited immediately adjacent to the right side of Battery Randolph, the left gun just 175-feet from the right gun of Randolph. It included a passageway/drain tunnel between the two otherwise disconnected structures. The concrete protection for the emplacement parapet wall of the No. 2 (leftmost) gun position was considerably enhanced, with the thought that it might be hit by large caliber shells being fired at Randolph. Otherwise, the battery conformed to the latest design of disappearing 6-inch guns as outlined in Engineer Mimeograph No. 59. It was a conventional

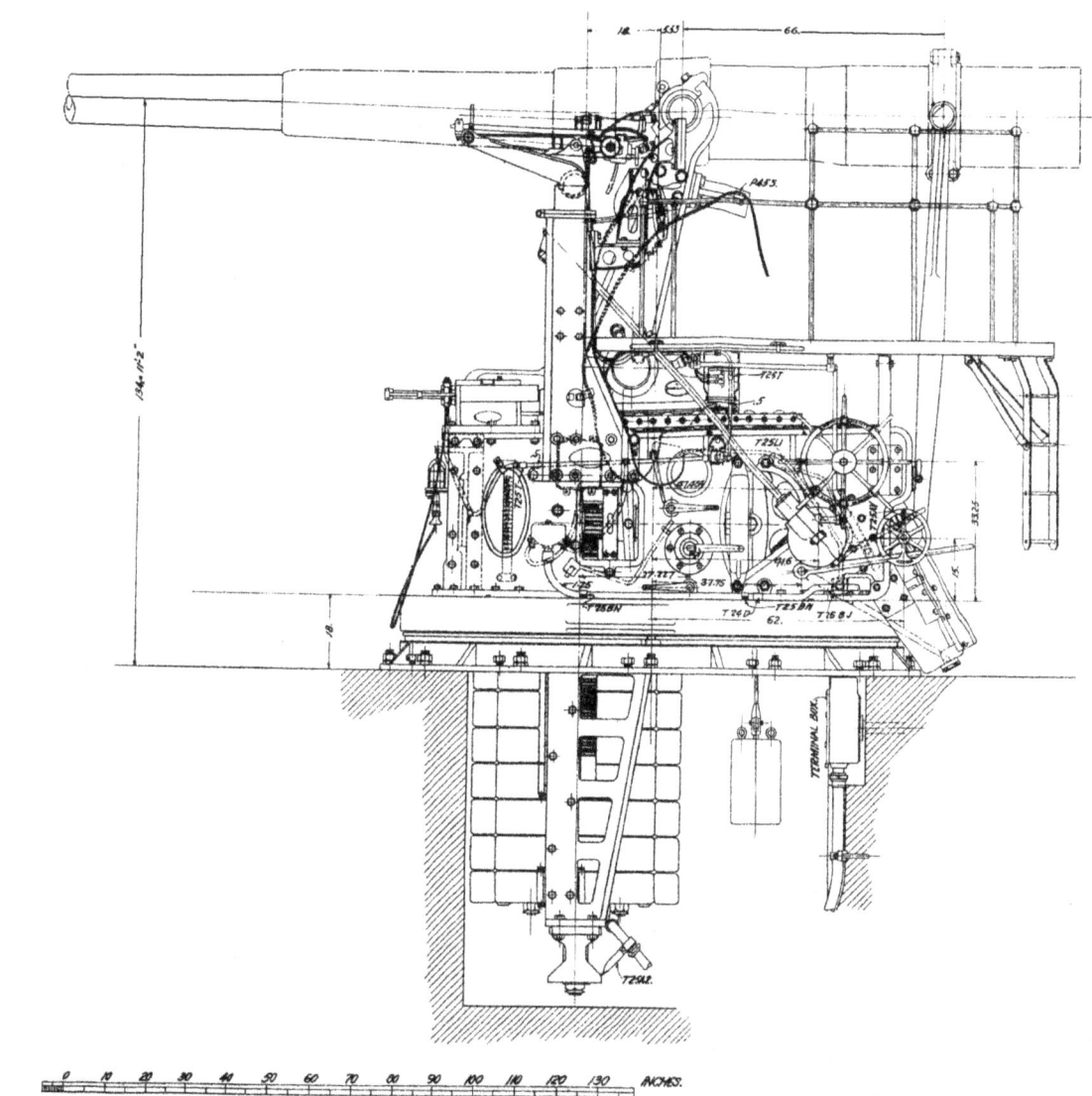

Ordnance Department side view sketch of a 6-inch gun on the Model 1905M1 Disappearing Carriage, such as was mounted at Battery Dudley, Fort DeRussy. *NARA.*

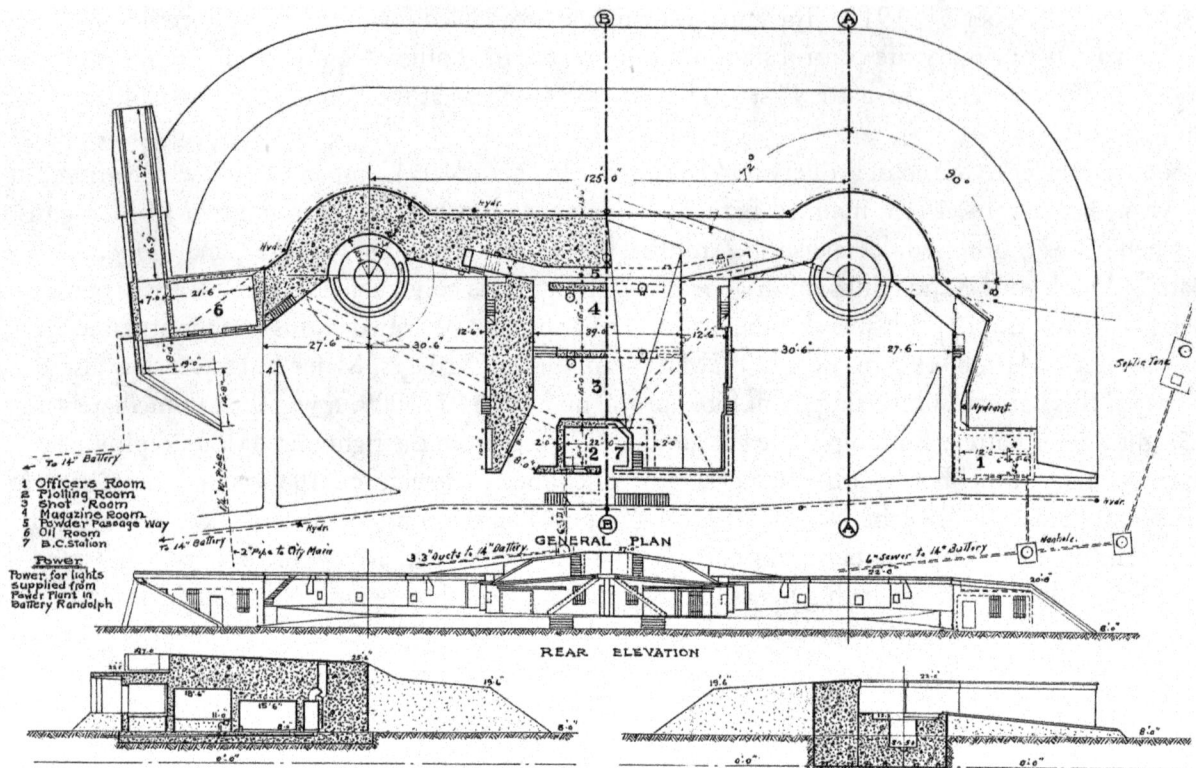

1919 Plan, elevation, and section view of Battery Dudley. *NARA*.

design with two separate gun positions, intervening traverse protecting the magazines. An expanded crow's nest station was built on top of the structure as a primitive sort of BC station, accessible by steel staircases. (36)

Work began a little slowly, in September 1909 only the foundation piles had been delivered to the site. As work concluded on Battery Randolph, manpower became available for work on Battery Dudley. In 1913, upon arrival of the ordnance aboard an army transport in Honolulu, the battery's guns and carriages were shipped to Waikiki on the flatcars of the Honolulu Rapid Transit and Land Company. The 55th Company, CAC, mounted the guns and carriages in July 1913. The battery upon completion at a cost of $75,000 was transferred to the Fort DeRussy garrison on August 5, 1913. The initial manning detachment for Battery Dudley (and Randolph), the 55th Company, CAC, which arrived in Honolulu May 13, 1913, and was assigned to the garrison until 1924. (37)

Battery Dudley's battery commander's station was atop the central traverse of the battery above the battery plotting room. This station also served as the battery's primary station (B'). There was also a 15-foot Bausch and Lomb Coincidence Range Finder (CRF) that was installed in the battery's CRF station at the northwest corner of the reservation. This station was built in 1931, and consisted of a 15-foot tall structural steel tower constructed around the concrete instrument pedestal on which the CRF was mounted.

The battle commander's command post and the meteorological station was housed in a three-story structure some 30-feet high at the rear of Batteries Randolph and Dudley. The two lower floors were built of concrete while the uppermost story was built of corrugated iron covered with asbestos. The lower floor contained a room for the senior officer at Fort DeRussy The second story housed the meteorological station and the upper story, measuring 14-feet by 16-feet served as the observing room for the battle commander and the mine commander.

Fort DeRussy received considerable criticism from the residents near the reservation. By 1910 the army's manner of sewage disposal proved to be less than adequate. The *Hawaiian Gazette* decried the manner used by the soldiers to get rid of their waste material. "Instead of digging cesspools," the newspaper reported, "the soldiers dump their refuse into the sea and unspeakable filth is to be seen at all times floating up and down the beach." Fortunately for the burgeoning hotel industry on Waikiki the tides tended to move the offensive

Construction at 6-inch Battery Dudley, this view shows the placement of the steel base ring for the disappearing gun mount. *USAMH*.

matter westward toward Ala Moana and Honolulu Harbor. Major Winslow refuted the claims of territorial health authorities that they had been prohibited from entering the reservation to investigate the complaints of the unhealthy condition, stating that they would require a pass signed by him, and noted that two concrete piers had been built out over the water on which two latrines had been constructed. The engineer went on to say that these conditions wouldn't change until the size of the garrison had been determined and congress had appropriated sufficient funds for construction of a projected pumping station that would carry the sewage out beyond the reef. (38)

In March 1911 Winslow received orders transferring him to Panama Canal Zone to design the canal's defenses. His successor, Major Wooten, arrived in the summer of 1911 and would remain in Hawaii until August 4, 1914. During his three-year tour, he would bring Fort DeRussy to near completion. Military affairs in general were a frequent topic of discussion in the Honolulu newspapers and Oahu's forts received a high level of coverage. The nature of Fort DeRussy's armament was of especial importance among the population of Waikiki. Rumors were rife as to the caliber of the guns initially and when it became known that a pair of 14-inch breechloading rifles would be emplaced at Fort DeRussy the date of their arrival was eagerly awaited by the media of the day. As early as 1911, when the United States Army Transport (USAT) *Sherman* stopped on its way to Manila, the two 14-inch guns on board the transport were thought to be for Waikiki. It must have been a disappointment when *Sherman* sailed on with its cargo still aboard. (39)

As early as 1910, plans had been formulated by the construction quartermaster on Oahu for the projected permanent garrison buildings on the island's posts and submitted for approval by the Congress. The initial plans had called for termite-proof concrete buildings and the first buildings built, those at Fort Ruger in 1910 and 1911, were built according to these designs. Their cost proved so prohibitive, however, that the decision

Practice firing of Battery Dudley in April 1938. *USAMH.*

was made to construct nearly all subsequent garrison buildings at Fort Ruger as well as those at Forts Armstrong, Kamehameha, and DeRussy of timber with slate roofs. The only concrete used was to be in the floors that were to be raised some four-feet above the ground with supports. Typically, the structures were single storied with peak roofs. The first government structure to be built on the Fort DeRussy reservation was the wireless (radio) station built in 1912. It was located on the south side of Kalia Road near the west end of the reservation at some distance from the batteries. Its antenna was sited across the road. (40)

Several more years of living under canvas passed before the long-anticipated funding for barracks and additional officers and NCO quarters was finally obtained. Construction of two, company-sized barracks at Fort DeRussy finally got underway in 1916 and were completed by February 3, 1917. These two barracks were built for $20,000 each were square one-story structures with a courtyard in the center. Each contained the dormitory for the enlisted men, toilets and showers, as well as a kitchen, pantry, commissary storeroom, and mess hall. The barracks also contained a day room, an office for the First Sergeant and the company office. Also built was a slightly smaller building containing the post headquarters, a post exchange, and theater, as well as the guardhouse and dispensary. This building cost $16,000. Much of the material used in the construction of these buildings was obtained by tearing down "old private houses located on the reservation and salvaged during the construction period." These three structures were ranged along the south side of Kalia Road that ran east and west through the post to the rear of Battery Dudley.

One set of field officers' quarters was built at a cost of $7200; four more sets of bungalow type family quarters for company officers were constructed at a cost of $6000. All were of concrete construction; and six sets of quarters for NCOs at $2500 each were built all were of the bungalow type frame construction. All were of similar design but were laid out as single-family dwellings. The officers' quarters were built in a row between the barracks and administration building and the two battery emplacements, while the NCO quarters were erected in a north-south line along Dealy Avenue on the west side of the reservation. (41)

Coast Artillery garrison barracks at Fort DeRussy in the 1920s. *USAMH.*

Once the size of the garrison had been determined and the sewage situation resolved by the employment of septic tanks, the water quality on the beaches was improved. In 1916, the garrison troops converted a part of the 37,000 square foot dredged area at the beach into an excellent swimming, and diving pool free of coral outcroppings and deep enough for all kinds of diving. A pier was built about 1912 and a diving platform erected in 1915. Bathhouses for officers and enlisted personnel and their families were built in 1922 for about $750 each. These facilities developed the fort's beachfront into one of the finest bathing beaches on the island. The diving platform was later expanded to three diving boards at varying levels and a third bathhouse was added in the 1920s for the ladies. On the beach a "tiki hut" was added for use by the troops. (42)

Construction of two additional single set family quarters for officers and three single sets of NCO family quarters were resumed at Fort DeRussy in 1919 and brought to completion November 3, 1920. Also built was a third barracks building similar in design but slightly smaller in size than the first two barracks at the post. This structure was built to house the Coast Artillery Recruit Training Center. With the end of the World War, basic training of newly recruited personnel was returned to various commands. In Hawaii, newly recruited personnel for the coast artillery were shipped out to Oahu where upon arrival they were assigned to the Recruit Training Center where they were instructed in the fundamentals of soldiering. Once his initial training was completed, the recruit was assigned to one of the coast artillery companies on the island for organizational training. (43)

The Fort DeRussy garrison had undergone reorganization in 1916 when the 10th and 55th Companies had been redesignated as the 1st and 2nd Companies, Fort DeRussy, respectively. A year later they were again redesignated, this time as the 9th and 10th Companies, Coast Defenses of Oahu. The early1920s saw major changes in the designations of the U.S. Army in Hawaii. The Hawaiian Division composed of infantry and artillery brigades was formed at Schofield Barracks in 1921. The Artillery District of Honolulu was redesignated the Hawaiian Coast Artillery District on April 5, 1921 and moved its HQ to Fort Shafter on June 21, 1921. Also on April 5, the defenses were separated into the CD of Pearl Harbor and the CD of Honolulu. The coast defense headquarters had initially been located at Fort Ruger, but when the garrison at that post was increased to four companies during the World War the headquarters of the defenses had moved to Fort Armstrong. When Fort Armstrong ended its primary coast defense role and was earmarked for transfer to the Hawaiian Branch Area Depots in September 1922, the coast defense headquarters was moved again this time to Fort DeRussy.

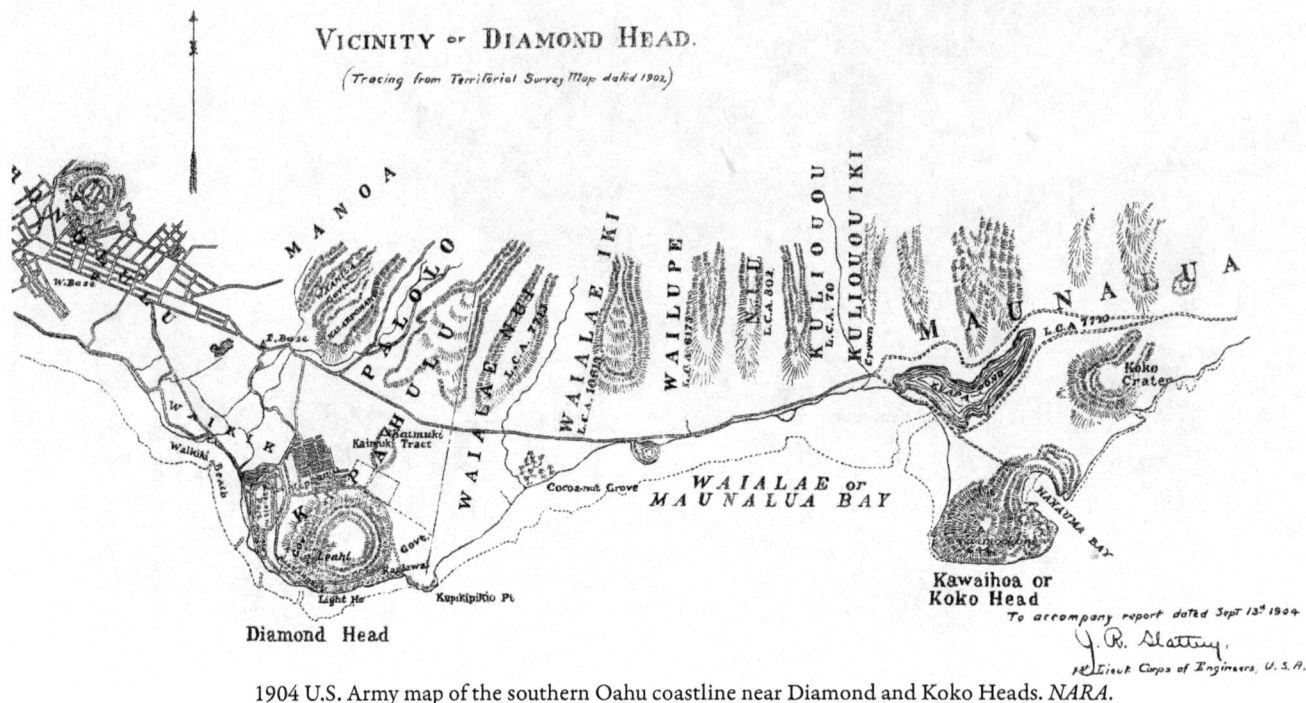

VICINITY ᵒᶠ DIAMOND HEAD.

*(Tracing from Territorial Survey Map dated 1902.)*

1904 U.S. Army map of the southern Oahu coastline near Diamond and Koko Heads. *NARA.*

Included in these crown lands acquired by the United States during the acquisition was Diamond Head, one of four prominent tuff lava cones created by ancient volcanic eruptions at the east end of the Waikiki section on the Honolulu side of the Koolau Range. Long dormant and likely extinct, Diamond Head is probably the most recognized topographic feature in the Hawaiian Islands. The Hawaiians called the Diamond Head crater Leahi. British sailors called the crater Diamond Head because calcite crystals found in the rock on the crater floor suggested diamonds. The southern rim of the volcano rises sharply at Point Leahi, 761-feet from the reef-strewn shoreline of Oahu's south coast. The remainder of the crater's hogback crest averages between 300 and 400-feet above sea level. Diamond Head, like much of Leeward Oahu, receives far less rainfall than the northeast or Windward side of the island. Consequently, natural foliage was sparse prior to the 1900s. This increased markedly as the U.S. Army planted and irrigated over the next seven decades, for aesthetic reasons as well as to camouflage the numerous military installations there.

In June 1898, Maj. William C. Langfitt organized the 3rd Bn, 2nd U.S. Volunteer Engineers, at San Francisco, and the battalion was ordered to Oahu on July 26. The location selected to quarter this unit was at the south end of the racetrack in Kapiolani Park, near Oahu's south shore. Here in the shadow of Diamond Head, Major Langfitt laid out the army's first establishment on Oahu, Camp McKinley, named for the incumbent President. The camp was soon relocated to higher ground some 400-yards toward the lower slopes of Diamond Head on what was known as the Irwin Tract. However, an inspection in the latter part of 1898, determined Camp McKinley to be in the "malaria belt." Consequently, the infantry was moved about three miles to a temporary camp on Waialae Beach, northeast of Diamond Head. They remained there until they embarked for the return to California on December 7, 1898. Throughout this occupation the engineers were engaged in extensive mapping of the island. (44)

In 1899, numerous other tracts were ceded to the Army in and around the City of Honolulu. One of these was the 2¼-acre "Barracks Lot" on the site of Old Fort Honolulu. Elements of the 10th Infantry and engineers continued to occupy Camp McKinley until 1907. The infantry battalion was projected to move to the newly established cantonment north of Honolulu on April 15, but they were ordered to Alaska on April 11, and the move to the new cantonment, named Fort Shafter in 1907, did not take place. (45)

Secretary of War William Howard Taft noted Diamond Head's military potential when he visited Oahu in 1905. On January 18, 1906, Diamond Head and Kupikipikio Point were reserved for military purposes by executive order of President Theodore Roosevelt. The initial 720 acres were designated the Diamond Head Military Reservation on January 27, 1906. On January 28, 1909, the reservation was named in honor of Maj. Gen. Thomas Howard Ruger, who served with distinction during and after the Civil War. Initially, the reservation encompassed all of Diamond Head crater, as well as the surrounding terrain, including Kupikipikio Point, called Black Point by the Army.

The first battery to be built at Diamond Head, two pits each for four 12-inch seacoast mortars, was begun in 1907. After completion it would be named Battery Harlow. Capt. Curtis W. Otwell, CE, obtained approval of the site located on the foot of the slope of the crater on its northern, or landward side in May. The design proposed was a standard two-pit mortar battery with flank and central traverse magazines. The rock foundation proved adequate for construction. As it was located behind the crater no forward earth protection was needed, though the flank walls were slightly extended. Another unusual feature was a basement on the left flank to house the two electric power generators. The cost estimate of $227,832.40 was unpleasant but accepted (Washington commented that the usual continental expenditure was just $120,000 for such a battery). (46)

The 12-inch mortar was a mainstay of American coast artillery. One of the first primary weapons developed for the Endicott Period, it was designed to complement flat-trajectory heavy guns with a high arc trajectory. It was a rifled, breech-loading howitzer intended to lob heavy shells, albeit at the sacrifice of range, onto the lightly protected decks of hostile warships. Usually emplaced in clusters of four mortars per "pit" the projectiles would descend on enemy fleets in batches of four, eight, or even 16 rounds. The standard weapon was the

Aerial photograph from 1938 showing the north, or backside of Fort Ruger's Diamond Head crater. *USAMH.*

Major General Thomas Ruger namesake of the fort surrounding Diamond Head. *Gaines Collection.*

Model 1890M1 12-inch mortar mounted on the Model 1896 carriage. It fired a 700-lb. projectile at a high elevation of 70° to a range of 15,291-yards, with a 360° field of fire. An improved Model 1908 mortar was used at Fort Kamehameha, and other M1890s were available in the 1920s on a railway carriage.

Construction progress was slow, as volcanic rock quarried from the battery site had to be crushed to make sand for the concrete. Maj. E. Eveleth Winslow arrived in late October 1908 and made some modifications to the battery plans. The two adjacent pits each were to be armed with 12-inch Model 1890M1 mortars on common Model 1896 carriages. Between the two pits was a reinforced-concrete central traverse topped with several feet of earth and rock. Inside were two powder magazines and two projectile rooms. The left and right traverses also contained projectile rooms, as well as powder magazines. The right traverse additionally contained a guardroom, an officers' bunkroom, and a storeroom on the main level. The left traverse design departed somewhat from the standard 1906 plan, extending to the rear of the emplacement and containing an enlisted latrine. Water was supplied from a connection to the city's municipal supply, though a steam pump was supplied to help overcome low water pressure.

A lower level contained four additional storerooms and two power-generator rooms, each with a 50 hp gasoline-powered four-cycle General Electric engine direct connected to a 25 kW 125-volt GE generator. This plant supplied electricity for both the battery and the intended fire control switchboard room at the north entrance to the Kapahulu Tunnel in the crater's north rim. Another departure from the standard plan was the battery commander's station (BCS) and plotting room in the upper level of the battery's central traverse. Since actual observation of the ocean approaches was blocked by Diamond Head's north rim, this station was more effective for the defense of Fort Ruger's land fronts. For the battery's seaward defense role, a second BC station was built in the Point Leahi fire control complex atop the south rim. The first 15-ton mortars arrived at Honolulu in June 1908; all eight had arrived at Fort Ruger by the end of the year and were mounted on their carriages by April 6, 1909.

The engineers pronounced Battery Harlow complete on February 18, 1910 and transferred it to the Fort Ruger garrison on March 16. Construction to that time had cost $205,000. An ordnance officer was sent from

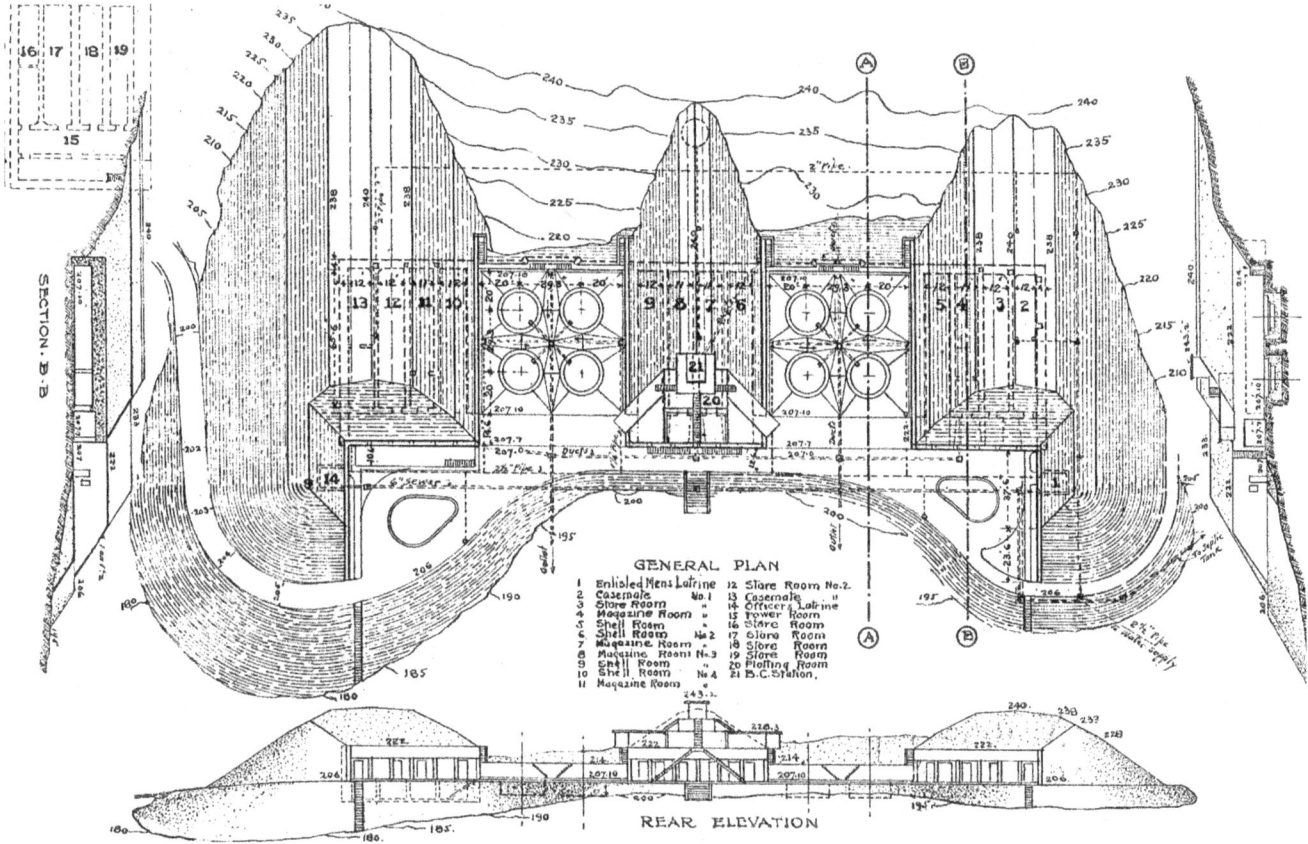

Plan, elevation and section from 1919 of mortar Battery Harlow. *NARA*.

Thanks to the camera of an ordnance specialist sent to Oahu, there is a good series of photographs showing the construction and erection of armament in Battery Harlow 1909-10. *McCarthy Collection, CSA*.

Here the base ring for one of the mortars awaits emplacement. *McCarthy Collection, CSA.*

Benicia Arsenal in California for proof firing the battery in July 1910. It did not go well. Widespread damage occurred to the gun carriages. Two racer rings were entirely broken, a pedestal for vertical traversing shaft was broken, numerous traversing wheels broke, and a multitude of small parts on the guns and emplacement itself (brackets and lamps) were smashed. Altogether three guns in Pit B were disabled, and one in Pit A. It would take over $15,000 and almost a year to make replacement parts and have them installed. The "best" remaining mortars were maintained in case an emergency required their use prior to the entire battery being restored. Beyond installation of the second 25 kW gasoline-powered generating set as a reserve in 1914, little was done to modify or improve Battery Harlow during the next two decades. Until 1915, Battery Harlow was Fort Ruger's sole seacoast battery. It covered a 15,000-yard arc from the Pearl Harbor Channel eastward to the waters off Makapuu Point, and if necessary, the mortars could fire over the Koolau Range into Kaneohe Bay. (47)

The picturesque magnificence of Diamond Head's crowing peak is still today synonymous with Oahu. It was inevitably an attraction to the army personnel in Oahu. Many found it almost obligatory to make the

Battery Harlow nearing completion in 1910. *NARA.*

View from about 1916 of Battery Harlow's A pit with model 1890MI 12-inch mortars on M1896M2 carriages. *Schmidt Collection.*

Battery Harlow had mechanical display indicators used to communicate range settings to the mortar gun crews. *NARA.*

climb up the trail—which admittedly became much easier and safer with the completion of the path the army built to get to the Leahi Peak fire control stations. Before then it was more of an adventure, as described once again by Anne Winslow in March of 1909: (48)

> Eveleth and Colonel Biddle invited the "family" to go along this morning on the trip to Diamond Head, where the mortar batteries and other interesting things are in the process of construction. You can see Diamond Head in almost all the pictures of Honolulu. The base of it shows in the little picture I put in this. It is an old volcano and towers up as clearly as a knife blade straight out of the sea. I have wanted to go in a mild way ever since I have been here, and Eveleth has been determined to get me to the top of it, so this morning the expedition came off. You have climbed mountains in a famous way, I know, but I don't think you ever climbed one that was built of dust and ashes without a spring of anything to catch hold of in case your hoofer slipped. To my mind it was a nightmare, and I don't know now what got into Eveleth to take us to such a place. It will always be a regret to me that I didn't take the camera and get some pictures of the children cavorting around the dizzy ledges of rock.

Once construction of Battery Harlow was well underway, the engineers began a second major project: a large center on Diamond Head's south rim to provide fire control and direction for many of the major seacoast batteries defending Honolulu. Captain Otwell began planning for this multi-tiered fire control complex when he prepared drawings for fire control stations on the island's south shore in 1908. In September, he submitted his plan for the Diamond Head stations to the chief of engineers. When Major Winslow succeeded Otwell, he reviewed Otwell's remarkable plan for a three-tiered group of primary stations in Leahi Point, Diamond Head's highest elevation. This obvious location offered an unobstructed view of Oahu's south shore. From its 761-foot summit on Diamond Head's south rim, one could see more than 30 miles to seaward and observe the south shore from Makapuu Point, the island's easternmost point, to Barbers Point west of Pearl Harbor. In addition

Some of the 12-inch mortar projectiles meant for Battery Harlow, Fort Ruger. Photo from about 1915. *Gaines Collection.*

Helicopter photo from 2009 of the "stacked" fire control stations at Diamond Head's Leahi peak. *Williford Collection.*

to starting construction of gun fortifications, Winslow revised and expanded Otwell's plan for the Diamond Head stations from three tiers to four, with a fifth station on the peak's summit that raised the elevation of the Leahi's summit three-feet.

Before Major Winslow could begin construction of the fire control installation, however, he had to bore a tunnel through the crater's north rim to move building materials to the construction site. Kapahulu Tunnel, as the modest tunnel was named, was some 200-feet below the 400-foot crest of the north rim and some 300-feet west of Battery Harlow. Initially, the tunnel was only five feet wide, seven feet high, and 580-feet long. After the tunnel was completed, a pathway was cleared across the crater floor to the interior slope of the south rim. About 350-feet up the interior slope of the south crater rim, a dormitory was built for enlisted personnel assigned to the fire control stations on the crater rim. This two-story structure was 48-feet long and 18-feet wide, constructed of cement plaster on wire mesh supported by a timber frame. Nearby, a 10 by 12-foot latrine was built of the same materials. Construction material was towed in small rail cars by mules or pushed by hand along a narrow-gauge railway through the tunnel and across the crater floor to the foot of the south rim. From there, a footpath wound up the interior slope of the south rim from the crater floor to the construction site.

For heavy loads that would be difficult for men or mules to manage on the narrow zigzag trail up the slope, a cable tramway was also built, powered by a generator in a building near the upper tram landing. This tramway stopped where the construction began, some 200-feet short of the summit. At the upper terminus of the tramway, some 250-feet above the floor of the crater and 642-feet above sea level, a small officers' dormitory 24-feet long and 13-feet wide was cut into the rock of the interior slope. From this level, a concrete staircase climbed another 40-feet to the entrance of a second tunnel. Like that through the north rim, this concrete-lined 225-foot tunnel also had a five-foot bore and seven-foot crown. The south end of this tunnel opened into a narrow ravine between two spurs of the interior slope of the south rim. Nearby were a switchboard room, 15-feet wide and 20-feet long, and a storage battery room 15-feet by 4½-feet. From the level of the switchboard room, a second concrete-lined stairway with 99 steps led up the ravine to a third concrete-lined tunnel, which entered the rim of the crater and afforded access to Leahi Point. Along this tunnel were storerooms, and at the midpoint, a spiral metal staircase led up to the higher levels of the fire control complex.

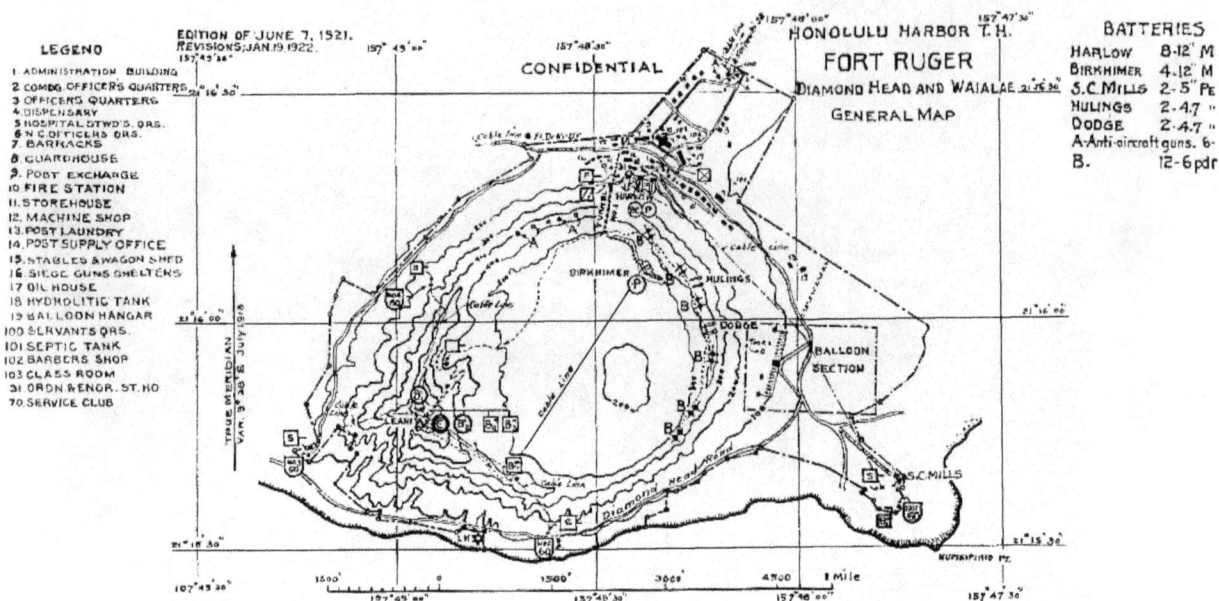

1919 Engineer plat of Fort Ruger, note the main cantonment area to the northwest of the Diamond Head crater. *NARA.*

From the top of the spiral staircase, a ladder gave access to the open-air station on the summit of Leahi Point. The complex and its supporting structures were completed in 1910 and transferred to the garrison on January 20, 1911. The first or lowest of the four interior levels of the fire control complex, with an instrument height 711-feet above sea level, contained storage and dormitory spaces on either side of the access passageway. At the south end was the secondary station (B") for the 14-inch disappearing guns of Battery Randolph at Fort DeRussy. This 12 by 16-foot observing room was equipped with an M1907 Lewis depression position finder (DPF). In 1934, the fire control scheme of the HD of Honolulu was revised, but this station continued as B" for Battery Randolph. The second level observing station, with an instrument height of 723-feet, was 12 -feet wide by 18-feet deep and initially assigned as the primary station for the Second Fire Command (F'2) at Fort DeRussy and the secondary station (B") for Battery Dudley at Fort DeRussy. The station was equipped with an M1908 seacoast observation telescope and an M1907 Lewis DPF.

In addition, two niches for emergency azimuth instruments were cut into the front wall of the station. The third-level station, with an instrument height of 734-feet, was the primary station for the First Fire Command (F'1) (which in 1915 consisted of Battery Harlow's two mortar pits) and B' Harlow. The station was also equipped with a class DM1 M1907 Lewis DPF and an M1910 Warner Swazey azimuth instrument. The fourth level, with an instrument height of 745-feet, was initially the battle commander's station, the C'1, often termed the "C station." In the early 1920s, this became the observation post for the harbor defense commander. The 12 by 18-foot observation room of this tier afforded a somewhat wider field of vision than the lower stations. It was also equipped with an M1907 Lewis DPF and an M1908 seacoast observation telescope. Behind this station were five telephone booths, each three-feet by three-feet, six-inches, and a private 12-foot by 10-foot, 6-inch bunk room and office for the battle commander, who commanded Fort Ruger and later the Honolulu Harbor Defenses.

Leahi Peak was excellent for observation stations, but it was also obvious to hostile fire, and couldn't be completely camouflaged. There is an interesting description of its vulnerability in a 1912 letter otherwise concerning the equipment allocated to the various stations: (49)

> The prominence of Diamond Head makes it an excellent mark for an enemy. Its height is no protection. We must assume that an enemy will have information regarding the use we make of Diamond Head; he will know of the existence regarding the use we make there; he will know of the existence of the range stations there; moreover he can see the tell-tale slots in the face of the cliff. Though we might make earnest efforts to conceal

them. Directing his guns upon this point, the fall of those shots where were short could be accurately observed, and, in a short time comparatively, one or two good hits would be made and the disruptive effect of a bursting shell filled with high explosive would get throughout all the stations; a few more such hits and instruments would be shaken and thrown out of adjustment, concrete pillars would be cracked, ceilings would crash down, instruments topple over and smashed, the entire top of the Head melted into a heap of ruins, burying telephones, cable and all other station paraphernalia in the depths of the wreck.

Fort Ruger was usually manned by three or four companies of coast artillery. The first two of these companies, the 105th and 159th Co., Coast Artillery Corps (CAC), under Maj. John Kirby Cree, arrived from San Francisco on August 12, 1909. After landing at Honolulu, Cree's command spent the night at Fort DeRussy and the next day marched through Waikiki to Fort Ruger. Reaching the barren mass of volcanic rock on the afternoon of August 13, they moved into the handful of timber shanties formerly occupied by Battery Harlow's construction crew and set up a tent camp on the heights north of the crater near Kilauea Avenue. The 105th Co. manned Battery Harlow's Pit A; the 159th took over Pit B, and Major Cree assumed command of the artillery district. The 105th Co., commanded by Capt. James R. Pourie, and the 159th Co., under 1st Lt. Chester H. Loop, were soon assigned to make Fort Ruger a livable post. In August 1909, the Quartermaster Department had not yet provided quarters for the new garrison and the construction shanties used by the construction gangs during the building of Battery Harlow were little better than the tents used by the newly arrived troops. While they awaited the completion of Battery Harlow, the two companies were placed on fatigue duty almost exclusively. (50)

The two companies cleared a parade ground of the lava rock scattered across the entire reservation, cleared underbrush, and built temporary barracks and quarters for the officers and their families. Trees, grass, and ornamental shrubs were planted. A water system for the temporary cantonment was also installed, as well as

Aerial photograph from 1932 of Fort Ruger's cantonment area. Battery Harlow can also be made out on the slope of the volcanic crater on the upper left. *Bennett Collection.*

Fort Ruger large barrack building as viewed from Battery Harlow in the late 1910s. *Schmidt Collection.*

a sewer line and septic tank. Other than administrative tasks, guard duty, and other necessary military duties, police duties were the order of the day from August 1909 until the next April. As no appropriations had been made for roads, the troops had to build their own, using what equipment and materials they could find. Major Cree borrowed a big road roller from the City of Honolulu in exchange for providing troop labor to the city. Much of the road-building material at Fort Ruger was quarried from the exterior slope of Diamond Head in the rear of where the NCO quarters would be constructed a few years later. The rock was hauled from this quarry to building sites on small cars the troops called the "Jumping Ruger Pacific" railway (the initials of the commander of the 105th Co. being J.R.P.). (51)

While the troops were required to live under canvas during those early months at Fort Ruger, plans were underway to provide permanent quarters for the garrison. The initial architectural drawings for the forts on Oahu completed in 1910 were finally approved so construction of permanent post buildings could be started in

One of the coast artillery company barracks buildings from the 1920s. *NARA.*

mid-1911. By June 30, 1913, $11,952.89 had been expended on temporary post buildings and $203,222.34 on permanent structures. Emphasis was first placed on constructing a barracks for two companies at the south end of the parade ground near the foot of Diamond Head's north slope, behind Battery Harlow. Construction continued at Fort Ruger through 1911 and all of 1912, when the first phase of post development was completed. In addition to the two-company barracks, an ordnance machine shop was completed in 1911. By the end of 1912, a concrete two-story field officer's quarters had been completed for the post commander on the bluff at the reservation's north end, near the intersection of Kilauea and 16th Avenues. Two more double sets of two-story concrete family quarters for officers were also completed along the west side of the parade ground in 1912, along with a bachelor officers' quarters. Coconut palms were planted around the edge of the parade ground, and a stable was built at the foot of Diamond Head's north slope west of Battery Harlow.

The post administration building, a mess hall, and permanent guardhouse were also built across Monsarrat Road (today's Diamond Head Road) from the barracks. Six sets of NCO quarters were erected along the south side of the road adjacent to the rock quarry east of Battery Harlow. Fort Ruger's garrison buildings, along with those of Fort Shafter, proved the only ones of that design built on Oahu. Being concrete, they proved very costly; thereafter, standard-design frame buildings were used at Forts DeRussy, Armstrong, and Kamehameha. (52)

The Cosmopolitan Boring Co. contracted to drill an artesian well north of the fort to provide reliable and secure water supply. When the boring was completed on June 29, 1911, the Lord-Young Construction Co. installed a pump to deliver 1,317,000 gallons of water daily. A pipeline was constructed from the well to the garrison area and thence through the Kapahulu Tunnel into the crater, where a large concrete reservoir was constructed high enough on the interior slope of the north rim to insure adequate gravity pressure.

On Tuesday, September 19, 1911, the repaired mortars of Battery Harlow were fired in the battery's first full-charge service practice. The navy tug USS *Navaho* towed two eight-by-eight-foot floating targets, each marked with large red flags, some three miles offshore. One target was anchored while the other was to be a moving target. As *Navaho's* radio was not functioning, signaling between the fire control station and the tug was by semaphore flags. Before the tug could reach the seven-knot towing speed, the targets became entangled and two hours were spent getting the two targets separated. When the practice was finally ready to be fired that afternoon, *Navaho* resumed course towards Koko Head. The mortars fired 10 rounds, and according to some observers, all 10 shots fell on the target or close enough to qualify as a hit. Maj. Edward J. Timberlake, CAC, commanding Fort Ruger noted: "Without having complete data at hand I am almost certain the mortars gave the Fort Ruger companies 75% hits out of the ten shots fired at a moving target. The shooting was quite satisfactory." (53)

A new mortar battery projected for Diamond Head's crater floor had its genesis in a pre–World War I plan to build mortar batteries around the island of Oahu to cover beaches that could be used by an invading enemy. The mortars would force enemy vessels to off-load landing forces much farther to offshore, in the open seas. Should an invasion force still attempt to land, it was believed that the mortars would make short work of the enemy transports and fragile landing craft. In 1912 a local board of officers chaired by General Montgomery M. Macomb considered the defenses of Oahu and examined the complexities of additional seacoast and land defenses. It specifically recommended two new 4-gun mortar batteries—one at Ahua Point at the eastern end of Fort Kamehameha, and one at Fort Ruger, inside Diamond Head Crater. The 15,000-yard range of the mortars covered the waters and beaches from Maunalua Bay on the east of Diamond Head to Waikiki Beach on the west.

Because of the protection afforded by the crater, it was thought that the standard mortar battery plan could be economically modified in this case. Its interior location would allow a strong defense against land attack, and to maximize its usefulness against land forces, it was to be supplied with extra high-explosive ammunition and a separate fire control system. The cost was estimated at $100,000 for engineering work and an additional $76,000 for the four mortars. (54)

Coast artillery soldiers at their quarters at Fort Ruger in 1916. *Schmidt Collection.*

One of the Fort Ruger NCO quarters in the 1920s. *NARA.*

Inside the Fort Ruger post exchange in 1916. *Schmidt Collection.*

SECTION. B-B

1 Store Room
2 Tool Room
3 Plotting Room
4 Shot Magazine
5 Powder Magazine
6 Shot Magazine
7 Officers Latrine
8 Men's Latrine
9 Store Room
10 Store Room
11 Engine Room
12 Fan Room

GENERAL PLAN

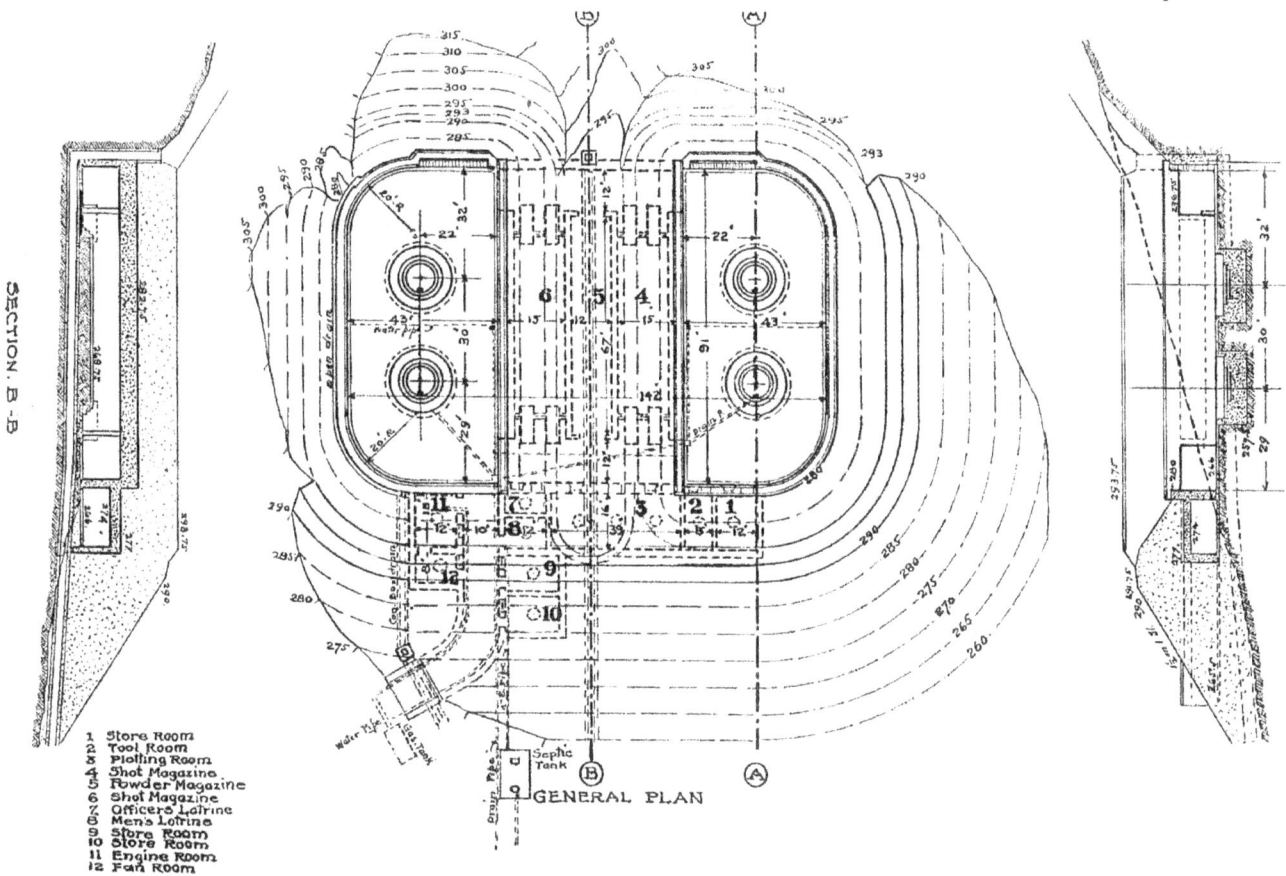

Engineering plan for mortar Battery Birkhimer, dated as of June 14, 1919, thus for the original design before rebuilding. *NARA.*

The suggestions of the board subsequently evolved before any funding was approved. In 1913 the War Department looked for ways to save money on the plan. The mortar battery for Ahua Point was dropped, and savings for the Fort Ruger battery were made by suggesting arming it with existing mortars from U.S. mainland batteries rather than new ordnance. The four mortars would eventually come from the battery at Fort Dupont, in the defenses of the Delaware River. The most expensive single item, the ammunition supply (some $145,000), would also come from domestic stores. Quickly approved, work was well underway by mid-1915. When the four mortars and carriages were delivered to Fort Ruger, the tunnel through the north rim proved too narrow, so it was necessary to drag them over the crater's north rim and lower them into the mortar pits. By April 1916, the battery was completed, but not transferred to the garrison.

Unfortunately, the design was seriously flawed. The four mortars were emplaced in small, deep pits, two on either side of a central traverse containing the usual powder and projectile rooms. On one end was an open gallery with offices, latrines, and the plotting room.

On July 3, 1916, engineer Maj. Robert R. Raymond penned a letter about proof firing the battery. Raymond obviously was not responsible for the design and was so leery of it that his proof firing trials were designed in anticipation of problems. The firings occurred on June 29, 1916. Raymond described the construction as a creditable piece of concrete work but expressed concern about the high vertical walls of the pits. Because of the confined character of the pits, problems with blast were anticipated, and the guns were to be fired over the entrance to the corridor in front of the plotting room. (55)

To measure the blast effect, special tin pans with recorded water levels were placed at key locations. Dummy men were prepared by artillery troops and placed near windows in the plotting room and latrine and live chicken hens were placed on leashes in the corridors. Each mortar was first fired twice, once with reduced charge and then with the full-service charge. The first firing of just one mortar with reduced charge destroyed the windows of the storeroom, tearing the sashes to pieces and hurling parts about. Fortunately, the hens were not injured, and assistant engineer Mr. Burbank, who volunteered to stay in the corridor, reported no stress. However the dummies were thrown down. (56)

For the second shot Raymond decided to risk staying in the plotting room. This time with a full charge the dummies were entirely thrown about and the men felt a violent shock. Subsequent firings of the mortar in the front of Pit A were even worse. There was no doubt that any board in the plotting room would severely displaced. In Raymond's opinion no plotting could be done with the battery in action. Subsequent firings destroyed the windows in the entrance gallery, and the water measurement pans were crushed, as was the hood of the light on the parapet. Stones and gravel were dislodged on the crater above the pit and would have interfered with ammunition trucks run to the guns, although no damage occurred to the actual mortars or gun blocks. It was generally agreed that "the confined pits so increase the effect of blast as to interfere to a considerable extent with service of the mortars and that the plotting room cannot be used in action." The team witnessing the proof firing at first thought a new plotting room outside the battery would be necessary.

It was obvious that the mortars could not be serviced as they were, with or without an adjacent plotting room, it was decided to move the mortars and leave the plotting room and magazines where they were. Moreover, in the original design, the pits formed shell traps. Since the battery's location inside the crater meant it would be exposed only to high-angle shells or aerial bombs, its funnel-like shape would concentrate hits and damage. If the mortars were moved outside the pits into the open, the dispersion would make the mortars safer than leaving them in the heavily protected pits, and were served by the existing magazines. This line of thinking provided the engineers with a convenient subterfuge to fix the battery without directly admitting that the original design was flawed. (57)

A reserve magazine had just been requested for Fort Ruger. As the eastern terminus of the land defense "redoubt" envisioned for Oahu, extra ammunition was needed to serve the field artillery pieces and new land defense batteries to be stationed at or near Fort Ruger. The plans also called for both mortar batteries, Battery Harlow and the new Battery Birkhimer, to have twice the ammunition supply of regular seacoast batteries—some at least to be high explosive or even hoped-for shrapnel shells. These needed protective storage convenient to the defense. Once the mortars were moved outside to new platforms, the old pits were roofed over and converted to additional ammunition storage at a fraction of the cost of a new magazine.

After some clarification with Washington, the project to relocate the battery's mortars was approved. Both Raymond (by now promoted to Lt. Col.) and Lt. Col. Eben E. Winslow in Washington were careful to not describe the process as "abandonment of the battery," but merely relocating the mortars more advantageously. Both the mortar relocation and the magazine conversion were ultimately approved, but there was a delay in funding. The 1917 appropriation did not contain money for the project, and apparently several more years went by without work. The mortars sat in their flawed pits, but the battery was not actively manned. It appears likely that the eight rounds fired (two per mortar) during the proof firing were the only ones ever fired in the original configuration. The report of completed works lists the battery as having been remodeled from February 1920 to June 1921 and transferred to troops (it appears the earlier configuration was never transferred) on June 17, 1921. Total cost, including the magazine enhancement, was by then $192,589—over twice the original projection. The four mortars now sat were in a line outside the southern pit, entirely exposed except for the protection offered by the crater. A new entry portal for ammunition was cut through the berm into the rear gallery and new handling tables built to serve the guns. Each of the former pits was subdivided into three long magazines. (58)

While Battery Birkhimer's initial mission was as a part of the land defenses of Fort Ruger that included coverage of the beaches east and west of Diamond Head, its secondary mission was to supplement the seaward defenses. To facilitate seaward fire control, Battery Birkhimer's reinforced-concrete battery commander's station was built atop Diamond Head's south rim. The battery was provided with two baselines, one for land defense of the area between the shore and the Koolau Range east of Fort Ruger and a second for seacoast defense. Four fire control stations served these two baselines. The primary station for its east (land) baseline was built in 1917 on the south rim of Diamond Head, about 1,480 feet southeast of the Point Leahi Station. The

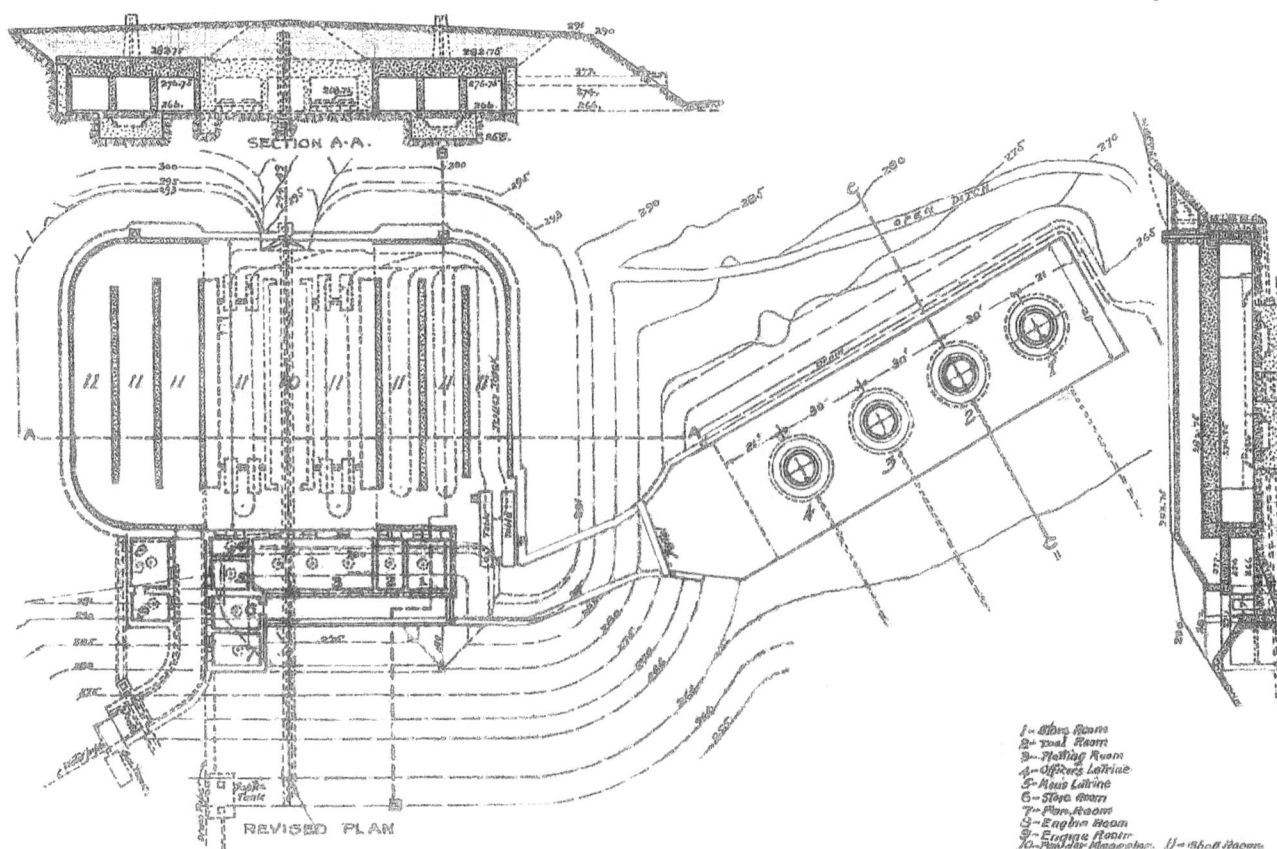

Engineering plan for mortar Battery Birkhimer, dated July 7, 1933, thus after thorough reconstruction involving moving the mortars outside of their original pits. *NARA.*

secondary station for the east was built on Puu-O-Kaimuki, at the southwest corner of the Kaimuki Reservoir. The observing instrument was on top of a concrete pedestal that extended up through a hole in a tower. The primary station for the battery's west (seacoast) baseline was also built on Diamond Head's west rim in 1916, with the other fire control stations at Point Leahi. The station was essentially the same as the east primary. Both primary stations were transferred to the garrison on March 9, 1917. The secondary station for the west baseline was 11,572-feet west of Leahi Point at the southwest corner of the Fort DeRussy reservation. Its observation room sat atop a 29.5-foot steel tower.

As early as 1910 installation of seacoast searchlights were planned for Fort Ruger. The first two of four searchlights at Fort Ruger were installed in 1911; one at Diamond Head and the other at Kupikipikio (Black) Point. A third light was emplaced at Diamond Head in 1913 and the fourth in 1920. Initially the searchlights at Diamond Head were manned by details from the coast artillery companies at Fort Ruger.

At the end of 1919, Fort Ruger had six officers and 210 enlisted men present for duty, distributed among four companies. The companies averaged about 49 enlisted men, approximately half nominal peacetime strength, and considerably less than wartime strength. During an inspection in December 1919, the command presented a less than outstanding appearance. Uniforms were faded, with considerable color variation; spiral puttees and leggings were worn interchangeably; equipment was only in fair condition, and men were missing articles of clothing, toilet, and mess kits. The troops required considerable individual instruction in the school of the soldier and military bearing, as well as in their physical appearance and training. Several men in each company had no identification tags; packs were not made up or adjusted correctly, and many shoes were unshined. The barracks were also less than outstanding. Window screens were missing or in need of repair, rifle-cleaning oil was found on the barracks floor, the basements flooded when it rained, the roofs leaked, and the interiors were generally unkempt. The gun and mortar batteries themselves fared only slightly better. Battery lighting was inadequate, telephone batteries were useless, and there was a general lack of attention to

5th Group Obs. Balloon Encampment, Ft. Ruger, B-952, 58.8-4.3 9-15-21, 10:04 AM. 1200.

Aerial photograph from 1921 of the new air service facility (balloons) at Fort Ruger. *USAMH.*

detail in the maintenance of the battery and its equipment. Further, the fire control stations and searchlights needed cleaning and maintenance. (59)

Fort Ruger became a hotbed of activity over the next few weeks as the deficiencies uncovered during the inspections were managed; by January 19, 1920, the commanding colonel was able to report that all areas of concern were being corrected.

The Army began active deployment of a new balloon-based aerial service early in the 1920s. The 3rd Balloon Co. of the Army Air Service arrived at Fort Ruger on May 6, 1920, to test fire control and direction of seacoast artillery from captive observation balloons. An Air Service depot was established on the east side of the Fort Ruger Reservation, about halfway between the main garrison area and Black Point. Here Various depot buildings and a balloon hanger were built there in 1920, along with a barracks for the 3rd Balloon Co. Three sets of officers' quarters were also built on the east side of the parade ground to accommodate the officers of the balloon company; on the west side of the parade a double set of officers' quarters and a bachelor officers' quarters were built for the coast artillery at the southwest corner of the parade ground. The YMCA building constructed during the war near the northeast corner of the parade ground was turned over to the service for use as a service club when all welfare work was placed under the Army. (60)

In 1921, the Coast Defenses of Oahu was separated into two separate coast defense commands, one for Pearl Harbor and the other for Honolulu Harbor. The 11th, 12th, 13th, and 14th Cos., CD of Oahu, that garrisoned Fort Ruger were redesignated the 4th, 5th, 6th, and 7th Cos., CD of Honolulu. Yet another renumbering in the Coast Artillery Corps occurred in 1922. The War Department again numbered the coast artillery companies in one list for the entire service. Finally in 1924 the CAC underwent another major reorganization. The serially numbered separate companies were abolished and either inactivated or redesignated as either headquarters batteries or lettered firing batteries of newly constituted coast artillery regiments. Sixteen regiments of harbor defense troops were created. The 15th and 16th Coast Artillery (Harbor Defense) Regiments were assigned to the harbor defense commands on Oahu. The 15th was assigned to Pearl Harbor's defenses, while the 16th was assigned to the Harbor Defenses of Honolulu. Companies of coast artillery that had been elements of the old regiments when they were broken up, were to be redesignated as elements of the new regiments.

In 1910, a fire control switchboard room, plaster on a metal lathe attached to a wood frame, had been built just outside the northern entrance to the Kapahulu Tunnel, Major Winslow's 1909 mule tunnel through Diamond Head's north rim. It had a manhole for the cables running to Harlow, the Leahi stations and on to Fort DeRussy and its batteries. However, this building was unprotected from bombardment, and plans were prepared about 1920 for a protected fire control switchboard room. In 1922, the old tunnel was enlarged to a width of about 15-feet with a 14-foot crown, to accommodate vehicular traffic. About 100-feet inside the tunnel, an area 32-feet square was excavated on the west side of the enlarged tunnel for a new fire control switchboard room. A 7 by 15-foot battery room supplied emergency electricity, and a seven-foot-square latrine was also provided. These rooms were transferred to Fort Ruger's garrison on January 5, 1923. The former switchboard building was converted into a guardhouse for the detail guarding the tunnel. (61)

12-inch Battery Selfridge at Fort Kamehameha. *McGovern Collection*

# CHAPTER 3
# THE OAHU LAND DEFENSE PROJECT

Once the United States annexed Hawaii, plans to protect the new territory were quickly developed. Certainly, there were challenges. The islands are geographically isolated, and a long way from the nearest friendly bases on the American West Coast. Even further away were the naval bases and fleet on the East Coast in the days before the opening of the Panama Canal. While the building of the Taft generation coast defenses might provide a deterrence to direct attack, the size and topography of Oahu meant that it was still possible for an enemy to land on beaches and attack the defenders from behind. Moreover, there was a likely protagonist in the Pacific Theatre—Japan. Hawaii had a large Japanese resident population; according to the 1910 census, 41% of the Hawaiian Island population was of Japanese ethnicity. Military planners always considered the possible loyalty to Japan of this population. (1)

It was quickly decided that only Oahu needed defended. There simply weren't any other significant cities, ports, or commercialism on the other inhabited islands to justify more than token defenses. Oahu itself was a tough proposition. The two harbors (Honolulu and Pearl Harbor) were adjacent to each other on the open and accessible southern coast. Beyond this region there were two rugged volcanic mountain ranges to the northwest and northeast. Coasts on all three "away" sides were mainly mountainous but did have numerous openings in reefs and beaches reachable by road or passes to attract landings, some of which were of considerable size. The agrarian plain stretching north of Pearl Harbor between the mountain ranges had maneuver room and acted as a potential funnel to approach the rear of the harbor defenses.

Besides the four coast defense forts built on the southern shore, the army staked out two other reservations for its island garrison. Fort Shafter bordered Honolulu but also stretched into the lower reaches of the mountains. Schofield Barracks was the largest reservation on the island and spanned a significant acreage in the island's center. It was established in 1908 as the base for Oahu's mobile defense troops. Located on the Leilehua Plain, between the Waianae Mountains and the Koolau range in central Oahu, the site commanded a strategic, central location. There was never any domestic resistance to American occupation, unlike in the Philippines, but still a substantial garrison of combined arms was always maintained on Oahu. Thanks to the sugar cane industry, the island did have a viable railroad network, though obviously it did not connect to all the sites of military potential. The quality of roads and trails varied considerably.

Numerous reports were submitted between 1908 and 1912 concerning the proper course to take in defending the island of Oahu. In 1908, there was a Joint War College study and the report of the National Land Defense Board. Three reports were made in 1910; a detailed letter by General Murray, Chief of Coast Artillery, and separate reports by the Moore Board (20th Infantry) and the 5th Cavalry Board, both of the latter based on Oahu. Two more reports came in 1911; A Report on Location and Construction of Posts on Oahu (known as the Cheatham Board), and another thoughtful letter from General Murray. In January 1912 an important memorandum on the issues was provided by Lt. Colonel H. Liggett to the War College. Ultimately a new comprehensive committee, the Oahu Land Defense Board under the chairmanship of Brigadier General Montgomery Macomb delivered a report on September 6, 1912 that set the recommendations the War Department finally endorsed and funded for action. (2)

Most of the debate between these various reports involved the wisdom and ability to defend against landings on both the western and eastern coasts of Oahu. One camp believed that every effort should be made to prevent any landing. It seemed the best way to do that was to have multiple batteries of 12-inch seacoast mortars emplaced around the island—plans included three four-gun batteries on the east coast, and two on the west coast. These could easily cover the islands coastline and outwards to 10,000-yards. Heavy plunging fire at that range could prevent any naval force attempting a landing.

The contrarians thought that there were only a few key passes or pinch points to get through or around either the Koolau or Waianae Range; that for less expense an expanded field army could stop any landing party from making progress into the northern plain—preventing access to Honolulu behind the coastal forts. A field army provided more flexibility for defense than fixed (and somewhat vulnerable) heavy seacoast mortar batteries. Other recommendations included providing improved topographical mapping of the island, the creation of a major military hospital, ordnance, and quartermaster depots to provide supplies to sustain a 30-day siege (the time estimate for the Navy to bring the battlefleet to the island). With either defensive scheme, there would be a need for prepared trenches, bombproofs, field gun emplacements to prevent enemy forces from approaching the southern fortifications.

What was finally approved was a plan that had added two new 4-gun mortar batteries to the Honolulu-Pearl Harbor defenses; one for the west at the Ahua Point reservation of Fort Kamehameha and the other for the east inside the Diamond Head crater. The more isolated separate batteries to the north were eliminated in favor of an expanded field army (and their accommodations primarily at Schofield Barracks). A defensive perimeter was to be established boxing in the adjacent harbors of Pearl Harbor and Honolulu. Into this perimeter the mobile garrison could retire in case of siege. This perimeter would run from the southwest at the eastern shore of Pearl Harbor, north to cross the Waipio Peninsula dividing Pearl Harbor between its west and east loch, then to Ford Island in Pearl Harbor, east to include the ridges to the north of Aliamanu Crater and along the Red Hill ridge further east, towards Diamond Head and then south to meet the sea again at Kupikipikio Point. (3)

At multiple locations, though generally within already established coast artillery posts, new gun emplacements were to be built to cover approaching enemy forces the from the west, east, or directly down from the north. These were to be smaller guns, firing high-explosive or Shrapnel against mobile artillery or troop formations. At certain key points fixed field redoubts were to be built, one on the Red Hill line, and two along with a ditch to seal off the Waipio Peninsula within Pearl Harbor. New searchlights, both fixed and mobile would be obtained to illuminate the approaches. With a nod to the (as it turns out unfounded) fear of local sabotage or partisan activities, obstructions and fencing of existing tactical structures were to be erected.

An April 1915 estimate of cost for approved projects summarized these as: (4)

| | |
|---|---|
| Kupikipikio Point 5-inch battery | $33,000 |
| Diamond Head 4.7-in & 6-pdr btty & misc. | $59,000 |
| Diamond Head mortar battery | $104,500 |
| Red Hill redoubt | $55,300 |
| Salt Lake field works | $110,000 |
| Ford Island 6 & 4.7-in batteries | $126,000 |
| Bishop Point 4.7 & 3-in batteries | $54,000 |
| Waipio Point redoubts & ditch | $122,575 |
| Searchlights | $105,000 |
| | |
| Total (engineering work only): | $769,875 |

Once the plan was approved, steps were almost immediately taken to reduce costs, The mortar battery at Ahua Point was stricken by the War Department. The Waipio Point redoubts were eliminated, and it is not clear if the field works near Salt Lake were ever actually constructed. The 4.7-inch battery on Ford Island was dropped (but not the 6-inch batteries), and the searchlight project was fulfilled by using older 36-inch mobile units and modifying them for use locally.

Overall, the biggest savings came, not from the engineering cutbacks, but cutting ordnance costs. The new guns, and as importantly the ammunition supply, for the project were obtained from existing emplacements in the continental U.S. By 1915 some of this ordnance had been emplaced for almost 20 years in various harbor defenses and batteries that were redundant or in harbor defenses of less importance. The older 4.7-inch and

6-inch guns were British Armstrong models purchased in the panic at the beginning of the Spanish American War in 1898. While serviceable guns, they had a low muzzle velocity attendant to this early generation of "Quick-Firing" guns. While that limited range for seacoast use, it was a perfect match for the new shrapnel shells recently developed, and ideal for use against land targets.

Altogether 34 guns were relocated from the continental U.S.:

> 4 x Model 1890M1 12-inch mortars
> 4 x Armstrong 6-inch QF guns
> 2 x Model 1900 5-inch RF guns
> 8 x Armstrong 4.7-in QF guns
> 2 x Model 1903 3-inch RF guns
> 6 x 2.24-inch (6-pounder) mobile parapet mount guns
> 8 x 1.57-inch (1-pounder) Vickers-Maxim mobile mount guns

The new batteries and tactical structures built under the Land Defense Project, are described following starting in a geographical order clockwise from the southwest (Fort Kamehameha) corner.

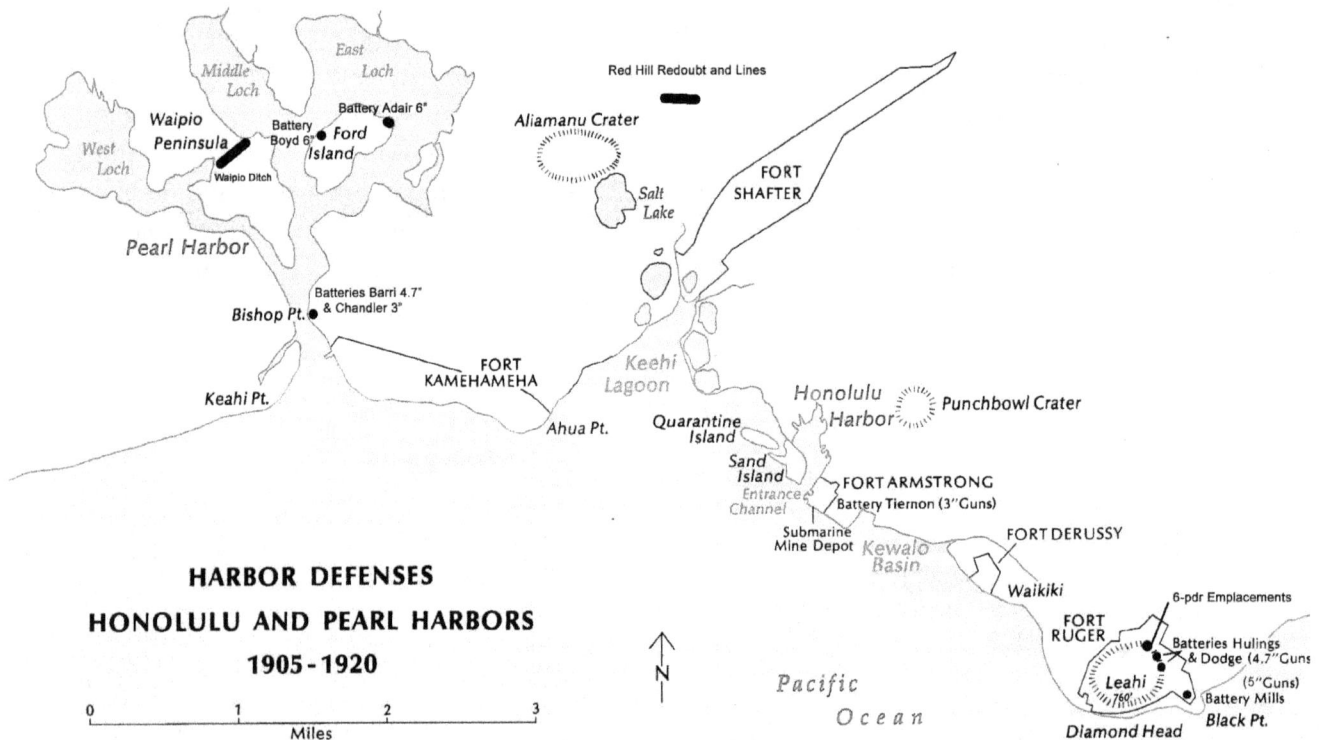

Map showing the major elements of the Oahu Land Defense Project actually completed. *Williford Collection from COE map.*

Bishop Point on the west side of Fort Kamehameha was selected for a new pair of batteries. They located north of the main post of Fort Kamehameha on the right (east) bank of the Pearl Harbor Channel. In 1914 this was part of the navy Pearl Harbor reservation, and it took several months to arrange the transfer of the tract to army hands. Tactically, the armament covered the western shore of Queen Emma Point and a good section across the channel on the Ewa Plain and Puuloa shore. Hostile units approaching from the west towards the channel could be brought under fire. The battery was built very close to the edge of the channel, behind a concrete retaining wall.

The design of the work was unique. the armament was casemated in gun chambers to afford increased protection to the manning details from the high angle fire of land-based howitzers and mortars. The battery structure had two faces: one to the northwest; the other to the southwest. The left flank of the southwest face

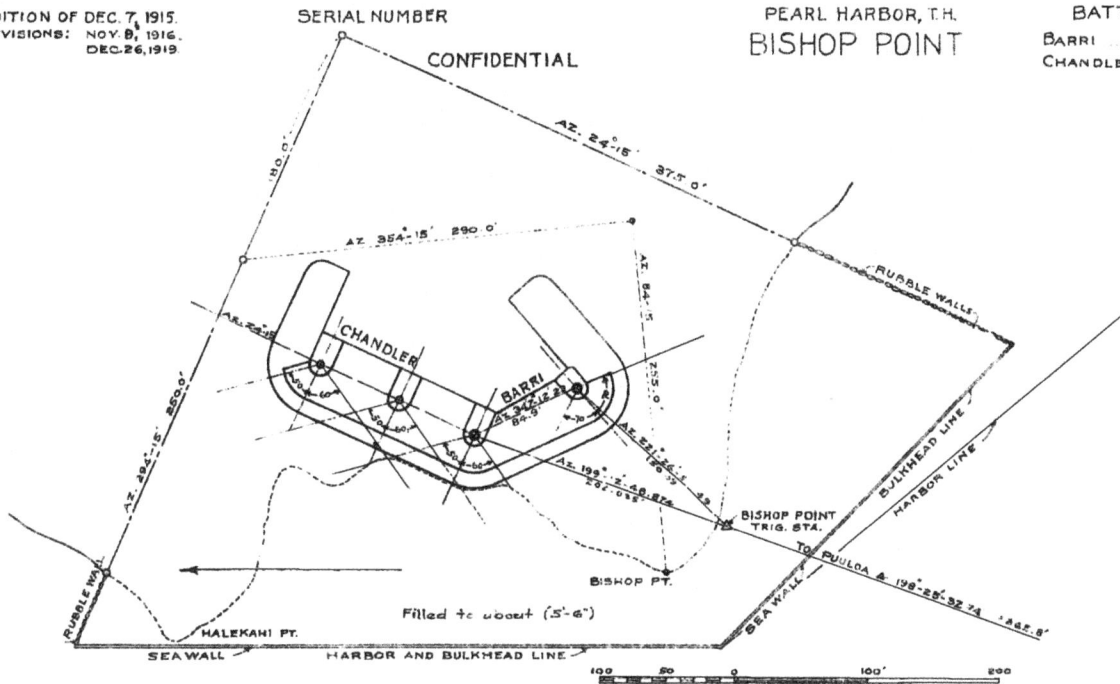

1919 Engineer plat of the army reservation at Bishop Point, used for the site of Batteries Barri and Chandler. *NARA.*

was refused in a sharp turn 90° and the right flank was extended to the southeast. These refused flanks nearly enclosed the structure in a defensive box and gave good protection against enfilading fire from flat trajectory weapons. Additionally, the occupied faces were built as a double story structure, with gun emplacements on the upper floors. While organized as two separate gun batteries, in the whole it was a single structure.

Two 4.7-inch guns of Battery Barri were emplaced 85-feet apart in their gun chambers or casemates in the southwest face of the structure. The Number 2 gun of the battery was emplaced at the juncture of the southwest face and its left flanking traverse and could be trained to cover the approaches across the salt flats that lay to the rear of Fort Kamehameha's coast defense batteries as well as the area of the Puuloa Flats across the Pearl Harbor Channel to the southwest of the battery. The Number 1 gun of Battery Barri and both of Battery Chandler's 3-inch guns bore directly across the Pearl Harbor Channel to the west, northwest to cover both the Puuloa Flats and the area south of the Pearl Harbor's West Loch.

The 4.7-inch rapid-fire (RF) guns had an interesting history. Thirty-four of these British designed and built Armstrong-type guns had been purchased early in the Spanish-American War in 1898. There was fear of Spanish attacks on the East Coast, and a lack of light RF guns needed for its defense. Batteries of two guns each were sprinkled around the various defenses. In 1913 eight of these guns were removed from their emplacements and sent to Hawaii for use in the Land Defense Project. The guns were older, but still serviceable items. Firing cased ammunition, they were mounted on pedestals for rapid traverse and protected with a box-like shield that shielded the front, top and sides. It fired a 45-lb. projectile about 9700-yards. The relatively low muzzle velocity was fine, in fact preferred, for use in firing shrapnel rounds against land forces.

The casemated room interiors of the emplacement measured 14-feet in width and 30-feet in length. The reinforced concrete roofs of the casemates were five feet thick and the front, or parapet walls were six to seven-feet thick. In front of these parapet walls was an earthen rampart some 34-feet thick. The casemates were located on the upper level of the structure and accessed from the battery parade by concrete stairs at the rear of the casemates. The lower level of the structure was divided into a number of rooms that were used as magazines, storerooms, personnel bombproofs, latrines, and power generator rooms. In event of engagement the casemates were to be covered by layers of sandbags laid on the five-foot thick reinforced concrete roof to provide an additional 15-feet of bombproofing protection.

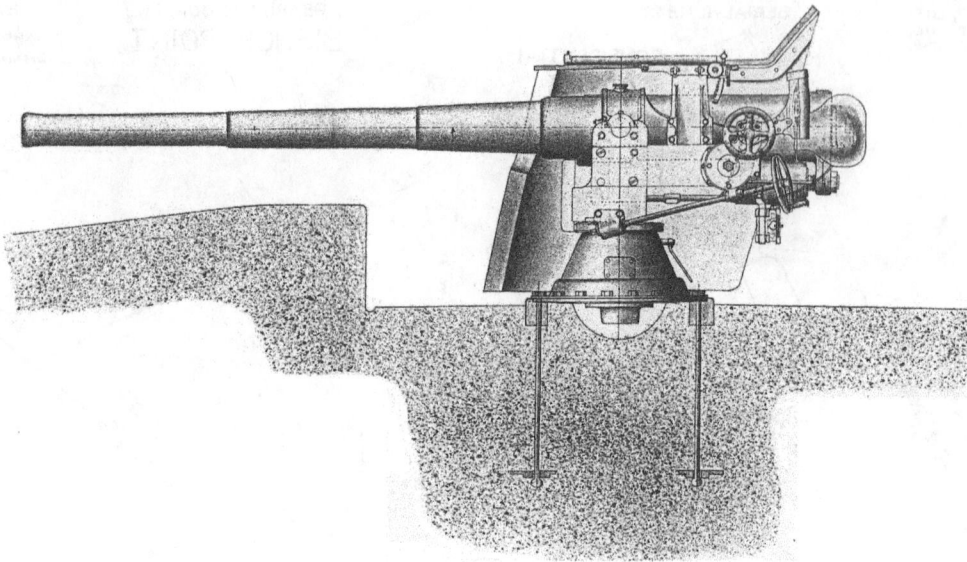

Armstrong-type 4.7-inch QF gun, eight of which were supplied to Oahu although only six were eventually placed in new emplacements.
*NARA.*

In the left flank traverse of the south face there were three rooms: a storeroom, the engine room and a fan room. The interior walls of these rooms were two-feet thick and the rooms measured about 12-feet deep and eight-feet wide. Between the Number 1 and Number 2 guns (in army terminology of the time, emplaced guns were numbered from right to left from a position facing the enemy) of Battery Barri on the south face were the two magazines of the battery; each measured about 20-feet by 20-feet. Outside each magazine a concrete staircase led to the rear of a casemate, the ammunition service was manual. Between Battery Barri's Number 1 Casemate and the Number 2 Casemate of Battery Chandler were three bombproof rooms. In the angle of the structure's right flank and southwest face separate latrines were provided for the officers and enlisted personnel.

The pair of 4.7-inch guns of Battery Barri had previously been mounted at Battery Griffin at Fort Hamilton in the Harbor Defenses of New York. The two 3-inch M1903 Rapid Fire guns had been made for the defenses of Guantanamo, Cuba. When that defense was cancelled the guns became excess and were sent to Hawaii for use in the Land Defense project. They were initially shipped to Oahu on June 6, 1913, with the

Aerial photograph from 1922 of Bishop Point, the concrete tops of the casemates for the two land defense batteries are quite evident.
*USAMH.*

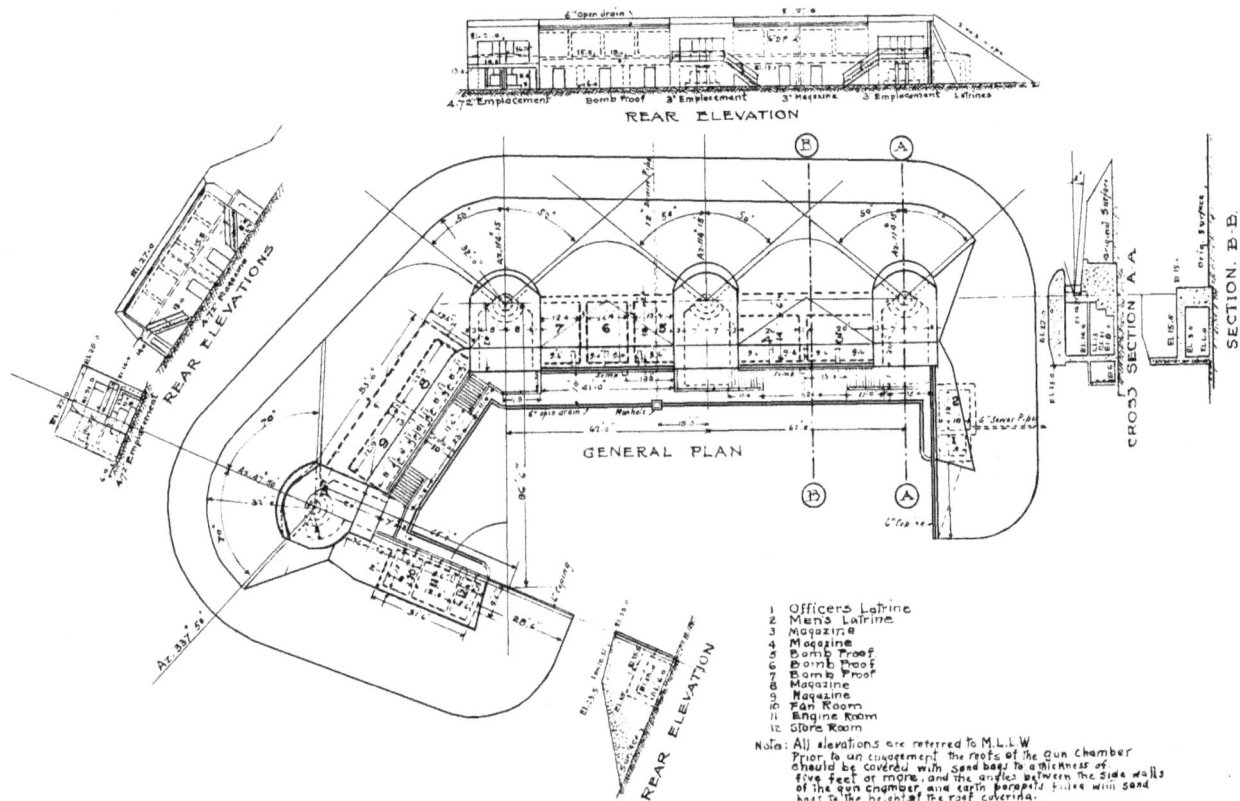

Plan, elevation and section of Batteries Barri and Chandler from 1919 engineer drawing. *NARA.*

Period photo of one of the land defense batteries. *Smith Collection*

intent of emplacing them at Fort Ruger. This plan was dropped in favor of emplacing them at Bishop Point on the Pearl Harbor Channel.

Plans for the battery were submitted in September 1914, and work was begun quickly, it was done within a year. Both batteries were turned over to the garrison on September 26, 1915, Battery Chandler's cost of construction was $26,923.37 and that of Battery Barri was $29,403.38. (5)

Following their completion, neither battery was manned on a regular basis. During the World War I the batteries were manned by companies at Fort Kamehameha on an irregular basis. Following the First World War, on July 22, 1919, the Adjutant General declared all the 4.7-inch guns of the Armstrong type obsolete and approved for scrapping. The command in Hawaii followed through and Barri was soon scrapped. By the early 1930s the Quartermaster Corps had erected five ammunition storage sheds at the Bishop Point reservation. In May 1935 a recommendation was made that Battery Chandler's armament, considered obsolescent locally, be declared obsolete by the Chief of Coast Artillery and that the two M1903 3-inch guns be stored as spares for Batteries Hawkins and Tiernon, the two remaining 3-inch gun batteries on Oahu.

The Macomb Board had recommended purchasing and deploying several searchlights for illuminating enemy positions and lines along the defensive perimeter. In the 1914 estimate, one 60-inch and five 36-inch fixed lights were requested, estimated to cost $65,000. The one fixed light was to go at or near the new Bishop Point reservation and be used to illuminate the opposite Puulau shore of Pearl Harbor. It was never funded or built. The new fixed 36-inch were also not provided. However the department was sent eight older 36-inch mobile lights. Also, funds amounting to $21,800 were allocated for providing new engines and controller cables to rehabilitate these lights. Apparently, the work was completed and the lights were stored pending need.

Waipio Peninsula is a large projection of land entering Pearl Harbor from the northwest. It divides the harbor into west and east "lochs" or bays. The East Loch needed to be isolated, a defensive perimeter was necessary to defend the southern end of the peninsula. A specific recommendation to dredge a ditch across the peninsula was made on June 9, 1913. defensive ditch across the peninsula was proposed, along with two defensive bombproof redoubts and a bridge over the ditch. This location would be swept by the new gun positions at Bishop Point, Ford Island, and the mortar batteries at Fort Kamehameha and Ahua Point. (6)

The two reinforced concrete bombproof redoubts proposed for the Waipio defenses were officially deleted from the project in June 1915. No specific reason was listed, but it is likely just that the funds were not forthcoming. It does appear, however, that $12,000 to finish the trench line on Waipio was expended. (7)

The Ford Island Military Reservation, named for Dr. S.P. Ford, a former owner, lies in the middle of Pearl Harbor's East Loch. Originally known as Mokuumeume, or "Island of game ume." It was also at one time in its history called the "Island of Strife." A low, flat island of just 10-15 height above mean low water, it had low bluffs all along its shores. In late 1914 the portion of land the army was interested in was owned by the John Ii estate. Arrangements could be made to purchase the land for $2191, and another $289.30 could be paid to the Oahu Sugar Company, current leaseholder who was producing sugar cane on the island. In 1916, two small tracts were acquired on the island for gun emplacements. (8)

The island was an ideal location for light guns to sweep west across the Waipio Peninsula, north against an assault across the Ewa Plain or east against ravines around Aiea and Makalapa. Initially eight 3-inch guns were considered adequate for the armament, to be deployed in two batteries of four guns each on the eastern and northern ends of the island. In early 1915 this plan evolved into using four 6-inch guns (in armored cupolas) and two 3-inch guns. Later in that year it was reduced to three positions: two 6-inch on the northwest corner, two more on the northeast point, and now two 4.7-inch guns on the northern point of the island. As with other land defense batteries, there was concern about protecting the guns from return fire from heavy field howitzers. It was determined that the largest opponent might well be a 6-inch (or 150-155 mm) howitzer and overhead and magazine protection should be sufficient (figured to be 7-inches of reinforced concrete). Comment was made that if the Japanese got some of their famous 240 mm howitzers ashore those weapons could potentially breech this protection.

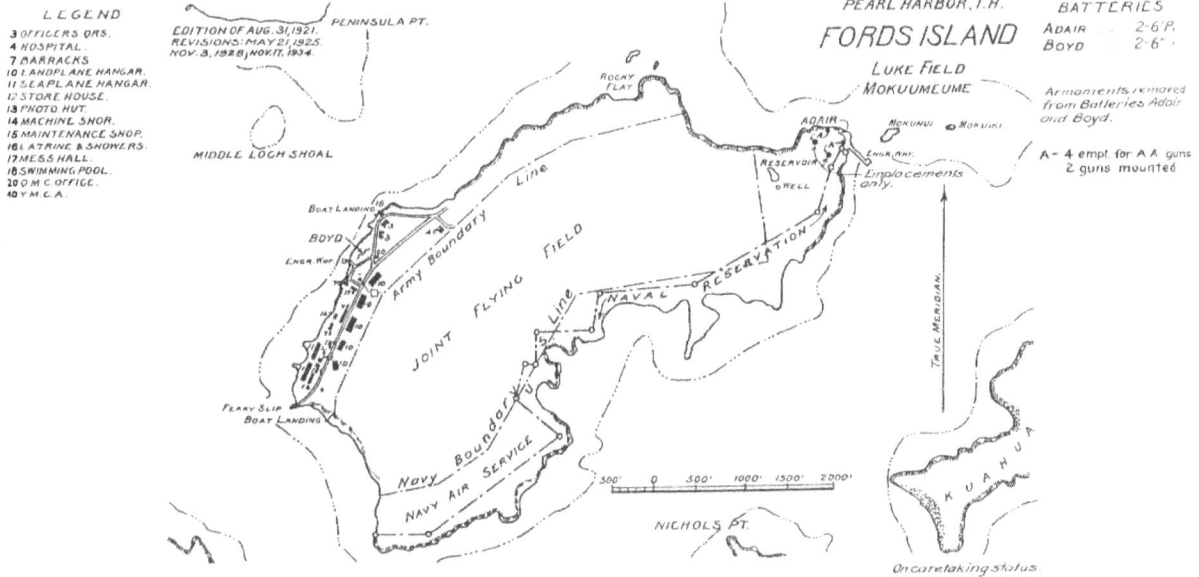

Engineer plat of the army reservation on Ford Island. Note the gun battery locations on the northeast and west-central points. *NARA.*

The guns used were part of an order of eight Armstrong-type 6-inch bore rapid-fire guns acquired during the Spanish-American War. They were being phased out in domestic coast defense installations. Being of relatively low muzzle velocity, they were ideal to use to fire shrapnel rounds against field armies and thought to be well suited for use in the land defense batteries. To provide a sufficient supply of projectiles and propelling charges for the Armstrong guns it was necessary to scavenge the magazines at Forts Williams, Maine; Adams, Rhode Island; Wadsworth, New York; and Dade in Florida. The gun was a 40-caliber rapid-fire type, firing a 100-lb. projectile about 12,000-yards.

1915 estimates included $9000 for land and rights of way, and $117,000 for battery construction. The battery for two 4.7-inch guns was soon dropped, and the project moved ahead for twin 6-inch Armstrong gun emplacements at the north and northwest positions. In May 1915 $3271.30 was authorized and paid for tracts of land totalling just 5.72 acres, but by the end of 1916 the entire island was acquired. It was shared with a narrow naval strip of property along the eastern shore (the famous battleship mooring row) and land for what would house the Army Air Service's Luke Field that was established on April 29, 1919. A naval air station for seaplanes was commissioned in January 1923, eventually becoming the Ford Island Naval Air Station. (9)

Battery Adair was the battery that ultimately occupied the northeastern point of Ford Island. Planning began in 1913 and went through several iterations. Actual construction was not begun until August 1916, and was completed in December 1917. The battery was armed with two M1898 6-inch Armstrong guns that had been removed from emplacements in coastal forts on the mainland (actually Battery Hobart at Fort Williams

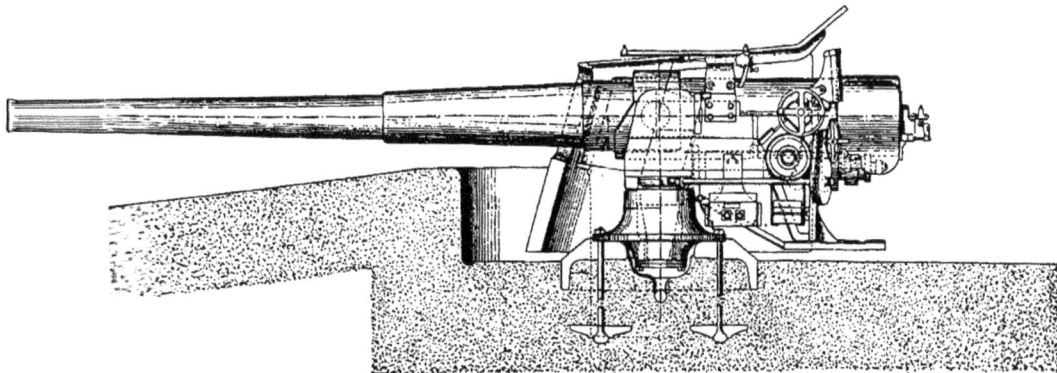

Ordnance Department side sketch of a British 6-inch QF gun, four of which were used to arm Batteries Adair and Boyd on Ford Island. *NARA.*

and Battery Bankhead at Fort Adams). Both guns and pedestal carriages were shipped to Hawaii from interim storage at the Benicia Ordnance Depot in 1917.

Designing the casemates and necessary overhead protection for the two Ford Island batteries was not easy. The gun selected, and its shield, did not conform easily to being held in a closely cropped casemate opening. As first designed the port or opening was a space five by 12-feet. Structural support for a thick, projecting overhead concrete cover was lacking. It took quite a bit of engineering to arrive at a satisfactory design. (10)

The two batteries on Ford Island were of different configuration. Battery Adair had a straight alignment with two casemated gun positions on either end. Their fields of fire (120° possible) were slightly splayed out from the perpendicular. Between the guns were separate shell and powder rooms for each gun and a central plotting room. A gallery, open above to the sky, ran between the guns and fed the magazine rooms, with an earthen parados protecting it from behind. On the sides were officer and enlisted latrines, and storerooms. The battery was not wired for electricity—none was required to operate the guns or light the interior. Work was done from August 1916 to December 1917. The battery was transferred to the custody of the garrison of Fort Kamehameha (for which Ford Island was a sub post) on December 17, 1917 for an engineering cost of $59,045.26. (11)

The battery was manned on a regular basis by rotating coast artillery companies from Fort Kamehameha during the period before and during World War I. Following the war, the battery was placed in maintenance status and by March 1925 the battery had been disarmed.

Battery Boyd, the second battery for Ford Island, was on the northwestern point of the island, and had a clear field of fire over the Waipio Peninsula and directions north and southwest. Although planning started for this battery in 1913 construction was not begun until April 1916. The battery was completed in February, 1917 and transferred to the garrison at Fort Kamehameha on February 27, 1917 for an engineering cost of $44,608.10. It was armed with two M1898, 6-inch Armstrong guns. The two guns, and its carriages had formerly been emplaced in Battery Bankhead at Fort Adams until 1913 when removed and shipped to the Benicia Arsenal in California. There they were stored until 1917 when they were shipped on to Oahu.

Aerial photograph of Ford Island in 1941. While the gun batteries were disarmed by this time, they can still be seen with careful inspection—otherwise the island is quite busy with its airfield and as an anchorage site for much of the battlefleet. *NARA.*

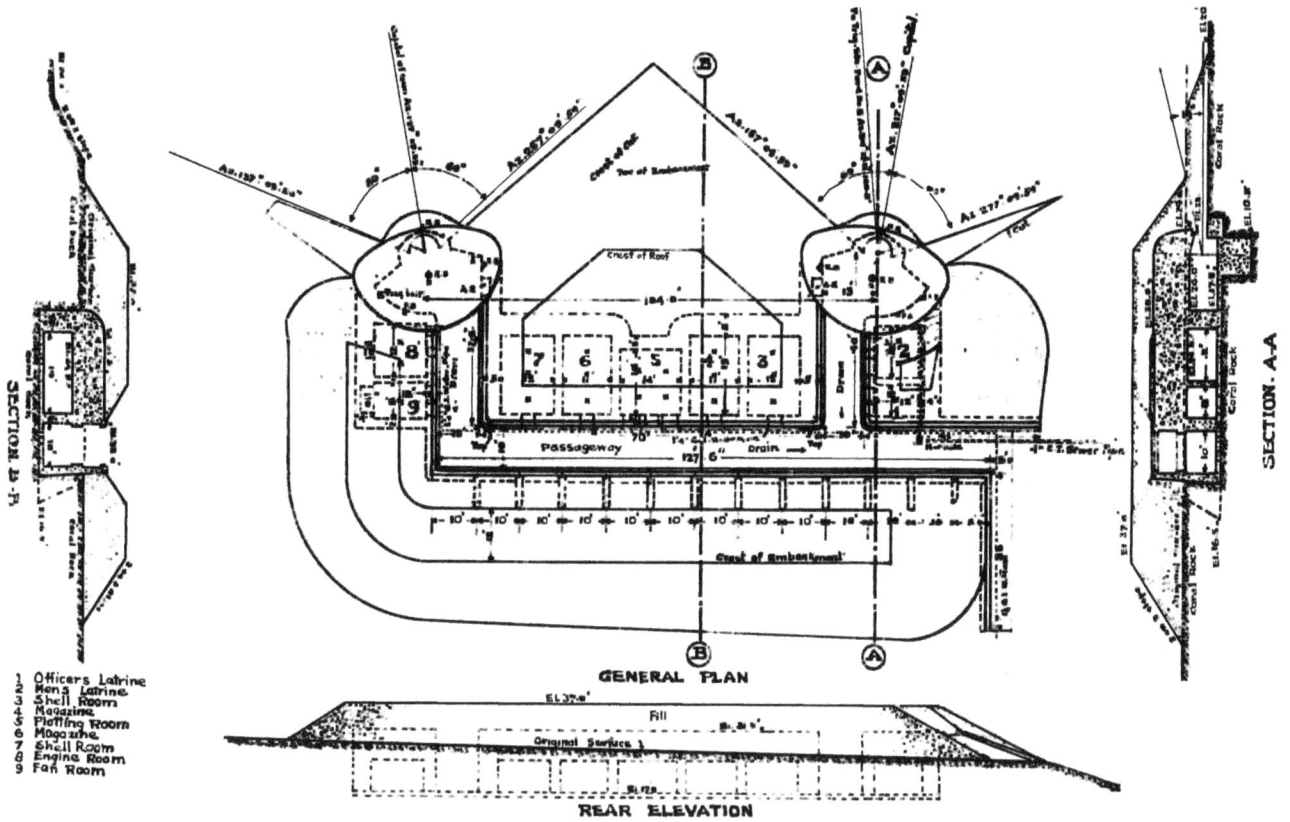

1 Officers Latrine
2 Mens Latrine
3 Shell Room
4 Magazine
5 Plotting Room
6 Magazine
7 Shell Room
8 Engine Room
9 Fan Room

Plan, elevation and section of Battery Adair, 1919. *NARA.*

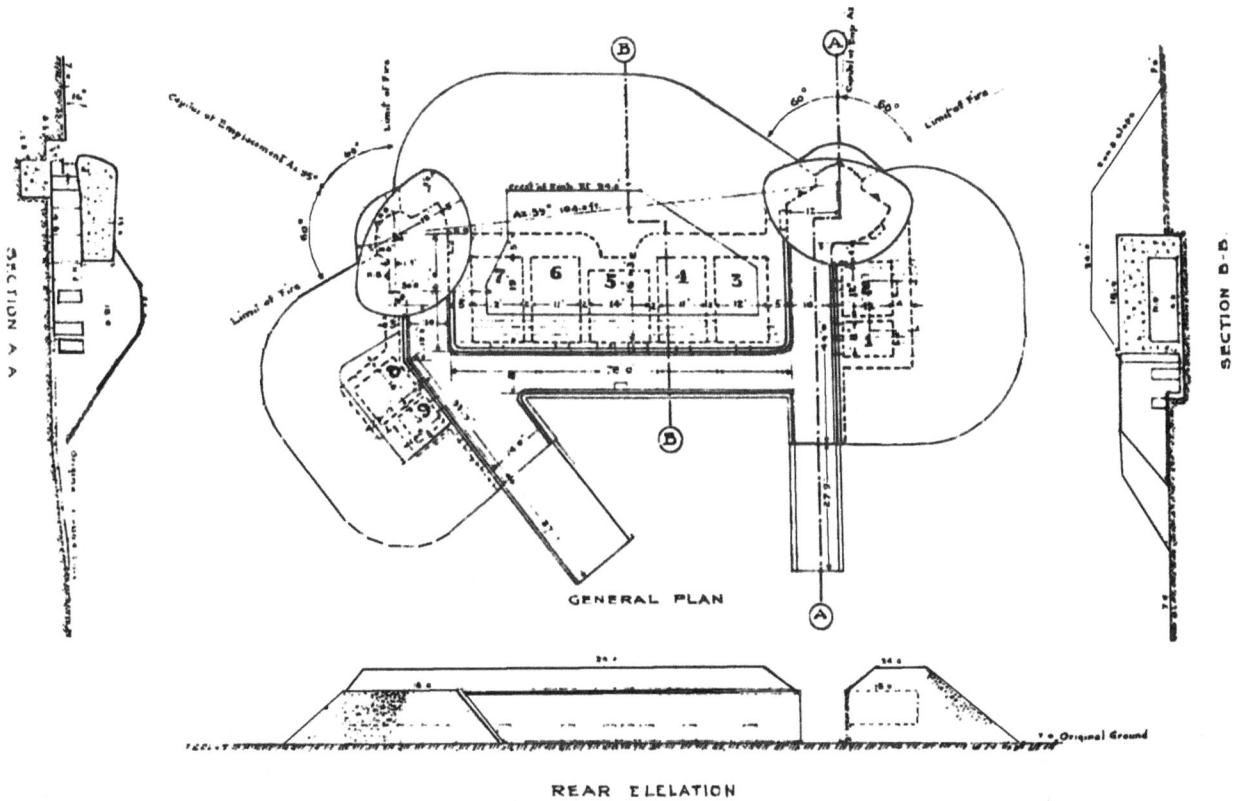

Plan, elevation and section of Battery Boyd, 1919. *NARA.*

Aerial view of Battery Adair in 1922—the gun tubes themselves can be made out projecting from the two casemates. *NARA.*

Battery Boyd in 1922 at the northwestern tip of Ford Island. *NARA.*

The emplacement was considerably different from Battery Adair. The No. 1 gun was in a casemated position, with a connecting gallery directly behind. Perpendicular to that gallery was a second gallery that branched to the left past shell and powder magazines for each gun and a plotting room between the sets. The No. 2 gun was on another shorter branch, and much more angled, splayed to a southern directrix. Another short gallery, turned heavily back to the right completed the design. There were latrines and storerooms, but no electric generator, the battery was not wired for lights. The battery was manned on a regular basis by rotating coast artillery companies from Fort Kamehameha during World War I. Following the war, the battery was placed in maintenance status, and by March 1925, the battery had been disarmed. At the end of World War I, the two gun batteries were placed in caretaker status. (12)

During the implementation of the Land Defense program a number of 37 mm, 1-pounder Vickers-Maxim (otherwise known as "pom-poms") guns on wheeled field mounts were supplied to the coast defense command. Not needing fixed emplacements, they were occasionally issued to forts for landward defense. At least in January 1916 a number of these guns were at Fort Armstrong where a new separate storage shelter was authorized, built, and transferred to troops on May 31, 1916. (13)

The perimeter line of defense was planned to snake along various ridges on the west-east line from Pearl Harbor through Fort Shafter to Fort Ruger. A couple of locations along this line were selected for pre-developed field works to act as potential strongpoints. One of these areas was that to the north of Aliamanu-Salt Lake (which was also needed as a site for fire control stations). It is not clear how many, if any, of these positions were completed. The statement of cost for April 1915 lists the Salt Lake-Makalapa Line as requiring $10,000 for sites, and another $100,000 for cost of engineer work. It was still listed as the same cost on September 15, 1917. The engineering notebook of about 1916 lists the site as "allotment made, negotiations in progress". The latter comment apparently referring to obtaining the land necessary. (14)

The Red Hill Military Reservation, a collection of military tracts lying between Fort Shafter and Pearl Harbor's East Loch, extended from the Red Hill spur of the Koolau range south to Honolulu Harbor encompassing Makalapa and Aliamanu Craters. This reservation was established in 1914 as the left flank anchor of the line of land defenses on Oahu. This line of land defenses extended from Diamond Head, which served as the right flank anchor of the line, to the crest of the Koolau Range and thence along the ridgeline of the mountain range northwestward until it reached the vicinity of Red Hill, one of a series of spur ridges running toward the sea from the mountains.

Numerous field works were constructed on the crater rims of Makalapa and Aliamanu, and along the crest of Red Hill itself. A redoubt was built on the northern end of Red Hill. This redoubt was located on the military crest of the steep south slope of Red Hill some 400-feet above the floor of the Moanalua Valley. In March 1915 it was described as being on the extreme right of the line, intended for the security of troops occupying the positions. It was to be a protected bombproof with latrines, storerooms and magazine for small arms ammunition. Construction commenced in March 1915 and was generally completed, except for electrical service and communications wiring, by September 1916. The redoubt was transferred to the Fort Shafter garrison on September 9, 1916 for an engineering cost of $39,117,00. (15)

The redoubt consisted of a concrete lined trench 274-feet long. In the rear of the trench line rooms of the cut and cover type were built into the side of the hill to serve as bombproof shelters, storage, etc. Five of these rooms (14 x 7-feet) functioned as personnel shelters and opened onto the trench line. Four other rooms were enclosed. Two of these served as latrines for officers and enlisted men and the other two as storerooms. A fifth room that adjoined one of the enclosed rooms served as a small arms magazine. The partition walls of the redoubt's rooms were composed of concrete two-feet thick, as was the structure's roof. These rooms were ventilated by vertical 8-inch diameter pipe vents with "star" tops. The roof over the structure's rooms was provided with an earth fill some 30-feet thick. At the centerline of the redoubt's rear wall a doorway closed by a thin slab of concrete was located to afford a possible exit from the redoubt through the ridge itself. The tunnel through the ridge to the north side of the hill was never constructed. (16)

The eastern end of the Oahu defensive zone received more attention and expenditure than any other location in the island. The closeness and vulnerability to a possible landing in the otherwise undefended Kaneohe Bay area was partly a factor. Fort Ruger's one coast artillery battery had mortars, but the site of the battery was on the gentle slope of the Diamond Head crater and not protected at all to the rear. The board recommendation called for an entirely new mortar battery to be built within the crater itself, emplacements for four 4.7-inch guns along the crater rim covering approaches from the northeast, plus six 6-pdr mobile guns, a concrete-lined infantry trench, and, at the adjacent Kupikipikio Point Reservation, a protected battery for two 5-inch guns, covering the Mokapu Peninsula to the east. Expenditures were estimated in April 1915, for just the engineering costs, to $197,000.

The new mortar battery, eventually to be named Battery Birkhimer shared a land defense and seacoast defense role. Its planned sister unit at Fort Kamehameha's Ahua Point was cancelled early in the planning process. Birkhimer was the only new mortar battery of the program. Its problematic construction is more fully described in Chapter 2. The other batteries, however, were built as planned. Two batteries for 4.7-inch guns were cut out of the Diamond Head rim and became Batteries Dodge and Hulings. On the rim emplacements were built for the (unnamed) 6-pounder guns. At Black Point (Kupikipikio Point) two 5-inch guns of Battery S.C. Mills was completed.

Battery Dodge was the southernmost set of two Armstrong guns emplaced about 40-feet below the crest of the east rim of Diamond Head between 411 and 425-feet elevation. Work was done in 1915. Transfer was made to troops on October 20, 1915 for a construction cost of $15,720.24. The gun casemates required long four-foot-wide access tunnels through the crater rim, 66-feet apart. Each access tunnel, varying between 60 and 90-feet long depending on the crater contour, provided access to a gun casemate and to the transverse corridors connecting each pair of casemates. The tunnel to the No. 1 casemate of the southern-most battery ran some 70-feet to a point where it intersected the 66-foot transverse corridor. At the interior end of Tunnel No.

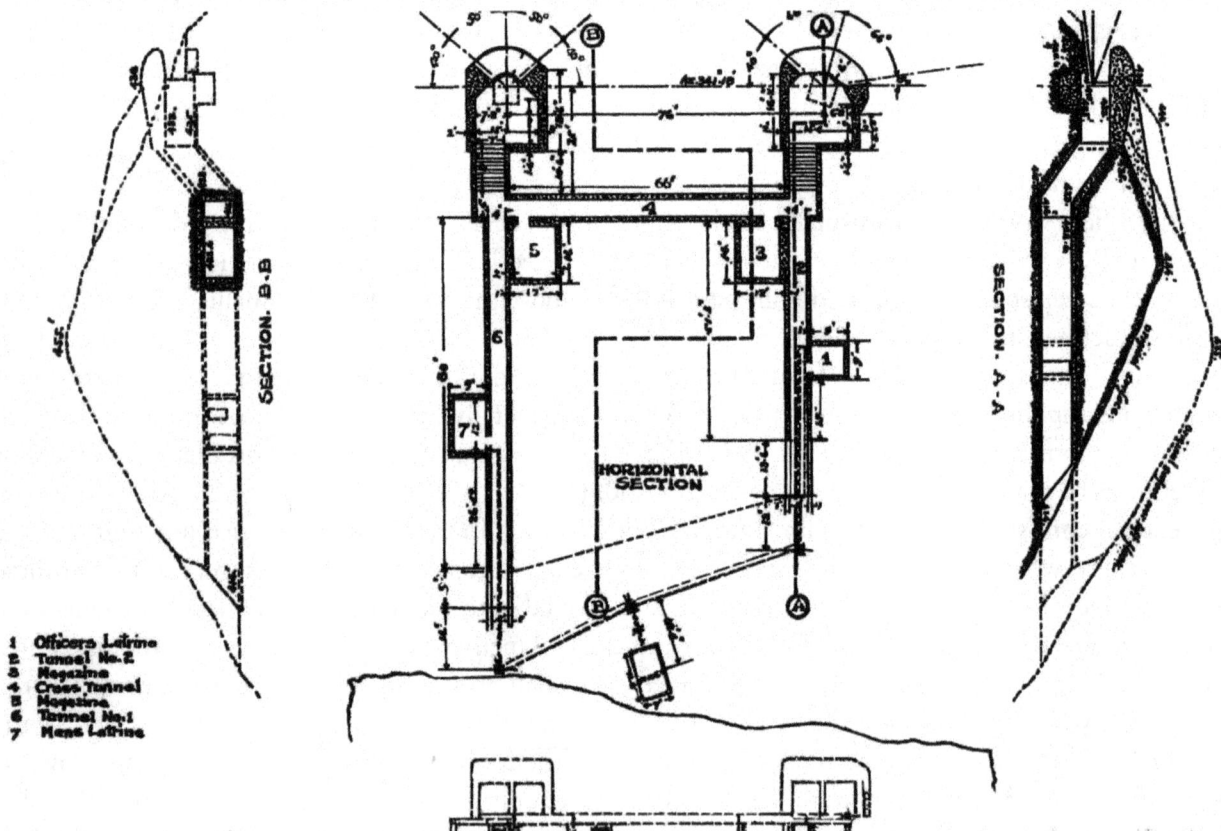

1   Officers Latrine
2   Tunnel No.2
3   Magazine
4   Cross Tunnel
5   Magazine
6   Tunnel No.1
7   Mens Latrine

Plan, elevation and section of Battery Hulings, 1919. *NARA.*

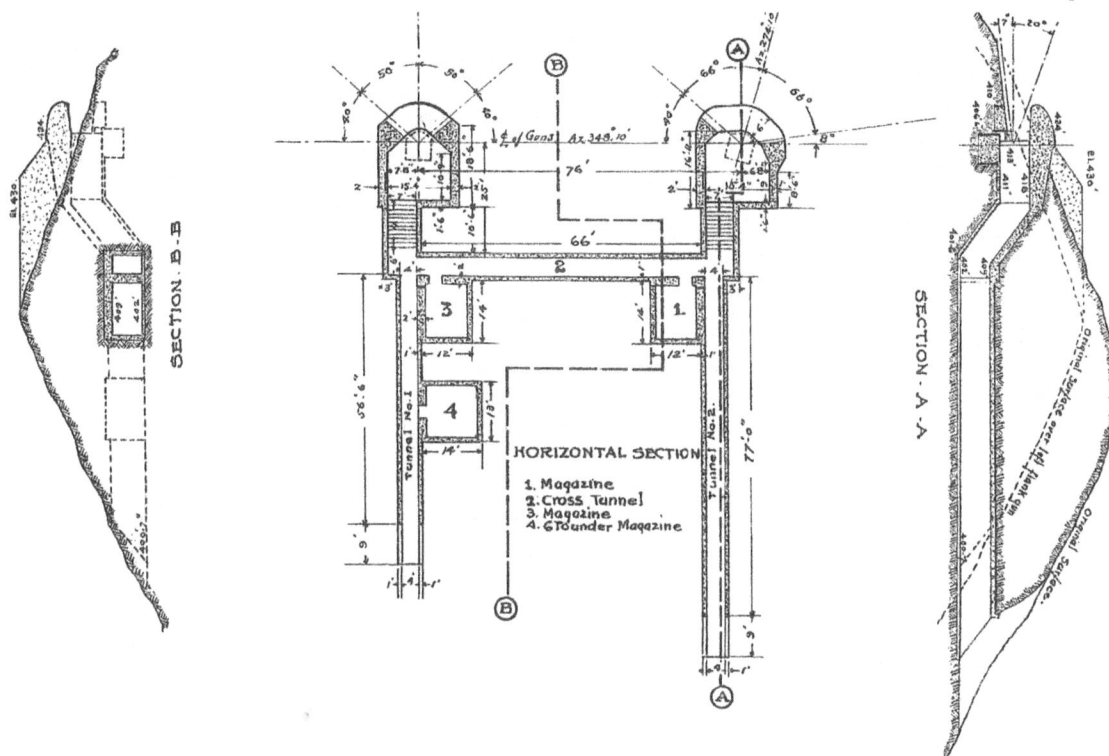

Plan, elevation and section of Battery Dodge, 1919. *NARA*.

1, a flight of stairs six feet wide led up about 9-feet to the gun casemate. The gun was served from a magazine at the foot of the stairway that opened onto the transverse corridor. A reinforced-concrete overhead canopy extended out about six-feet over the front of the casemate. The gun's pedestal mount was well suited to the role, since it allowed the gun to be depressed 7°. The No. 1 (right) gun casemate was angled slightly to the right to cover the flank and rear approaches to Battery Mills located below. The battery's second casemate was accessed through a 57-foot tunnel. Midway along its length, a 14 by 18-foot magazine housed the ammunition for the 12 6-pounder RF guns on wheeled carriages that could be placed along the top of the crater rim for close-in defense. The magazine for 4.72-inch Gun No. 2 was at the junction of tunnel No. 2 and the transverse tunnel. No electric plant was contained with the battery, reserve power came from the plant at Battery Birkhimer. (17)

The battery was disarmed in the early 1920s, the adjutant general had declared all 4.7-inch obsolete and authorized their scrapping. Early in the Second World War the battery was rearmed using two ex-navy 4-inch Mk IX low-angle deck guns. They originated with a Navy loan of 25 such guns to the army to bolster land defenses immediately after the Pearl Harbor attack. Modifications were needed to the casemates to allow the guns to fire to their maximum elevation, as well as to allow one gun its maximum traverse by cutting a special niche in the flank wall for the gun's projecting sight. These guns served from mid-1942 to early 1944.

Battery Hulings was the northernmost of the two batteries was constructed on Diamond Head's east rim in 1914 and 1915, about 525-feet to the left (north) of Battery Dodge. Its two 4.72-inch Armstrong guns were also emplaced about 40-feet below the crest of the rim at 425-feet. The battery's layout varied only slightly from that of Battery Dodge; its No. 1 Tunnel had an officer's latrine at its midpoint, and the No. 2 Tunnel had a larger latrine for enlisted personnel assigned to the two batteries. Its two four-foot wide access tunnels were excavated through the rim 66-feet apart. Each tunnel provided access to a gun casemate and to the transverse corridor connecting the two casemates. On October 20, 1915, Battery Hulings was transferred to the Fort Ruger garrison, although some finish work and the electric supply remained to be completed. Cost to date of transfer was $15,720.24. The 4.72-inch ammunition for the two batteries had been gathered from numerous coast artillery posts along the Atlantic and Gulf coasts where guns of that caliber had been emplaced. The 2nd Co., CAC, manned Batteries Hulings. The battery was disarmed in the early 1920s, the adjutant general had declared all 4.7-inch obsolete and authorized their scrapping. (18)

Shields for the 4.7-inch guns, probably taken at Fort Ruger just prior to installation in Batteries Hulings and Dodge. *Gaines Collection.*

Six pairs of emplacements for M1898 6-pounder (2.24-inch) guns on M1898 wheeled carriages were built along the crest of Diamond Head's east rim, near and both between and flanking either side of Batteries Hulings and Dodge, for the close-in defense of the eastern approaches to Diamond Head from the vicinity of Maunalua Bay. The 12 emplacements consisted of simple concrete slabs and epaulements. The guns, stored at Fort Ruger, were to be brought to the emplacements when needed. Small concrete magazines were included in the construction, one for each four guns. Work was done from October 1914 to August 1915. Also there was a larger magazine purposely included along the northern entry corridor into Battery Dodge. The emplacements (never named) were transferred to troops on August 10, 1915 for $11,620.50

The guns for these land defense batteries were received with only the most basic fire control equipment (essentially just the sights included on the carriage), and as late as November 1918 this equipment was still incomplete. Inspection reports called for a 15-foot coincidence range finder (CRF), plotting board, and telephone connections with the battery commanders for each battery. There is no indication this fire control equipment was ever provided. This type of gun was declared obsolete in June 1919, and they were soon discarded. (19)

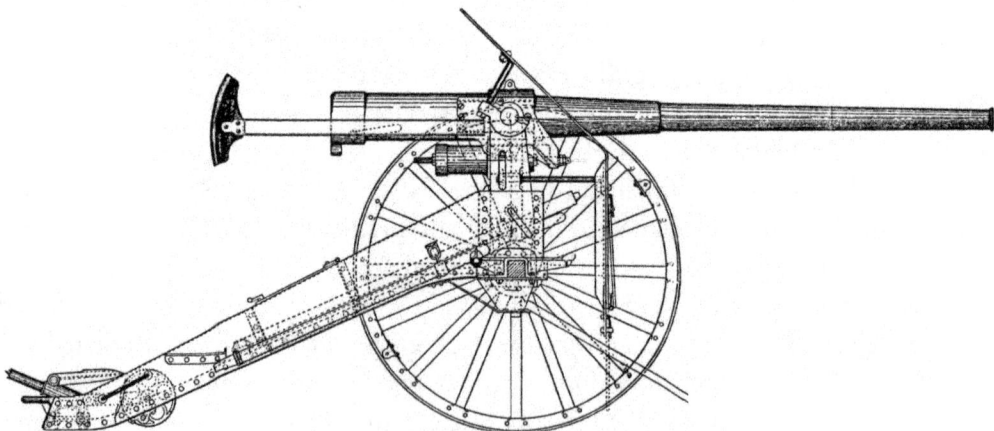

Army 6-pounder (2.24-inch or 57mm) light gun. While three slightly different types were acquired in the late 1890s, this drawing is representative of the style supplied to the Oahu land defenses. *NARA.*

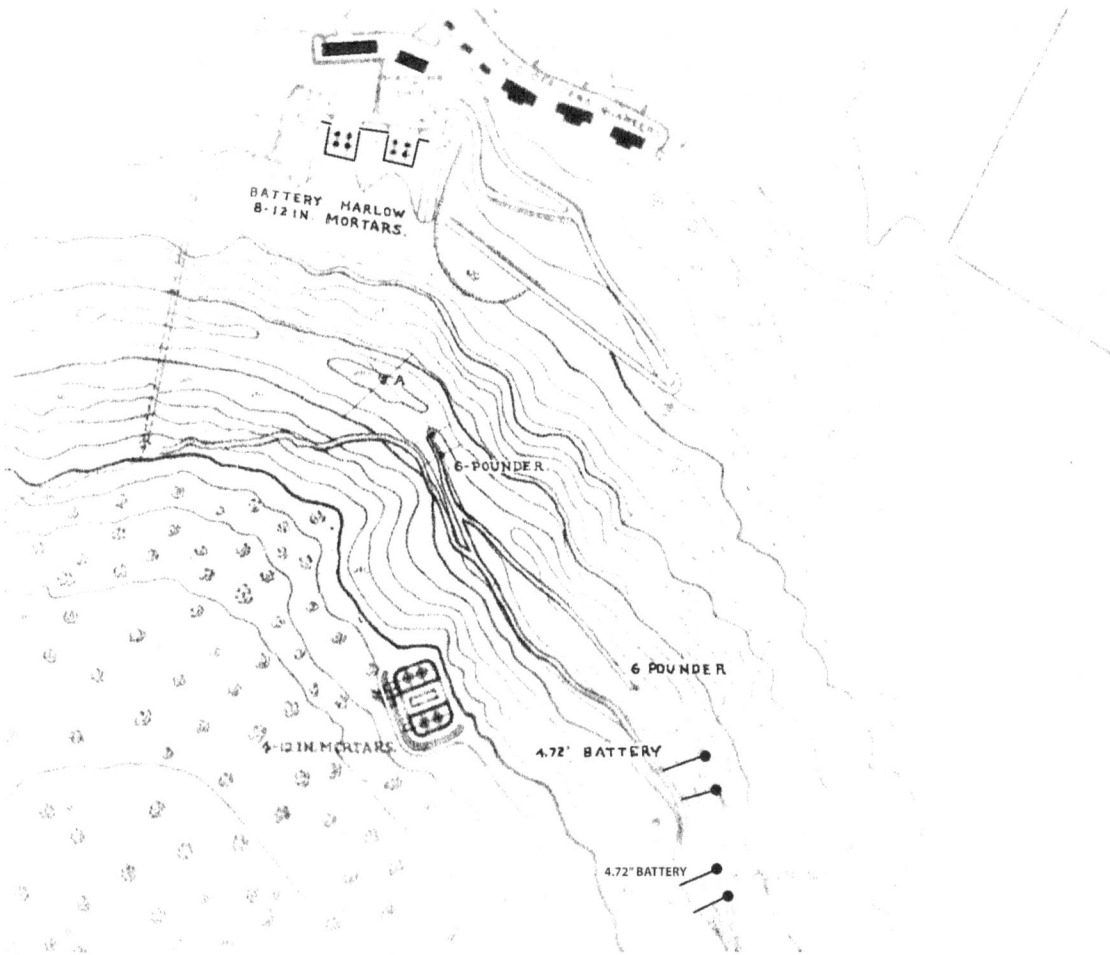

1919 map of the northeastern side of the Diamond Head crater, showing relative locations for Battery Birkhimer, the 4.7-inch guns of Battery Hulings and Dodge, and for the northernmost four positions for 6-pdr guns. *NARA.*

This real photo postcard illustration shows the six 6-pdr of the Fort Ruger land defense battery lined up on the post in 1915.
*Schmidt Collection*

Plans for the land defenses of Fort Ruger had always included a relatively lengthy protected infantry trench. The engineering notebook of 1916-17 reports 2800-feet of infantry trench was completed. (20)

Included with the weapons supplied for land defense, were four Vickers-Maxim automatic 37 mm guns on wheeled carriages. These also had been originally acquired right after the Spanish-American War and were generally obsolete for mobile armies. Some of these guns were stored at Fort Armstrong, and another four were allocated to Fort Ruger. On July 23, 1917, a shelter was authorized for construction inside the crater portion of the reservation for properly storing these guns. The guns were removed within a short time after being received with the widespread declaration of obsolete material right after the end of the war, and the shelter re-purposed. (21)

The position at Kupikipikio Point (subsequently known by the army at Black Point) was one of the earliest elements foreseen with the Oahu Land Defense Project. Plans were submitted by Maj. William P. Wooten on September 8, 1914, and it was reported as being near completion by late November. The reservation had been a part of the Fort Ruger property intended as the site for a fixed 60-inch searchlight. About 100-feet above sea level, Battery Mills faced almost due east. It was armed with two standard Model 1900 5-inch seacoast guns on Model 1903 pedestal carriages relocated to Oahu for use in this project. The carriages had previously been mounted at Battery Gregg in the Delaware defenses of Fort Mott. This 50-caliber gun fired a 58-lb. projectile at 15° elevation to a range of 10,431-yards. The choice of armament seems peculiar when there were still un-used 4.7-inch and 6-inch guns in the U.S. Transferring of just two guns with unique parts and ammunition for this type of secondary use doesn't appear to be sensible.

As the battery was subject to enfilading fire from seaward, a 20-foot traverse extended to the rear some 115-feet from the right (south) front of casemate No. 1. In this traverse, engine and fan rooms furnished re-serve or standby electricity to the battery; the primary source was the nearby power plant for Searchlight No. 1. The magazines and bombproofs for the men were on the battery's lower level, partially below grade, with

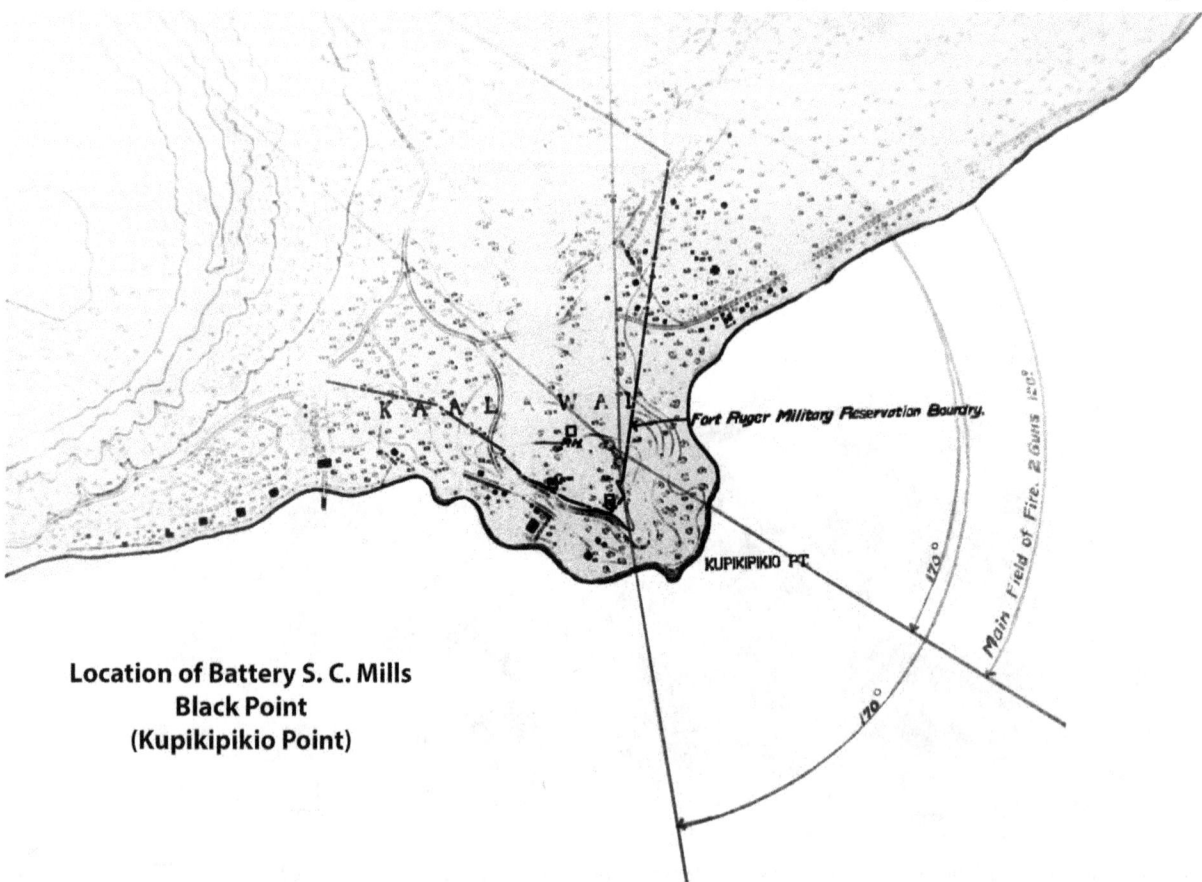

Map showing location and field of fire intended for the 5-inch gun battery at Black Point. *NARA.*

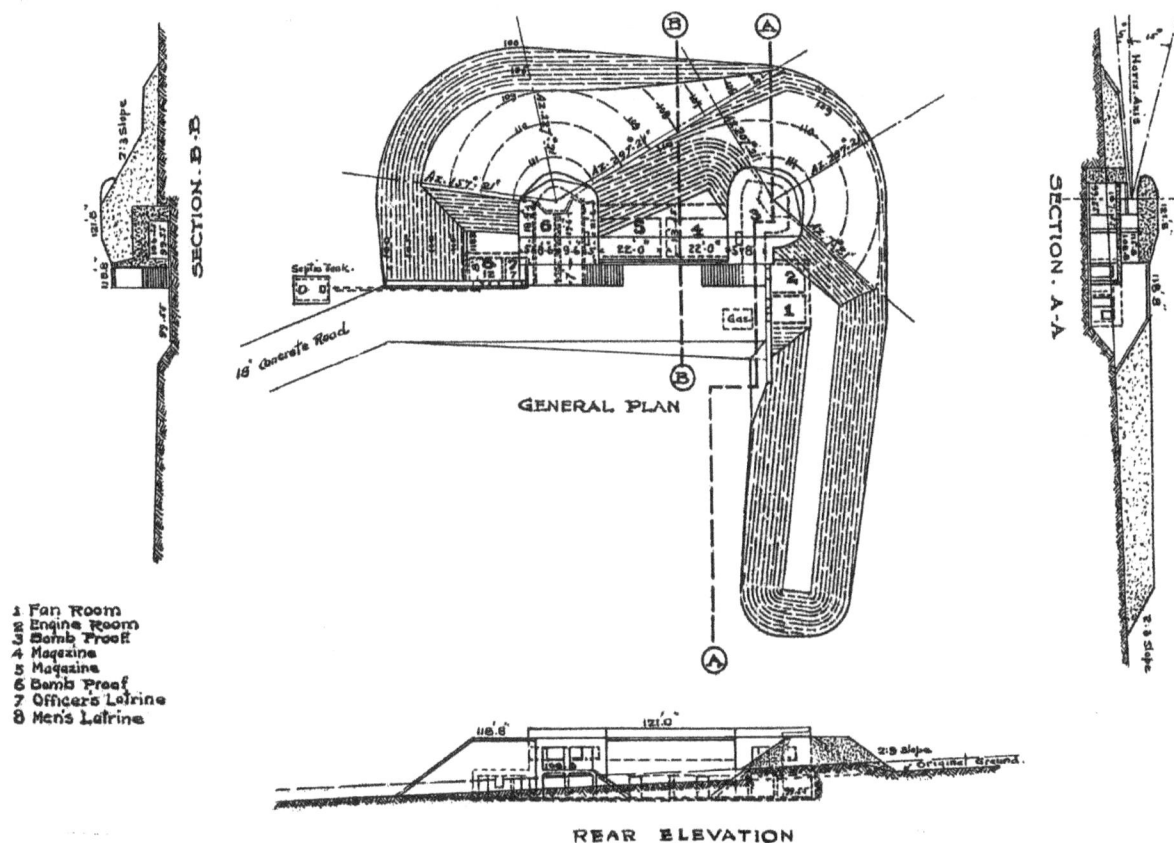

## GENERAL PLAN

## SECTION. B·B

## SECTION. A·A

1 Fan Room
2 Engine Room
3 Bomb Proof
4 Magazine
5 Magazine
6 Bomb Proof
7 Officer's Latrine
8 Men's Latrine

## REAR ELEVATION

Plan, elevation and section from 1919 for the 5-inch Battery S.C. Mills at Black Point. *NARA.*

the latrines for both officers and enlisted men. The battery was transferred to the coast artillery on May 15, 1915, before it was finished, electric power connections from the nearby searchlight power plant not having been completed. Consequently, the battery's electrical power was initially provided by the reserve plant. An inspection in July 1916 revealed more than a few problems with the material used in the protective parados and ammunition service path and were eventually corrected by appropriate troop labor. When completed in late 1916, it was judged to be the "most effective" land defense battery then in service. Battery Mills was formally transferred to the Fort Ruger garrison on December 12, 1916 for a construction cost of $30,560.58. It was also manned by the 2nd Co., CAC. (22)

The Army Adjutant General declared all 5-inch seacoast and siege guns obsolete in June 1919. The command in Hawaii removed Mills' guns but was authorized to keep them as post ornaments. For many years the two distinctive shields of the 5-inch Model 1903 carriages were positioned, upside down, as entry guards outside the gate into Fort Ruger. It is a view well accounted for in private photographs and public postcards.

By early 1916, a third company of coast artillery was needed for the Fort Ruger garrison for manning the land defense batteries being constructed at Diamond Head and on Black Point. Consequently, augmentation of the garrison was authorized. On July 12, 1916, the 2nd Co., CAC, some 108 men strong, disembarked from U.S.A.T. *Sheridan* and was temporarily quartered at Forts DeRussy and Armstrong. On July 14, the company marched to Fort Ruger. Pending the completion of a new frame barracks at Fort Ruger, the company was quartered under canvas. Upon settling into their tent camp, the company was assigned to Battery S.C. Mills' 5-inch guns at Black Point and Battery Dodge's two 4.7-inch Armstrong guns. The company was also provided with four 0.30 cal. Vickers-Maxim machine guns. The 2nd Co. finally moved to its new permanent quarters in December 1917. (23)

The garrison, however, was still short one company if all batteries were to be manned simultaneously, so a fourth Fort Ruger company was activated using cadre personnel from the other three companies and filler troops from the mainland. This new company, constituted in July 1917 was initially assigned to the mortars of Battery Birkhimer, and later to the 5-inch guns of Battery S.C. Mills, which it manned through the World War. After a few weeks of familiarization with the newly assigned 4.7-inch armament, the Fort Ruger garrison conducted its annual war-condition period from August 12 to 19, 1916. The 1st Co., Fort Ruger, conducted its first target practices with Battery Dodge's 4.7-inch guns on August 18, followed by additional practices on August 24 and 28. A service practice on October 12 with two 6-pounder RF guns atop the east rim of Diamond Head, near Batteries Dodge and Hulings completed training with the new batteries.

As mentioned previously, the loyalty of the large first and second generation of Japanese residents in the Hawaiian Islands was frequently questioned by the army defenders. As it turns out this population was loyal to the United States and was not infiltrated with agents or provocateurs by Japan. But that wasn't known in 1910-1940. One of the suggestions of the Macomb Board and included in subsequent funding requests was for obstacles or at least protection for the important tactical structures; ways to protect them from saboteurs or even mobs. Funds did not allow for concrete walls or ditches, but there was enough for fencing. On July 29, 1915, authorization was granted to protect essentially all the gun batteries and mining structures with "non-climbable wire fencing". An interesting series of battery and fence outline plans was issued describing the new fences. Just how complete the provision was carried out, or how long they lasted is unknown. (24)

Several diplomatic disputes erupted between the United States and Japan in the early 1900s. On April 15, 1913 the lower house of the California State legislature passed a bill forbidding foreign aliens (at least those ineligible for citizenship—like Japanese and other Orientals) from owning property. While the Wilson administration sent Secretary of State William Jennings Bryan to the state to prevent the bill's passage, that effort failed, and the Webb bill was passed on May 4. Japanese response was immediate and negative. They felt it contrary to the spirit of the negotiations that had ended a 1907 dispute, and distinctly racial in nature. There were vocal demonstrations in Tokyo and heated diplomatic exchanges. Both the Navy and Army took steps to prepare for possible hostilities. The Navy wanted to move their three cruisers of the Asiatic Squadron to Manila from Shanghai, and to prepare and concentrate the six armored cruisers of the Pacific Fleet in Hawaiian waters. General Funston in Hawaii appealed for reinforcements. Actually, on Oahu all eight Taft-generation batteries were in full service and alerted for the crisis—which soon passed.

After the 5-inch guns were removed from Battery S.C.Mills, they were turned into gate entry ornaments for the Fort Ruger post; though placed upside down for some reason. This view from the early 1940s. *Williford Collection.*

# CHAPTER 4
## THE FIRST WORLD WAR AND AIRPLANE DEFENSE

The First World War, despite this terminology, was primarily a European, or maybe an Atlantic/European War. Besides cleaning out small German detachments in the Far East, the Pacific Ocean was generally unaffected, which also meant that the American defenses at Oahu were generally unaffected. By the country's entry into the conflict in April 1917 there was no viable threats to the nation's Pacific bases.

The German auxiliary SMS *Geier* had taken protective haven in Honolulu early in the war. An unprotected colonial cruiser dating from 1895, she had departed German East Africa before the war started, but was still at sea when hostilities began. Trying to work her way to Admiral von Spee's Squadron crossing the Pacific, the ship was plagued with engine problems and lack of coal and water. The vessel entered neutral Honolulu on October 15, 1914, and was eventually interned on November 8th. Diplomatic relations were suspended with Germany on 3 February 1917. On that day three officers and 64 enlisted men of Fort Ruger's 1st Company of Coast Artillery were ordered to Fort Armstrong near the harbor's commercial anchorage. Later in the afternoon and evening, two Vickers-Maxim machine guns and four more men arrived from Fort Ruger. The following morning the detachment was ordered to the army coal pile on Allen Street, with the Maxim guns on the right flank to command the German ship, whose crew had attempted to destroy the vessel by setting it afire and disabling its engineering equipment. At 4:45 p.m. after being relieved by elements of the 1st Infantry from Schofield Barracks, the 1st Co. returned to Fort Armstrong overnight. That was the extent of the garrison's direct contact with the Central Powers. The ship was seized, and the crew removed to the nearby Territorial Immigration Center. Finally at the start of war the ship was taken into full U.S. custody, and in fact repaired and reused as the USS *Schurz* later in the war.

German colonial cruiser SMS *Geier* interned in the port of Honolulu during the First World War. *NHC*.

Naturally there was great excitement in both the civil and military communities when the United States entered the Great War in April 1917. It was, however, not totally unanticipated. Hawaii was far from any likely theatre of combat, the German surface fleet already bottled up in the North Sea and the range far too great for any submarine activity. No hostile attack on the island was anticipated. But the soldiers assigned to the defenses knew that their personal lives might well be affected as the country's need for its professional military personnel was urgent.

For the local defenses about all that was done was a tightening of security against possible acts of sabotage and a general heightening of awareness and alertness. Some of the logical steps had already been taken. Since early 1916 Brigadier General John Wisser, Hawaiian Department commander, had ordered a general strengthening of guards at all army posts. With the declaration of war in April liquor was banned at all the post clubs. At Forts Kamehameha, DeRussy, and Ruger, one battery was nominally designated to be manned around the clock to have an immediate response to any possible need. Troops were actively used to plant and cultivate gardens in anticipation of food shortages, which never occurred. Censorship was increased, and mail contents were officially monitored. In October 1917, a typhoid outbreak occurred at Schofield Barracks, though tied to local water supply issues that were not repeated at the coast defense posts. (1)

There was a moderate augmentation of the coast artillery garrisons on Oahu in the latter half of 1917, but no large reinforcements were sent there during the war. In fact, most of the Regular Army organizations, except for a few companies of engineers, the Signal Corps, the 6th Aero Squadron, and the coast artillery companies were withdrawn from Hawaii early in the war. The Regular Army infantry was replaced by the 1st and 2nd Inf., Hawaii National Guard.

The extensive use of military aircraft in the First World War encouraged the U.S. Army to examine its own preparations for this new mode of warfare. While some efforts had been made beginning in 1910 to develop anti-balloon guns and spur some thoughts about how to organize and deploy them, no comprehensive plans were authorized. As unusual as it might seem that an isolated location like the Hawaiian Islands would be threatened by the early types of short-ranged aircraft, there could have been a threat from airships. The expanded German use of these lighter-than-air instruments against French and British colonial possessions in Africa and the Pacific was also a distinct possibility. Airships could conceivably operate on a global scale.

In October 1917 the War Department assigned responsibility for a new antiaircraft branch to the Coast Artillery Corps. The rationale for this was that it had the most experience in hitting moving targets, many of the technical and practical aspects of tracking and predicting targets were similar. Antiaircraft defense would remain with this branch for many years. The Ordnance Department was challenged to come up with both a "fixed" 3-inch antiaircraft gun for static emplacements, and a 3-inch mobile gun to accompany a field army. However, the critical needs for the American Expeditionary Force (A.E.F.) sent to the war theatre in France had to be met with a series of limited-production French and American stopgaps. (2)

The first American fixed antiaircraft gun design was finalized in April 1916, and it was issued the next year as the Model 1917 3-inch antiaircraft gun. The tube was a long 55-caliber, with a drop-block breech mechanism. Continuous-pull type of firing was used, achieved by a firing handle. The relatively simple pedestal mount allowed elevation of 0 to 90°, and a full 360° traverse. There was no shield or other crew protection. The carriage with its pivot yoke and racer were designed to be emplaced on fixed platforms. For a foundation a concrete block 12-feet in diameter held 16 hold-down bolts emplaced in concrete. The gun fired a 15-pound projectile with a muzzle velocity of 2600-fps. This gun and the subsequent M2 fixed 3-inch gun used the same ammunition, also in common with the Model 1903 seacoast gun. On March 15, 1918, the Army Adjutant General reported that 159 new Model 1917 3-inch AA guns were under manufacture for use at static coast defense forts in the United States and overseas possessions. Based on suggestion from the Chief of Coast Artillery the previous December, Oahu was initially assigned 15 of the new guns. The same correspondence states that as of the end of January 1918, 28 emplacements for the new 3-inch carriage had already been completed in Oahu. With the Chief of Engineers' backing, the decision was made to delay delivery of guns to the U.S. West Coast to provide the full allotment of guns to Hawaii. (3)

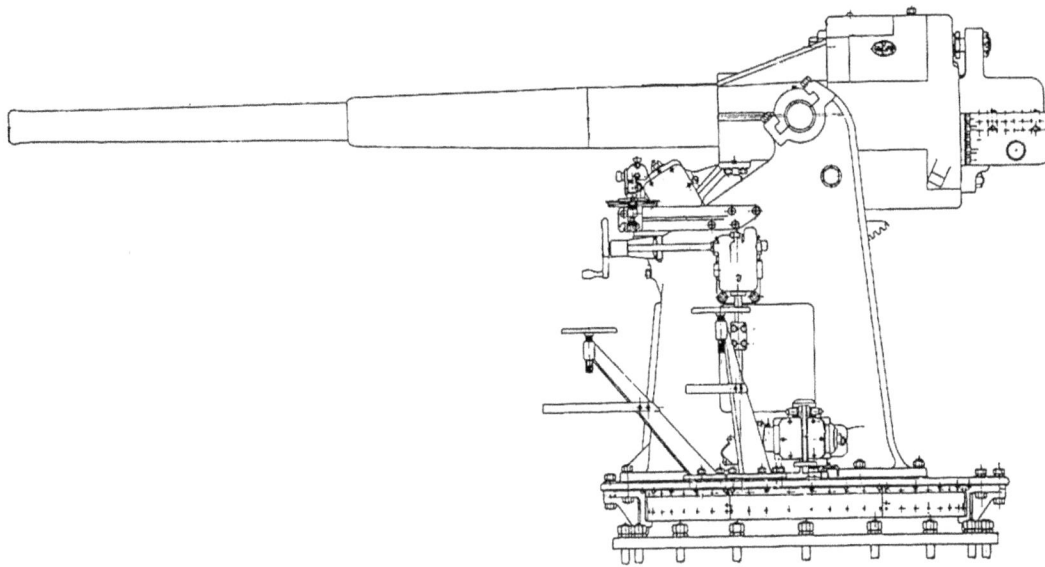

Ordnance Department sketch of the fist standard American fixed anti-aircraft gun, the 3-inch Model 1917 type. Over twenty of these guns were emplaced in Hawaii about 1920. *NARA.*

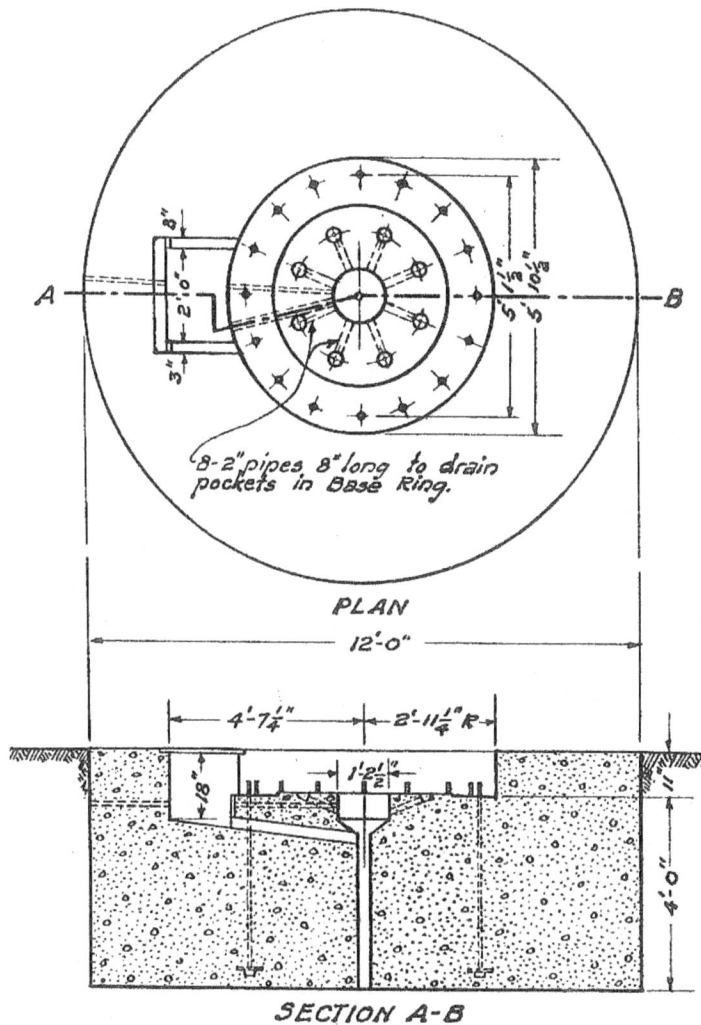

Plan for the concrete block to hold the 3-inch AA Gun Model 1917 pedestal. *NARA.*

The Oahu sites for the AA guns had been projected in June 1916 by a local Board of Officers under Brigadier General Frederick Strong. Prevailing concept for deployment envisioned pairs of guns, separated by 150-200 feet. Because the guns couldn't cover the space directly overhead, a second pair located about 400-feet away was often included. Emplacements were simple gun blocks with hold down bolts and occasionally a protected magazine for the fixed ammunition. The locations recommended were: (4)

| | |
|---|---|
| Fort Ruger | Two dual gun sites |
| Fort DeRussy | Two dual gun sites |
| Punchbowl | One dual gun site |
| Sand Island | One dual gun site |
| Fort Shafter | Two dual gun sites |
| Fort Kamehameha | Two dual gun sites |
| Ford Island | Two dual gun sites |
| Schofield Barracks | Two dual gun sites |

Twenty-eight guns were requested, and the 1925 Defense Project states that 26 3-inch Model 1917 guns were present in Hawaii. Initially just nine guns were shipped to the islands in 1920, and the final seven shipped in December 1922. It appears that altogether 26 were delivered to the Hawaiian Islands. No more fixed AA guns would follow. It appears the guns were emplaced immediately as they arrived, starting in 1920.

Map showing location of the fixed AA gun blocks at Fort Ruger in 1935. *NARA.*

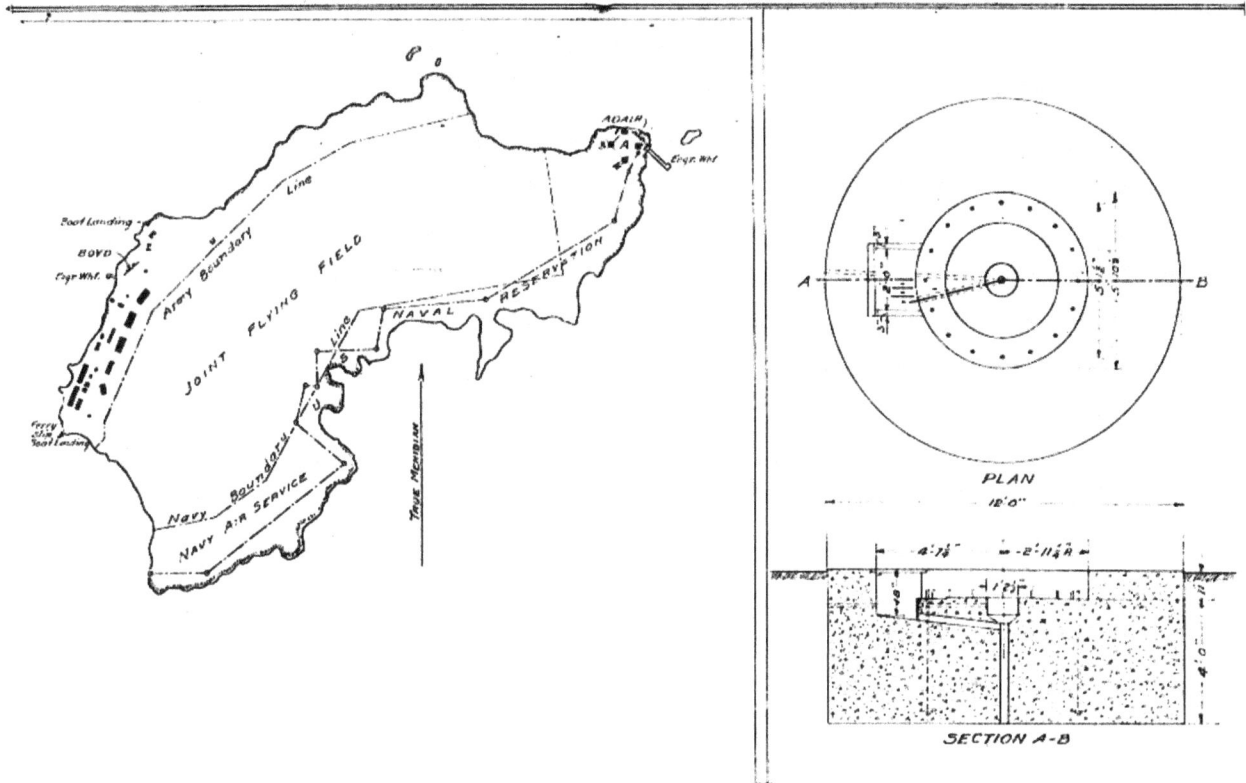

Map showing location of the fixed AA gun blocks at Ford Island in 1925. *NARA.*

The initial 14 sets of emplacements (for two guns each) were begun in the summer of 1917 and soon finished. Then the re-shuffling began. The original location for four guns on the crater ridge of Fort Ruger proved ill-conceived, and the guns were relocated to new blocks in the fort's cantonment area about 1925. In 1927 Fort Weaver's completion required protection with a new battery of four guns. Problems at one of the Fort Kamehameha sites forced a relocation of the blocks in 1928. Then in 1934 two of the Fort DeRussy blocks were moved closer to the other pair to form a quadrangle of a single four-gun battery. New seacoast batteries also required closer protection. In 1937 the new 8-inch battery at Black's Point necessitated the relocation of some of Fort Ruger's guns. Likewise, Battery Closson at Fort Kamehameha received a new battery (at this point the emplacements almost always had four gun block positions) relocated from the older positions that same year. Finally, Fort Barrette got a battery of four 3-inch Model 1917 guns by relocating the existing armament from Fort Shafter, again in 1937.

The later batteries usually were designed with guns around a square, with 200-feet between corners. The blocks were flush with the ground, though often protected with a berm of sandbags. Usually a small concrete magazine supported a pair of guns. The AA predictor (the primary method of directing fire from the battery) was kept in a small concrete storage room. Where necessary for remote locations, a small power generator room was also required. (5)

For a description of the fixed, 3-inch Antiaircraft emplacements built on Oahu, see Appendix III.

Also produced late in the First World War was what would be considered the standard mobile AA gun. This was also a 3-inch bore tube, carried on a four-wheeled trailer. It could be both transported and fired from this carriage, though for accuracy it was usually stabilized by outriggers or even held steady by a previously

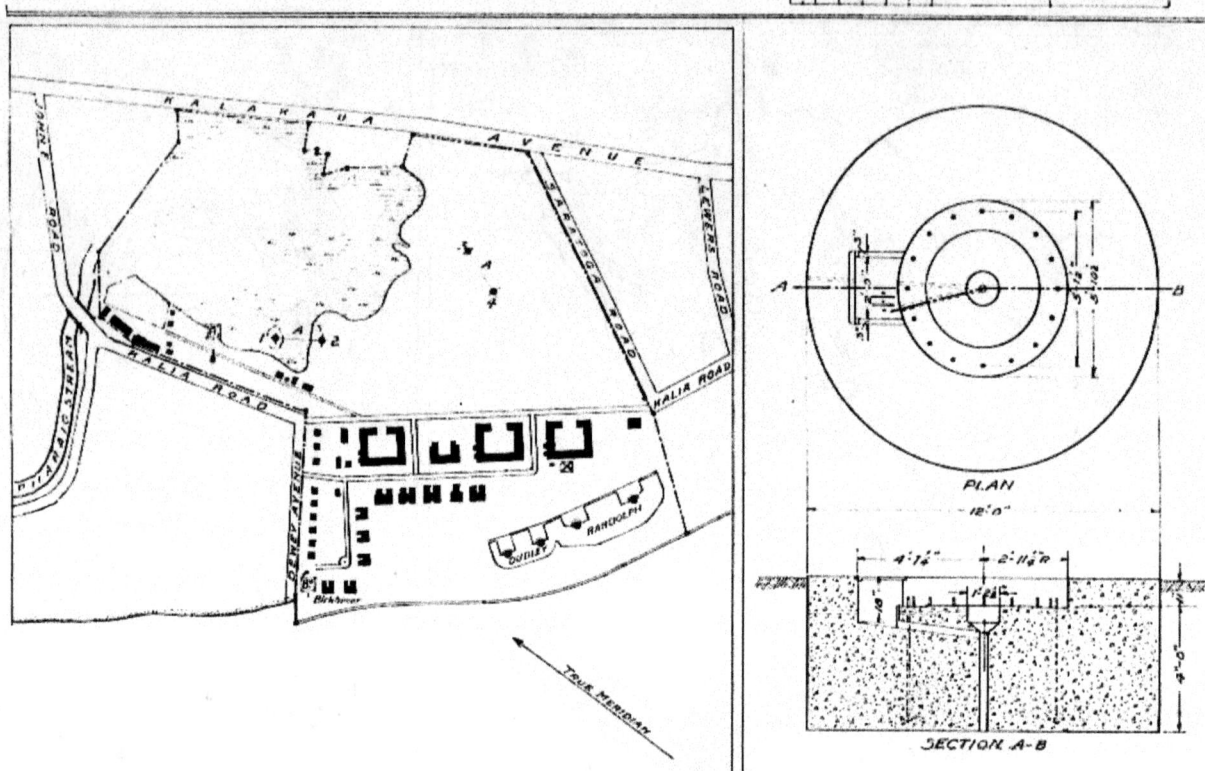

Map showing location of the fixed AA gun blocks at Fort DeRussy in 1925. *NARA*.

Photograph, probably taken in the early 1930s, is of a fixed 3-inch Model 1917 AA gun being exercised for target practice by its assigned gun crew. *Gaines Collection*.

emplaced guide platform. Assigned to Coast Artillery Antiaircraft Regiments, they were truly mobile guns intended to be used where needed, and not deployed as fixed batteries. It was designated the 3-inch Model 1918 gun (and trailer). It fired the same ammunition as the fixed Model 1917 gun. The 40-caliber barrel could be elevated to 85° and fire its 15-lb. projectile to a vertical height of 8530-yards. In 1920 nine initial Model 1918 3-inch AA guns were supplied, and before long 24 of these guns were assigned and sent to Hawaii. They were destined to be the primary gun armament for the new Coast Artillery AA regiment being formed in Oahu.

Over the next few years 16 additional guns, initially at least held in reserve, were sent to Hawaii, bringing the total inventory to 40. They were popular parade items for the big reviews the army staged at Fort Shafter in the 20s and 30s. In practice though, the gun proved not to be a technical success. It was never stable enough, and the size and range became too limited as targets became more advanced. Of course, as a mobile weapon, it was not assigned to permanent emplacements, but over time likely locations on military reservations were identified and prepared with sunken pits or elevated berms for potential occupation.

On June 3, 1921, an organization known as the Hawaiian Antiaircraft Regiment was created at Fort Ruger. Unfortunately, its creation didn't come with soldiers, its initial two battalions, 32 officers and 802 enlisted men, had to come from companies manning the fixed seacoast batteries on Oahu. The 1st Bn, formed from Fort Kamehameha companies in the CD of Pearl Harbor, was organized as an antiaircraft gun battalion with three batteries, each with four mobile M1918 3-inch guns and a searchlight battery with 12 36-inch high-intensity barrel-type searchlights on Mack trucks. The 2nd Bn, organized from inactivated companies at Forts Ruger and DeRussy, was a machine-gun battalion with four batteries, each equipped with eight AA machineguns. The two battalions perfected their organization and on June 23, both battalions were assembled at Fort Ruger. Initially, the Hawaiian Antiaircraft Regiment was quartered in temporary frame structures built back during construction phases in 1909-1910. Some troops were housed in a tent encampment east of the main garrison area, practically on the slopes of Diamond Head, a location that proved ill suited. (6)

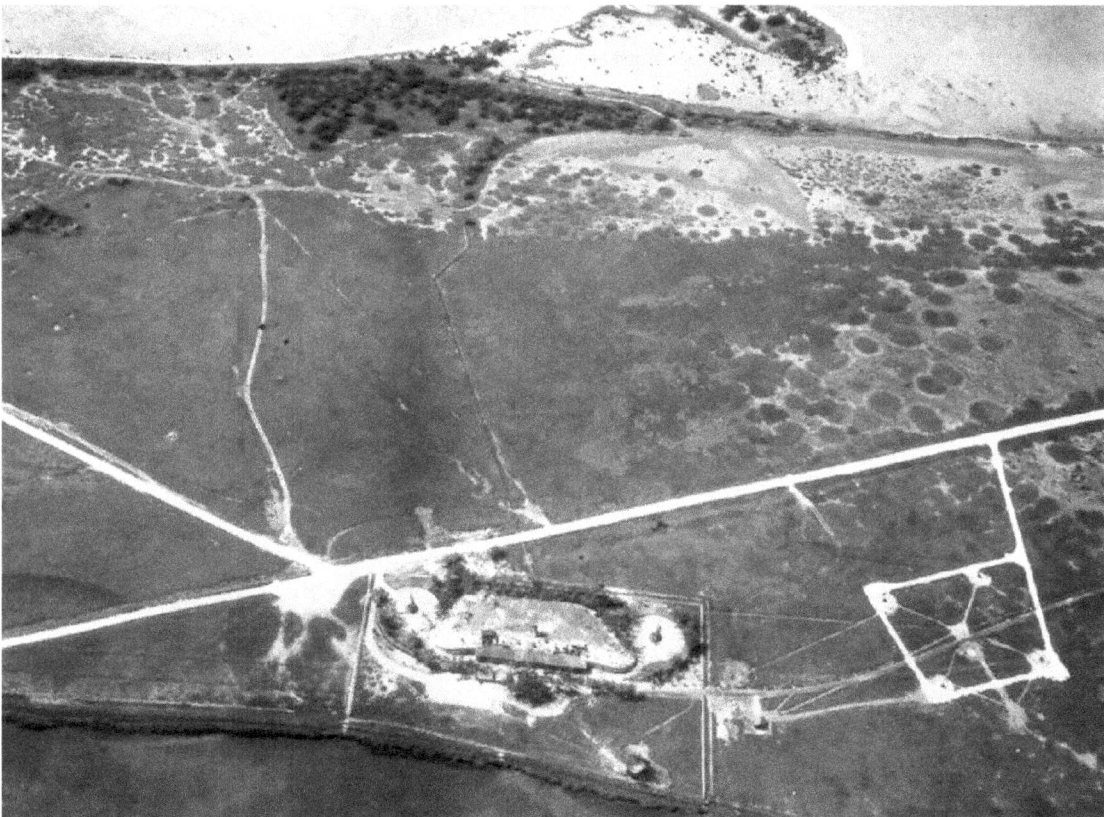

At the close of the First World War Fort Kamehameha gained the services of Battery Closson for two 12-inch barbette, long-ranged guns and a protecting fixed 3-inch AA battery on the eastern side of the reservation. *USAMH.*

Obviously mobile guns allowed movement of units to different geographical locations as needed, here a column from the 64th CA (AA) is on the road in 1925. *Williford Collection.*

The initial battalions were assembled for the first time near Diamond Head at Fort Ruger, under the command of Major Harrison on June 23 of 1921. The Hawaiian Antiaircraft Regiment spent the next three weeks completing its tent camp, constructing mess shelters, bathhouses, latrines, etc., as the regimental staff began to assemble the regiment's initial equipment. Among the first heavy pieces to arrive were 12 Mack searchlight trucks. As soon as the encampment was prepared, orders were issued on July 11 to march to Schofield Barracks for 10 days of training. As the regiment's only available transportation was the 12 searchlight trucks, the next

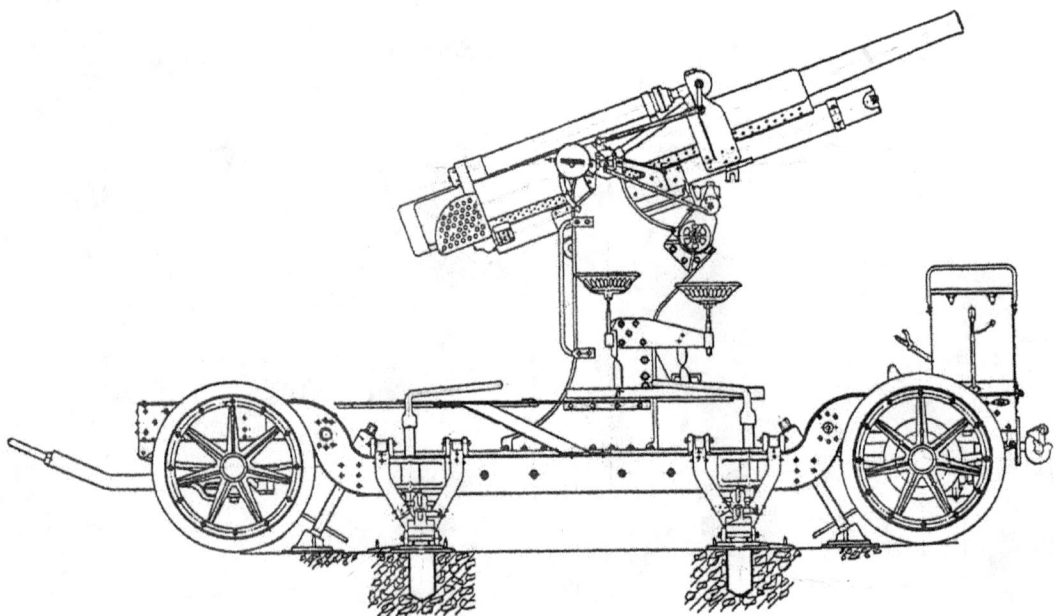

Mobile AA guns, such as this 3-inch Model 1918 gun and wheeled carriage, began being supplied to the coast artillery defenders on Oahu in 1922. *NARA.*

four days were spent borrowing motor vehicles from the 11ᵗʰ Field Artillery Brigade at Schofield Barracks so that the AA regiment could move to the central Oahu base. While at Schofield Barracks, the regiment conducted road marches, field exercises, and machine gun school. At the end of the training period the regiment was transported back to Fort Ruger. (7)

The remainder of 1921 was spent in training, and in reorganization in accordance with the new table of organization. The number of officers authorized for the regiment was increased to 37 in August 1921, and in December a further revision of the table of organization and equipment changed the peacetime strength of the regiment to 36 officers and 828 enlisted men. The revised T/O&E (Table of Organization and Equipment) also altered the regimental organization: After a scant five months, the machine gun battalion, armed with 0.30 cal. AA machine guns, was determined to be unnecessary and reorganized as a second gun battalion. Battery H became the regiment's second searchlight battery, equipped with mobile 60-inch searchlights on Cadillac trucks. The Cadillac units were somewhat of an improvement over the Mack trucks of Battery D, having slightly better range. They did not, however, prove as efficient in terms of continuous usage. (8)

The Hawaiian Antiaircraft Regiment's encampment at Fort Ruger proved to be poorly sited and generally unsatisfactory, by the following January Fort Shafter was selected as their new home. In early February the regiment had moved to its new station in the uplands behind Honolulu and occupied the World War I cantonment barracks on the east side of the former infantry post. In addition to the barracks, single-story bungalow-style field officers' quarters and NCO quarters were occupied. The temporary World War I barrack were little better than shacks and contrasted markedly with the excellent permanent barracks and quarters across the gulch around Palm Circle, occupied by the HQ and special troops of the Hawaiian Department. Intensive training and familiarization with equipment followed once the regiment was settled in at Fort Shafter.

By March 8, 1922, the adjutant general notified the regiment that the secretary of war had approved the regimental coat of arms and its motto, "We Aim High." On June 1, 1922, the Hawaiian Artillery Regiment was redesignated as the 64ᵗʰ Artillery (AA) CAC. (9)

From 1921 until mid-1924, the 64ᵗʰ Artillery, CAC, was the only antiaircraft regiment in the U.S. Army. Due to it being in an overseas location, the 64ᵗʰ was one of the few CAC units maintained at near full peacetime strength. The 64ᵗʰ was extensively involved in evaluating antiaircraft materiel and tactics. As a mobile unit it was also engaged in testing various military vehicles. In the spring of 1924, the regiment conducted a series of firing tests and demonstrations for the McNair Board, appointed to evaluate the relative tactical efficiency of the Army Air Service and antiaircraft artillery. The increasing role of antiaircraft artillery in the CAC was

3-inch Model 1918 anti-aircraft guns on their familiar trailers on maneuvers at Fort DeRussy in 1934. *USAMHI.*

The 64th also had mobile, or at least transportable, searchlights and sound locators to accompany its guns, here shown in the 1930s during a review. *Gaines Collection*.

reflected in November 1925 when a third battalion was added to the 64th CA. Each battalion was to consist of a battalion headquarters and three firing batteries.

The improvement of the 64th's cantonment continued in 1926, when the roadways were hard surfaced and a parade ground was laid out. On July 1, 1928, the 64th Coast Artillery was expanded again. The machine gun battalion (a single battery, Battery I), was abolished and Batteries K, L, and M were constituted. Batteries K and L were activated as gun batteries, but Battery M (AAMG) was inactive. The regiment would now consist of three battalions, each with a searchlight battery, two gun batteries, and a machine gun battery. Firing practice with the 3-inch guns had to be canceled in 1930 because of the lack of Air Corps planes equipped for towing. The necessary searchlight exercises and machine gun practices, however, were carried out.

The battalions of the anti-aircraft regiments also had batteries of light AA guns for use against low-flying strafing aircraft. The preferred weapon in the 20s and 30s were water-cooled machineguns on tripod mounts. *Gaines Collection*.

The men of the 64th turned out to have a strong affinity for sports. Several years of effort were expended on the development of athletic facilities. By1930, a running track had been laid around the playing fields, cement bleachers were built into the northeastern slope of the bowl, and a boxing arena was formed by terracing a horseshoe shaped indent in the southeastern slope. By early fall the entire sporting complex was in full use by the men of the 64th. This use lasted only a few weeks, for about one o'clock in the afternoon of November 18, 1930, according to the official history of the regiment: "...it began to rain hard, and this developed into a cloudburst around 3:00 P.M., resulting in the worst flood here in many years. A tremendous volume of water came suddenly down the upper gulch, wiping out a large part of the upper gulch road and all the walls, bridges, and revetments in the upper stadium. Immense boulders were carried along like chips of wood. The main water pipelines, gas mains, and sewer lines were broken and the stadium proper covered with mud and in places with ten feet of rocks and debris." All organizations suffered more or less from water rushing through the stadium. The Sector Garden was practically washed out. (10)

The 64th CA assisted the civil authorities in dealing with the widespread damage around the island. Searchlight detachments assisted the Honolulu Police and Fire Departments in Kalihi, where nine persons drowned. Other parties from the 64th worked through the night of the 18th rescuing people and property from the floodwaters in the Moanalua Gardens and Kalihi Districts just outside Fort Shafter. Numerous heroic acts were performed by the men of the 64th CA. One example was the efforts of Pvt. Jesse Compo of Battery B to rescue Col. James P. Barney, of the 8th FA, his sister, and his chauffeur from their automobile after it was swept off the Moanalua Bridge. After a civilian had been swept to his death trying to rescue the colonel's party, Private Compo "plunged into the raging waters" in an attempt to carry a rope to the three victims who had gained temporary safety in the branches of some trees. Compo barely escaped with his life and was later awarded the Soldier's Medal. Three other members of Battery B and one from Battery I received written commendations in connection with the rescues. (11)

Antiaircraft defense in the Army received even greater emphasis during the early 1930s, and training coast artillery troops on Oahu in antiaircraft gun drill was intensified in 1931. The number of AA regiments in the Regular Army had been increased to seven in a major reorganization in 1930, but only two were maintained close to nominal peacetime strength. Both were overseas: The 60th CA (AA) Regiment in the Philippines and the 64th CA (AA) on Oahu. Between the two world wars, these two regiments served as the primary overseas antiaircraft training and equipment testing organizations.

A new generation of mobile guns, the 3-inch M3, was introduced in the early 1930s. Over a couple of years the Oahu defenders gradually replaced the M1918s with the newer model. Finally in 1940 the older Model 1918 gun was declared Limited Standard. Initially it was decided that 28 of these guns would be kept in the Department's strategic reserve, but eventually. as there was no other use intended for the obsolete model, all 40 currently in inventory were placed in reserve in mid-1940. Some of these were in fact utilized after the December 7 attack. In December 1941 16 M1918 guns were sent with small detachments to outlying islands. A battery of four guns each went to the big island of Hawaii, Kauai, Maui, and Christmas Island. (12)

The M3 was a more modern development for a 50-caliber anti-aircraft gun of improved performance. The breech mechanism was semi-automatic. It developed 2800-fps muzzle velocity with HE rounds and 2600-fps with shrapnel and could fire to a vertical ceiling of 14,200-yards. It was fired using a continuous-pull mechanism and short lanyard. Maximum rate of fire was calculated at 25-30 rounds/minute. For its time and bore size, it was considered a successful weapon. This newer type of AA gun featured a much more stable firing platform. Large folding outriggers supported the gun when deployed, and then could be folded in prior to attaching sets of wheels for towing. The first allotment of eight M3 AA guns arrived in April 1939. By April 1940, 40 were assigned to units, with 60 reported by December 7, 1941. As reinforcement poured in after the December 7 attack, in early 1942 30 batteries of mobile 3-inch guns (120 mounts total) were reported available to the defense. As coast artillery units were chronically short of personnel in the island, more than a few of the guns were kept as secondary assignment for various batteries. When deployed the carriage usually sat in a dugout location, revetted with timber and with pipes holding a frame or chicken wire holding camouflage netting.

Sketch of the second generation mobile AA gun, a new 3-inch M3 gun which was a towed rig deployable on a steady ground platform with outrigger legs. *Army Technical Manual.*

A mobile 3-in AA gun and a pair of 155mm field guns on display in 1941 during a review at Fort Kamehameha. *Williford Collection.*

In 1920 there was a clear need for two types of anti-aircraft weapons. High-flying level bombers approaching or dropping bombs should be engaged by heavy guns capable of having their shells reach the necessary altitude and to carry explosive shells with an appropriate fuze. For dive bombing or strafing attacks, a lighter gun capable of rapid change in azimuth and elevations delivering a rapid-fire volume was needed. Early Coast Artillery AA Regiments were generally organized with batteries for heavy 3-inch guns, either fixed or mobile. They would usually also have at least one battery per battalion of lighter AA machine guns (either the Browning 0.50-caliber water-cooled gun, or its smaller 0.30-caliber type) on tripods mounts for use against low-level attackers. As aircraft increased in speed and developed new combat techniques, most nations developed intermediate "automatic" light cannon for improved defense. For the U.S. this weapon was initially the 37 mm automatic antiaircraft M1A2 gun, later replaced by the improved 40 mm M1 gun during 1943-44.

1932 was an important year for transition in terms of the 64th regiment's armament and equipment. The regiment's old M1918 AA guns were replaced by the new 3-inch M3 AA guns; the unit received 11 new Sperry Mk VI AA searchlights and generators on trailers supplanted a portion of the old Mack and Cadillac truck-mounted searchlights; as well as more modern AA machine guns and fire control equipment. Organi-

The M3 guns began first arriving in Oahu in April 1939. Here some of the guns are lined up during a period of training. *Bennett Collection.*

Camp Malakole in southwest Oahu became the home for the 251st CA (AA) in 1941 with its 3-inch M3 guns. *NARA.*

zationally, the 64th continued to evolve. Battery I was converted from a searchlight battery back to an AAMG battery, giving the regiment two searchlight batteries, six gun batteries, and a machine gun battery. Parades and reviews were a regular part of coast artillery soldiering on Oahu between the two world wars. Regimental parades were held frequently, often tied to the arrival of a visiting dignitary. The presentation of military proficiency or athletic awards to one of the organizations in the Hawaiian Department seems to have a sufficient justification for a departmental review, or at least a parade by the Hawaiian Separate Coast Artillery Brigade.

On December 16, 1932, the 15th CA, manning the defenses of Pearl Harbor, was honored with a brigade review at Fort Kamehameha for winning the U.S. Coast Artillery Association plaque for the highest target practice scores in 1932. All the regiments in Hawaii assembled on the Fort Kamehameha parade ground for the awards ceremony and the review by Maj. Gen. Briant H. Wells, commander of the Hawaiian Department. During a visit to Hawaii in 1933 by Newton D. Baker, the highly respected former secretary of war, a full-scale departmental review of some 12,000 troops at Schofield Barracks, with all the "materiel that could be moved economically," was provided in his honor. The 64th took part, its guns, searchlights, and other mobile equipment passing in review in a broad battery front. (13)

A new firing point for extended practice firings was established on the beach at Fort Weaver in the early 1930s. It included the four fixed 3-inch AA guns, and a similar number of prepared emplacements for the mobile guns. The 64th CA had regularly conducted its annual service practices at this facility. In July 1934 however, the regiment went to the newly established Bellows Field at Waimanalo on Oahu's Windward Side for the annual gun and machine gun service practices. Here Battery L, in conjunction with the Air Corps, demonstrated AA tactics for a large group of Army and Navy officers and civilians. The exercise was the air defense of a troop column moving through a series of defiles. Battery L, commanded by Capt. William G. Brey, CAC, marching at the rear of an advance guard, pulled out of the troop column and took up a firing position between two defiles and was ready within 30 minutes to repulse a bombing raid from the sea. Battery B's 3-inch guns joined with those of Battery L and at the end of the series of air raids, the AAMGs of Battery I were added to the antiaircraft fire on the towed sleeves of the "attacking" planes. Although still partially equipped with increasingly obsolete ordnance and equipment, the regiment turned in a very credible set of scores in this practice. (14)

Mobile anti-aircraft searchlights, such as this Sperry 60-inch type, accompanied the M3 mobile battalions. *Gaines Collection.*

The 64th CA arrived back at Fort Shafter in time to take part in the largest military demonstration since the World War, when President Franklin D. Roosevelt paid a three-day visit to Oahu on July 26-28, 1934. President Roosevelt arrived on the field and passed in front of the troops in his car with 30,000 spectators giving him a great ovation. The president took his station in the reviewing stand and watched over 15,000 troops pass before him. In 1935, the 64th Coast Artillery was now partially equipped with more modern equipment, the new M3 3-inch AA guns as well as new model Vickers M1A1 and Sperry M-2 AA gun directors. These in combination with the T-2 and T-9 height finders enabled the firing batteries to earn the "E" for excellence, and for Battery B to win the coveted Knox Trophy. (15)

As the 1930s ended and war in Europe and Asia loomed, AA training in Hawaii intensified even more, and the 64th CA was tasked to train other units in the department. The regiment was also reorganized once again. The six gun batteries were grouped together as the 1st and 2nd Bn, three batteries to each battalion, while the two searchlight batteries, A and E formed the 3rd Bn. As a precautionary move, the staff reexamined the AA defense of Oahu and selected new positions for the existing AA batteries of the 64th as well as those projected to reinforce the island in the future.

A system of 12 numbered antiaircraft intelligence stations (AAIS) were established around the perimeter of the island, frequently in conjunction with the existing lettered fire control stations. These were observation stations for spotting enemy aircraft, manned with personnel trained for the purpose; tied into the command cable network. To facilitate the regiment's expanded training mission, a permanent regimental firing point was established in 1939 at Malakole, about a mile and a half northeast of Barbers Point. Here, among the algaroba trees near the beach, ground was cleared for a firing point for mobile antiaircraft guns. Target practices at the new firing point were interrupted in October 1939, when torrential rains fell on the relatively dry southwestern portions of Oahu. The camp of the regimental HQ Battery at the firing point was flooded some three feet deep. By fall of 1940, expansion of the firing point into a permanent camp was carried out by fatigue parties of the 64th CA. (16)

The 369th CA (AA) (Colored) was one of the first units assigned to the defenders with a predominantly Afro-American roster. Most of its service was with the North Shore Groupment. *Bennett Collection.*

On November 4, 1940, advance elements of the Federalized California National Guard 251st CA (AA) Regiment, CAC, arrived in Honolulu and was sent to Malakole. The camp was deluged by another heavy rainfall on November 19 forcing the newly arrived personnel out of their tents to higher ground. The tents were set up on the floors of the roofless buildings still under construction, fortunately elevated some two-feet above ground. When the remainder of the 251st arrived in Oahu on November 23 they were temporarily quartered at Forts Shafter and Ruger pending completion of the quarters at the firing point. The new antiaircraft firing point and post was named Camp Malakole on January 9, 1941. As completed it contained some 96 "theater of operations" buildings: 48 barracks, 12 mess halls, nine magazines and storehouses, five officers' quarters, seven showers and latrines, dispensary, officer's mess, post headquarters, post office, regimental day room, theater, laundry, motor repair shop, gasoline station, fire house, guard house, and photo laboratory, as well as quartermaster and engineer buildings. With the arrival of the 251st at Camp Malakole, the 64th only irregularly used it thereafter for firing practice. The 251st CA had three 3-inch firing batteries, fully equipped with the required 12 3-inch M3 guns. (17)

New AA regiments were organized in place with unassigned replacement troops already in the islands. Several thousand coast artillery replacements, most with less than six weeks of training, began arriving in Oahu in 1941. On July 10, 1941, the 98th CA (AA) Regiment was partially organized at Schofield Barracks. Three months later, on October 9, 1941, the 97th CA (AA) Regiment was partially organized at Fort Kamehameha, again with cadres from other units and recent trainees. Command of the newly-named HSCAB (Hawaiian Separate Coast Artillery Brigade) changed in August 1941 with the arrival of Maj. Gen. Henry Burgin. The coast artillery on Oahu was reorganized as the Hawaiian Coast Artillery Command (HCAC) and the newly-formed 53rd CA (AA) Brigade was constituted to command the antiaircraft regiments on Oahu. War plans called for the deployment of antiaircraft regiments to various points on the island. The 1941 revision to the Hawaiian Defense Project, which was to be submitted to Washington at the end of the year, envisaged defending eight distinct areas: Fort Ruger and Diamond Head area, Fort DeRussy Honolulu Harbor area, Aliamanu Crater area, Pearl Harbor-Hickam Field-Fort Kamehameha area, Fort Weaver, Fort Barrette, Schofield Barracks-Wheeler Field, Kaneohe Bay NAS-Camp Ulupau. (18)

Soldier with 3-inch anti-aircraft fixed rounds at Schofield Barracks in 1925. *Williford Collection.*

# CHAPTER 5
# TO FIGHT AT LONG RANGE

Naval weapons evolved rapidly in the early 20th century. By mid-decade of the 1910s the caliber of battleship main batteries had increased to 15-inches, and their maximum elevation was increasing, resulting in a significant increase in the maximum range. The 12-inch disappearing guns and mortars of the Endicott generation, and even the 14-inch guns of the Taft period were in danger of being out ranged. A War Department initiative resulted in a board of review in 1915 to recommend improvements in seacoast defense armament for American domestic and insular territories.

This board, publishing its report in 1916, recommended installation of new types of longer range guns and carriages. The board recommended that several major harbors, already having defenses from the past generations, get just a couple of these new guns to enhance their ability to engage at long distance. The primary new weapon was the proven Model 1895 12-inch/35 gun. While widely used on disappearing carriages, it was now to be emplaced on a new heavy long range barbette carriage. This was a 360°, rotating carriage capable of elevating the gun to 35°. The carriage was designated the Model 1917 and referred to as a BCLR mounting for Barbette Carriage Long Range. As mounted, it could send its 1070-lb. projectile at 2250 f.p.s. out to 29,300-yards. That was more than twice the distance of the Model 1901 12-inch gun on the disappearing carriage, and 12,000-yards more than that carriage newly modified to 15° elevation. Several harbors were also to get a much more powerful 16-inch guns and mortars on a similar sort of barbette carriage, but they lagged in development and were still several years away.

Recommended for Hawaii were six of the new long-range 12-inch guns emplaced in three new emplacements. At this stage no 16-inch were suggested, though once informed the Hawaiian command requested the substitution or the larger gun for the smaller, but to no avail. Incidentally the 16-inch weren't much more expensive for emplacement, but the cost of the gun and carriage and ammunition were, and would require lengthy procurement times. The Board of Review report suggested the 12-inch guns go to emplacements at Ahua Point (eastern end of Fort Kameheha), Diamond Head Crater (soon changed to Barbers Point) and

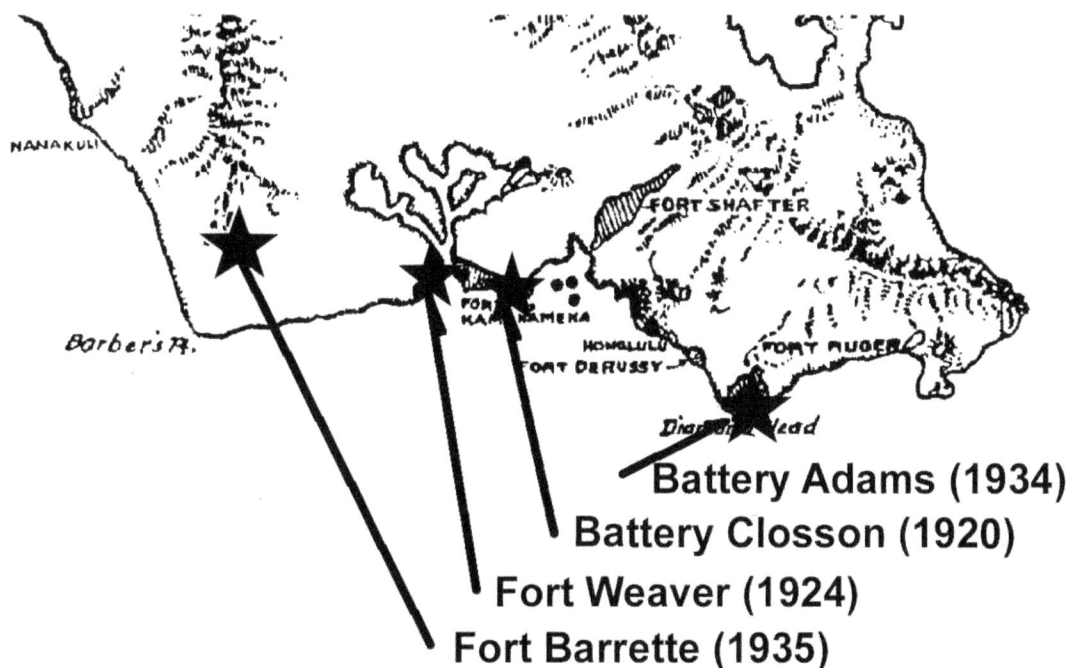

Between the wars four new long-range gun batteries were added to the defenses on Oahu. Two were in new posts at Fort Weaver and Fort Barrette, and two were at existing forts at Kamehameha and Ruger's Black Point. *Williford Collection from NARA map.*

Schofield Barracks (soon changed to Waialua Bay on the north shore). It was estimated that these emplacements could be built with a total expenditure of $1,439,900. The report's exact words were: (1)

> Oahu: Six 12-inch guns mounted for long-range fire to cover water areas not covered by existing armament from which the naval utilities can be bombarded by a hostile force, and to afford a means of defense against hostile warships attempting to support landings on the island.

For the new carriages the Army's Corps of Engineers designed a new type of emplacement. Each new emplacement was for two of the new 12-inch on BCLR. The gun carriages were placed within a pit with just the sides of the top carriage and gun mounted above ground, the rotating mechanism and space for the breech to pivot when elevated being below ground level. The two guns were to be 420-feet apart, each pit being surrounded with a concrete apron. Between the guns and set a little behind was a shared protected magazine. In this structure were two separate projectile and powder magazines, one set for each gun. Inside the magazines were rails in the ceiling for transferring shells, outside was a dedicated rail path for carts taking the shells and powder around to the guns. Over the magazine was a 6-foot concrete roof covered with 11.5-feet of earth for protection from shells or bombs up to 1000-lb. Within the battery structure were the usual operation rooms—plotting room, power room for two 25 kW gasoline generators, latrines, aid room, storerooms. On the top of the structure were two enlarged crow's nests to serve as a commander station for each gun. (2)

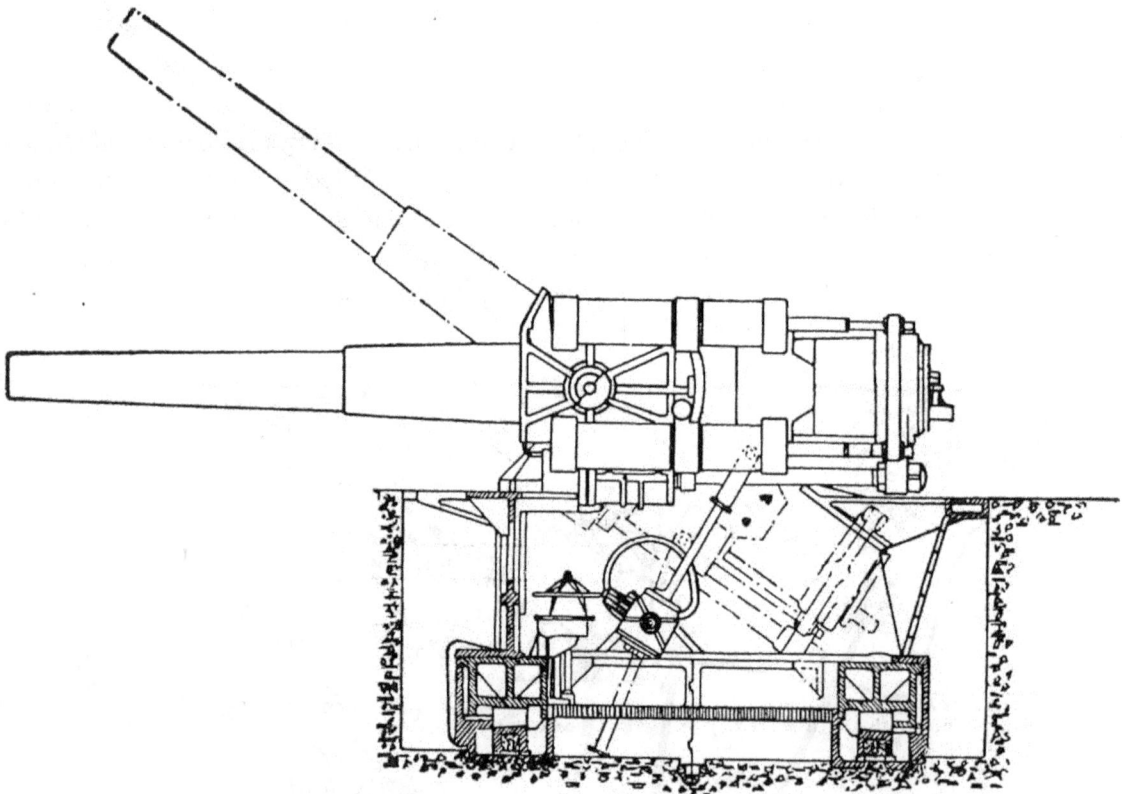

As proposed by the 1915 Board of Review, the standard coast artillery 12-inch gun was coupled with a special type of barbette carriage permitting a higher angle of elevation and thus range. *NARA.*

One serious drawback to the design was that the gun and carriage were entirely unprotected and in the open, protected by "dispersion"—one hit would not disable both guns. However the circular shape of the rotating platform and surrounding concrete apron looked exactly like a bull's eye from the air. The design allowed for 360° fire capability, something highly desirous on a compact island like Oahu, but at a cost for survivability of both the gun and crew. Of course, camouflage could be used to help hide the gun, but the distinctive magazine "hillock" on a flat tidal plain of just 6-foot elevation over water level still made the emplacement obvious. Eventually this would have to be remedied.

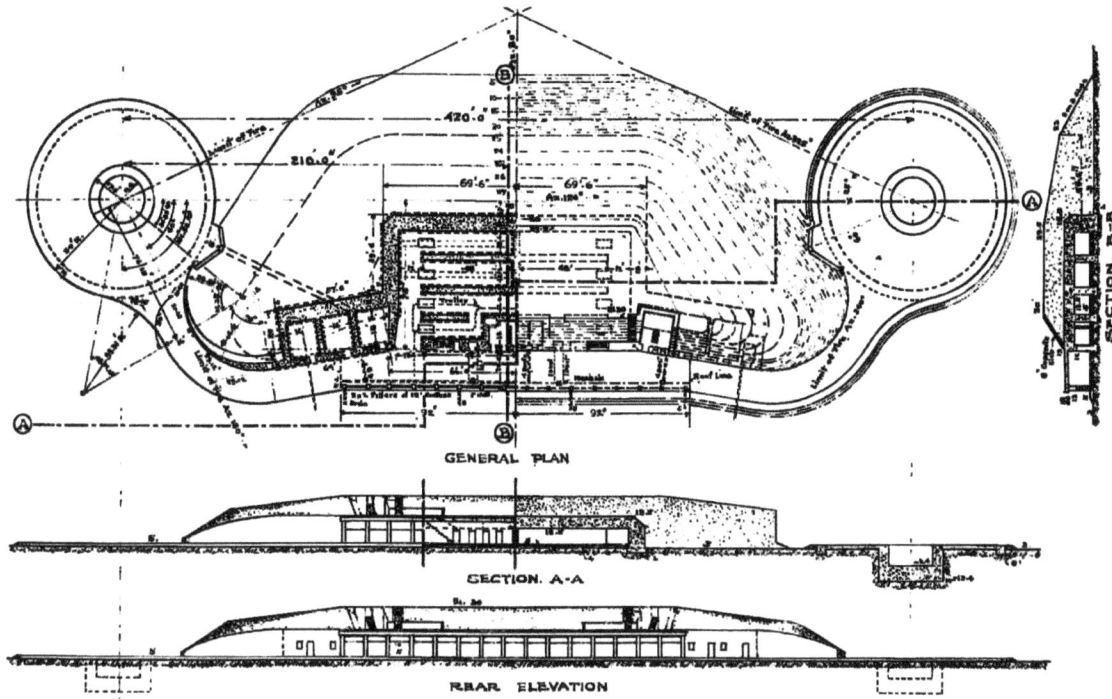

Plan, elevation and section for new Battery Closson built on the Fort Kamehameha reservation between the old cantonment and Ahua Point. *NARA.*

Aerial photograph from 1938 showing Battery Closson somewhat isolated on the flat land extending to Ahua Point. *NARA.*

Another aerial from the late 1930s of Battery Closson. Without any attempts at camouflage, the two gun platforms stand out very clearly for friend or foe alike. *NARA*.

Battery Closson's 12-inch gun being readied for the next round to fire in 1932. *NARA*.

Battery Closson at Fort Kamehameha practice firing in 1941. *Williford Collection.*

During 1916-1917 correspondence between Washington and Oahu discussed the precise locations for the new batteries. However, when it came actual funding only one new battery was authorized. The location on military land already garrisoned by the army prevailed, and the construction of the new 12-inch long-range battery was begun at a site to the far east on the existing Fort Kamehameha reservation, on the sub-post at Ahua. This was still flat scrub land, ultimately the emplaced guns would have a trunnion height of just nine-feet above mean low water. It was a withdrawn away from the southern beach, perhaps a thousand feet north of the direct cable line which ran from Battery Selfridge to the fire control complex on the eastern point. Local engineers closely followed the standard engineering designs of the current Engineer's emplacement mimeograph. The two guns were 420-feet apart, the standard splayed plan allowed significant fields of fire on either side of the emplacement such that all the southern shoreline and much of the western shore of Oahu was covered. Construction began in September 1917 and reported to be finished in April 1920. The battery was transferred to troops on May 4, 1920 for a cost of $300,249.42. (3)

The guns were initially manned by the 75th Company CAC, until the summer of 1924. After the major Coast Artillery reorganization, Battery B, 15th CA Regiment took over manning this powerful addition to the Oahu defenses.

The original Board of Review for 1915 had not recommended any 16-inch batteries for Oahu, but just the three dual 12-inch batteries. However, it seems that a desire for placing such a powerful installation in southwest Oahu soon emerged. The War Department initiated active planning for one of the new 16-inch gun and barbette carriage sets for Oahu in late 1920, even before any of the new guns or carriages had been produced. Eventually this became a project for a new emplacement to be built at the Puuloa reservation of Fort Weaver. Once in production, one of the new guns went to the final disappearing carriage mount emplaced, the next two to New York City defenses, and then the next pair to Oahu.

The 16-inch Model 1919 gun was a very potent weapon, arguably the most powerful conventional gun ever put into series production. Of a long, 50-caliber size, using a combination or wire-wound and built-up construction, as completed it could fire a 2340-lb. projectile at 2700 f.p.s. to an impressive 49,140-yards at full elevation. Only nine Model 1919s were ever built, mainly due to developments that led to a reasonable replacement with navy MkII guns that will be discussed with Fort Barrette's gun battery. (5)

The real key to the success of this weapon was the barbette carriage. Designing, producing, and then accurately placing the carriage was a major technological challenge. The sheer weight and size of the apparatus was daunting. This may not sound particularly difficult but placing a 385,000-lb. gun tube of over 800 inches long precisely into a sleeve and then rotating and elevating it with ease required precision engineering and exacting tolerances. A report done right after the Second World War is worth quoting: (6)

> There is no type of manufacture in America which involves more development design and manufacturing problems than modern 16" barbette carriages for seacoast defense.

Development of the new carriage began with specifications written in 1918, with the first prototype tested at Aberdeen Proving Ground in 1921. As emplaced, it was motor driven for changing azimuth and elevation. It had on-carriage recoil and counter recoil cylinders, and power ramming. Added after initial installation, the mounts also had breech air-scavenging systems. Protection of the mount and crew came from either the emplacement design or a steel splinter shield placed around exposed elements.

The Puuloa Flats, a 322.33-acre tract of land on the west side of the Pearl Harbor Channel across from Fort Kamehameha, was acquired on December 20, 1904 from the Dowsett Company, Ltd. and ten other landowners of smaller tracts. Initially reserved by the navy, a major portion was transferred to the army in 1921. Subsequently named the Puuloa Military Reservation, it was later designated the Iroquois Point Military Reservation. The acreage remained essentially undeveloped through World War I; but that changed significantly in the years following the war. On March 27, 1922, the War Department in its General Orders Number 13, named the reservation Fort Weaver in honor of Major General Erasmus M. Weaver, Chief of Coast Artillery from 1911 to 1918, who had died November 13, 1920. Here, a battery of two long-range 16-inch guns as well as emplacements for a 155-mm gun battery, antiaircraft batteries, searchlight installations, a power plant, and an anti-motor torpedo boat battery were built over the span of the next 20 years. Fort Weaver served as a sub-post of Fort Kamehameha throughout its period of service. In many cases the crews for the batteries here commuted from the well-developed garrison cantonment at Kamehameha. (4)

The site for the battery was on the reservation's central section inland about 800-feet from the shore at Keahi Point. Following standard emplacement engineering instruction, the guns sat on two large circular concrete slabs measuring some 73-feet across. In the center of the that apron was the gun block with a diameter of

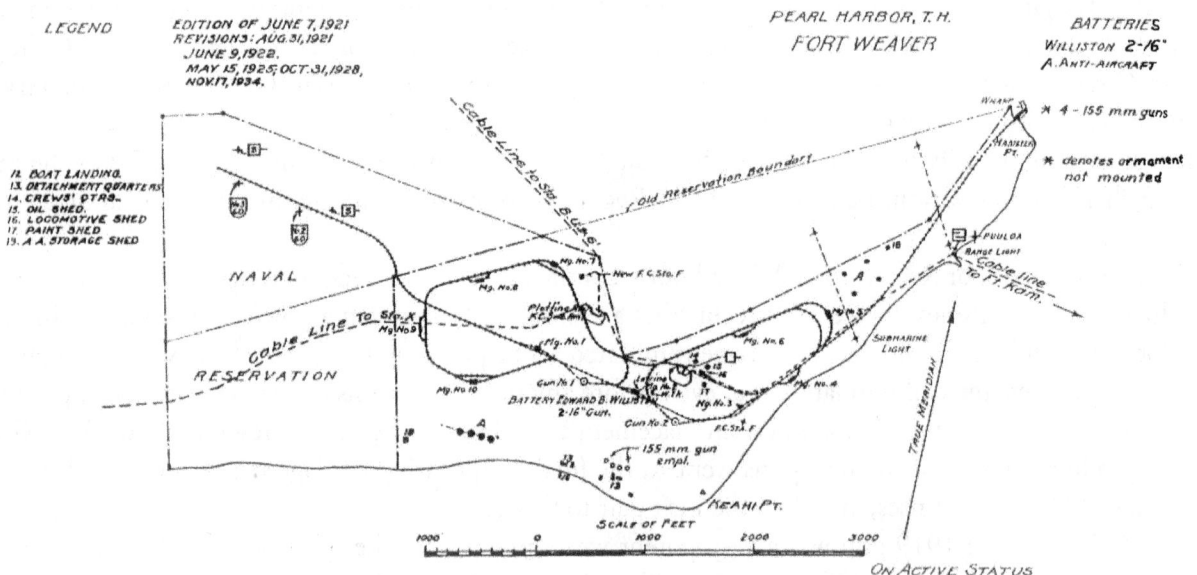

Engineer map of the full Fort Weaver reservation, dated to November, 1934. *NARA.*

51-feet. The two blocks were separated by 900-feet. The pair was arranged for a firing directrix of south, but of course they were in no way prohibited from a rotation and fire direction of 360°. The guns on their carriages were completely open and exposed. A narrow-gauge railroad track on which the ammunition service trucks moved, extended around the circumference of the gun platform, extending between them and connecting to initially three separate magazines, one between the mounts and one on either side. The magazines were gable-roofed, using hollow-tile brick walls with iron roofs and measured 32-feet by 61.5-feet, but not otherwise protected from gunfire or bombs.

Additional supporting structures were generally spread to the rear of the guns. These included a concrete, sand covered plotting room with fire control switchboard about 600-ft. to the rear of gun No. 1. When covered with earth this had the effect of making the structure appear to be just another of the sand dunes on the reservation when viewed from seaward as well as increasing the protection from bombardment. Only the rear of the building that faced toward the landside or north was uncovered. Upon completion of the structure, the whole area was planted in native algaroba trees that grew profusely in the area. In addition to the plotting

Fort Weaver was named for Major General Erasmus Weaver. A former Chief of the Coast Artillery Corps, he passed away in 1920. *Gaines Collettion.*

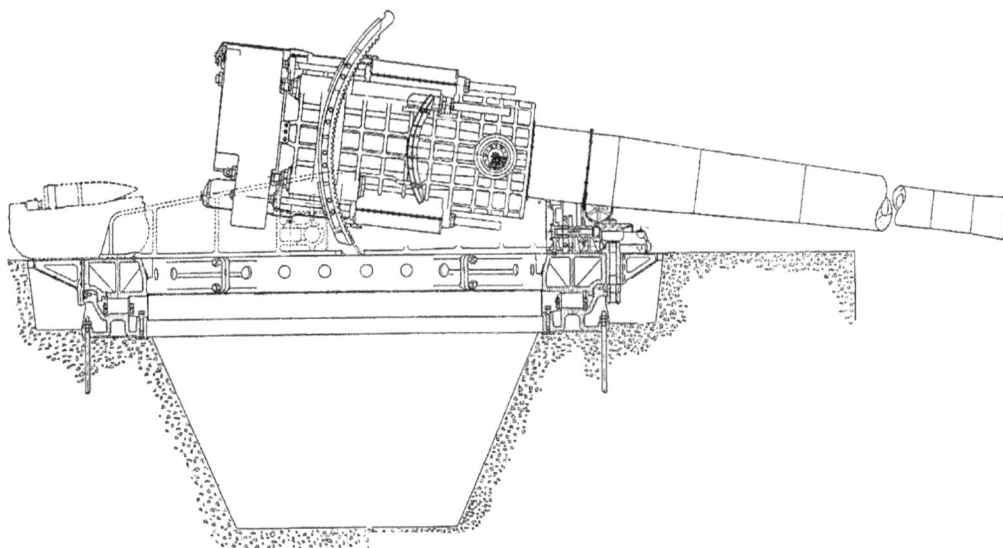

Ordnance Department sketch of the new 16-inch gun on Model 1919 barbette carriage, the type of gun mounted at Fort Weaver. *NARA.*

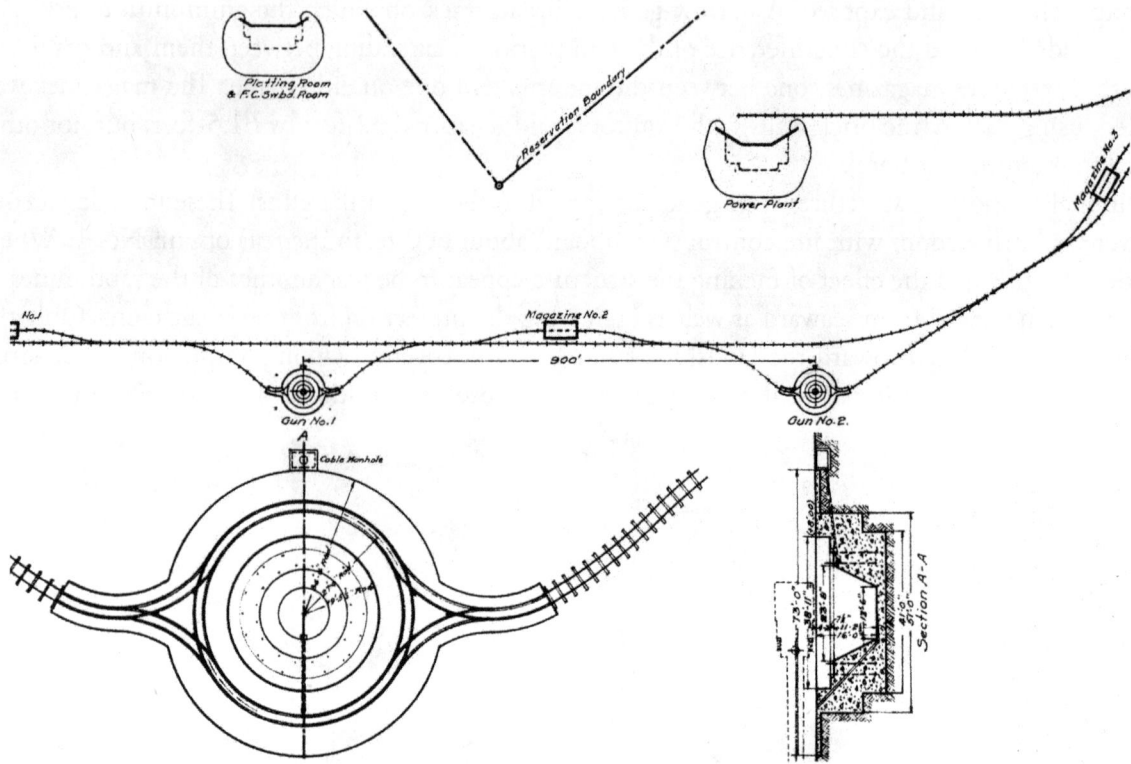

1924 engineering plat showing details of the gun blocks for 16-inch rifles at Fort Weaver. *NARA.*

Navy tug USS *Navaho* with a coal barge and one of the 16-inch gun tubes in transit to Fort Weaver after its arrival in Oahu. *CDJ.*

One of the 16-inch guns of Battery Williston, Fort Weaver. Note shells at the ready on the holding table behind the breech. *USAMH.*

16-inch gun of Battery Williston at full elevation, note the ammunition rail leading to the gun. *USAMH.*

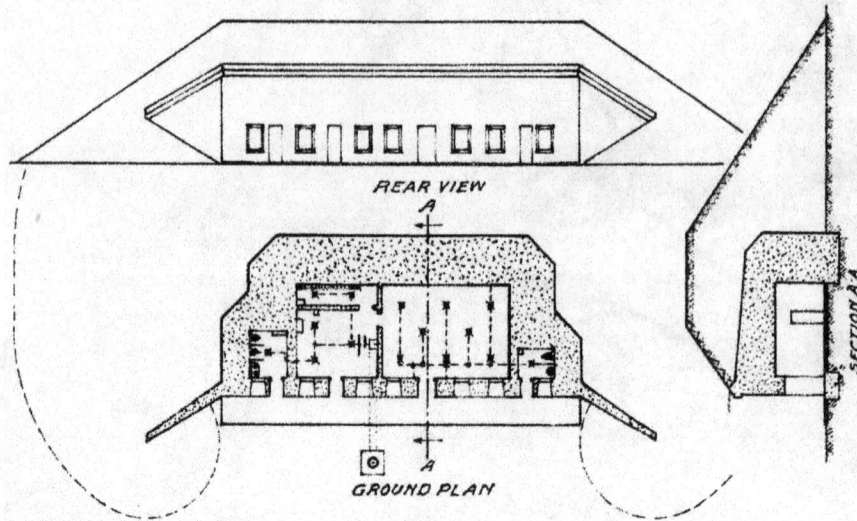

1925 elevation plan of the separate plotting room for the gun battery at Fort Weaver. *NARA.*

and switchboard rooms in the building there was a room containing a bank of electrical batteries to provide an emergency power supply as well as latrines for the officers and enlisted personnel. It was equipped with an M-1 plotting and relocator board. A concrete, earth-covered separate powerplant was built about 400-feet behind gun No. 2. It was equipped with three 90 kW Winton diesel generating sets. Rounding things out were a steel and iron paint room, a separate wooden latrine, and the post got a new wharf on its northern parcel projecting into the Pearl Harbor Channel. No permanent troop quarters or facilities (other than the latrine) were provided. Manning personnel would be quartered at Fort Kamehameha, or in tentage on post if required in an emergency.

Concrete work was done from October 1921 to September 1924. It was transferred to troops on September 19, 1924 for a cost of $121,549.72. For the separate structures, the plotting room was transferred Sep-

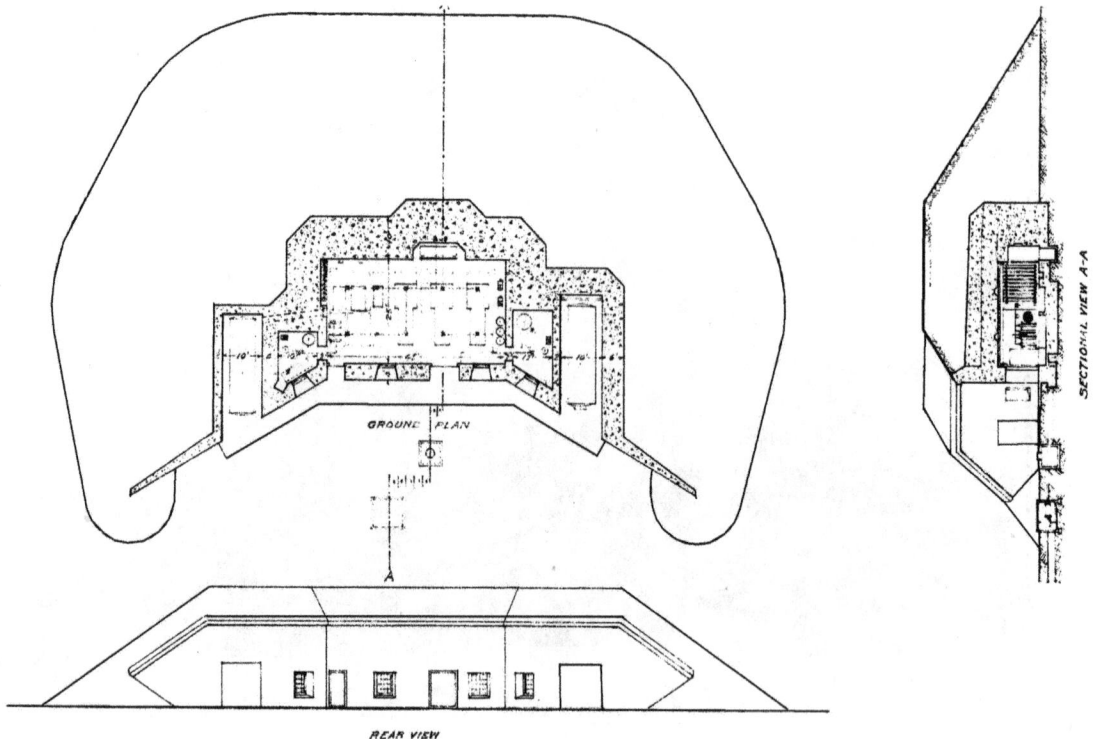

1925 elevation plan of the power supply room for Fort Weaver. *NARA.*

tember 21, 1923 for $65,271.65. The three service magazines were transferred together on that same date for $11,346.32. The wharf transferred on September 19, 1924 for $64,627.41. (7)

When completed the new battery was named by General Orders as Battery Williston. The guns were mounted during the summer of 1924, and on September 19, the battery was transferred to the care of the 15th CA (HD) Regiment that had been organized during the summer of 1924, as the coast artillery garrison of the Harbor Defenses of Pearl Harbor.

Even with the completion of the 16-inch Battery Williston at Fort Weaver, the need for additional long-range emplacements on Oahu continued. A 1925 Board in Hawaii recommended a new 16-inch battery originally projected for Schofield Barracks be moved to the Salt Lake/Honolulu area. In December 1927 another board recommended a project for two guns and an estimated budget of $514,000 for an emplacement in the southwest of Oahu. Eventually these projections coalesced with an approved project to emplace on Oahu a second set of 16-inch guns on barbette carriages at a new reservation at Kapolei.

Thus a decade after Fort Weaver, a second set of long-range 16-inch guns was approved for installation on Oahu. Puu Kapolei is a volcanic cone at an elevation of 166-feet on the Ewa Plain at the southeastern end of the Waianae Range. It overlooks the former sugar cane fields of the Ewa Sugar Plantation that lay between Pearl Harbor and the south shore of Oahu south and west of the Pearl Harbor Naval Station. According to Hawaiian legend, Kapo was a sister of the volcano deity Pele and lei means beloved. Hence the small hill's name--Kapolei, means beloved Kapo. In the latter part of the 19th century, James Campbell, a pioneer land owner and businessman, gained possession of the Kapolei Hill area. He leased the land in the vicinity of the puu to Benjamin F. Dillingham, a prominent land developer and railroad magnate. Dillingham, in turn, leased the land to the Ewa Plantation near Kapolei. Other than a small reservation at Gilbert, a few miles to the southwest of Kapolei that served as a firing point for the coast artillery's railway mortars, the area remained largely one of sugarcane fields through much of the early 20th century.

The Kapolei Military Reservation, was established in 1931, and was composed of two tracts. Tract Number 1 occupied the volcanic rise of Puu Kapolei itself. This tract was selected as the location for two 16-inch gun emplacements, their powder and projectile magazines, a power plant, command and control facilities, and bar-

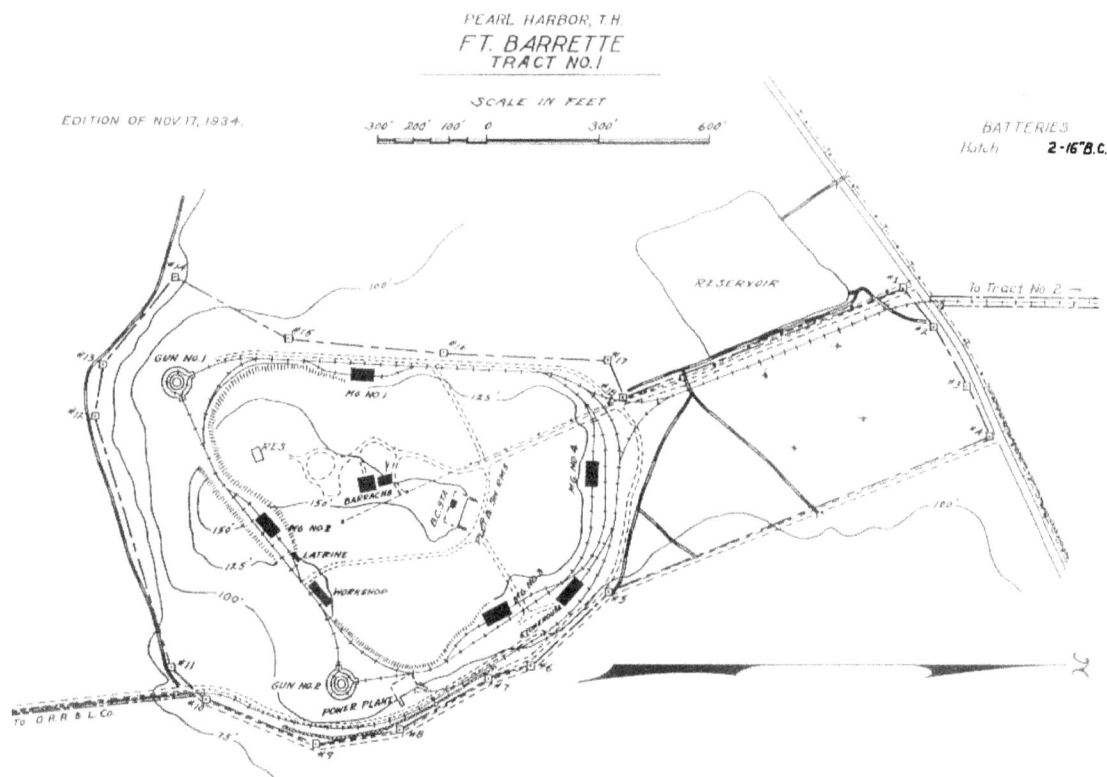

Engineer map of the main reservation of Fort Barrette (Tract 1) in 1934. *NARA.*

racks for the guns manning details. Tract Number 2 was located northwest of Puu Kapolei in a ravine between two spurs of the Waianae Range. Tract Number 2 contained four additional magazines and was connected to Tract Number 1 by a narrow-gauge spur of the OR&L line that continued on southward past Kapolei for three quarters of a mile where it joined the OR&L main line that served the agricultural interests of the island as well as other commercial, and military functions. (8)

The 1922 Washington Naval Treaty, more properly known as the Treaty for the Limitation of Naval Armament (and also as the Five-Power Treaty) grew out of a wave of popular and political sentiment to control military competition following the world war. Initiated by the United States' Harding administration, it attempted to moderate a new naval race, particularly between the U.S., Britain, and Japan. The conference, held in Washington D.C., began in November 1921, and concluded with a draft agreement on February 6, 1922. Among the significant provisions were a moratorium on new capital ship construction, limitation on the size of retained battlefleets, which forced the scrapping of excess ships, and restrictions on building new fortifications in the Pacific. The latter proviso carefully exempted the Japanese homeland, the main regional base of Britain (Singapore) and the United States (Hawaii). Hence it did not directly impact the existing or new harbor defenses in the islands.

However, the signing of the treaty led to the cancellation of the construction of numerous of capital ships for the treaty signatories. The United States agreed to cancel building of seven battleships and six battlecruisers already funded and under construction. All were to have been armed with the navy's MkII 16-inch gun. Though war-delayed in some cases, many of these guns were already finished or at least partially forged or fabricated. The Navy was initially a little guarded about disposing these assets, unless until the treaty was fully ratified by each signatory's home government. Articles of ratification were formally exchanged on August 17, 1923. After confirming that the Navy Act of July 11, 1919, permitted exchange of equipment between services without compensation, the Navy offered the guns to the Army.

In anticipation of the transfer, the Army suspended all work on its own 16-inch Model 1919 gun in June 1923. In several lots between 1923 and 1941, 112 guns were transferred. This was an excellent gun, the product of a thorough development by the Navy's Ordnance Bureau. It was a built-up tube of 50-caliber length. Somewhat lighter (almost 40 tons less) in weight and construction than the army's contemporary Model 1919, it was roughly comparable in performance. It used a typical navy interrupted-screw breech. At a 47° elevation is could fire a 2240-lb. projectile at 2650-f.p.s. to a range of 45,100 yards. While not quite as powerful as the army's own model, the cost savings were irresistible. Quick modification was made to the design of the current barbette carriage and a new ballistic table composed. This gun became the army's primary weapon for the final decades of coast artillery. (9)

The carriage used to mount this gun was simple modification of the Model 1919 type, originally made for the army's own 16-inch gun and emplaced locally at Battery Williston. The "thinner" gun required side frames to be placed closer together. The base ring was bolted together and emplaced on a concrete foundation. Atop the ring were the racer ring and traversing rollers. The top carriage had two side frames with trunnion beds to hold the cradle. The cradle had trunnion extensions, and completely encircled the gun tube. A single recoil cylinder and piston were underneath the cradle, and three recuperator cylinders were mounted atop the cradle. Electrically powered traverse could rotate the gun 360° and powered elevation by rack could deliver elevation of -3 to +47°. The heavy shell weight required power loading and ramming—generally at a 4° elevation. They were equipped with compressed-air gas ejector systems. (10)

Construction at the fort, known initially as Battery Kapolei, was begun in secret by the Corps of Engineers in July 1931, despite some protests by local civilian contractors. General Orders Number 10 of the War Department dated November 23, 1934, designated the post Fort Barrette in honor of Brigadier General John D. Barrette who had commanded the Hawaiian Coast Artillery District from April 1921, to August 1924. General Barrette died on July 24, 1934. The Barrette family donated $623 in late 1940 for the purpose of establishing a memorial to General Barrette at the post bearing his name. Troop labor was used to construct the

Major General John D. Barrette, namesake for Fort Barrette.
He had been an acting Chief of Coast Artillery in 1918 and died in retirement in 1934. *Gaines Collection.*

memorial with the donated funds being used to defray a part of the cost of materials. On February 21, 1941, the memorial to General Barrette in the form of a stone gateway was dedicated by Brigadier General Fulton Q. C. Gardner, commanding general of the Harbor Defenses. A bronze plaque in the gateway structure was unveiled by General Barrette's daughters. The gateway was graced with a heavy steel decorative gate of two ten-foot sections for vehicular traffic and two three-foot sections for pedestrians It was flanked by two sentry boxes measuring built of local lava rock. (11)

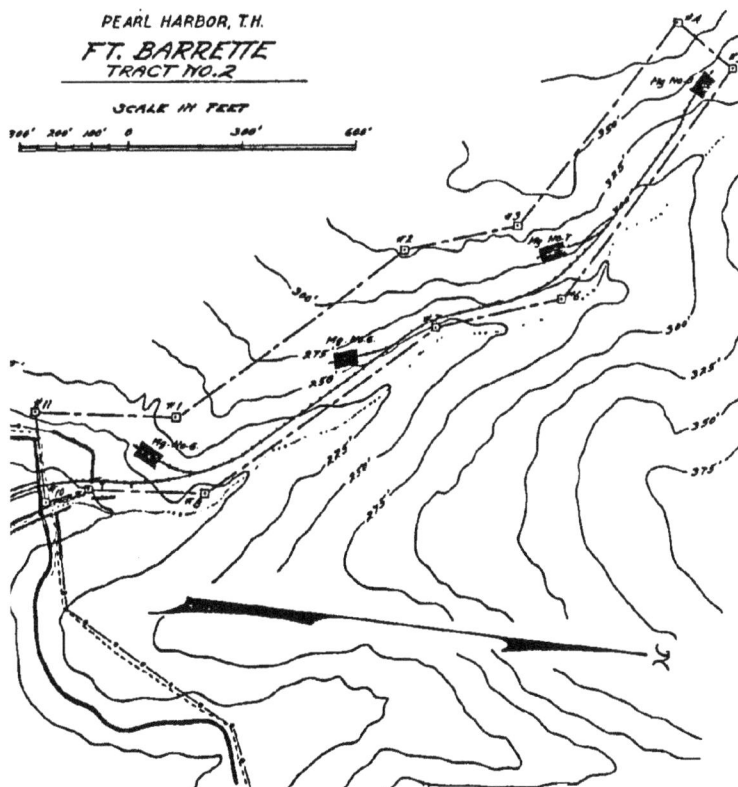

The additional Tract 2 of Fort Barrette, which was used for the majority of the ammunition storage of the fort. *NARA.*

Construction of the two concrete gun emplacements at Fort Barrette began on July 5, 1931, and continued until mid-July 1935, when the battery was pronounced finished, and the guns mounted. It was transferred to troops on July 6, 1935, for a construction cost of $730,558.64, In keeping with the adopted harbor defense doctrine of dispersal, no attempt was made to provide overhead protection to the two guns of the battery. It was felt at the time, that separation of the guns themselves and wide dispersal of the various support elements of the battery would provide a certain amount of security to the guns, as they would be reasonably safe from aerial, or sea borne bombardment. Also being on this hillock or remains of a volcanic cone meant that any major structure projecting above ground would be conspicuous at a distance, and not be able to be concealed.

Very similar to the emplacements of Battery Williston, the battery had two heavy reinforced concrete gun wells of 51-foot diameter to hold the barbette, roller path, and ultimately carriage and gun. These were emplaced 900-feet apart defining a line that pointed perpendicular to the southwest. They were slightly to the southwest themselves on the rather small military reservation. Now a proven technique, the emplacements were supplied ammunition by a set of dedicated ammunition railway tracks.

There were four unprotected, lightly sheltered magazines on this line, one between the guns, one north of each gun emplacement, and the fourth on the north side of the reservation where the track curved to ultimately encircle the central part of the reservation. The four magazines located at Kapolei were well dispersed and three of them were on the reverse slope of the hill. Magazine Number 2 although on the forward slope was placed in defile between the gun emplacements and was sheltered from any direct fire that might come from seaward. Each of these structures was built of reinforced concrete and had a metal louvered overhead door at each end. The railway tracks ran through each of the magazines before rejoining the main spur. Each magazine roof was camouflaged with a covering of several inches of earth making them difficult to discern from the air. There was also a storehouse along the line. (12)

1933 aerial photograph of what will become Fort Barrette. The elevation of the terrain over the adjacent fields is evident. *NARA.*

Delivering and placing one of the 16-inch gun tubes into the carriage at Battery Hatch. *NARA*.

16-inch gun Battery Hatch at Fort Barrette soon after installation. *USAMH*.

One of the ex-navy MkII 16-inch guns at high elevation of Battery Hatch. *USAMH*.

Battery Hatch was armed with two MK II Mod 1 16-inch naval guns. The two massive gun tubes, each of which weighed 153 tons, were hauled from Pearl Harbor to Fort Barrette on the right of way of the OR&L by a pair of that rail line's steam locomotives. To facilitate the move a special ¾-mile spur had been laid north from the main line to Puu Kapolei. While enroute one of the gun tubes slipped from its cradle almost derailing the entire train. Once repositioned by the accompanying wrecker car the gun train continued to Fort Barrette. The guns were mounted on Army M1919M1 long-range all-around fire barbette carriages (Serial Numbers 7 and 8). From its location on Puu Kapolei, Battery Hatch's guns could attain a maximum range of some 45,100 yards, sufficient to cover the waters all around the island of Oahu except at the island's easternmost point at Makapuu Head. However, the usual effective range was somewhat less at 44,670 yards. (13)

Nestled into the east slope of Puu Kapolei was another bombproof building—the battery power plant, built of reinforced concrete, and located about 25-feet below, and to the left rear of, Emplacement Number 2. Here three 100 kW General Electric Company generators were installed and powered by three 150 BHB 6-Cylinder Winton Diesel Engines. A fourth 25 kW generator also of General Electric manufacture, was powered by a 45-54 horsepower English Electric gasoline engine that provided additional electric power at the post. Additional structures built with the original 1935 reservation work were a storeroom, water storage, a concrete detached latrine, two concrete tool and paint sheds, a ten-man barracks and a combination kitchen/mess hall/office/quarters facility. A concrete reservoir with a capacity of 50,000 gallons of water was located near the barracks. Water was pumped from a water tank on the higher reservation to this reservoir. (14)

The battery plotting and fire control switchboard rooms were combined into a single structure of reinforced concrete and built into the reverse slope of the hill at a point between the two gun emplacements. It was transferred to the Coast Artillery on July 31, 1935, at a cost of $41,521.16. At the west end of the structure a concrete staircase provided access to earth-covered roof of the building. Located atop the plotting room was

Protected, separate plotting room for the guns of Battery Granger Adams, Fort Ruger, as completed in March 1934. *NARA.*

an elevated battery commander's station (BCS). This station was a reinforced concrete structure supported on concrete pillars. The observation room of the station, some 20-feet above the plotting room roof, measured about 15-feet by 15-feet. The BCS also had a concrete splinter-proof roof that was also covered by several inches of earth to camouflage it from aerial view. (15)

One of the magazines for shell and powder of Battery Hatch. While sturdy, its protection lies in its location in a ravine rather than by overhead concrete. *NARA.*

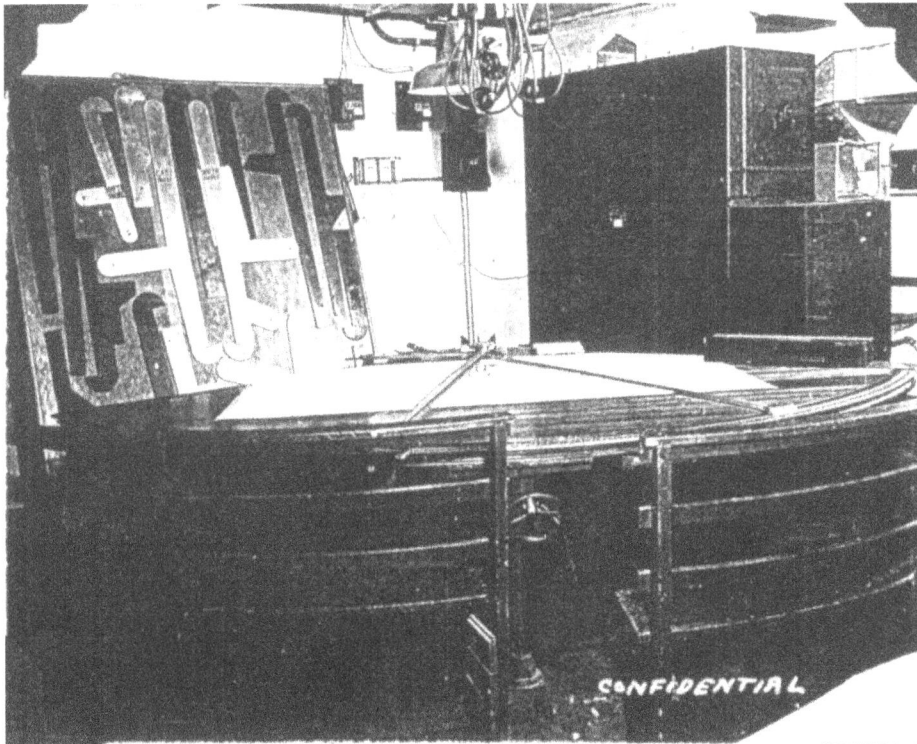

Photograph taken inside 16-inch Battery Williston plotting room at Fort Weaver. This is where information obtained at distant end stations would be used to calculate firing solutions. *Gaines Collection.*

The plotting room was equipped with a very large, 360° Cloke plotting board, a fire adjustment board, a range correction board, and a deflection board, each of the M1 type. In addition, the battery was provided with an M2 Spotting Board. In May 1940, a study called for the replacement of the spotting board with one more suitable to the requirements of the battery. This same study also found the M1 plotting and relocating boards to be unsatisfactory, and the brigade recommended on November 25, 1940, that they be replaced with the M1 Gun Data Computer. Delivery of this computer was scheduled for January 1942.

As mentioned above, there were two sections to the original Fort Barrette reservation at Kapolei. The second tract was a narrow piece of land that stretched to the northwest of the main reservation. It followed the line of a narrow gulch that ran between two adjacent ridges. The ammunition rail line from the post was extended up this ravine. On it were built four additional magazines to hold the reserve ammunition for the battery. This seems to have been a much better solution than that chosen for Fort Weaver, that for several years had its reserve magazines at an entirely different facility at Fort Shafter.

All eight magazines at Barrette (the four around the battery and these four on Tract 2) were built to the same plan. They were 41 by 71'6"-foot structures with the tracts of the railway running through the center and seven bays on each side. All eight were transferred together on July 6, 1935, for $114,095.52. The sides were built of reinforced concrete, and the concrete roof had a dirt cover. While adequate for splinter protection, their protection from direct shell hits was by defiladed location rather than earth and concrete. When Battery Hatch was given overhead protection, magazine space was figured into the casemate design, there was no need to add a new protected magazine for the fort during the war.

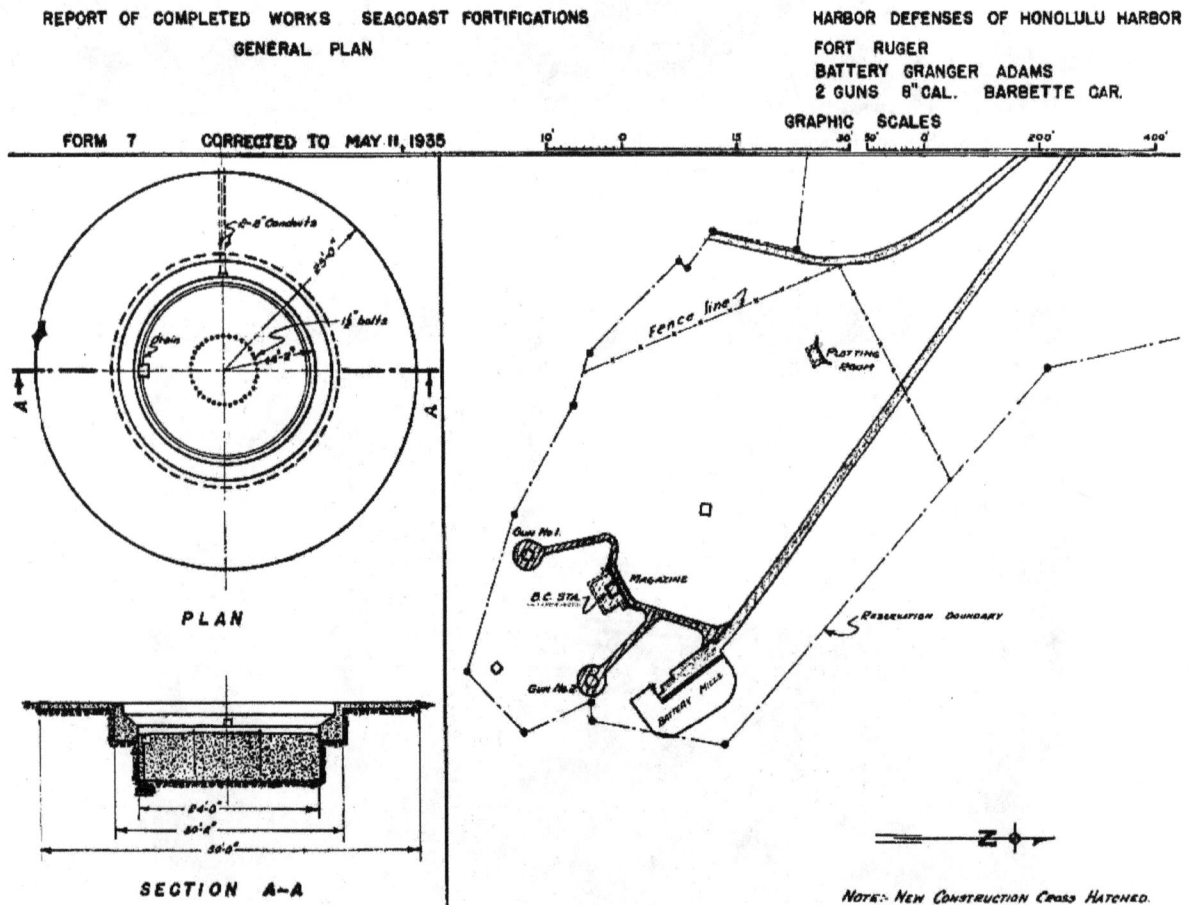

Engineering map and plan of gun block for the new 8-inch long-range battery to be emplaced at Black Point, Fort Ruger in 1934. *NARA*.

One final new long-range battery was erected on Oahu before the Second World War. The small reservation on Kupikipikio (Black) Point was acquired with the rest of the Fort Ruger property for military purposes in 1909. Southeast of Diamond Head, the point projected into the Pacific along the southern shore of Oahu. Previously it had been first used for a searchlight and later for one of Battery Birkhimer's fire control stations. In addition, it was selected for 5-inch Battery S. C. Mills of the Land Defense Program prior to the World War. The point was considered following that war for one of the early batteries of 155 mm guns on Panama mounts. This plan was ultimately rejected by the early 1930s in favor of a pair of longer-ranged 8-inch guns.

In 1931, the Hawaiian Department ruled that civilian contractors would no longer be allowed to bid on fortification construction, primarily to help maintain security about new defenses. Both the contractors and the Honolulu Chamber of Commerce filed strong protests with the War Department, but to no avail; bidding by civilian contractors was prohibited. Thereafter the Honolulu District Engineer was placed in charge of all fortification work in the islands. The construction at both Kupikipikio Point and Kapolei (Fort Barrette) were some of the first projects conducted under these guidelines and attracted a lot of newspaper print on this subject.

There is little existing correspondence relative to the rationale for this emplacement at Kupikipikio Point. This was early in the depression years when military budgets were sparse. There does not seem to be any urgent reason to suddenly need to enhance the coverage to the south of Diamond Head during this decade. Nonetheless this new battery was constructed, though of minimal elements and using pre-existing armament no doubt helping to keeping costs in line.

1938 aerial photograph of the new 8-inch battery at Kupikikio Point (Black Point). To the left (east) of the emplacement can be still seen the overhead casemates of Battery S.C. Mills. *USAMH.*

The weapon chosen for this new position was the old Model 1888M2 8-inch gun that had been removed from seacoast disappearing carriages during the First World War and placed on a Model 1918 barbette top carriage normally married to the Model 1918 Railway Car. Eight of these railway guns, with cars reset to the island's 36-inch rail gauge, had arrived in Oahu in early 1934. While the gun was certainly an older model (the oldest heavy seacoast gun still in U.S. deployment), its railway top carriage did allow an elevation of 42°. This allowed it to fire the 160-lb. AP projectile at 2450-f.p.s. to a range of 23,900-yards. This was further than the range of the old Taft disappearing batteries or the 155 mm GPFs used in field or Panama mount emplacements. Also, the top carriage could traverse on the rail car base fairly quickly allowing the gun to track moving vessels. (16)

Newly emplaced 8-inch gun, still on the Model 1918 railway top carriage, at one of the positions of Battery Granger Adams, Fort Ruger in 1934. *NARA.*

The drawback for using the mount like this was the awkward size and height of the unit on the rail car, and the problematic stability of the mount when emplaced. It was found that the top carriage and gun could be removed and bolted to a much more stable ground platform and still retain the favorable range and traverse. Never used in extensive numbers, this type of mounting was used at select locations by American forces in Newfoundland, the Philippines, and on Oahu. It appears that the first location utilizing this arrangement of a Model 1918 railway top carriage on a fixed ground platform was this battery at Fort Ruger.

On October 5, 1933, construction began on this battery for two 8-inch M1888M2 guns on M1918 barbette carriages at Kupikipikio Point, immediately to the rear of disarmed Battery S.C. Mills. Like most other batteries built between the world wars, Battery Granger Adams was constructed in accordance with the dispersal concept then in vogue. The two circular gun blocks for the guns were sunk into the ground and surrounded by large concrete service aprons some 65-feet in diameter. Each emplacement was constructed in the open, unprotected and readily visible from the air. The Model 1918 top carriage was fixed to the block and carried at an elevation above sea level calculated at 112-feet. The Report of Completed Works states that the work was begun on October 5, 1933, and completed on May 11, 1935. It was transferred to troops on March 2, 1935 (sic) for $116,832.70.

Between the two emplacements, and to the rear, was a large bombproof reinforced-concrete magazine. Ten-foot-wide ramps extended from the loading platforms to each side of the magazine, where they entered the 100-foot corridor extending the width of the structure. Some ten feet inside the entry at each end of the corridor was a tool room. The magazine had four other rooms: Two projectile rooms, each 40-feet long and 10-feet wide, were between the corridor and the two powder magazines, each 28-feet long and 16-feet wide. The exterior walls of the magazine were five feet thick, the interior walls two feet. Overhead, the concrete roof varied between two and five feet thick, covered with several feet of earth. On top of this earth covering was a

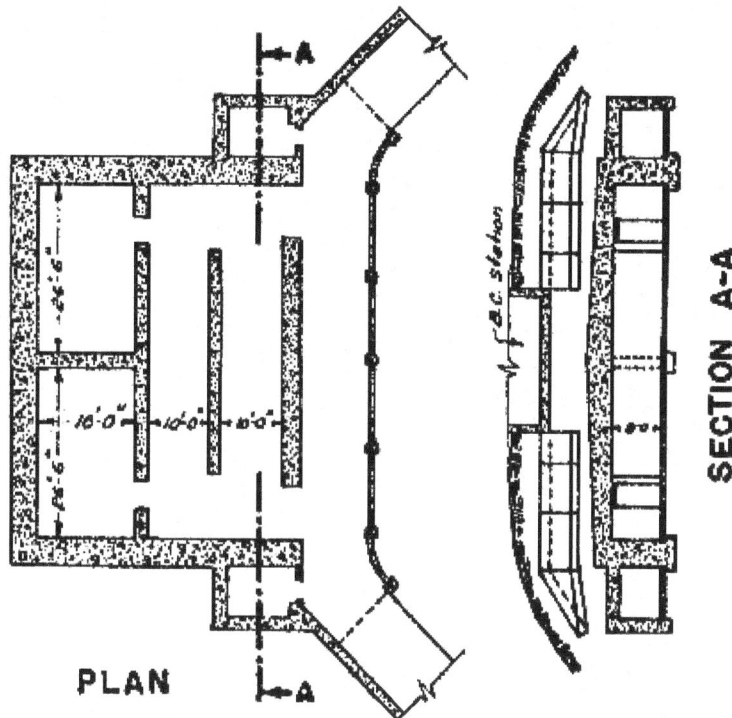

Plan for the central traverse magazine of Battery Granger Adams. *NARA.*

1934 construction photograph of work building the Granger Adams projectile magazine. *NARA.*

burster course of concrete about a foot thick, covered with several more feet of earth and sod. The magazine had a small 5 kW gasoline electrical generator set. Reserve power was supplied by the adjacent, but otherwise abandoned Battery S.C. Mills. The magazine was transferred on March 2, 1935, for $42,936.85.

Built into the top of the magazine was a splinter-proof battery commander's station of reinforced concrete, with a foot-thick concrete-slab roof. The 10-foot-wide roof of the magazine's corridor formed a walkway that extended the width of the structure. At each end of the walkway, a stairway curved to the ground. The BC station was transferred on March 2, 1935, for a construction coast of $2532.17.

The separate battery plotting room, some 600-feet northwest of the magazine and battery commander's station, was constructed like the magazine, only smaller. Its exterior walls were five feet thick and its interior walls 18-inches. Its roof, two to five-feet thick, also had a concrete burster course, along with 10-feet of earth. The entire 35-foot-square structure was entered through a short passageway some ten-feet long and three-feet

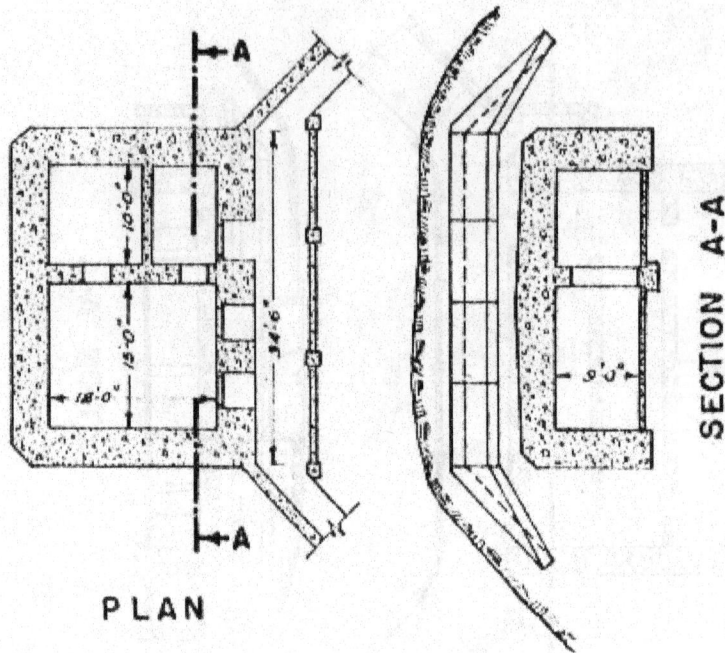

PLAN

SECTION A-A

1935 plan for the separate, protected plotting room for Battery Granger Adams, emplaced almost directly behind the gun battery.
*NARA.*

wide that opened into a six-foot wide rear corridor extending the full length of the structure. There were two other rooms inside; a fire control switchboard room 10-feet square, and an officer's bunk room about 10-feet long and 6 feet wide. This plotting room was transferred on March 2, 1935, at a cost of $19,914.23.

The two 8-inch M1888MIIA1 guns and M1918 barbette carriages were specifically supplied to Oahu for this use; they did not come from the armament of the 41st CA's 8-inch railway batteries discussed in the next chapter. The new weapons arrived aboard USAT *Republic* on June 2, 1934 and were mounted by early 1935. (17)

In addition to the battery commander's station atop the magazine, Battery Granger Adams used three fire control stations. The primary station (B') atop Diamond Head's south rim was begun in 1934, near the combination emergency station for Battery Birkhimer and the base end station for Battery Closson. This structure was a dug-in type, entered through a manhole in the earth-covered concrete roof. The station was equipped with pedestals for two M1907 DPF instruments, one of which was installed on the east pedestal in October 1934 as the battery was prepared for its October 29 transfer to the garrison. The two secondary (B") stations were located on Koko Head and on Makapuu Point on Oahu's east end. All the stations were equipped with DPFs and azimuth instruments.

As noted in the previous chapter, each of the new long-range, barbette mount batteries of the interwar period received a new fixed antiaircraft battery; often relocated from a position of lower priority. Battery Closson at Fort Kamehameha, Fort Weaver, Fort Barrette, and finally Black Point of Fort Ruger received these batteries in the 1920-30s. For Black Point two gun blocks for the familiar 3-inch Model 1917 fixed AA gun were emplaced to the northwest of Battery Granger Adams in 1937 and guns moved from their previous emplacement near the tennis courts at the Fort Ruger cantonment. They continued to be mounted here, though only occasionally manned, until late in the war.

# CHAPTER 6
## EQUIPPING THE DEFENDERS WITH MOBILE GUNS

As noted in Chapter 2 there was concern by the military planners that an enemy could avoid the conventional harbor defenses along the southern shore and land parties, large or small, at a variety of beaches on the northern and eastern coasts of Oahu. The Army acknowledged the unique tactical situation of this substantial, but isolated, mountainous island. The deployed army forces for Oahu were projected to include a heavy field artillery battalion, at this time usually having 12 guns or howitzers of 4.7 and 6-inch bore; but larger landings, perhaps supported by substantial naval forces might require stronger opposition. The 1910-13 Macomb Board, authoring the *Land Defenses of Oahu*, recommended two new mortar batteries for Oahu to fire on an enemy preparing to land on the coast. (1)

While a start, the need for more heavy artillery for remote locations, persisted. Although the natural obstacles to an enemy force were considered to be formidable, additional plans for the defense of the north and northeastern coasts were developed. During the First World War plans called for seven fixed mortar batteries, each armed with four 12-inch mortars. These batteries were to be located where the mortars could place vessels standing off the beaches under fire and force them to either retire or off-load their landing forces in the open seas much further offshore. Should an invasion force still attempt a landing, the mortars would make short work of the unarmored decks of the transports and destroy the fragile landing craft. Each of these projected batteries was to have two bays or pits, separated by a large bombproof magazine. Two 12-inch breechloading mortars were to be mounted in each of the pits. Two of these batteries were proposed for Oahu's North Shore: one in the vicinity of Mokuleia to cover the waters of Waialua and Kaiahulu Bays, and the other in the Waialae area to cover the potential landing beaches of Kawela and Waimea Bays. On the island's windward side, three more mortar batteries were planned. One was projected to cover the beaches of Laie Bay, the second on the Mokapu Peninsula was to be positioned to cover the waters of Kaneohe and Kailua Bays, and the third was to be built at Waimanalo to place fire on the waters of that Bay. The remaining two batteries were those already suggested by the Land Defense Board mentioned above. World War I intervened before these plans could be implemented. War priorities and the period of conservative fiscal responsibility following put an end to these plans, at least in the form of fixed emplacements. (2)

The development of a new generation of mobile guns during the First World War offered a possible solution to the Oahu dilemma. The use of railway artillery for seacoast defense by the United States Army began with discussions and experiments in the years immediately preceding World War I. The theories expounded at the time argued that heavy guns mounted on railway cars could be sent to threatened points along the nation's seaboard and thus avoid the high construction costs of fixed permanent fortifications. Fewer guns and crews would be needed to create the same deterrent as fixed defenses. Beyond study, however, nothing was done at this time to implement the use of railway artillery in the United States. Soon the efficacy of mobile heavy artillery to pound field and fixed fortifications, and logistical targets further behind was demonstrated during the European War. By 1917, when the United States entered the fray, railway artillery was playing a major role on the Western Front. In the last 17 months of the World War, the U.S. Army fielded no less than eight regiments of railway artillery. Because of its experience with large-caliber weapons, the Coast Artillery Corps was assigned to provide the men for these organizations. Following the war, four of these regiments, the 42nd, 43rd, 52nd, and 53rd Artillery (Railway), CAC, were retained in an active status for training, although at reduced strength. Then in the summer of 1921 the army's railway artillery was reduced just to a handful of batteries in the 52nd Coast Artillery. This regiment was divided between Fort Monroe, VA, where one battery was available to work with the Coast Artillery School and Board, and a reduced strength service battalion posted at Fort Hancock, NJ. (3)

Oahu already had in place an operating commercial rail system that would seem to favor the use of railway artillery to solve the tactical challenges mentioned above. By 1921, the 36-inch-gauge Oahu Railway and Land

(OR&L) Company line extended from Honolulu, past Pearl Harbor, and across the Ewa Plain to the island's Leeward Coast, with a branch line north to Schofield Barracks. The railroad tracks also extended along the Leeward Shore of the island to Kaena Point and thence around the point to the North Shore. The line ran close to the shoreline to the island's northernmost point at Kahuku, where it connected to the Koolau Railroad that operated between Kahuku and Kahana Bay on the Windward Coast. As a result, railway artillery using this rail system could cover all but a quarter of Oahu's coastline.

In addition to the OR&L and Koolau lines, there were many miles of sugarcane and pineapple plantation railroads on the island that could be utilized in time of war. The War Department decided to authorize the formation of a battalion of railway artillery for Oahu in 1920. The battalion of two batteries was to be armed with 12-inch mortars and carriages on cars modified for the narrow-gauge tracks of the OR&L. The mortars had been removed from fixed coast defense batteries during the World War to be mounted on railway cars for service by the AEF in France. The war ended before any of this armament could be sent to the front, but some were used to equip one of the regiments retained on active status following the war. Others were simply stored for future use. Eight M1890M1 12-inch mortars and railway mounts were shipped to Aberdeen Proving Ground from Camp Eustis and Watertown Arsenal between late December 1920 and early 1921 for disassembly and preparation for shipment to Hawaii. (4)

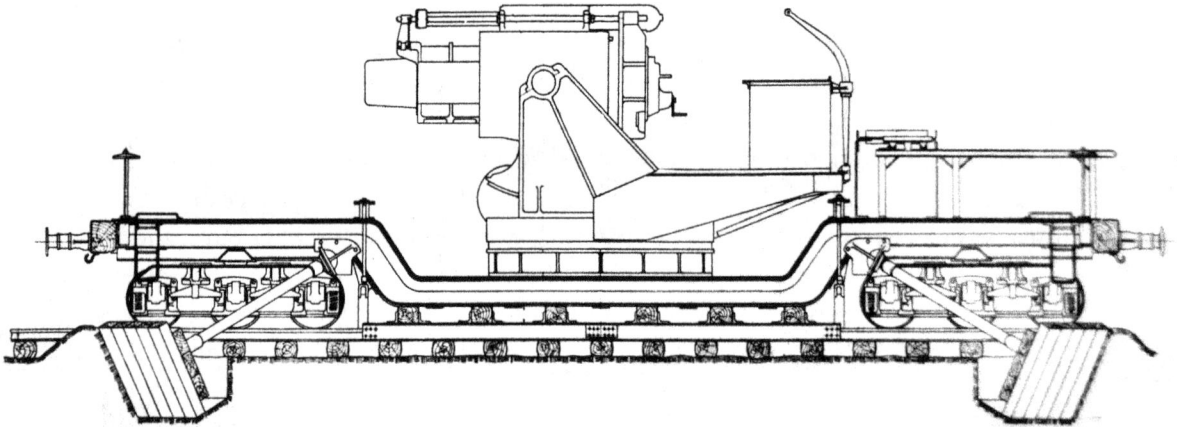

The abundant 12-inch seacoast Model 1890M1 mortar was married to the simple Model 1918 railway mount and car for expeditionary use during the First World War. *NARA.*

One of the eight 12-inch railway guns sent to Oahu in 1922 and placed on cars modified to fit the narrow gauge of the island's sugar railway network. *McGovern Collection.*

The Hawaiian Railway Battalion was created by the War Department on January 15, 1921. It was not physically organized at Fort Kamehameha, however, until very late that year. The battalion was initially composed of a battalion headquarters and two lettered firing batteries. The personnel were obtained by inactivating organizations manning the fixed batteries. On November 23, 1921, the eight M1890MI 12-inch mortars, eight disassembled M1918 mortar carriages, and eight M1918MI railway cars were shipped from Aberdeen Proving Ground to Pearl Harbor. Upon arrival they were moved to nearby Fort Kamehameha in January 1922. Assembly and modifications were required to fit the rail cars to the narrow gauge. While the railway ordnance was being prepared, the Hawaiian Railway Bn was redesignated 41st Artillery (Railway), CAC. The small battalion headquarters detachment received the additional designation of 198th Company, CAC, while Batteries A and B became the 199th and 200th Companies, CAC.

Firing a 12-inch mortar of the 41st CA Regiment, perhaps at the Fort Kamehameha firing point in the 1920s. *Bennett Collection.*

Meanwhile surveys along the rail lines of the island's perimeter were underway to ascertain the best positions from which to cover the invasion beaches. Consideration was also given to using the many miles of sugarcane and pineapple plantation railroads in addition to the OR&L trackage, especially on the North Shore and Ewa Plain. Numerous cane hauling branches of the Waialua Agricultural Company's plantation railroad spread from the main OR&L line inland like tentacles across the North Shore's sugarcane fields reaching nearly to the Leilehua Plains in the central part of the island. Schofield Barracks, the Army's sprawling reservation occupying the center of the island had direct rail service to Honolulu, but this rail service stopped at Wahiawa. Consequently, during the period between the two world wars the coast artillery gun and mortar trains traveling to the North Shore and the upper reaches of Windward Oahu, had to travel via the southwest coast and Kaena Point. Not until after America's entry into World War II, would the necessary rail connection between Wahiawa and Haleiwa on the North Shore be established.

Most of the firing positions along the coastline of Oahu were selected with a view to cover the likely invasion beachheads from the flank, as there were few locations where the mortars could be retired into the

Another view of training on one of the railway 12-inch mortars at the Fort Kamehameha siding. *USAMHI.*

interior. As the overhauling, assembling, and mounting of the Hawaiian railway mortars dragged on until September 1922, the process of facilitating the mobility of the battalion was begun. The engineer railroad at Fort Kamehameha was upgraded and connected with the OR&L. Two firing spurs were constructed at Fort Kamehameha and the battalion held its first target practice on September 20, 1922. The practice proved far from satisfactory, due in large part to inadequate fire control equipment. An Engineer Department locomotive and a locomotive crane were borrowed to maneuver the guns on the Fort Kamehameha reservation until a dedicated locomotive could be provided for the battalion from army stocks (a suitable army locomotive for off-post travel was not obtained until World War II). (5)

The OR&L cooperated fully with the Army, making their system and a locomotive available free of charge. When the first off-post firing point was established at Gilbert in the Ewa District, the railroad provided a switch connecting the mainline to the spur. The battalion conducted its first target practice at the new Gilbert firing position on November 4, 1922, using two mortars. Firing results were improved; three of the four rounds fired were "on target." While several issues were identified, the target practice had been generally satisfactory.

Early in 1923, battle position locations were selected along the south and north shores of Oahu. One site was the North Opaeula Gulch, a few miles inland from Waialua. Access to this position used parts of the Waia-lua Agricultural Co.'s plantation rail system. However, it was found the plantation railroad beds would not bear the weight of the trains, so the Opaeula Gulch position was only to be used in the event of war. Nearer the shoreline at Waialua a position on the OR&L right-of-way was selected for exercises and maneuvers, although it was not suitable as a battle position. Beyond the firing point at Gilbert, no actual off-post firing spurs were prepared for the mortars. By the late 1920s, a firing point had been planned at Maili on the southwest shore and some provision had been made for support of firing practices in the form of a fire control station atop the cliffs at Puu O Hulu, and a fire control switchboard at nearby Maili. Both structures were built of reinforced concrete. (6)

During the latter 1920s the 41st CA remained posted at Fort Kamehameha, where it participated in peri-odic maneuvers and exercises, and fired its mortars in annual service practices from various positions on the island. The off-reservation positions were simply selected places on the OR&L right-of-way—no firing spurs other than those at Gilbert had been constructed for want of funds. The 41st CA continued as a two-battery battalion attached to the 15$^{th}$ CA until the early days of World War II. With ten years of experience, it had been demonstrated that had been found that the heavy mortars were too cumbersome and slow to deploy with too many site load restrictions to efficiently perform their wartime assignment. In 1933, after two years of consideration and study, the War Department decided to replace the 12-inch mortars with the lighter 8-inch railway guns in response to the recommendations of the Hawaiian Separate Coast Artillery Brigade (HSCAB).

Like many other weapons deployed to Oahu, once there they seldom left. The overseas departments of Philippines and Hawaii both had large strategic reserves of weapons, ammunition and equipment stored on-site in case of war. The railway mortars were not required elsewhere, so they were carefully stored and given occasional maintenance in the warehouse depots of the department's Ordnance Section. Subsequently the 41$^{st}$ CA had just one active battalion with eight 8-inch railway guns. The inactive 2$^{nd}$ Battalion would be formed when needed with a second batch of such guns that arrived in 1940 and stored. The regiment's third battalion, also to be composed of two batteries was, if required, to be equipped with the eight obsolete 12-inch railway mortars. These outmoded weapons were still being held in storage at the Army's Ordnance Storage Facility near Fort Shafter in 1941. Large stocks of ammunition were also still available as 12-inch seacoast mortars were still active at Forts Kamehameha and Ruger.

On December 24, 1941, the department engineer suggested to the Hawaiian Department that as the obsolete 12-inch mortars then in the war reserves storage facility were no longer part of the defense project, they be could be deployed in dummy battery positions. General Burgin noted that he intended to use the mortars in the event of damage, destruction, or other loss of any of the 8-inch guns or the heavy caliber armament at Fort Kamehameha. On January 1, 1942, the engineers reiterated their recommendation that the mortars be moved to the unoccupied railway positions as dummies. Burgin relented and agreed that four of the mortars be placed at Waianae and the other four be positioned at the Puuiki battery. The following day he noted that since there were no coast artillery or infantry personnel available to guard the armament in the unmanned batteries, the "breechblocks, sights and similar parts...be removed before emplacement." By January 15, however, the decision to place the mortars in the two unused railway batteries had been reversed and the construction of special dummy guns for the positions recommended instead. Also the HCAC was still considering the expansion of the 41$^{st}$ CA to three battalions, the 3$^{rd}$ Bn would need the eight 12-inch mortars. (7)

The preparation of dummy batteries was not restricted to the firing spurs at Fort Kamehameha and the Puuiki and Waianae batteries. On January 4, 1942, the 34$^{th}$ Engineers were ordered to build a complete dummy railway artillery position near Browns Camp and thoroughly camouflage the actual Browns Camp position as soon as possible. The War Department considered the Browns Camp position too close to the shoreline and difficult to camouflage. In fact, there was a significant growth of Algaroba trees between the firing spurs and the shore and the spurs themselves were generally well hidden by trees. Camouflage was also ordered for the positions at Kahuku and at Camp Ulupau. Eventually, dummy 8-inch guns disguised by dummy houses, manufactured at the Kalihi Camouflage Factory, were positioned at the Waianae Battery and four 12-inch mortars were placed on the firing spurs at Puuiki. The remaining four 12-inch mortars were held in reserve on the spurs at Fort Kamehameha, but never used again. Finally the 12-inch mortars were ordered scrapped on May 1, 1945. (8)

The First World War Railway Gun project had involved several different types of guns. In addition to the seacoast 12-inch mortars, 8-inch, 10-inch, and 12-inch guns were taken from continental U.S. emplacements in 1917 and married to newly-made railway mounts and carriages. Even the navy contributed 7-inch broadside guns and 14-inch tubes for railway batteries. Only a very few of these were completed in time before the war ended in November 1918. Probably the most successful of these extemporaneous weapons was the 8-inch. While not initially considered for Hawaiian deployment, the troubles with the 12-inch mortar project prompted the decision to replace them with 8-inch railway artillery. The Hawaiian Department Engineer Office at Fort Shafter was instructed to prepare detailed specifications for the acquisition of land for battle positions for eight 8-inch M1888 guns that were to replace the mortars. In the summer of 1931, eight of these guns were collected at Aberdeen and prepared for shipment to Hawaii. For a variety of reasons this shipment did not occur, however until the early summer of 1934. (9)

The weapon chosen for this mount was the "old" 32-caliber 8-inch seacoast gun Model 1888. During the First World War, fifty 8-inch Model 1888 guns were removed from seacoast batteries and subsequently mounted on Model 1918 Railway Mounts. The carriage had the usual base ring, traversing rollers, side frames, cradle with a single hydraulic recoil cylinder mounted underneath and four distinctive counterrecoil spring cylinders surrounding the tube. It had ammunition cranes on the rear of an attached steel working platform. Essentially a complete barbette carriage, it was originally simply bolted on an appropriate railway car. The gun could be elevated to 42° and fire a 240-lb. HE round to 15,700-yards or an AP round to 23,900-yards. It could also be bolted to a fixed emplacement ground platform. The railway equipment never made it to France, but 47 complete units were in inventory with the Coast Artillery Corps in 1919. The top carriage mount had all the same ballistic performance as the railway mount. Mounted as a fixed barbette it was an awkward, exposed weapon, but one with a good useful range and payload that could prove important for the interim.

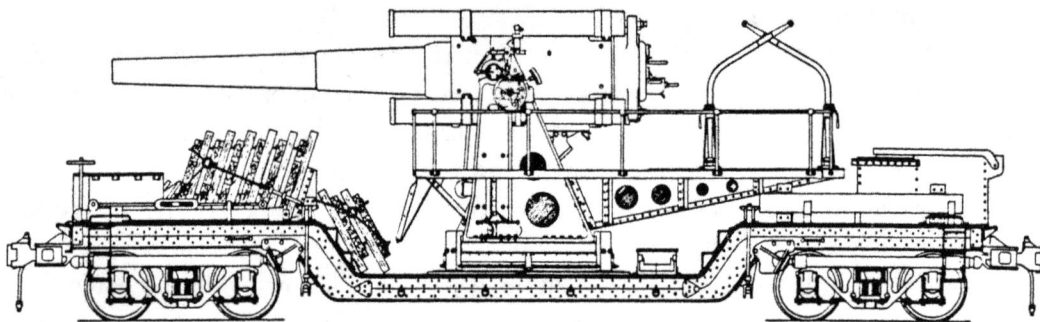

Developed at the same time as the railway mortar, the seacoast 8-inch Model 1888 gun tube was placed on a new barbette carriage atop a Model 1918 railway car. *NARA.*

Four 8-inch M1888MIIA1 guns and M1918 barbette carriages were shipped from the Aberdeen Proving Ground on May 14, 1934. These four arrived aboard the USAT *Meigs* on July 15. These were followed on August 14, with another shipment of four more guns and carriages of the same type. The eight guns and carriages were then mounted on the M1918MII railway cars fitted with the narrow-gauge trucks required for use by the 41st CA's two active firing batteries on the OR&L's right of way. Eight more guns and carriages were to be shipped to Hawaii when transportation funds became available, but these did not arrive immediately. (10)

Finally, in 1940 the augmentation of Hawaii's railway ordnance was authorized, and the additional 8-inch railway guns were earmarked for Hawaii. Eight guns, barbette carriages, and railcars, were shipped to Oahu on April 3, 1940. On April 24, a ninth gun was shipped, giving the island a total of 17 8-inch railway guns, though damage to one unit forced it to be returned.

The change in gun (and weight) size required selection of new battle positions. Maj. Ira B. Hill, commanding the 41st CA outlined the requirements for new firing positions:

Guns should have all around fire

While the primary targets would be naval vessels, the batteries should be positioned to cover adjacent landing beaches and defiles in the main routes of communication on land.

The positions should be located where they could be concealed from the air and sea by the use of camouflage and be far enough from the shoreline to avoid easy ranging from hostile warships bracketing the beach line.

The position should be selected with a view to economic and rapid construction with an ease of withdrawal if required. (Major Hill noted that placing the battery positions in the sugarcane and pineapple fields should be avoided because of the high cost of land acquisition in peacetime and the difficulty of concealment from the air. The plantation railway system was too lightly constructed to be used by the guns.)

The positions should be removed from material masks far enough so as not to restrict long-range fire.

The selected locations should have good roads nearby to facilitate the replenishment of ammunition and supplies.

Positions should have suitable fire control stations. (11)

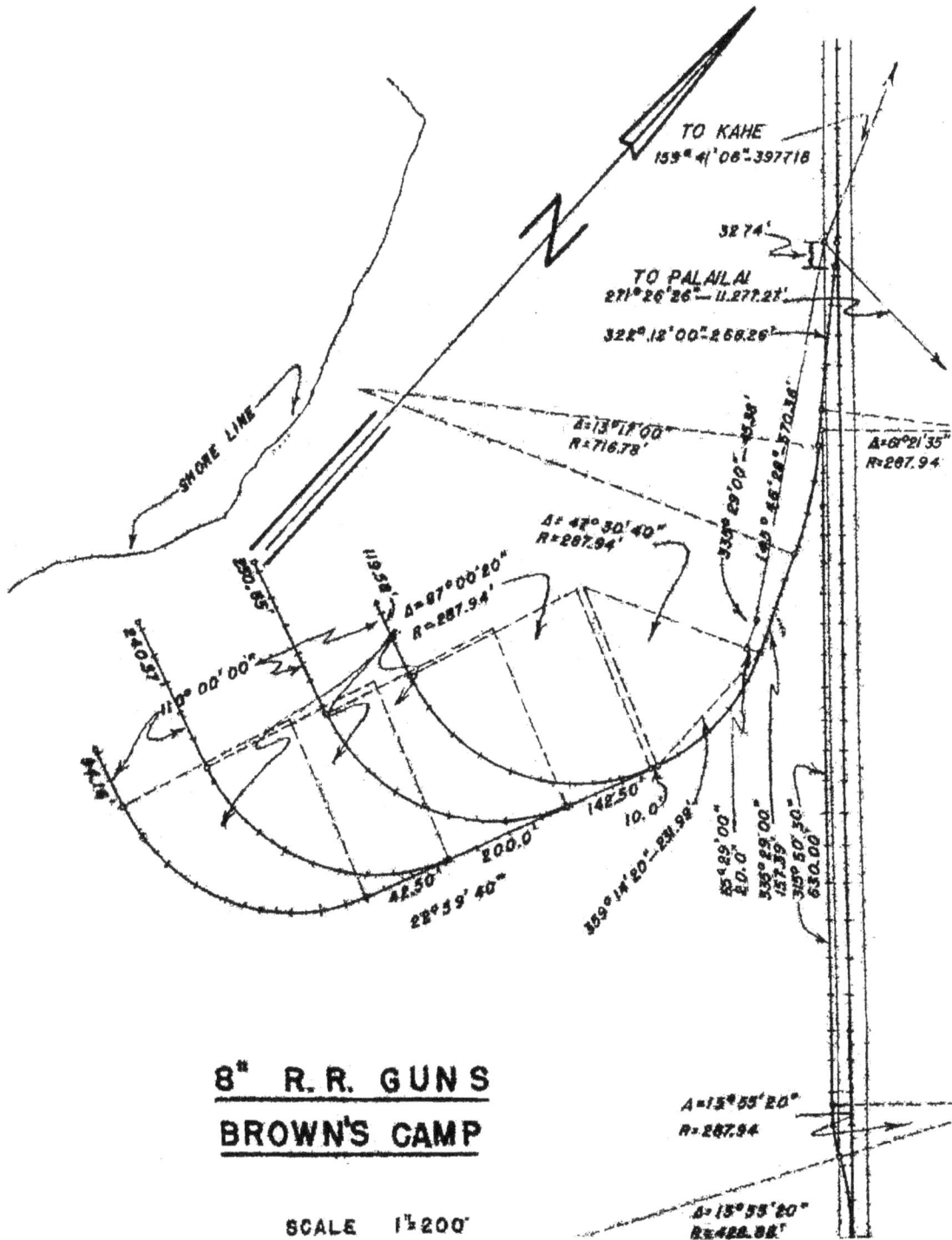

**8" R.R. GUNS**

**BROWN'S CAMP**

**SCALE 1"=200'**

Sidings with separate tracks for each of the four guns of an 8-inch Railway Battery were prepared at several strategic locations around the island, like this one at Brown's Camp covering many of the northern beaches on Oahu. *NARA.*

Each of the new positions was to consist of firing spurs for four guns adjacent to the main line of the OR&L. The end of the spurs were the firing locations. They were usually arranged in a trapezoidal configuration; two positions in the front (towards the enemy) separated by 200-feet. The outer positions on each side were withdrawn another 200-feet at an angle "out". Additional surveys were conducted between 1935 and 1940, to select the new firing positions. None of these battle positions had been constructed when World War

Loading an 8-inch railway gun emplaced at the Haleiwa position prewar. *Smith Collection.*

Special tracks were laid to allow the 8-inch Model 1918 railway cars to be displayed at the big reviews held at Schofield Barracks. Here in 1935 is a unit of Battery B, 41st CA, reviewed by Secretary of War George Dern. *Welch Collection.*

Periodically the 8-inch guns were also put on display at Fort Kamehameha itself, as is this unit in mid-1941. *Williford Collection*

II erupted in Europe in 1939. The firing spurs established at Gilbert on the Ewa Plain in the 1920s (originally for the old 12-inch mortars), continued in use during the 1930s, as the sole off-post firing point. After new positions were developed between 1939 and 1942, this position was abandoned as a battery firing position. (12)

Part of the efficiency of the 41st CA was attributable to the frequent movements undertaken by the battalion. At first the battalion used the firing point at Gilbert, but after the completion of the Browns Camp position it frequently used that position. Army frugality prevented acquiring a locomotive with air brakes

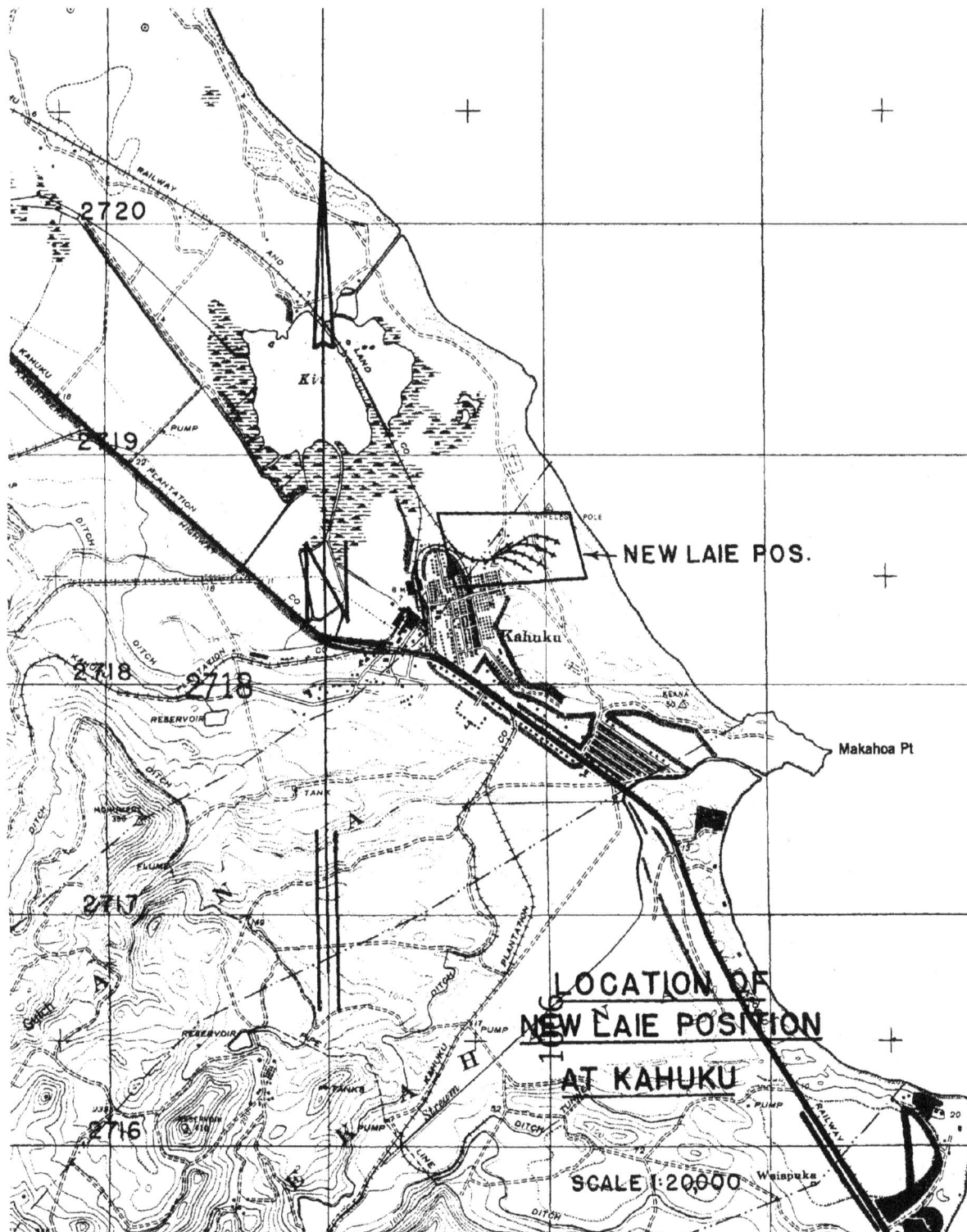

A new position started during the war at "New Laie" on the far northeast Oahu coast was eventually developed into what was known as Battery Kahuku. *NARA.*

The Gilbert Firing Point on the Ewa plain in southwest Oahu was developed to facilitate practice firing of both the 12-inch mortar and 8-inch gun railway guns. Here a battery of four 8-inch is practiced with in the mid-1930s. *Gaines Collection.*

and the capacity to pull the armament trains fast enough, the OR&L consented to provide a locomotive for the battalion at $10.00 per hour. In addition to providing for the training of the troops, the armament train itself required frequent "exercising" to preserve journals, otherwise hot boxes were prone to stall the train. This happened almost every time the guns were moved. Not until the early months of World War II were suitable army locomotives provided.

Overall the railway guns proved to be less effective than hoped for. Faulty rotating bands on the 260-pound naval projectiles sent with the guns rendered the firing of the guns highly inaccurate when target practices were held. The inaccuracy of the railway guns prompted the Hawaiian Department to suggest that delivery of the remaining M1888 guns be deferred and possibly replaced with newer materiel and ammunition if the problems could not be resolved. Discussions between the Hawaiian Department, the Ordnance Department and the Chief of Coast Artillery eventually resolved the ammunition deficiencies paving the way for delivery of the final eight 8-inch guns. In April 1940 these eight guns were to be used to arm two additional batteries of the 41st.

Eventually ten 4-gun firing positions were developed in nine locations. Fort Kamehameha itself had two battery locations, more to facilitate training than as an actual tactical emplacement. While several positions were intended as emergency or alternate locations rather primary ones, they all were functional from early in the war.

See Appendix IV for a description of the 8-inch railway firing positions.

As for constructing the new firing positions, after lengthy consideration in Washington, a modified project was resubmitted in February 1935 by Maj. Gen. Halsted Dorey. A two-phased project was developed. The first phase, or peacetime project, was to construct the two sets of firing spurs at Fort Kamehameha and to procure and ship all the material required for construction of the remaining eight firing locations. The projected layouts of the battery positions were also downsized and simplified. The second, or wartime, phase called for construction of the eight firing positions and construction of the long-awaited inland railway connection from Wahiawa to the North Shore. (13)

Initially only the on-carriage sights were used for fire control, or at most portable coincidence range finders on train cars for position finding, With the new firing positions, proper range-finding stations were also emplaced. At first the railway batteries of the 41st CA had only the two 50-foot rangefinding towers. These were to be equipped with 15-foot CRFs and used at Browns Camp and the Maili and Waianae positions. It was not until April 16, 1941, that two more 65-foot steel battery commanders' towers were finally approved for use when operating from Gilbert, Kahuku, or Laie. In February 1935, plans called for using preexisting fire control stations used by Batteries Closson, Williston, and Hatch as well as three projected new locations: Puu Manawahua in the southern section of the Waianae Range, Kahuku at the north end of the Koolau Range, and further to the east on that range near the town of Hauula on the windward coast. While the demountable towers were used for battery commanders' stations, base end stations were also projected. These positions were treated as fully functional emplacements, they were assigned equipped position-finding stations either through new construction or assignment to share an existing station. In nearly all cases these fire control stations were to be equipped with M1 depression position finders (DPFs) of various classes depending upon the elevation. (14)

The U.S. Navy's plans to establish a major land and seaplane base at Kaneohe Bay on the Windward Coast of Oahu in the late 1930s would involve the 41st CA. The proposed permanent project for the defense of this installation called for a battery of two 8-inch naval guns on barbette mounts. Pending the arrival of the naval guns, the Hawaiian Coast Artillery Command was called upon to provide a battery of four M1888 8-inch railway guns to defend the airbase. As there were only 16 8-inch railway guns on the island, this would reduce the mobile railway guns to three instead of the projected four batteries. The emplacement of the 8-inch guns on the Mokapu Peninsula at Kaneohe Bay would provide the coverage of Windward Oahu that had been partially provided by the proposed battery at Laie. General Henry Burgin, new commander of the HCAC recommended in August 1941, that the proposed railway artillery positions at Gilbert, Kahuku, and Maili be eliminated and that the Laie position be moved about 1,500 yards northward, close to the Kahuku Plantation Golf Course near Kahuku Village. General Short's recommendation to the War Department concurred with Burgin's.

In the last months of 1941, an additional alternate battle position was established at Waianae on the southwest coast, bringing the number of completed firing points to seven. The Waianae position was built under the jurisdiction of the District Engineer, and when inspected by a board of officers the technical aspects of the construction were found to be very well done. However, there had been no attempt to maintain any security at the site or to camouflage it during construction. An obvious right-of-way revealed the position, and rock and soil were shifted making it obvious in aerial photographs. Working roads had been cut in from the main road to the end of the firing position which further increases the hazard of disclosing the position. The situation highlighted the lack of coordination between the engineers and the coast artillery as well as the lack of communication between the District Engineer, the Hawaiian Department Engineer, and the engineer units in the field during early weeks of the war. After several weeks of discussion and the impetus of the Pearl Harbor raid, the Waianae Battle Position was finally accepted, but there is no indication that the site was ever used except as a dummy battery location. (15)

This affair helped illuminate an ongoing debate about the relative value of railway batteries. It was determined that the ability to successfully move a railway battery from one position after they were committed to action was unlikely, especially under enemy air activity. There is little doubt that these issues had an impact on the continued construction of railway artillery positions on the island. After the Japanese air raid, it was decided to keep the Kamehameha spurs in service until Battery Closson's 12-inch guns were casemated. On January 8, 1942, General Delos C. Emmons, the new Hawaiian Department commander, accepted the completed firing spurs at Fort Kamehameha.

Another class of mobile weapons deployed on Oahu were the tractor drawn 155 mm G.P.F. guns. The modern, yet relatively simple 155 mm gun developed and produced by Puteaux in France was known as the Grande Puissance Filloux, or GPF for short. It had only been recently introduced into French service and was considered quite modern. It was one of the first large field guns to utilize a split carriage. In August 1917 the U.S. ordered 48 guns from the French for direct delivery to the A.E.F. in France, many others followed. Thanks to quick French supply, it was issued and used towards the end of the war in Europe. In the U.S. extensive plans were made for licensed production, actively aided by government-funded plants to be operated by ar-

SECTION ON A-A

SCALE $\frac{3}{8}$=1'-0".

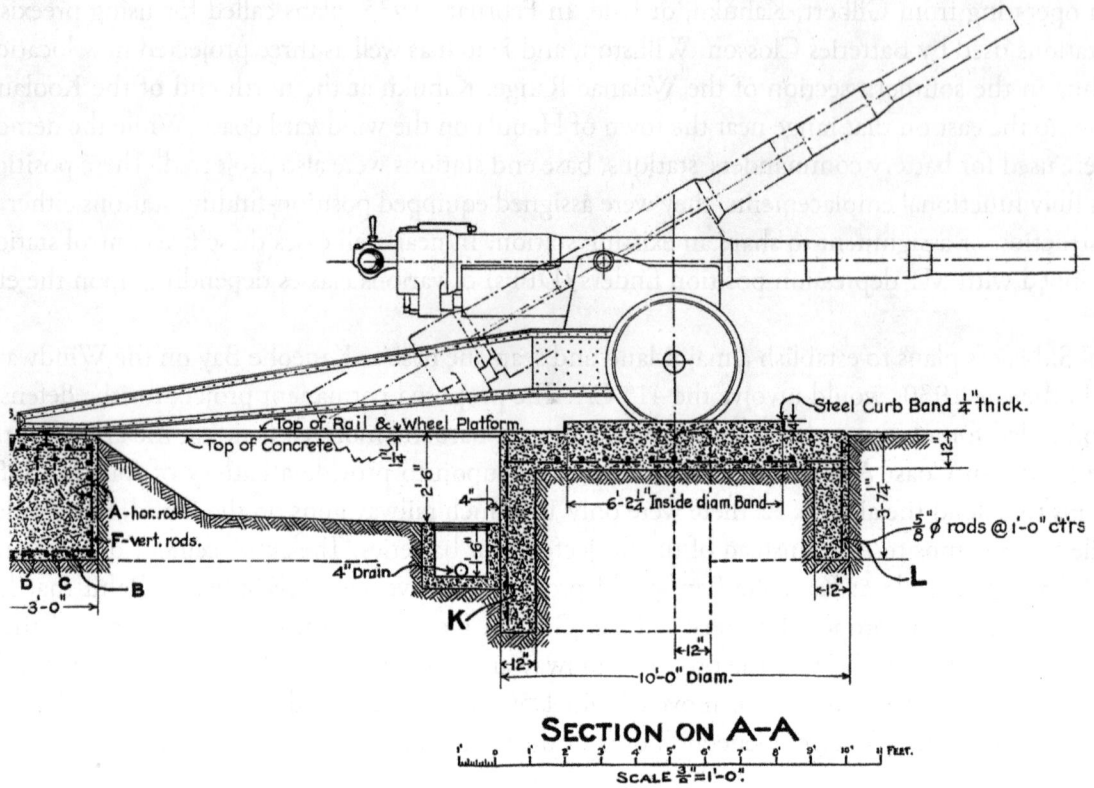

A special concrete mount was developed for the army 155 mm Model 1917/18 gun. Called a "Panama Mount", they could be built without guns at various locations and allowed for a stable, rapidly traversing platform for seacoast use. *NARA.*

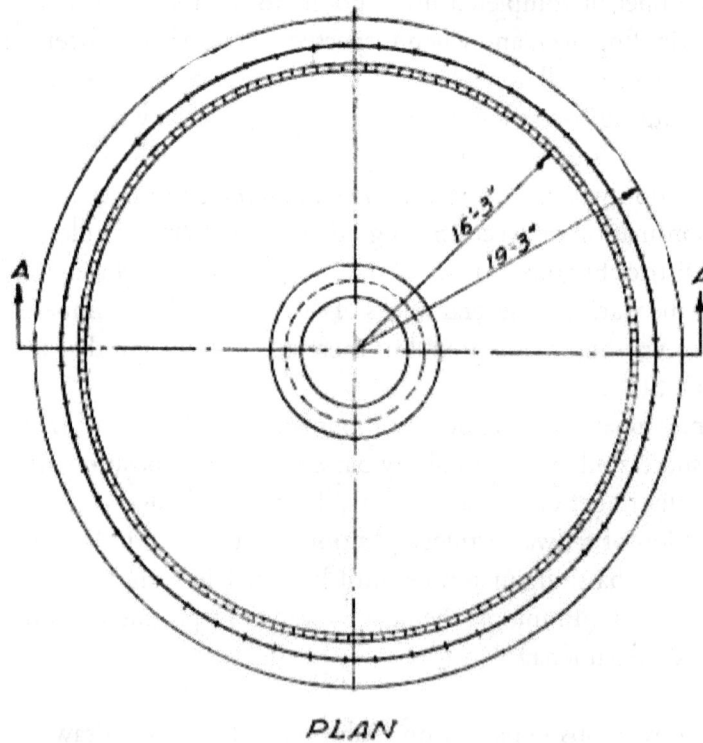

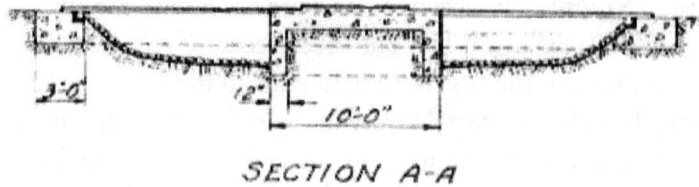

PLAN

SECTION A-A

Plan for a typical Panama Mount, this is one of the individual gun platforms built at Fort Ruger. *NARA.*

senals or contractors. Sources vary, but about 997 155 mm guns were acquired from all sources by the U.S. up until 1920. It was issued and used by both field and coast artillery units throughout the U.S. and overseas possessions well into the Second World War. (16)

The gun was a 38.2-calibre modern tube on a split-trail carriage that allowed a 35° elevation. It could fire a 95-lb. HE round or 100-lb. AP projectile at a muzzle velocity of 2400-f.p.s. about 19,100-yards. (17)

The 55th Artillery, CAC, a regiment of tractor-drawn artillery had been organized in the Defenses of Boston during the World War and saw action in France. This regiment was one of nine mobile regiments of that had been retained in active status at the end of the war. The regiment, equipped with thirty-six 155 mm (GPF) guns, was transferred to Oahu in 1921 on the United States Army Transport *Buford*. The regiment arrived in Honolulu on May 20, 1921.

After its arrival on Oahu, the 55th Artillery had been reorganized to three firing batteries in each of its two battalions. In Oahu it entered service with just 24 guns, the other dozen was in storage pending activation of a third battalion. Soon the regiment selected and surveyed numerous potential battery sites along the north, northeast, and southwest shores of Oahu for its guns. In order to familiarize themselves with the fields of fire and terrain, the regiment's firing batteries carried out exercises at numerous locations around the island on a regular basis. When first operated in the coast defense role in field emplacements, the guns that had proven so effective against stationary targets did less well against moving naval targets. Having only 60° of on-carriage traverse, they were unable to track fast moving vessels quickly enough without moving the 14,560-pound split trail gun carriage.

By the late 1920s, mostly due to personnel shortages, the 3rd Battalion of the 55th had been inactivated. The remaining six active firing batteries of the 55th, like most other coast artillery units on Oahu, had difficulty maintaining their nominal peacetime strengths through most of the inter-war period. Further manpower reductions plagued the 55th. By 1935 the personnel available for the defense of Oahu's shores were stretched so thin that manning responsibilities for the coast artillery's 155 mm gun batteries on the North Shore were transferred to the 11th Field Artillery Brigade of the Hawaiian Division. (18)

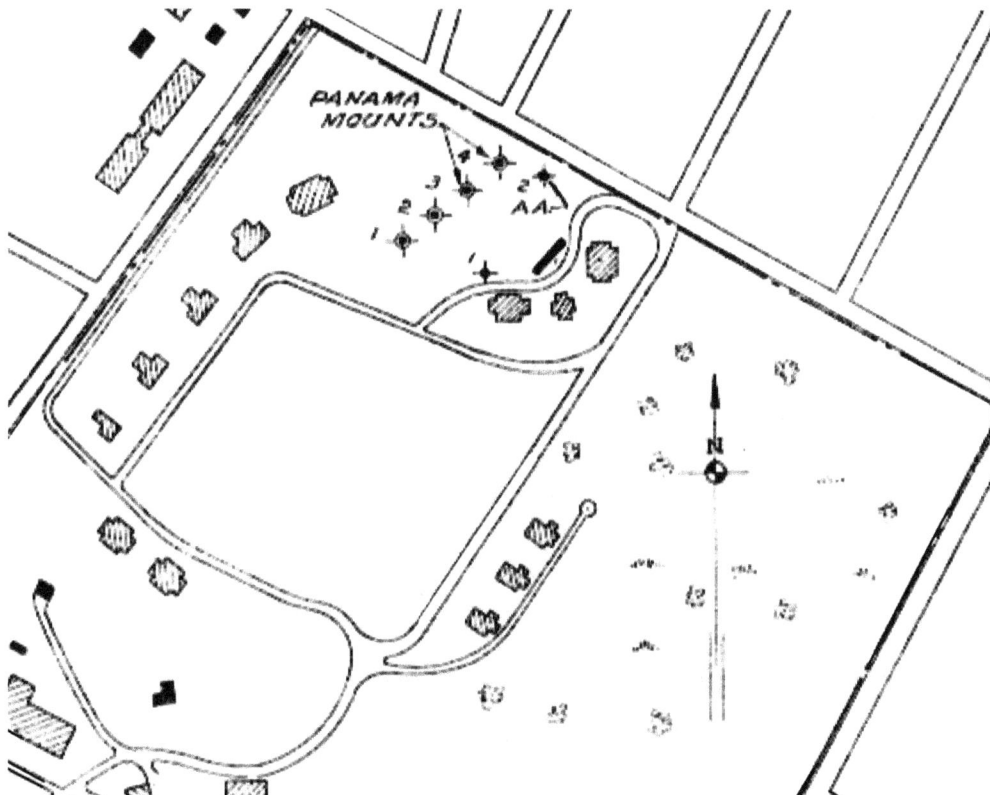

Map of Fort Ruger showing where that fort's Panama Mount emplacements were built, in fact almost adjacent to the older 3-in fixed AA gun blocks. *NARA.*

A second regiment of tractor drawn artillery was slated to be deployed to the Hawaiian Islands in the event of national emergency. In anticipation of this augmentation, an additional thirty-six 155 mm GPF mobile seacoast guns were shipped out to Hawaii in the mid-1930s and added to the Hawaiian Department's war reserves. In addition to their peacetime training activities, the mobile 41st and 55th CA Regiments were given the wartime mission of providing the seacoast artillery defenses of Windward Oahu and the North Shore. Implementation of a defense plan for the North Shore that had begun in the early 1920s continued, albeit slowly, into the late 1930s before the pace quickened. Waialua Bay was specially considered a potential location for invasion by a hostile force and the area received considerable attention during the period between the two world wars. Tests were conducted there with underwater obstacles to landing craft. The defense scheme for the North Shore was further revised in the early 1930s. In 1940, the clearing of cane fields to improve fields of fire in time of war was arranged with the plantation owners and planning for a large fortified base camp on the Pupukea Plateau on the North Shore was carried out. (19)

Unoccupied Panama Mount emplacements (No. 3 and No.4) built at the Punchbowl Military Reservation in the late 1930s. *NARA.*

155 mm GPF guns belonging to the Coast Artillery at Fort Kamehameha in 1941. *Gaines Collection.*

The 55ᵗʰ Coast Artillery regiment initially shared crowded quarters and barracks space with the companies manning the fixed batteries defending Pearl Harbor. As a consequence, a physical move was planned for early 1922. Under this plan, the Hawaiian Antiaircraft Regiment would relocate to the World War I cantonment barracks on the east side of the gulch at Fort Shafter. The 2ⁿᵈ Bn, 55ᵗʰ Artillery, Batteries D, E, and F, was transferred from Fort Kamehameha to Fort Ruger to occupy the buildings left behind by the Antiaircraft Regiment.

Fort Ruger provided the home post for the 155 mm GPF guns of the 2nd battalion of 55th Coast Artillery Regiment in November 1932.
*NARA,*

The usual towing prime mover for the 155 mm gun was the Holt 10-ton tractor, shown here being inspected at a review in Oahu.
*USAMH.*

In the 1920s the Coast Artillery Corps had developed a fixed emplacement design for the 155 mm carriage. It consisted of a central concrete pedestal to support most of the weight of the gun and allow the wheels to rotate. A separate circular concrete ring surrounded the pedestal and held the extended gun trails. A 155 mm gun could be towed to such a prebuilt site and emplaced on it relatively quickly. On this stable firing platform its accuracy against moving naval targets was improved, it could traverse quickly to fire at moving targets (ships) and could be tied into an already established system of fire control stations and plotting room. Batteries with four, three and two gun emplacements were constructed. Also, they were usually built with a complete circular ring to allow 360° traverse, even if the practical field of fire was less. Because they were developed from prototypes perfected in the overseas army's Panama Department, they became known as "Panama Mounts".

Normally the guns remained in gun sheds at a post and deployed to the positions only when a need existed. Of course, that meant that the number of emplacements could exceed the number of actual available guns. Many of the 155 mm battery emplacements were active (i.e. armed or at least immediately available for armament) for several years. With the gun emplacements were often built protected magazines, plotting room, battery commander's station, aid stations, encampment for the troop detachment. Also not infrequently positions for AA guns (particularly light automatic machineguns of 37 mm and 40 mm types) and searchlights were emplaced near Panama Mount batteries.

The Hawaiian coast artillery command heartily embraced the new 155 mm Panama mount. It offered a practical solution of a low-cost way to prepare in advance for island protection and combat, increasing the accuracy and thus effectiveness of the guns. Over 30 battery emplacements, most for four guns, were constructed on the island of Oahu between 1934 and 1943. The first 155 mm Panama mount battery was erected at Fort Weaver, between the 16-inch battery and the shore in 1934. Before the war, between 1937 and 1941, ten new 4-gun emplacements were built. They were sort of sprinkled around Oahu at locations that had been selected and previously used as field positions, including Browns Camp and Barbers Point in southwest Oahu. The two forts that hosted the battalions of the 55th each received an emplacement very convenient for training—Battery Kam at Fort Kamehameha and Battery Ruger at Fort Ruger. Also on the south shore was a battery on Sand Island. For the northern beaches were batteries at Ashley Station, Kahuku, and Kawaiolu. Then two batteries were built for the new Kaneohe Bay base; East Beach and North Beach/Pyramid.

155 mm guns were of course capable of being used also in conventional field positions not needing the previously emplaced Panama Mounts. This unidentified position is simply captioned "somewhere in Oahu". *McGovern Collection.*

After the war started, authority was granted, and funds released for building 15 new alternate sites. All were to be built under the supervision of the 34th Engineer Regiment. The rationale for this series is interesting: (20)

> Early in the war it was decided to construct additional Panama Mount 155 mm battery positions because maps taken from Japanese aviators indicated that several of the then existing battery position were known to the Japanese. Also, it was thought that most of the existing locations were built too close to the water. Fifteen new positions authorized by Department and 34th Engineers were instructed to construct them.

Known emplacements for this generation sort of echoed the initial distribution, with locations in the southwest (Kahe, Palailai, X-Ray), south (School, Round Top, Aliamanu, Wili), Kaneohe Bay (Loko, Kapoho, Papaa), and north shore (Pupukea, Waimea, Pine, Mokuleia). Most of these were never permanently armed and usually maintained by only a minimal detachment. Sometimes they would be used for temporary deployment training of a unit. Most were abandoned by 1944.

The final generation of Panama Mount emplacements was emplaced in late 1943-early 1944. The light 4 and 5-inch guns loaned by the navy to the army after the start of the war and emplaced in temporary or expedient batteries were declared obsolete and either returned to the navy or scrapped after a survey in October 1943. Several locations still needed defensive guns, so new Panama Mount batteries were constructed, in some cases immediately adjacent to the old gun positions. Another batch of twenty-five 155 mm M1917/18 GPF guns was shipped from the U.S. to Hawaii specially for this purpose. Five positions received new Panama Mount batteries: Homestead, Nanakuli, Kahana, Dillingham, and Kaena.

For Descriptions of the individual 155 mm Panama Mount emplacements constructed on Oahu see Appendix V.

One other type of mobile weapon was deployed on the Oahu defenses. The US Army's 240 mm howitzer was developed in World War I as a heavy regular army siege gun. While the American army initially borrowed 8 and 9.2-inch howitzers from British sources, this was a domestically produced gun made from a purchased Schneider design. It was a stout, heavy, high-angle howitzer intended to be placed on static ground platforms and relocated in sections only with some difficulty. Plans were made and orders issued for manufacture of 1200 complete howitzers for the war effort. It took time to construct and equip plants, of course along with correctly rendering French plans and specs into manufacturing guides. Only the pilot howitzer was completed by the time of the armistice, but production, though at a reduced rate, continued in the immediate postwar period. A total of 330 240 mm M1 howitzers and carriages were made and delivered from 1919 through 1921. At a full elevation the howitzer could hurl a 345-lb. projectile at 1700 f.p.s. a range of 16,390-yards. That was impressive performance, but the relatively static, 41,206-pound gun and carriage were ponderous and certainly not easily mobile. (21)

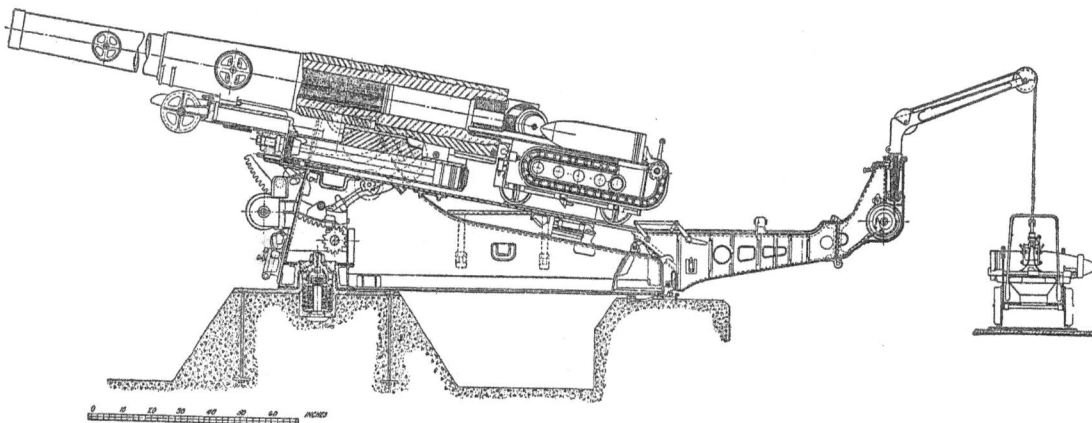

Produced immediately after the First World War, the Army's 240 mm howitzer was delivered to Oahu in 1922. Like the 155 mm, a special concrete firing platform was designed for emplacement in potential battery sites in advance of need. *NARA.*

Twelve of these 240 mm howitzers were to go to the Philippines to form a heavy counterbattery force for the Harbor Defenses of Manila Bay. However, the Washington Treaty intervened, and while onboard at sea the howitzers were diverted and turned over to the Hawaiian defenses. The treaty had been signed by the American representative, but not yet ratified, on February 6, 1922. A week later, on February 13, the army's Chief of Staff approved a memorandum on "The Interpretation of Article XIX". Two days later the 12 howitzers on USAT *Wheaton* were scheduled to arrive in Honolulu. The Secretary of War cabled the Hawaiian Department concerning the ordnance supplies: "circumstances now render these unnecessary in the Philippines." Honolulu was offered the chance to remove what ordnance they needed, with the understanding that in no circumstance would there be any increase in personnel to man it. Hawaii grabbed the howitzers and their ammunition. (22)

Despite the lack of permanently assigned manning detachments, these powerful howitzers were considered an important augmentation to the coastal defenses of Oahu. Emplaced near the previously unprotected coastlines, their range made them especially valuable in covering the waters that lay offshore from the potential landing beaches along the southwest, northeastern, and North Shores. At first, the howitzer batteries were emplaced on field platforms of timber grillage used in maneuvers and exercises, but the limited 20° of traverse permitted by the carriage proved to be inadequate when attempting to track the moving ship targets. By the latter part of 1922, the Hawaiian Department Engineer, Major W. A. Johnson, was assigned the task of modifying the howitzer carriages and providing a platform that would provide all around traverse ability as well as possess the necessary structural strength to withstand the stresses to which such a platform would be subjected. By March 1923, Major Johnson had designed a large circular concrete emplacement that in later years would be described to be similar in appearance to an "oversized Panama mount." (23)

The Department Engineer with 55th Coast Artillery provided the necessary working parties and constructed an experimental emplacement at Waimanalo on Oahu's northeast side during the summer of 1923 and a howitzer, mounted on a modified carriage, was emplaced on the concrete mount. A proof firing was carried out on September 14, 1923. Seven rounds were fired at various azimuths and elevations. The emplacement proved to be very stable and there were no failures of the emplacement's concrete work. This was followed with an additional 32 rounds that were fired to determine the weapon's accuracy on the experimental emplacement at ranges that varied between 5,000, and 15,000-yards. These tests also proved to be an unqualified success, and the modified howitzer carriage was ultimately adopted by the Army as the M1918M1 Carriage. Johnson's prototype emplacement at Waimanalo was later abandoned in favor of another site for a pair of howitzers nearby. (24)

The initial prototype fixed platform for a 240 mm howitzer was emplaced to cover Waimanalo Bay on the southeastern Oahu coast.
*NARA.*

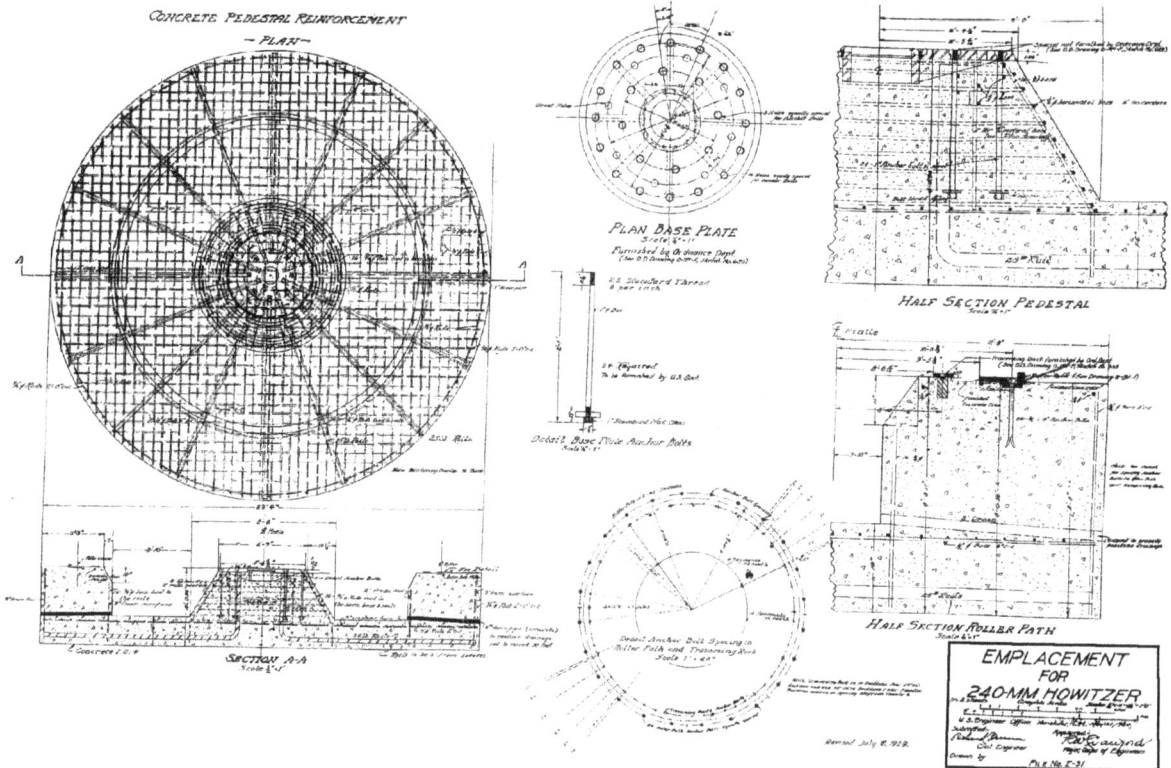

1929 plan of the 240 mm platform with its central pivot point and encircling rail for the trails. *NARA.*

Good view showing tests evaluating the new howitzer platform with its gun. *NARA.*

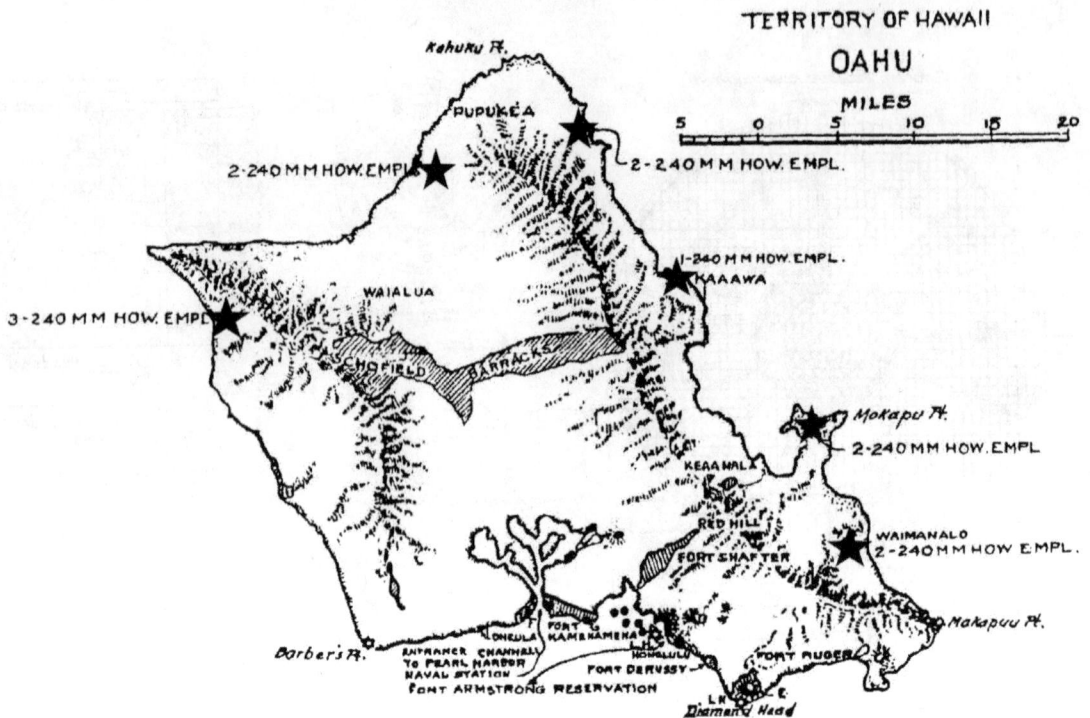

Map showing the locations of the final six prepared emplacements for the 240 mm howitzer. *NARA.*

During the next few years, six sites, each for a pair of 240 mm howitzers were on permanent concrete emplacements based on Maj. Johnson's design. They were at locations along the windward, north, and southwest shores of the island. Authority to proceed on constructing emplacements for ten howitzers was given on April 15, 1924. Four of these permanent installations were located on Windward Oahu. The first of these, completed April 7, 1927, was a single emplacement at Kaaawa just east of Puu O Mahie and Kahana Bay. On May 20, 1927, the second battery site was completed for two howitzers at the Kuwaahoe Military Reservation

1930 photograph showing one of the 240 mm positions under construction. *NARA.*

in the shadow of Ulupau crater's west rim on the Mokapu Peninsula. A third set of two emplacements was constructed on the Waimanalo Military Reservation in October and November 1929 these were reconstructed and completed by May 29, 1931. A similar pair of emplacements was constructed near Laie on the island's eastern shore during the spring of 1931. Two more sets of emplacements were also constructed to cover less well-defended portions of the Oahu shoreline, one of two emplacements was built at Pupukea on the north

Various efforts were made to assist camouflaging the 240 mm sites, this photo clearly shows some of the overhead nets and material to break-up sighting from the air. *NARA.*

Another early war (1942) view of a camouflaged 240 mm. By this time the howitzers had been turned over the field artillery units primarily tasked with beach defense. *NARA.*

side of Waimea Valley. Construction of the two howitzer emplacements began on May 23 and was completed by June 30, 1927. The Pupukea Battery was the only battery of that type built on the North Shore during the inter-war years. A sixth battery, consisting of three emplacements, was constructed on the Makua Military Reservation south of Kaena Point on Oahu's southwest shore. (25)

A fuller description of the constructed 240 mm fixed howitzer positions is given in Appendix VI.

As the 240 mm howitzers were portable weapons, they were retained in storage except during exercises. The howitzers were broken down into three loads to facilitate transportation to a site and reassembled at the prepared emplacement. These three loads were: the gun tube, the top carriage, and the gun cradle. The field platform that could be used at an unprepared site constituted a fourth load. Each load could be carried on a truck or towed on a trailer to the firing location. Additional vehicles were required for the manning detachment and ammunition. Periodically the howitzers were taken out into the countryside by the coast artillery, and later by the field artillery, to practice mounting the powerful weapons in their prepared emplacements. At the battery site the gun and carriage could be assembled and emplaced ready for use in about three hours by a well-trained manning detachment. (26)

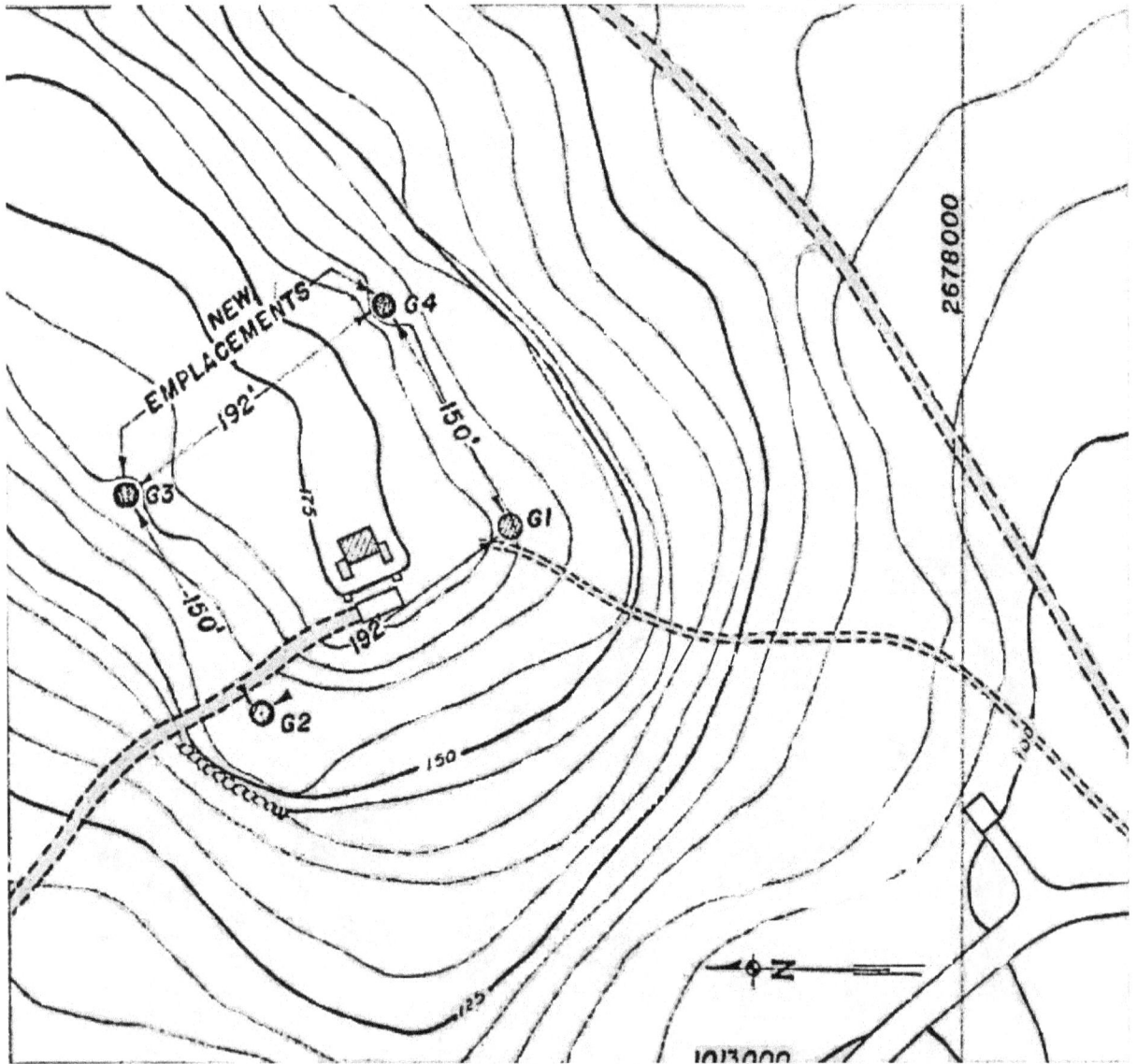

Besides the prepared gun blocks and circular traversing track, little else permanent was built at the howitzer location. Look at how bare this finished emplacement appears—just the two new blocks (re-sited a little from the initial placement) and a central magazine. *NARA.*

As the war progressed the 240 mm howitzers were used less and less. This firing by the 90th Field Artillery Regiment in late 1942 could well have been the final time this gun was ever fired. *USAMH.*

Driven by the chronic manpower shortages of the 1920s and 1930s, eventually the howitzers became a shared assignment with the field artillery. The howitzer emplacements at Makua and Pupukea were initially assigned to the coast artillery companies at Fort Kamehameha. However, soon the 2ⁿᵈ Battalion, 55ᵗʰ Artillery, CAC, assumed the collateral responsibility for manning the howitzers on Windward Oahu while the 3ʳᵈ, and later, the 1ˢᵗ Battalion of the 55ᵗʰ provided the manning detachments at Makua on Oahu's southwest coast and the position at Pupukea. The 55ᵗʰ was still providing manning detachments in the early 1930s, when Battery F set a record for emplacing two of the howitzers in just two hours and 40 minutes. As the 55ᵗʰ CA's personnel level was maintained far below its authorized peacetime strength, the regiment could do little more than pay lip service to its howitzer manning responsibilities. (27)

However, the howitzers were found to be ill-suited for the coast defense mission. Their projectiles were unsuitable; proper fire control equipment for firing on moving targets was not available, their rate of fire was slow, and their maximum range was too limited. Because of these inadequacies the decision was made by the mid-1930s to change their mission to that of heavy field artillery in support of the field forces in the event enemy landings were made on the island. The howitzers were reassigned to the 11ᵗʰ FA Brigade of the Hawaiian Division posted at Schofield Barracks. The 8ᵗʰ, 11ᵗʰ, and 13ᵗʰ FA Regiments composing the brigade each provided manning detachments for two howitzer batteries.

In 1936 the commanding general of the department wanted to eliminate the 240 mm gun contingent. He wanted to keep four guns and mounts for emergency field employment but requested to return the other eight stateside. Washington determined that as they had no other use in mind, all 12 might as well be kept in hand at the department's war reserve. As a result of the changed mission, six new emplacement locations further inland were eventually authorized and selected for six two-gun batteries of howitzers in the late 1930s and constructed between 1940 and 1942. In selecting these new firing positions for the howitzers, consideration was given to multiple fields of fire, so that when emplaced, one or more of the six batteries could place almost every acre of Oahu under the fire of at least one of these powerful high trajectory weapons. Four of the new howitzer battery sites were located to provide coverage of the North Shore. One was located at an elevation of about 1,500-feet on the summit of Quadrupod, a peak in the Koolau Range, another was located on the Anahulu Flats east of Kawailoa, the third was the initial battery site retained at the Pupukea Military Reservation and the fourth was emplaced at the Kalihi Military Reservation at the Kalihi Pali The remaining three howitzer sites were located to cover the south and southwest portions of the island. In most instances the base rings were removed from the old howitzer emplacements, however, in one case, the emplacements at Pupukea were retained as an alternate battery position. While the howitzers were manned during the early months of World War II by field artillery units, the 1st through 6th Provisional FA Batteries, were constituted in the Hawaiian Department to man the howitzer batteries. The 240 mm howitzer batteries would continue in service until the summer of 1944 when they were disarmed, and their manning detachments disbanded. (28)

8-inch Railway gun on display during a review. *Smith Collection*

8-inch Railway gun being loaded. *Smith Collection*

# CHAPTER 7
## THE INTERWAR YEARS

The original forts defending Pearl Harbor and Honolulu continued to be the centers of coast artillery activity during the interwar years. The new posts at Fort Weaver and Barrette, while they had the most modern and powerful weapons of the island, did not have the infrastructure to support their own garrison. Of course, the growing importance of the mobile batteries of both the seacoast and antiaircraft branches resulted in many new geographical sites and locations, but most were just prepared positions for arming if necessary and had few or no permanent structures.

Soon after fortification construction began, units of coast artillerymen began arriving in Oahu. They were soon joined by other ground forces that would ultimately constitute the defensive mobile forces. Finally, the Army Air Corps (and subsequent Army Air Force) units would also greatly grow in importance. In 1933 there were somewhat over 15,000 army troops in the Hawaiian Department. About 3,000 were in the seacoast units, 1,100 in the AA defense, and just over 11,000 in the mobile army and service functions. Over 25% of the total military forces were assigned the coast artillery harbor defense, antiaircraft and mobile units. (1)

Thereafter the troops assigned to all the army branches grew substantially as the 1940s approached. The manpower more than tripled by 1941, though the relative proportions by function stayed fairly constant. By December 1941 the breakdown of forces by corps or service was as below: (2)

Army Manpower Strength by Function December 22, 1941

| | |
|---|---:|
| Department Headquarters | 683 |
| Mobile Army Units | 23,550 |
| Army Air Force | 8,802 |
| Harbor Defense | 6,220 |
| Anti-aircraft Defense | 8,993 |
| Service Functions | 5,811 |
| Hospital/Medical | 3.009 |
| Service Command | 83 |
| | |
| Total | 57,241 |

Personnel staffing by fort or camp was also interesting. On November 30, 1941, the following army personnel distribution was reported: (3)

Army Manpower Strength by Post November 30, 1941

| | |
|---|---:|
| Bellows Field | 409 |
| Camp Malakole | 1,495 |
| Fort Armstrong | 818 |
| Fort Barrette | 133 |
| Fort DeRussy | 542 |
| Fort Kamehameha | 2,171 |
| Fort Ruger | 897 |
| Fort Shafter | 3,425 |
| Fort Weaver | 346 |
| Hickam Field | 5,380 |
| Honolulu | 51 |
| Schofield Barracks | 22,173 |

The 15th CA Regiment (Harbor Defense) had been constituted in February 1924 by the War Department which reorganized the Coast Artillery from a corps of separate companies manning the fixed coastal defenses into regiments of harbor defense, railway, mobile tractor drawn and antiaircraft artillery regiments. The 15th CAR was organized at Fort Kamehameha and headquartered at the fort for many more years. The 15th along with the 41st Coast Artillery Regiment (Railway) and 1st Bn of the 55th Coast Artillery Regiment (Tractor-Drawn), were all posted at Fort Kamehameha from 1924 until the United States entered World War II in December 1941.

Small alterations were made to guns and emplacements in the 1920s and 1930s. At Kamehameha's Battery Selfridge, following the First World War the old-style powder hoists were recommended for removal, but it wasn't until March 1928 that the work was carried out. After Battery Closson began service in the early 1920s Battery Selfridge was placed in a reduced manning status. In 1930, the Battery Commander's Station atop the central traverse of the battery was provided with a splinter-proof concrete roof. On May 5, 1935, it was recommended that Battery Selfridge be made a Class B battery or alternate assignment, for Battery B, 15th CA Regiment. In 1937, the battery's operational spaces such as the plotting room, fire control switchboard room, rest rooms, and latrines were provided with gas proofing collective protectors and airlocks by the Chemical Warfare Service.

Coast artillery units were annually subject to evaluation during their annual Service Firing. Regimental batteries deployed to their evaluation locations (either at a gun battery or practice firing point). Then using approved ammunition and target techniques the batteries performed a firing exercise and were scored for accuracy and time performance. Pre-war this was a major event. Competition was keen for the awards and cash bonuses. Even for the average serviceman the special activity provided a sense of adventure and certainly broke the routine. Many of the memories recorded in letters home and diaries revolved around this annual service firing. Here's a comment by Hope F. "Buck" Wilmer, Battery "A" 55th Artillery, CAC, dated about 1939: (4)

When they were out on maneuvers several battalions of 3-inch Anti-Aircraft guns would gather. All batteries would fire for record and they had plenty competition. "We would be out about two weeks, training and firing for record, and if we did well, we got a big party. In order to be eligible for the Expert Gunners Badge, and $5.00 a month bonus for a year, (if the money don't run out) we had to qualify on all of the weapons, 155 mm,

For crew members of active batteries, the annual "firing for record" was a major activity and consumed weeks if not months of careful preparation before the moment of discharge. This is such a firing by Fort Kamehameha's Battery Selfridge in the 1930s. *USAMH.*

Anti-Aircraft, machine guns, etc. I think the 64th Artillery, C.A.C. was stationed in downtown Honolulu at Ft. Shafter. Back at Ft. Kam we had the 1st Battalion of the 15th Artillery, 41st Artillery, and the 55th Artillery. I don't have any idea of where the rest of the 55th was during this time. At various times I was a member of the Gun Crews, Cook, and Mess Sergeant. Back then we were trained to take over about any position.

Careful records were maintained during the annual service firing. Information was recorded during target practice relating to the gun, carriage, conditions of loading, laying, etc., which would be of value in calculating future firings. A special recorder for each battery was assigned to this task and the records kept in dedicated Gun and Battery Record Books.

As a major coast artillery post, Fort Kamehameha served during the interwar period as a host to mobile 155mm GPF gun battalions. *USAMH.*

8-inch Railway Guns firing from the rail firing point at Fort Kamehameha in the 1930s. *Williford Collection.*

For much of their service lives, the gun batteries at Fort Kamehameha (as well as the other forts on Oahu) were not provided with any camouflage. Then reducing the visibility of the fixed batteries received the attention of General Walter Short, who voiced his concern on this matter soon after he assumed command of the Hawaiian Department. Numerous concealment methods were to be incorporated to conceal the gun batteries including overhead cover with garnished nets, tone-downed painting, transplanting bushes and shrubbery, as well as the preparation of dummy positions and the extension of roads past the batteries. Cost of the project for Battery Selfridge alone was estimated at $5000.

Battery Hasbrouck served as an integral part of the fort's armament until the mid-1930s, when its eight mortars were judged to be obsolete, and the battery was relegated to Class C status. In spite its obsolescence, some of its magazines and other spaces were gas proofed in 1938 when portions of the battery structure were chosen for adaptation to use as the Harbor Defense Command Post (HDCP) for the Harbor Defenses of Pearl Harbor. Despite its new mission, Battery Hasbrouck's armament was retained in place. In the weeks just prior to the Japanese air raid on Oahu the battery was reactivated for active service. Battery B, 41st CA had begun training with the mortars as an alternate battery assignment during the week before the Japanese attack. During that attack, the magazines of the structure were utilized as a shelter for civilian personnel at the fort. (5)

Following that war, Battery Jackson was not manned on a regular basis because of personnel shortages, though it retained its armament throughout this period. Similarly, Battery Hawkins was manned through World War I, but in the post war years was placed in maintenance status due to the lack of personnel to provide a detachment.

When the mine project for Pearl Harbor was eliminated following World War I, the mining casemate building was converted into the Harbor Defense Radio Station and radio operator's school. In 1937 the radio room itself was gas proofed and provided with an airlock by the Chemical Warfare Service. The other rooms in the structure, the power generator room, and two schoolrooms, were not gas proofed. The building was returned to service with these modifications on December 1, 1937.

At long-range Battery Closson, only minor improvements were made to the battery during the period between the two world wars, one of which was the provision of a concrete splinter-proof roof over the battery commander's station atop the central traverse. In 1937, the Chemical Warfare Service installed collective pro-

While a small post with limited space, Fort DeRussy was periodically used for maneuvers. Here 155mm GPF guns of the 55th Coast Artillery Regiment are deployed in 1935. *USAMHI.*

Almost all coast artillerymen started their Hawaiian experience with the Army transport journey from San Francisco. Here soldiers of the 251st CA (AA) arrive in 1941 on USAT *Leonard Wood*. *Bennett Collection*.

tectors and airlocks to make the plotting room, first aid room, fire control switchboard room, and latrines on the interior of the battery traverse gas proof. Since completion, Battery Closson had become a key element of the defenses for both Pearl Harbor and Honolulu. A full set of fire control end stations were completed and operational. The primary station was erected at the Punchbowl reservation, with secondary stations at Makapuu and Puu Palailai.

Camouflage of the Battery Closson's two exposed 12-inch barbette emplacements received serious attention immediately before the war. The reduction of the battery's visibility from hostile aerial observation began in late February 1941. Numerous concealment methods were utilized, including overhead cover of garnished nets and toned-down painting. Cost of the camouflage project just for Battery Closson was estimated at $6000.

The fort's C-1 station was remodeled and gas proofed in 1937. About this time the third story observation post was rebuilt in concrete and a fourth story of smaller dimensions added to the roof of the observation post as the harbor defense signal station. In 1941 a second steel frame tower surmounted by a two-story splinter-proof concrete house was built directly adjacent to the "C" Station. It served as the Harbor Entrance Control Post (HECP) Observation Post and Signal Station and housed the command function of the Pearl Harbor Groupment.

No new searchlights were reinstalled at the Fort Kamehameha location until 1941, when two 60-inch portable seacoast searchlights, Numbers 7 and 8, were positioned near Battery Hawkins and the Pearl Harbor Channel. These lights were mounted atop a pair of 50-foot towers that were procured in 1941. When the portable 60-inch lights were installed in 1941, they were powered by their own mobile power plants, and controlled from a nearby fire control tower. The lights were manned by a detail from Headquarters Battery, 15th CA Regiment. (6)

The cantonment building complex at Fort Kamehameha continued to expand, with support buildings being added to or replaced gradually. These included a new hospital, guard house, post administration building, post exchange, garage, bakery, cold storage house, laundry, carpenter shop, incinerator, telephone exchange, fire station, post office, and vocational school. There were specific post sections for functional specialties—shops and storehouses for quartermaster, engineer, ordnance departments. Sports facilities were developed, including a swimming pool, bath house, hand ball and tennis courts, grandstand and ball field, officer and

NCO clubs, YMCA for enlisted men. Fort Kamehameha was renowned in army circuits for its pool and running track in particular. One thing that wasn't needed was a central power plant. Commercial connections and service was more than adequate, though individual gun batteries and searchlights had emergency power generators. During the interwar period, Fort Kamehameha was quite an attractive post. Army wives worked hard on an extensive number of outdoor gardens. Lots of sisal from the Old Queen Emma estate were replanted on the post.

New 155mm GPF gun Panama Mount positions were constructed on post for Battery Kam, which was established at Fort Kamehameha during the Department maneuvers of May 1939. A company of the 3rd Engineers was assigned to build four Panama mounts. They were located on the shore between Batteries Hawkins and Jackson and covered the approaches to the Pearl Harbor channel. Magazine bunker revetments were built of corrugated iron, placed behind vertical pipes, and backed up with a large amount of dirt. The battery position was transferred to manning detachment of the 55th CA on March 25, 1940. Battery A of the 55th manned their four guns here on the morning of December 7, 1941. (7)

In these years the 15th Coast Artillery was the primary unit for the fixed gun and mortar batteries assigned to the Harbor Defenses of Pearl Harbor. These defenses were centered around the Fort Kamehameha reservation on the east side of the Pearl Harbor Channel and included the defenses of Fort Weaver established in 1922 on the left-hand side of the channel. In 1934 Fort Barrette at Kapolei Hill on the Ewa Plain was added to the Pearl Harbor Defense command. Both western side forts were organizationally sub-posts of Fort Kamehameha, which served as the headquarters post of the Pearl Harbor defenses. The regimental commander of the 15th also functioned as the harbor defense commander for Pearl Harbor. He had administrative and tactical control over the 41st Coast Artillery (Railway) Battalion and the 1st Battalion of the 55th Coast Artillery (Tractor Drawn) Regiment, both of which were posted at Fort Kamehameha.

All active firing batteries of the 15th functioned as training units for the thousands of coast artillerymen who served one of more tours of duty on Oahu during the interwar years. The 15th and the other coast artillery units in Hawaii were the primary training units for the entire Coast Artillery Corps, along with similar units in the Philippines and in the Panama Canal Zone. Since Fort Kamehameha had no adjacent civilian development, it was also one of the principal locations for live fire target practice. The Harbor Defenses of Pearl Harbor had the most modern heavy seacoast gun batteries—a long range 12-inch battery, and two long range 16-inch gun batteries. In 1935, a modest augmentation was authorized, and the 15th Coast Artillery was able to increase the number of men assigned to each of the active firing batteries of the regiment to nearly 200 men. Each of the active firing batteries were assigned to one of the long-range gun batteries: Battery A manned the 16-inch Guns at Fort Weaver Battery B the 12-inch Battery Closson at Fort Kamehameha and Battery C manned the 16-inch guns at Fort Barrette. (8)

There were generally not sufficient troops available to permit a battery organization to have only one tactical assignment. Battery A manned a battery of 155mm guns at Fort Weaver, while Battery B manned the 12-inch disappearing guns of Battery Selfridge as its alternate assignment. Battery C also manned a battery of 155mm guns. During the interwar period antiaircraft defense became increasingly important tactically and strategically. The batteries of the 15th Coast Artillery had a tertiary mission manning fixed 3-inch antiaircraft guns at each of the posts in the harbor defenses. All three batteries of the 15th were well versed in beach defense infantry tactics and in the preparation of field fortifications. The efficiency of the 15th Coast Artillery was high, and the various batteries were frequently recognized for their performance, winning numerous departmental awards. The years of 1939 and 1940 saw an increased intensity in the training program of the 15th Coast Artillery. Battery manning complements were gradually increased as well and by the latter part of 1941 the number of troops in the batteries had increased to well over their wartime strength levels. In October 1941 a number of the men in the 15th Coast Artillery were reassigned to the 97th Coast Artillery (Antiaircraft) Regiment when it was organized at Fort Kamehameha. By the end of November, the 15th had 28 officers and 712 enlisted men assigned to it. Although only a part of the 97th Coast Artillery had been activated, sufficient batteries had been organized to provide manning details for the fixed Antiaircraft gun batteries at the three forts in the Pearl Harbor defenses.

The 55th CA regiment with its towed 155 mm guns was headquartered and resided at Fort Kamehameha when it arrived in Hawaii in 1921. Its original complement was 19 officers and 586 enlisted men with twenty-four 155 mm guns. While this was sufficient armament for two battalions, there was not enough personnel to man all of them simultaneously. The 55th struggled with this handicap for many years. A serious accident occurred with one of the 155 mm GPF guns of this unit. During the standard record practice on April 29, 1929, of the 1st battalion (almost always the fully staffed portion), an accident killed the gun sergeant and three gunners.

With the same logic the newly formed Hawaii Railway Battalion was also assigned to the Fort Kamehameha post when created on December 22, 1921. It was armed with eight 12-inch mortars on Model 1918 railway cars. A year later its designation was changed to the 41st Coast Artillery Regiment, with just one battalion. It certainly made sense to locate this unit here where it had convenient rail connection to the main Oahu Railway & Land Company network. As noted previously the 55th and 41st had designated firing positions elsewhere on Oahu. These regiments were just quartered and in readiness for deployment at Fort Kamehameha, though there were some developed firing points at the fort mostly for the convenience of training with the armament.

After World War I, the stations near Ahua Point were provided with an additional fire command station atop the existing structure. These four stations continued in service through August 1944, manned by personnel of the range sections from the 15th CA Regiment until they were placed in caretaking status, and maintained by elements of the 53rd CA (HD) Battalion through the end of the war. In January 1933, a plan for a second 60-inch searchlight on a disappearing tower was projected for the point. The cost for a disappearing tower was nearly three times that of a rigid steel frame tower, consequently an 85-foot tower was built on the point in 1937. The searchlight shelter atop the tower measured 15-feet by 12-feet. The light was powered by a 54 h.p. General Electric gasoline powered engine and a 25-kW generator that was housed in a splinter-proof concrete powerhouse about 295-feet from the tower. The searchlight, its tower and power plant were transferred to the garrison on August 19, 1937. This light was designated SL No.10. The light's distant electric controller was located in the B" station for Battery Hasbrouck. Searchlight Numbers 9 and 10 remained in service through World War II.

The Army's aviation facility on Ford Island saw increased use during the 1930s. On January 23, 1939, the Army designated Luke Air Field a sub-post of the newly activated Hickam Air Field. However, on October 31, 1939, the Army turned the entire island over to the Navy, moving all of its Ford Island aviation activities to Hickam Field. When the Navy took over the Ford Island in 1939, the old land defense batteries were converted into a storage warehouses. On the eve of World War II, the Army established new battery sites for automatic weapons on Ford Island. During the Japanese air attack on Pearl Harbor, the protective casemates, magazines and passageways of the old batteries were used as shelters by a large number of dependent women and children of the naval personnel quartered on the island.

For several years after the First World War the submarine mine defenses of the Honolulu Harbor Defenses were retained in deployable status. At Fort Armstrong there were changes in the assignment of army mine planters, with the planter *Whistler* taken out of service in 1921 and replaced by the USAMP *General Royal T. Frank*. There never seemed to be a lot of enthusiasm for the mining projects of Pearl Harbor and Honolulu. It was assumed that any attacking fleet could directly bombard the naval and commercial facilities of both harbors directly from deep water south of the Oahu coastline The possibility that small raiding parties would try to penetrate the harbors seemed minimal. In any event, in 1922 the submarine mine project was placed into a sort of inactive status, and then in 1927 eliminated entirely from the defenses.

On September 20, 1922, Fort Armstrong was redesignated as the Army's Hawaiian Department's General Depot, except for the mine complex area around Battery Tiernon which remained under the aegis of the HD of Honolulu until February 15, 1927, when the torpedo storehouse, loading room, cable tank, dynamite house, mine dock and boathouse, and trackage pertaining to the above, were transferred to the Quartermaster Corps. Even the 36-inch searchlight No. 5 was moved and used on army tug *Cebu* by 1929. The mine planter *Frank's* mission was changed to that of serving as a target towing vessel and an Army inter-island transport. The vessel continued to use the old mine wharf at Fort Armstrong into the early days of World War II. It was torpedoed and lost in January 1942 by a Japanese submarine while underway between the islands of Hawaii and Maui. In October 1937 Fort Armstrong became the principal port of embarkation for the Army on Oahu. Battery Tiernon was retained armed as an inactive element of the Harbor Defenses of Honolulu. For a few years the battery was utilized as the saluting battery for Honolulu Harbor. During the inter-war years, some of its interior spaces were converted for use as a morgue. Although portions of the battery structure had been adapted for other uses, the two 3-inch guns, though unmanned, remained emplaced at Battery Tiernon into the 1940s.

The land-defense field works were mostly abandoned after World War I. The redoubt at Red Hill was eventually stripped of removable equipment and essentially abandoned by 1930. It was considered in late 1931 as an alternate location for the command post of the Army's Hawaiian Department. No further record of this structure exists. Overtime, various other military reservations, and installations have been carved out of the original Red Hill Military Reservation. Among these are the Makalapa Naval Reservation, the Army's Aliamanu Military Reservation. The U.S. Navy's Red Hill Underground Fuel Storage Facility was begun in December 1940 and completed in August of 1943, with the first of the tanks being placed in service in September 1942.

During the early 1920s the first efforts were made to drain the swampy northwest quarter of the Fort DeRussy reservation. A canal was built to drain the wet backwaters and in March 1923 $100,000 was appropriated for the reclamation of the remaining low marshy and generally unsanitary area of the reservation. When the draining and filling of the swamp was completed around 1928, the AA emplacement in the swamp had been pretty much buried, the gun was dismounted, and the emplacement abandoned reducing the battery to three guns.

For much of the early 1920s, the 10th Company served as the manning detachment for both the seacoast gun batteries at Fort DeRussy as well as the post's antiaircraft battery. The Waikiki area saw considerable development in the years following the World War. By the early 1920s, the encroachment of Waikiki's commercial and residential development had reduced the use of the big coastal guns and the post's antiaircraft battery for target practice. The concussion and noise of a full-service target practice with the 14-inch guns was too aggravating to the civilian populace not to mention the cost of repairing damage and replacing broken pottery. Consequently, full-service charge firing of the 14-inch guns had ceased by the mid-1920s and would not be fired again until war erupted in December 1941. While sub-caliber practices and loading drills continued, the troops at Fort DeRussy were required to travel to the more isolated Pearl Harbor defenses for their annual service practices.

The duties for the 10th Company, CAC were generally uneventful from 1920 until 1924. In addition to regular training with the armament, the company performed the usual post duties that included the care and maintenance of the 6-inch and 14-inch guns as well as the 3-inch AA guns. They also performed guard duties for the small post. The recruit center staff shared in some of these duties but concentrated most of their energies on the basic training of the recruits that arrived from the mainland.

On February 28, 1925, the five regiments of coast artillery on Oahu (15th, 16th, 41st, 55th, and 64th), were organized into the Hawaiian Separate Coast Artillery Brigade. The brigade headquarters initially occupied space at the Alexander Young Hotel pending the preparation of suitable offices and quarters for the brigade staff at Fort Shafter.

In the Harbor Defenses of Honolulu, in June 1924, Battery A, 16th CA Regiment became the sole active firing battery posted at Fort DeRussy. It provided manning detachment for both Battery Randolph and Battery Dudley during the remaining years between the two world wars. At Battery Randolph in the late 1930's structural modifications were made at the rear of the central traverse to install the collective protectors of the gas-proofing project for the protection of personnel assigned to the battery's plotting room.

Colonel Harold L. Cloke arrived in Hawaii to assume command of the 16th Coast Artillery in the late 1920s, about the same time that the regimental headquarters was moved from Fort DeRussy to Fort Ruger. Cloke worked to improve the post which had somewhat shabby appearance when he arrived. Over the next few years considerable changes were made in the appearance of the Waikiki installation. As Lieutenant Clarence M. Mendenhall noted in 1934: (9)

> Lovers of Fort DeRussy wouldn't know the old hole now. That waste of Algaroba, date-palm, coral and swampland "Mauka" (toward the mountains) is nearly metamorphosed.
>
> First, some 900 Algaroba, date-palm and opuina[sic] trees and scrub were yanked out by tractors and cleaned up by C.W.A. workers. A grove of large date-palms was left along Saratoga Road. Sixty-three large date-palms were transplanted, screening the backside of the buildings on Kalakaua Avenue.
>
> Second, Topsoil from Punchbowl was hauled in and spread by C.W.A. workers to a beautiful level grade—5,000 cubic yards in five months.
>
> Third, six lines of a semi-automatic sprinkling system were installed, covering a little more than half the area designed for the Hawaiian Separate Coast Artillery Parade Ground.
>
> Fourth, Kalia Road (main road through the post) was widened to a four-lane boulevard.
>
> Fifth, a work of art in the form of a cut-rock portal and wall on Kalakaua Avenue was built. At the same time the baseball diamond was Regraded, grassed, packed, and roads and parking spaces built.
>
> Sixth, the boxing arena and carpenter shop were razed. The boxing arena is being rebuilt with adequate drainage and parking facilities in the vicinity of the corral (removed in 1932), and the carpenter shop is taken care of by a large addition to the back of the Utilities Building (in rear of Randolph-Dudley enclosure), thereby concentrating all utilities.

Mendenhall went on to note that construction of a quarter-mile racetrack and athletic field, salvaging and rebuilding the lower warehouse and truck shed, gasoline station and old stables, into a group of three buildings that would border the Waikiki Amusement Park were projected for the future. Also planned were a new gas station, a searchlight truck shed, a storehouse and a dressing room for the athletes. A new entrance and stonewall from the John Ena-Kalia Road Junction to the north gate was also on the drawing board. Also planned were the completion of the grading and the application of topsoil to some 75 acres of the parade ground.

The coast artillery garrisons of the Honolulu harbor defenses were frequently called upon to render honors to visiting dignitaries. When Prince Kaya Tsunenori and his wife of the Japanese royal household passed through Honolulu on a world cruise in 1934, a battalion of the 16th Coast Artillery commanded by Major Carl S. Doney rendered the appropriate honors on the couple's arrival and departure.

While the 14-inch guns were not fired after the early 1920s because of their blast effects, Battery Dudley was fired on a more frequent basis, although rarely were more than a few rounds at any one practice. In May 1934, Captain Rodney C. Jones' Battery A, 16th Coast Artillery put on a performance for the tourists along the beach firing two salvos at a high-speed target being towed some 11,000 yards offshore. Both salvos hit the expensive target and it was necessary to bring in a second target to finish the practice. The target practice "season" began about the first week in March each year and Fort DeRussy was the scene of considerable activity as the troops prepared for their annual service practices. In 1934, both Batteries A and D of the 16th Coast Artillery conducted their practices at Fort DeRussy giving the tourists wintering on the sands of Waikiki an entertaining show as the AA shells burst over the waters offshore. (10)

For officers, Fort DeRussy was a pleasant post to be assigned to. Servants were the norm for officers' families and Japanese maids who performed all manner of cleaning except washing could be obtained at salaries ranging between $35 and $60 per month. In addition to the "best swimming in the world," at Waikiki beach, officers assigned to Fort DeRussy could use the practice golf course at Fort Ruger and memberships could be

Sporting competition was thought to be helpful for morale, physical fitness, and competitiveness. Posts and units fully participated in leagues and annual tournaments, like this baseball team from Fort Ruger. *NARA*.

At times the sporting playoffs even rose to inter-service levels. Here is a photo of the Oahu Army-Navy football game of 1924.
*Bennett Collection*.

obtained at both Waialae Golf Course and the Oahu Country Club. There were also monthly dances at both forts Ruger and DeRussy as well as bridge tournaments. Those officers with school age children usually sent them to the Punahou School where tuition was $150 per year. (11)

The entire year was filled with military training of all types for Honolulu elements of the Hawaiian Separate Coast Artillery Brigade. In addition to the seacoast and antiaircraft gunnery, the men of the 16th Coast Artillery also trained in infantry tactics participating in the annual Hawaiian Departmental exercises as well as periodic brigade exercises. When the troops were not fully engaged in military training of one kind or another they were keeping in shape through various athletic competitions. Each post had baseball, swimming, and football teams, and there were boxing competitions at the battery level. The seacoast artillery units competed with the infantry at Schofield Barracks, the antiaircraft regiment at Fort Shafter and the Navy and Marine

Polo was popular among the Army officer corps. Here are the participants at the Kapilolani Park field with Diamond Head looming in the back-ground. *Bennett Collection.*

With the nearness of tropical beaches, swimming was naturally popular. Here is the diving station at Fort DeRussy takes advantage of the dredged area once used to bring the gun tubes of Battery Randolph close to shore. *USAMHI.*

Corps in almost all sports events. Trophies were awarded at the regimental, brigade, and department levels and unit pride played a major role in the competitions.

A major element of both work assignments and entertainment for the units were the periodic army reviews. All the coast artillery forts had frequent inspections. Both the performance of the troops and the condition of the elaborate emplacements and their important guns were subject to regular reviews and evaluations. However, every once and a while the entire regiment would turn out for a military review. In most cases that meant a massive parade past a reviewing stand with some presiding dignitary, followed by inspection of displays of weapons and skills, overflights of squadrons of aircraft, and refreshments and celebrations. The first major review in Oahu on record is one in October 1909 for Governor Walter Frear at Schofield Barracks (where most all seemed to be held). Then in March of 1916 there was a review of the Department's field units at Schofield Barracks for the territorial governor. In 1921 the newly designated Hawaiian Division provided a major formal review at Schofield. A few massive command-wide reviews seem to stand out in the historical record. On August 14, 1928 Secretary of War Dwight F. Davis reviewed the Hawaiian Division. The mid-30s seemed to be particularly prevalent for these reviews. Three were held in 1934 and four more in 1935.

If a soldier was directly involved in the Review (and in many cases most were—the 15,000 marching for the President amounted to almost the entire army garrison at that time) weeks were spent in careful instruction on the march, logistics of maneuvers, timing, practicing, etc. Spacious Schofield Barracks was well-suited for this type of activity. Before and after his particular moment to shine the soldier would become a spectator in the stands or on the grounds to see how his compatriots did. From the written accounts of the 20s and 30s these reviews are seen as big highlights, almost in a fair-like attitude. Participation in the parades is proudly reported home in letters. Specially appointed photographers recorded the units parading by and sets of these prints sold to soldiers at the post exchange. Almost every surviving interwar photo scrapbook by the soldiers stationed on Oahu have a set of photos of "his" big review.

One of the high points in the period between the two world wars came on July 26, 1934 when the heavy cruiser USS *Houston,* escorted by sister cruiser USS *New Orleans,* entered Honolulu Harbor bearing the President of the United States Franklin D. Roosevelt, and the Army on Oahu was geared up to give him a warm welcome. Lieutenant John R. Lovell, CAC recalled: "The sky was literally black with Army and Navy aircraft; everything that would fly was in the air to take part in the welcome." (12)

The Commanding General of the Hawaiian Department, Major General Brian H. Wells, had ordered a massive review of the troops to be held on the newly turfed review field of the Hawaiian Division at Schofield Barracks. All elements of the Coast Artillery Corps on the island participated. Instead of marching as infantry in the review as in previous years the coast artillery units were to man either primary or secondary armament. Batteries A and D from Fort DeRussy and Battery C from Fort Ruger along with Batteries A, B, and C of the 15th Coast Artillery from Pearl Harbor formed an antiaircraft "regiment" for the occasion, passing in review mounted on trucks with their secondary assignment of antiaircraft machine guns. Lieutenant Lovell described the event: (13)

> Everything that could march or roll was in the gigantic formation, which extended as far as the eye could see in every direction. It actually required over five minutes to call all the elements of the command to attention. Mr. Roosevelt arrived on the field and passed in front of the troops in his car. The crowd of spectators numbering approximately 30,000 people gave him a great ovation. The president took his station in the reviewing stand and watched over 15,000 troops pass before him.

On the occasion of the arrival of Major General Hugh A. Drum to take command of the Hawaiian Department in March 1935, reviews of the various commands on the island were held. The review for the Hawaiian Separate Coast Artillery Brigade was held on the newly completed parade ground at Fort DeRussy on April 25, 1935. The brigade, some 2500 men strong (about 25% of the entire Coast Artillery Corps) was commanded by Brigadier General Robert S. Abernethy. Three coast artillery regiments participated: the 64th CA (AA) Regiment, the 15th CA (HD) Regiment from the Defenses of Pearl Harbor and the 16th CA (HD) Regiment

Big Reviews by units, post, and even the entire Hawaiian Department for new commanders or visiting dignitaries became centerpieces of ceremony and a welcomed break in mundane routines. They are well documented in soldier photographs, as here at Schofield Barracks in 1934. *Olbrych Collection.*

Commanders (Major General Herron & Brigadier General Gardner) review the Seacoast Command at Fort DeRussy on February 10, 1936. *Bennett Collection.*

The usual procedure was to carefully arrange and either march or drive each unit's motor equipment past the reviewing stands. Here the 64th CA (AA) drives its towed sound-locating apparatus and searchlights in March, 1927. *Olbrych Collection.*

Eventually even special rail spurs were erected into Schofield Barracks to allow the railway artillery and their newly-painted steam engine to be "paraded". *Olbrych Collection.*

Equipment and armament displays were usually staged with the reviews. Here 155 mm GPF guns are displayed at Kapiolani Park by the 55th CA Regiment. *Bennett Collection.*

from Honolulu. As one observer noted: "It was a fine show. The uniforms and equipment looked splendid, the marching excellent. Officers executed the saber manual and gave their commands in the best Infantry manner. It was a review to make all Coast Artillerymen proud of their Corps." (14)

In the Departmental Review held September 2, 1935, to honor a congressional delegation, the brigade again traveled to Schofield Barracks where for the first time in the history of the United States Army, the 8-inch guns of the 41st Coast Artillery Railway Battalion passed in review. That required specially laid railroad tracks for the purpose emplaced by the Engineers. Another feature of the review was the appearance of the massed Coast Artillery Band of 160 pieces. The band with its three drum majors led by Warrant Officer George W. Dahlquist, and the Drum and Bugle Corps of some fifty pieces led by Technical Sergeant Stanley H. Walker marched down the field to the strains of coast artillery marching song "Crash on Artillery" drawing a thunderous ovation from the 10,000 spectators. The band was garbed in their regulation uniforms, but wore white leggings, red and gold cross belts. The helmets worn by the bandsmen were painted a brilliant red. (15)

The congressmen had scarcely left the islands when the Secretary of War George H. Dern arrived at the end of September. He too was treated to a mammoth review of the Hawaiian Department troops and the massed coast artillery band that, once again, receive high praise from all who witnessed the parade. Secretary Dern was soon followed by the former Chief of Staff General Douglas MacArthur who was on the way to the Philippines Islands to assume his military duties with the fledgling Philippine Government. Although the general was not afforded a review due to brevity of his stay, the brigade provided a guard of honor for MacArthur on his arrival and departure.

Battery B, 64th Coast Artillery won the Knox Trophy for 1935 for having the best annual service practice record in Coast Artillery Corps. In acknowledgement of this achievement a special brigade review was held for Major General Drum, commanding the Hawaiian Department, at the Fort DeRussy parade ground on the evening of February 17, 1936. It was a spectacular event. Forty 60-inch AA searchlights, each providing 800,000,000 candle power of illumination, were arranged in a circle around the parade ground their beams focused on a point above the center of the field providing a brilliant canopy of light as the brigade passed in review. Between five and eight thousand persons attended the review while thousands more watched the spectacle from the roofs of nearby buildings and the hills behind Waikiki. The 64th CA received the Coast Artillery Cup from General Drum. In 1937 two reviews were held, the second in August for Secretary of the Treasury

Competition for, and hopefully winning the annual Coast Artillery Knox Trophy (as here in 1931 by the 55th CA) invoked much ceremonial activity. *Bennett Collection.*

Henry Morgenthau, Jr. (16)

The local Honolulu Chamber of Commerce began hosting an annual Army Day in 1935. An exhibition camp at Fort Armstrong was equipped with displays of army equipment and expertise and opened to the public. The chamber of commerce hosted regular free concerts and entertainment for all island army servicemen. Apparently, it was quite popular and even helped somewhat in community relations. (17)

As the decade of the 1930s drew closer to an end, the pace in the training schedule continued to increase. In addition to the regular artillery drill and service target practices with both the seacoast and antiaircraft armament, considerable attention was also given to infantry tactics and the local defense of battery positions. Joint exercises with the Navy were also carried out. In 1939 with the approach of the spring, Captain Donald B. Herron's Battery A, 16th Coast Artillery fired its annual antiaircraft service practices at a new firing point for 3-inch AA guns on Sand Island.  On Army Day, April 2, 1939, Battery D, 16th CA, commanded by Captain Milo G. Cary, fired two salvos from the guns of Battery Dudley. The firing was carried on an Army Day radio broadcast heard all over the mainland United States. These events were followed by the annual Hawaiian Department maneuvers that included a joint Navy-Coast Artillery exercise and a joint antiaircraft-Air Corps exercise. (18)

The garrison of Fort Ruger had settled in its new quarters and assignments by the end of 1923. In February 1924 units of the 16th Coast Artillery Regiment (Harbor Defense) were assigned to the Coast Defenses of Honolulu. At this time, only three active batteries of coast artillery manned the fixed batteries in the CD of Honolulu. The three firing batteries of the 2nd Bn, 55th CA, Batteries D, E, and F, remained assigned to Fort Ruger, with only a slight augmentation as part of the 1924 reorganization. (19)

Fort Ruger's fire control complex served additional batteries as the Honolulu defenses grew. In 1935, this station became the command post for the Honolulu Groupment command that included the Leahi, Punchbowl, and Diamond Head Gun and Mortar Groups. On the summit of the Diamond Head rim at 764-feet elevation, a supplementary station for the First Fire Command (F'''1) was established. A simple dugout station with concrete walls and pedestals, it could be equipped with portable fire control instruments. Why the Fort Ruger Fire Command had both its primary and supplementary station at the same point is unclear; perhaps the primary station viewed to the west and the supplementary station viewed to the east. (20)

After construction of Battery Granger Adams in 1934 at Kupikipikio Point, it was necessary to move the

Barrack life, particularly in the early years, was adequate but certainly not private. Here in 1917 is a view inside of one of the Fort DeRussy enlisted barracks. *USAMH.*

Diamond Head was probably the most photographed site on Oahu, almost every surviving photo album by Army soldiers from the interwar years include this view. *Olbrych Collection.*

searchlight there to the southeast slope of Diamond Head where it was eventually renumbered as Searchlight Number 16.

When the headquarters of the Harbor Defenses of Honolulu came to Fort Ruger in January 1927, Diamond Head was an obvious location for the command post (termed the "H" station). The "H" station would serve as the operations center for the harbor defense commander. A bombproof command post with rooms adjoining the Kapahulu Tunnel was created in the existing fire control switchboard room, which was lined with gunite. The old tunnel was enlarged, and its 14-foot crown was raised to 17-feet. Between the fire control switchboard room and a point 50-feet inside the tunnel's north portal, a large cavern-like lateral was excavated on the tunnel's west side. This 31 by 156-foot area, encompassing an additional 3,912 square feet, was lined with gunite. It was filled with eight rooms partitioned with concrete-block walls. The "H" station was completed in 1934. (21)

The additions to Fort Ruger were not limited to the fortifications. The transfer of the headquarters of the 16th CA in 1927 to Fort Ruger and a new regimental commander hastened the improvement of the post's appearance. Monsarrat Avenue was dressed up – repaved, curbing and sidewalks added, royal palms were planted along both sides of the avenue, and additional plantings were carried out. Cloke's successor continued his predecessor's work. By 1934, the parade ground had been "turfed, and what was once an ugly pile of stones, grass and debris" near the south end of the parade had been converted into a "beautiful garden." The old theater was being converted into a gymnasium and replaced by new theater across the street from the band barracks. (22)

As early as the 1920s, residential expansion around Fort Ruger curtailed target practices at Diamond Head. By the mid-1930s, target practice with the mortars was halted, as well as firing practice for the 3-inch AA guns atop the crater rim, at the north end of the garrison area, and on Black Point. Consequently, much of the target practice by Battery C, 16th CA, in the 1930s was done in the Pearl Harbor defenses, using the 12-inch mortars of the 41st CA at Fort Kamehameha and the fixed AA guns at Fort Weaver. With the completion of Battery Granger Adams, Battery Harlow joined Battery Birkhimer in caretaking status. The mobile 2nd Bn, 55th CA, also fired its annual service practices at Fort Kamehameha and conducted some target practices at Fort DeRussy and elsewhere along Oahu's shoreline. In addition to its 155 mm GPF guns, the 55th fired 240 mm howitzers in the "wilds of Oahu." In addition to their primary assignments, coast artillery units at Fort Ruger also had supplemental assignments, usually manning antiaircraft matériel.

By the mid-1930s, the commanding general of the HSCAB considered the 12-inch M1890MI mortars at Diamond Head obsolete. With the completion of the 8-inch battery at Black Point in 1934, the mortars became an alternate assignment for Battery C. In July 1935, Battery Harlow was recommended for downgrading to class "C" - out of service in caretaking status. When the decision was made in 1937 to gasproof the plotting rooms and latrines of the gun and mortar batteries on Oahu, the gas proofing was done at both Batteries Harlow and even the decommissioned Battery Birkhimer. At Harlow, collective protector units were installed for the remodeled plotting and operating rooms, and the latrines in the right-hand traverse. The battery was returned to service troops on September 27, 1937. Battery Birkhimer served as reserve magazine and was again remodeled in conjunction with the gas proofing. The original powder magazines in the old central traverse were converted into a first aid room and bunkroom. Where they opened onto the rear corridor, an airlock was provided for the altered spaces. Across the corridor at the rear of the battery, the plotting room was reduced in size and an airlock made it possible to gasproof the plotting room and latrines. After this work was completed, the battery was transferred back to the 16th CA on September 27, 1937.

An augmentation of the Coast Artillery Corps in 1935 brought significant reinforcement to the garrison in Hawaii, increasing the enlisted strength of the brigade from 2,958 to 4,182 men. While most of the increase went to the 15th CA at Pearl Harbor, additional men were assigned to the HQ Battery, 16th CA, and the HQ Detachment and Combat Train, 2nd Bn, 55th CA, improving their ability to operate the harbor defenses' motor transport. There were enough additional enlisted men to provide antiaircraft range sections for Batteries A and C, 16th CA, enabling these batteries to change from their primary seacoast mission to their secondary

AA mission more efficiently. There was also a small increase in manpower to help staff the searchlights of the harbor defenses. (23)

The 16th CA was further organizationally expanded. An eighth battery, G, was authorized in the 1930s, but

Visiting a pineapple plantation and sampling the fruit or even sending some home was a popular soldier activity. *Olbrych Collection.*

Posing soldiers from Fort DeRussy sitting on a lava flow of the Kilauea Volcano. *Bennett Collection.*

Major posts offered regular bus service to servicemen to points on other posts and major Honolulu stops. This is the Fort Kamehameha bus in the mid-1930s. *Olbrych Collection.*

Obligatory sightseeing destinations in Honolulu included the Iolani Palace built in 1882 by King Kalakaua. *Olbrych Collection.*

like Batteries B, D, E and F, it too remained inactive. Despite the recent increases, regimental strength in early 1940 was still only 14 officers and 336 enlisted men. Through 1940 and 1941, there was still a great shortage of commissioned and noncommissioned officers in the rapidly expanding army. It was not uncommon for the batteries of the 16th CA to have only one officer assigned, frequently a lieutenant rather than a captain, the usual rank for a battery commander. After war broke out in Europe in 1939, the number of enlisted men increased more rapidly. As a result, the strength of the active harbor defense and mobile tractor-drawn batteries gradually increased to the levels called for in their peacetime tables of organization. By January 1941, the firing batteries of the 16th CA averaged about 100 officers and men, while the regimental HQ Battery numbered about 170 officers and men. By mid-1941, the batteries of the 16th and 55th CA had reached, and in some cases nearly doubled, their authorized peacetime enlisted strength.

Battery C of the 16th CA posted at Fort Ruger was the primary unit assigned to Granger Adams. In February 1941 measures were taken to camouflage the installations in the Hawaiian defenses. The cost of these efforts for Granger Adams was estimated at $2000. On November 27, 1941, in response to alerts from the Hawaiian Department, Battery C of the 16th CA had posted anti-sabotage guards at Black Point to safeguard Battery Granger Adams and the military property and materiel there. Battery C had begun "digging in" on Black Point even before the outbreak of war, blasting holes in the lava rock to construct machine gun positions, slit trenches, and a number of ready-service ammunition pits and personnel shelters near the AA emplacements. The pit interiors were provided with wooden linings and heavy timber roofs topped with layers of sandbags.

During the interwar years some steps to modernization were taken in the late 1930s at Fort Weaver. In addition to the service magazines at the post, five additional reserve magazines intended to hold ammunition for the 16-inch guns at Fort Weaver had been built. For security reasons they were placed at the Fort Shafter reservation outside Honolulu. Of reinforced concrete, though not earth covered, they were transferred on September 6, 1923. Reconsidering the wisdom of placing the reserve so distant for the guns they were to supply; seven new magazines were subsequently built at Fort Weaver to the west and east of Battery Williston along with extensions to the ammunition railway tracks. They were of reinforced concrete and measured 71.5 by 41-feet, but not buried or earth protected. The set was transferred on January 14, 1934, for $101,800. After transferring their ammunition, the Shafter magazines were used for general ordnance storage.

Work was undertaken to gas proof the operating spaces for Battery Williston while also increasing the size and the amount of protection to the battery's plotting and fire control switchboard facilities. The modifications of the plotting room were undertaken in 1937. An addition to the existing plotting room structure was constructed at the open rear of the building. Four rooms and a corridor running parallel were added to the outside of the structure. The new entry passage opened into an airlock and protective collector room equipped by the Chemical Warfare Service (CWS). A central corridor separated the old and new parts of the building. The first room was an officers' bunkroom, then the first aid station. At the rear of the first aid station was an anteroom that adjoined a second plotting room. The enlarged and gas proofed plotting and switchboard facility was provided with additional bombproofing. A new burster course a foot thick topped with an additional six-feet of earthen cover was laid over the original roof, another six or seven-feet of earth topped the structure and the entire hill, now about 30-feet high, was planted in algaroba trees. With the completion of these improvements on November 30, 1937, the engineers transferred the complex back to the custody of the 15th CA with a newly figured construction cost of $108,519.65. (24)

The proof firing of the two new 16-inch rifles of Battery Hatch was scheduled and carried out on April 24, 1936. Each gun was fired four times at different elevations and with powder charges varying from 75% of normal to 15% above normal. These tests were conducted by Battery C, 15th CA and various ordnance personnel and engineers who observed the test. No problems with either the gun or the emplacements were encountered and further tests to determine the accuracy of the guns were scheduled for later in the year. (25)

When the battery was completed in 1935, the powerful 16-inch guns were assigned to Battery C of the 15th CA (HD) Regiment. Battery C only staffed the post during formal exercises in the years prior to World War II, although a small combination guard and caretaking detachment from the battery occupied the small barracks atop Puu Kapolei. Battery C held a number of target practices with the 16-inch gun at Fort Barrette, but more frequently they practiced using the 155 mm gun as a substitute for the larger caliber weapon.

The barracks building built in 1934 was used to house the caretakers. In 1936 WPA funds were used to raze the vintage wooden Marine Corps barracks at Pearl Harbor and the salvaged materials were used to construct three temporary board and batten huts with a capacity of 16 men each at Fort Barrette. These temporary structures had tongue and grove flooring and five-foot high walls surmounted by a continuous screened window on each side that could be closed with shutters. These were used to quarter troops when at the fort for exercises or target practices. The original kitchen/mess hall building former graced with a gable roof and chimney was initially occupied by the caretaking detachment during the 1930s, but after the commencement of hostilities became the post headquarters. (26)

There are some interesting surviving accounts of what a typical workday was for a general enlisted man at one of posts in the 1920s. This one about life at Fort Kamehameha is from an account now at the US Army Museum of Hawaii was published in the *Coast Defense Journal:* (27)

> Typical Day on Post Our day began with reveille at 0600, the bugler at the guardhouse playing "Can't Get Em Up." By 0615 all the troops were formed in front of their barracks, roll was called, the gun was fired, and the colors were raised. The officer of the day (the "O.D."), the only officer around, received reports from the battery first sergeants. The next half-hour was calisthenics until "Chow Call" sounded at 0700. You could always tell what day of the week it was by what was in the mess hall – "S.O.S," "gas check pads" (hotcakes), "slum," "buckshot," "goldfish," "coldcock," etc. "Drill Call" sounded at 0800 and everyone marched off to the parade ground for infantry drill. The men were armed with .30 cal. M1903 Springfield rifles, cartridge belts, campaign hats, and O.D. woolen shirts. First sergeants and cooks were armed with pistols. Officers carried drawn sabers during drill and pistols during field exercises.
>
> When "Recall" was blown at 0930 the troops returned to barracks to change into fatigues. The fatigue utility uniform was the "most hated uniform in the army." Nothing matched. The materials were all different and appeared to have been made from "mill ends at cotton mills. You could draw a blue denim floppy hat, a disgusting colored blouse, which had two side entry pockets at the chest, and a baggy pair of cotton pants that would have made a Georgia slave run away!" Malone recalled, "One corporal in my battery traded and purchased around until he had a couple of full suits of blue denim fatigues. But he was a perfectionist." The men hated the floppy fatigue hats and would wear campaign hats with their fatigues. The officers were understanding and overlooked this infraction of the dress code. If you saw "motley colored clowns coming towards you in Fort Kam, it was an artillery battery heading for their guns." Artillery drill was held from 1030 to 1130; the plotters received hypothetical courses by telephone and calculated elevation and deflection; the gun crews would ram dummy shells into the breech and practice sighting and aiming. Recall sounded at 1130, with "all the mules tearing off for the corral. Troops march back to barracks to wash up for 1200 chow." Malone detested the frankfurters and sauerkraut served on Wednesdays, so once a week he spent 35 cents on ham and eggs at the PX.
>
> "Fatigue Call" sounded at 1300 and everyone but the noncoms and "trumpeters" (buglers) headed either to the balloon hanger to haul barbed wire or to the dock to unload barges of artillery shells brought out by the USAT Meigs. The NCO swimming pool, where trumpeters practiced, mostly while floating in the water, had been blasted out of the coral reef. The next formation was at 1700 fronting the barracks, when the band played "Retreat." They mostly played field music, and on occasion, the national anthem. The gun was fired, the colors were lowered, and the bugler played "Chow Call." The troops were free to attend a movie or do whatever they wanted after chow until "Tattoo" at 2145, then all lights were extinguished except in the sergeants' rooms, day room, barbershop, and library. "Call to Quarters," "the most beautiful bugle call in the Army," sounded at 2245. After "Taps" at 2300 hours

Garrison posts, particularly those in foreign or territorial regions, often have problems with soldier morale. If not engaged in active military operations, troops just being present in case of need, or manning fixed defenses against some theoretical threat can get bored with the tedium. The service tried to equip its posts in overseas territories with a full range of recreational facilities. The army posts in Oahu, including the three major coast artillery forts were no exception. There were movie theaters that showed films nightly. Chapels and religious meetings were provided for all faiths. Service clubs, with food and drink service were open for segregated officer, NCO, and enlisted men. A YMCA provided athletic and recreational facilities for enlisted men, usually including card rooms, pool tables, library and reading rooms.

During the war many of the tourist hotels were contracted by the military as rest and recreation venues. *Olbrych Collection.*

The most popular destination in Honolulu for much of the time of active coast artillery was Fort Street—location of restaurants, bars, and clubs catering to the needs of young servicemen. *Williford Collection.*

Each of the posts with major garrison quarters had their service clubs. This is the 1920s Officer's Club at Fort Kamehameha. *USAMH*.

There were even recreational camps further afield in the islands. Some posts had reserved or private beach areas specially for servicemen. For both officers and men from 1916 there was Kilauea Camp, on the Big Island with cabins available for reasonable rent at an altitude avoiding the tropical heat and humidity. On the slopes of Mauna Loa, it was the closest habitable camp to the spectacular volcano. It was run by a small detachment of four officers and 44 enlisted men all on detached service from other units. In the 1930s each soldier could take part of their annual two-week vacation here if desired. Special travel rates by boat were made available for those wishing to avail themselves. Of course, on Oahu officers (social status, rank, and income were more closely class related in these early decades of the 20th century) could participate in the well-developed Honolulu social scene. Dances, luaus, charity events and membership in social and country clubs were open to the officer class. About the only activity missing in the islands was an extensive hunting opportunity.

The military purposely promoted organized sporting activities on its posts and for its personnel from at least the mid-19th century. It was good for physical conditioning of all ranks, and it encouraged team building and competitiveness. It was also considered a much healthier recreational activity than idle time, and to prevent too much alcohol drinking or worse amusements. All the fort posts with large cantonments (primarily Fort Kamehameha, DeRussy, Ruger, Schofield Barracks, and Fort Shafter) had army-supplied sports facilities including tennis, handball, and basketball courts, baseball field with viewing grandstands or stadiums, running tracks, boxing rings. Depending on the date and location there were also bowling alleys, swimming pools, bath houses. Unit teams were formed, and annual competitions held between them with elaborate trophies and the associated bragging rights. Sporting victories were a big deal during the interwar years. Tournaments at time seemingly became as important as military readiness. Schofield had some of the finest boxers anywhere and bouts could attract many thousand spectators. In 1935 the army sent a Hawaiian boxing team on an exhibition trip to Manila and other spots in the Orient.

Swimming and beach sports were a natural fit for the coastal forts like DeRussy and Kamehameha. Surfing had been a popular activity for the native locals long before the arrival of American settlers and continued to be practiced by soldiers. At Fort DeRussy they found that the channel dredged in the coral rock to allow barges to bring in the 14-inch gun tubes left a wonderful deep hole for swimming and diving.

Officers were segregated for activities, though they were often the professional "coaches" for enlisted men teams. Officers (and dependent wives) had access to both military and community country club golf courses and tennis courts. Horseback activities, both trail riding and polo teams were also favored by the higher ranks. Being a coach or sponsor of a successful team could be of value to a professional career. Over time more and more facilities became open for the enlisted men. For example, in 1917 the golf course at Schofield Barracks opened for enlisted men participation. (28)

The June 1930 *Coast Artillery Journal* had an instructive article about what to expect if you were transferred to Oahu. It described the different quarters, facilities, and ambiance at the various posts: (29)

Fort Ruger is located six miles from the center of Honolulu, just under Diamond Head, at an elevation of about three hundred feet. It is the coolest post on the island of Oahu and is a very desirable and sought-after station for both officers and men. Quarters: Officers' quarters at Fort Ruger-eighteen. (Including one set for four bachelors). Four sets are now being used for field officers. There are one single and three double sets of concrete quarters, the balance are of frame construction, and all are two-story. All are in excellent condition. N. C. O. quarters at Fort Ruger: sixteen. All double sets. Three sets are concrete, balance are of frame bungalow type. All are occupied by enlisted men of first three grades. A practice golf course is being laid out at Ruger. In addition, Ruger is only ten minutes' ride from the Waialae Golf Club where the annual Hawaiian Open Championship is held. Green fees in this club are two dollars per week and two dollars and fifty cents on Saturdays, Sundays and holidays. The Oahu Country Club is open to service membership. Dues are nine dollars and fifty cents per month.

Fort DeRussy is on Waikiki Beach, three miles from town. Officers' quarters at Fort DeRussy: ten. Five sets are now being used for field officers, three of which are assigned to Headquarters Hawaiian Separate C. A. Brigade-Commanding General and two of his Staff. All are of frame bungalow type and single sets and very desirable quarters. N. C. O. quarters at Fort DeRussy: nine. All are single sets, and of bungalow type. Eight are occupied by enlisted men of first three grades and one by a warrant officer assigned to duty with H. S. C. A. B. There are excellent post exchanges at both Ruger and DeRussy. Servants: Japanese maids may be obtained at wages of thirty-five dollars to sixty dollars per month, depending on the degree of training. As a rule these maids do all work except washing. Automobiles. An automobile is almost a necessity. A bus line, connecting with a streetcar, runs through the post, but distances in Honolulu are so great that the officer and his family are greatly handicapped without a car. There is a garage for each set of quarters. When you arrive, turn your car over to a service station agent who will be at the dock. Recreation. Fort DeRussy has the best swimming in the world The Coast Artillery posts near Honolulu usually have one dance every month and a bridge party about once every two months. Uniforms. There is a standard cloth in use in the Hawaiian Department that differs somewhat from that in general use throughout the Army. If you are ordered to Hawaii and there is sufficient time before you sail it might be well to have a friend send you a supply of this cloth if you have your uniforms made on the mainland. A mess jacket is necessary. A wool uniform will be very comfortable during the winter months, especially after your first year. Civilian clothes. Worn almost entirely when off the post, linen suits in summer and white trousers with odd coats in winter. School bus. The posts operate school busses which take children back and forth to the different schools in the city, most officers send their children to Punahou School. The tuition here is one hundred and fifty dollars a year. The public schools are crowded with mixed races. A number of officers here are on a commutation status. They rent houses near Ruger and DeRussy at reasonable rates, less, generally, than commutation allowance. There are excellent tennis courts and athletic fields at both Ruger and DeRussy.

Fort Kamehameha is situated on the Island of Oahu at the entrance of Pearl Harbor, an important naval base, about nine miles from the city of Honolulu. It can be reached via automobile only. The road is paved, and the journey is made in about twenty minutes from the center of Honolulu. Quarters: The officers and non-commissioned officers' quarters and the barracks for enlisted men are, of the bungalow type. There are four field officers' sets. thirty-three company officers' sets and a bachelor officer's set. The field officers' sets contain four bedrooms, 2 baths, a large lanai, living room, dining room, kitchen and servant quarters with bath. The company offices' sets contain three bedrooms, one bath, living room, dining room, lanai, kitchen and servant quarters with bath. It is unusual for officers to be on commutation from this post. In the bachelor officer's set a small mess is operated by a caterer. This provides excellent service at reasonable expense to the individual. There are twenty-five single sets and one bachelor set for the noncommissioned officers of the first three grades. There is no noncommissioned officers' mess operated at the present time. Troops. There are stationed at Fort Kamehameha the following troops: 15th Coast Artillery, 41st Coast Artillery, 55th Coast Artillery (less 2nd Battalion), Detachments of Medical, Ordnance, and Quartermaster, and Signal Corps

Fort Weaver: There are no officer quarters at Fort Weaver. An outpost guard only is maintained for safety of government property. The Officers' Club maintains a delightful bathing beach and picnic grounds at this post, which is accessible by a short boat trip of about five minutes from Fort Kamehameha. Sports. Baseball, basketball, volleyball, boxing, track and field, bowling, tennis, swimming, football. Library. There is an excellent library room in conjunction with the enlisted men's Service Club. Post Exchange. The Post Exchange on this post has a complete line of staple articles. In connection therewith is an excellent restaurant, gasoline filling station, barber shop and tailor shop. Splendid motion pictures are furnished daily. Schools. During the school season busses are run each school daily, transporting the children of the post to and from the schools in Honolulu.

Fort Shafter. Fort Shafter is about three miles from the center of Honolulu and on the main road leading to Schofield and Kamehameha. It was formerly an infantry post. It is now occupied by the 64th C.A., Tripler General Hospital, and in addition, is the headquarters of the Hawaiian Department. Officers' quarters are available on post. Headquarters Area (Staff post): Ten large, two-story, wooden sets, of the type found at most Coast Artillery posts. These quarters are occupied by the Commanding General and such of his staff as he determines upon, they being assigned through Post Headquarters on information received from Department Headquarters. Four small, two-story wooden sets, which at one time were noncommissioned officers' quarters, but are now occupied in accordance with assignment orders as above. These quarters are very comfortable. One large, two-story set, assigned permanently to the Commanding Officer, Tripler

General Hospital. Regimental Area (64th): Five field officers' sets. These are not of a standard pattern. In general, they are one-story bungalow type, with three bedrooms, besides maid's room. All quarters in this area were at one time the quarters of civilians employed in building the Headquarters Post but have since been remodeled and enlarged so as to make them presentable and fairly comfortable. Fourteen sets of battery officers' quarters. These are of the same type as the field officers' sets, but with the exception of two they have only two bedrooms, exclusive of maid's room. As in the ease of the field officers' quarters mentioned above, these are fairly

The set-up for the Christmas Dinner at a post in December, 1934. *Olbrych Collection.*

Much was made of holiday dinners for the troops, particularly with most all of them being thousands of miles from home.
*Olbrych Collection.*

presentable, and comfortable, their main disadvantage being the insufficiency of bedroom space and the fact that they are temporary quarters. Noncommissioned officers' quarters. There are thirty-four sets of noncommissioned officers' quarters in the regimental area, Fort Shafter. These are assigned to enlisted men of the first three grades only. They are occupied almost exclusively by noncommissioned officers of the 64th Coast Artillery, the exception being four new sets occupied by key men of the 9th Signal Service Company. With the exception of these the noncommissioned officers' quarters at Fort Shafter are very unsatisfactory as to condition. For those officers, warrant officers and noncommissioned officers, who are required to live on rental allowance, but who are on duty at the post, thus being required to commute back and forth by automobile, there is an utterly inadequate number of garages. Applicants are placed on a priority list maintained at Post Headquarters, but in some instances, garages have not been available until a year after request. Miscellaneous. Commissary-none at post, daily service from General Sales Store, Honolulu, T. H. An excellent Post Exchange is conducted as well as a branch fruit and vegetable store. Medical service. Small dispensary for first aid treatment. Post surgeon. Recreation. Nine-hole golf course. Three tennis courts. Fresh water swimming pool. Riding horses, with access to good bridle paths in the cane fields. Schools. School bus service operated between post and schools of Honolulu.

Ordnance stores in remote overseas garrisons offered a special challenge. Overseas departments like Hawaii, the Philippines and Panama needed to be self-sufficient, with ordnance supplies capable of lasting through any attack or a subsequent campaign or siege. They needed enough equipment and ammunition to arm local forces (militias, national guard, reserves, and in the case of the Philippines, another nation's armed forces) created at the start of a conflict. Collectively these stores were known as War Reserve Materiel. Substantial reserve quantities of munitions were authorized and planned for the Hawaiian Department. Funding and facilities were also needed for the proper care and storage of these munitions, and their protection from sabotage. The first Hawaiian storage facility was simply sets of wooden warehouses located in the Ordnance Department's compound at Fort Shafter. Arrival of a large stock of excess war supplies following the First World War prompted plans to construct a dedicated storage facility.

After years of discussion, the Army finally moved to acquire a site and protected facility for a departmental ammunition magazine. In late 1928 land was appraised in the Salt Lake district south of Red Hill for the new Aliamanu Crater Military Reservation. Following the Congressional appropriation, that land was purchased for $73,000 from the Damon Estate. The first construction work was done in 1932-33. A tunnel was driven into the north rim of the crater through to the internal crater floor—of about 675-feet length. It was for vehicular traffic and 11-feet wide. At the junction of this tunnel to the interior of the crater were bored three parallel galleries. When built in 1932 provision was made for eventual conversion of this complex into the emergency conversion to a command post, they had entries from both ends and a latrine room was built into one of the galleries. Inside the crater a roughly circular perimeter road was built on the crater floor just inside the crater walls. Into these walls eight "blocks" or sections of tunnels were excavated. Each section had three to five individual munitions storage tunnels. The tunnels were of a uniform 10.5-feet width, and 117-feet long. The ceiling was arched and either nine-feet 11-inches or eight-feet five-inches in height. Walls and ceiling were sealed with sprayed gunite. They were built by private contract and had water lines and commercial electrical connections. Lessons learned from these first sets of magazines were carried into the second set. While some of these first tunnels had rear exits and were connected laterally to their sister units in the set, that was not reproduced in subsequent units. Also, it was found unnecessary to include vertical ventilation. The floors were sloped so that any infiltrating water would flow out through floor gutters.

Building Aliamanu Ordnance Facility Magazine No. 5 in 1934. *NARA.*

In late 1935 General Hugh Drum requested an additional $1,580,000 from the congressional appropriations committee to complete the required number of magazines. By mid-1937 some 166 numbered galleries were reported completed, all of similar width but with varying lengths of from 52 to 156-feet. On the east rim of the Aliamanu Crater, a large ammunition workshop and a guardhouse were built where the access road from the Kamehameha Highway entered the reservation. Two free-standing wooden, 12 by 12-foot black powder magazines were constructed. There were two gatehouses, warehouses, carpenter shop, garage, and gasoline station and even a large separate two-story 136 by 159-foot barracks built in 1937 for $43,556.17. Water was supplied by main from commercial systems, but also a 15,000-gallon reservoir and 8500-gallon tank provided a reserve. The storage complex was also served by a spur branch of the Honolulu Plantation Railway. (30)

The ammunition tunnels housed the reserve stores of munitions for the field artillery units at Schofield Barracks as well as the war allowances of powder charges and projectiles for the coast artillery gun and mortar batteries. The Hawaiian Ordnance Depot maintained the fixed ammunition for the 64th Coast Artillery (AA)

Regiment posted at Fort Shafter. A major portion of the war reserve ordnance matériel was also stored at the Aliamanu Crater. In many ways this magazine represented the same functionality as Malinta Tunnel did for the Philippine Department.

Salt Lake Hill (north rim of Aliamanu Crater) was also selected over time for a variety of fire control stations. Initially it had the secondary fire control station (B") for Battery Harlow's 12-inch mortars at Fort Ruger. In succeeding years additional fire control stations were built into the crest of the crater's north rim at elevations varying between 475 and 480-feet for the long-ranged Batteries Closson, Williston, and Hatch. In 1937, a fifth station to serve as the observation post for the Hawaiian Department's forward echelon command post was dug into the crest of Salt Lake Hill at an elevation of about 477-feet above sea level. This station was directly above the department's ordnance storage tunnels designated for use as a command post.

As early as 1931, the Hawaiian Department wanted to establish a protected command post and had briefly considered the adaptation of the old infantry redoubt on Red Hill. In 1936 the development of a command center at Aliamanu Crater began. Construction started with a vehicular tunnel bored through the crater's north rim in 1937, then a bombproof operations center for the department staff was constructed. At 40-feet inside the crater floor portal of the tunnel was a kitchen and a refrigerated storage room with its own exit to the south slope of the crater's north rim. Forty feet further into the tunnel was the facility's emergency power generator room and the complex's airlock and gas proofing collective protectors as well as a first aid station. Through this area another tunnel branched off at right angles to the vehicular access tunnel. This lateral served as a six-foot wide corridor for a length of 70-feet leading to the operating spaces of the center itself.

The operations center of the departmental command post consisted of a 40-foot-wide tunnel 100-feet long that had three laterals, each of which were 110-feet long and 20-feet wide that branched off the first tunnel to the right at an angle of about 70°. The three laterals were each connected by a ten-foot-wide communication gallery. A shorter tunnel branched off the main tunnel to the left at about its midpoint. This branch contained the installation's latrines. The main tunnel and its laterals were designed to provide quarters, messing and workspace for the departmental staff during wartime.

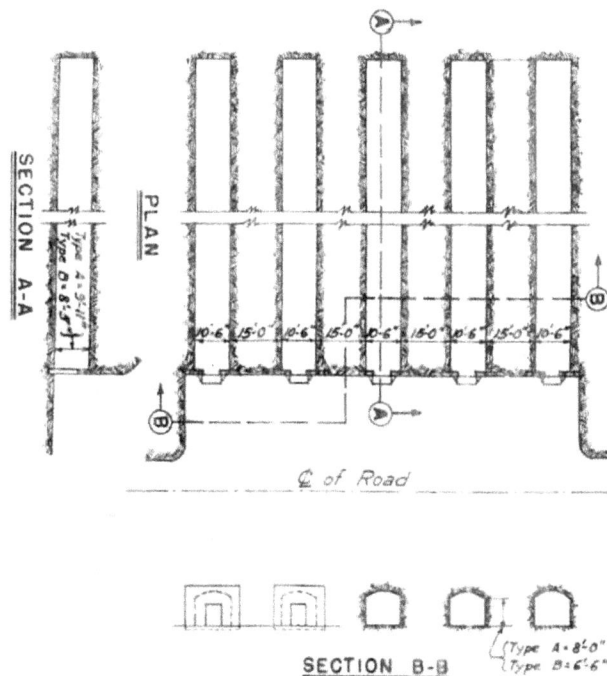

Plan of a typical munitions tunnel section of the Aliamanu Ordnance Facility 1930s. *NARA.*

# CHAPTER 8
# THE JAPANESE ATTACK ON PEARL HARBOR

On November 27, 1941, the War Department issued its famous "war warning" message. The primary consequence of this warning was the implementation of Alert Status No.1 for the Hawaiian coast artillery command. This initiated efforts to prevent acts of sabotage and deal with uprisings, with no additional provisions for external attacks. Locally, a subsequent anti-sabotage alert bolstered guard detachments throughout Oahu. With this alert some units were deployed to their readiness assignments. Battery F of the 55th CA was dispatched to Sand Island and Battery E reinforced its guard detail at Camp Ulupau. Battery D, 16th CA, was deployed to man the Ulupau Searchlight Group's 12 seacoast searchlights along the northeast coast of Oahu from Makapuu Point to Punaluu, just north of Kahana Bay. The specific missions under this alert for the Hawaiian Coast Artillery Command were: (1)

1)    To protect all seacoast and antiaircraft armament, searchlights, observation and fire control installations, and other elements of the seacoast and antiaircraft defenses.
2)    To protect all vital installations on posts and reservations of the command.
3)    To protect the radio beacon on Sand Island
4)    To provide a guard for the rear echelon of department headquarters and Tripler general hospital.

The Japanese attack on the navy base at Pearl Harbor has been written about in great detail over the 80 years since it occurred. This text will not revisit the causes and events of that attack, but rather highlight the role of the Harbor Defenses during and following the attack of that day. Suffice it to say that a Japanese task force of six aircraft carriers (HIJMS *Akagi, Hiryu, Kaga, Shokaku, Soryu* and *Zuikaku*) launched two waves of a devastating air attack on the military facilities on Oahu. Their primary target was the ships of U.S. Pacific Fleet, but to achieve air superiority they also attacked the Army Air Force airfields at Bellows, Hickam, Wheel-

Lieutenant General Walter Short, commander of the Army's Hawaiian Department at the time of the Pearl Harbor attack. *USAMH.*

Major General Henry T. Burgin, commander since August 1, 1941 of the Hawaiian Coast Artillery Command. While it was renamed a number of times, Burgin essentially retained this position for the rest of the war. *Gaines Collection.*

One of the Japanese A6M2 "Zero" attack planes for the aircraft carrier Akagi taking off for the Pearl Harbor attack operation on December 7, 1941. *NARA.*

er Fields and Navy air facilities at Ewa and Kaneohe Bay. Other army bases were not targeted, though the antiaircraft facilities at bases adjacent to those listed above certainly became involved in opposing the attack.

American radar, developed in the late 1930s, was still an experimental and top-secret technology in 1941. The secretary of war authorized construction of installations in the Hawaiian Islands on December 13, 1939. The Department Headquarters quickly formed an Aircraft Warning Service (AWS) Board and work began in January 1940, to establish potential sites for permanent radar installations. But it was not until late May 1941 that a priority list of radar equipment and a delivery schedule was established. On May 28, 1941, the War Department authorized the acquisition and development of sites for SCR-270A and SCR-271 air-warning radar equipment that were to begin arriving on Oahu in August 1941. Among the 12 radar sets allotted to the Hawaiian Islands, three were slated for installation within Oahu's North Sector, and two more in the South Sector of Windward Oahu. The remaining sets were to be installed on the south shore of Oahu and on other islands in the Hawaiian chain. The SCR-270A and SCR-271 radars were capable of detecting an aircraft at ranges up to 150-miles away under ideal conditions. (2)

One of the stations on the northeastern coast was to be located at the Kaaawa Military Reservation near Kahana Bay several hundred feet from the former site of an old howitzer emplacement. About 100-feet southwest of Cable Hut "L" a mobile SCR-270A radar was to be set up to cover the northern approaches to Oahu. Although termed "mobile," these SCR-270A sets could better be termed "transportable" in light of the time required to set one of them up. The SCR-270A radars were the first to be installed, because the installation of the fixed location SCR-271 radars required regular barracks, and other buildings, roads, and all the usual support facilities.

The mobile SCR-270A units employed on Oahu consisted of a K-30 operating van, which contained the electronics equipment including the linear oscilloscope or "A-Scope" serving as the radar screen. The power supply for the radar was provided by a K-31 power generator truck. A K-32 prime mover towed the antenna trailer that supported a huge vertical array antenna and its supporting tower and rotating mechanism. Additional parts and equipment of the unit were carried on a five-ton stake body truck. Two of the stations in the North Sector were to have SCR-271 fixed-installation radars. One of these was to be permanently emplaced at the Opana Triangulation Station on the heights above Kawela Bay south of Kahuku Point; the other set was to be west of Schofield Barracks on Oahu's highest peak, Mount Kaala, in the Waianae Range. An SCR-270A radar was also to be set up at Kawailoa, near the North Shore Groupment command post. (3)

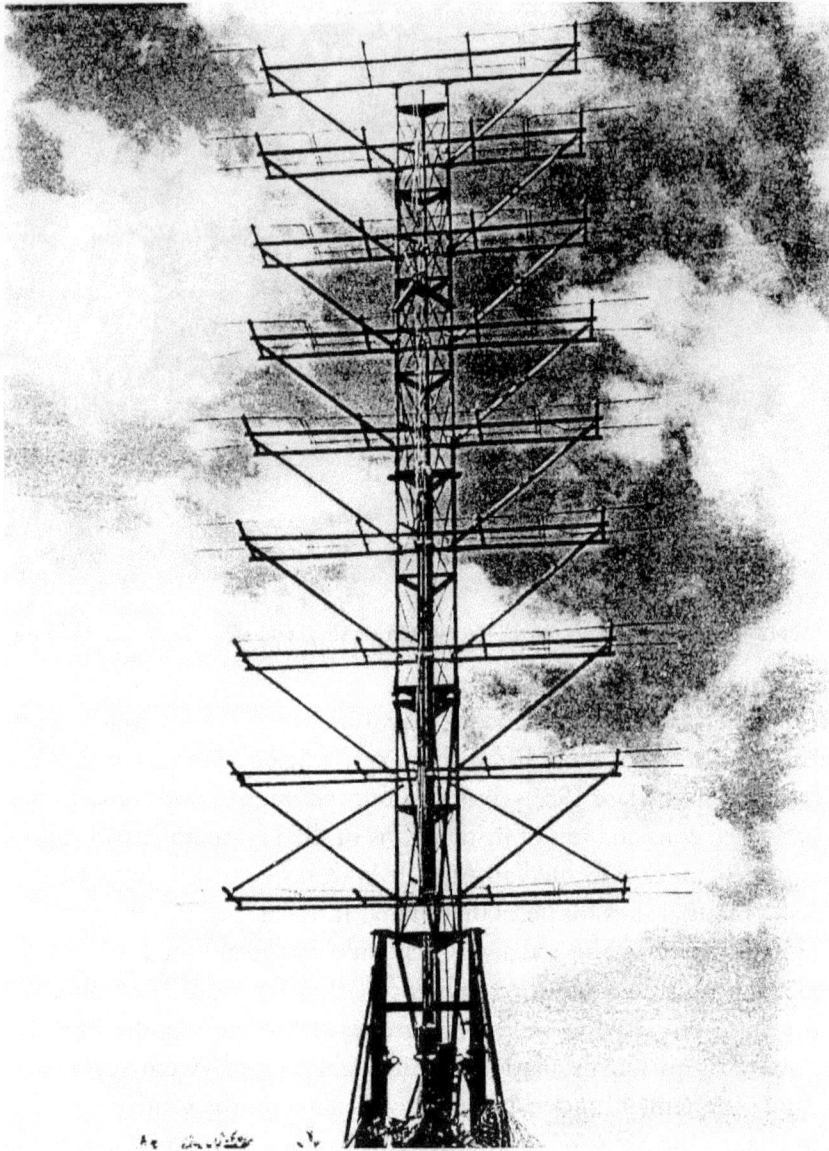

The infamous tower for SCR-270 Air Warning Radar at the northern Oahu station at Opana. *Bennett Collection.*

The set assigned for Kawailoa arrived in Honolulu early in August 1941, and on August 15, its emplacement was approved by Headquarters. This set was installed and operating by September 1941. A second SCR-270A set, originally slated for installation at Fire Control Station "B," atop Puu Pailailai at the south end of the Waianae Range, was retained at Schofield Barracks for training until November 27, 1941. It was then moved to Opana on the North Shore, where the permanent SCR-271 radar installation was still under construction. The mobile radar was to occupy Opana until the permanent station was operational. From the time of their installation until November 27, the radars were operated on training basis about eight hours a day, Monday through Friday, and for about four hours on Saturdays. On November 27, however, the Hawaiian Department was placed on alert and directed to operate from four o'clock to seven o'clock every morning, seven days a week. (4)

On Sunday morning December 7, 1941, the U.S. Army forces on the north and northeast shores of Oahu were few in number and widely scattered: Platoon-sized guard detachments at Bellows Field, Camp Ulupau, and the Kawailoa area, and the half-dozen men at the Kaaawa radar station constituted the bulk of the army personnel on all of Windward Oahu. At about 6:44 a.m., just before its operators secured the mobile SCR-270A radar set at Kaaawa, two aircraft were detected flying north of the island. These two aircraft were Zero type floatplanes that had been launched from the Japanese heavy cruisers *Chikuma* and *Tone* at 5:30 a.m., to carry out a reconnaissance flight over the naval anchorage at Lahaina Roads and of Pearl Harbor itself. *Chikuma*'s plane flew over Pearl Harbor about 7:30 a.m. and noted the abundance of naval targets below and reported the current mctcorological conditions. At around 7 a.m. the SCR-270A installed at Opana detected planes approaching from the north. Although the incoming planes were reported to the Army Air Corps AAIS Center at Fort Shafter, the report was discounted, and the Japanese were able to obtain tactical surprise.

As the Japanese first attack wave crossed from the Pacific Ocean to Oahu on the south shore, they encountered two private Piper Cubs of the K-T Flying Service. They had recently taken off from the John Rodgers Field on a Sunday morning rental by three off-duty soldiers to hone their flying skills. All three were enlisted men from the 251ˢᵗ Coast Artillery Regiment. Corporal Clyde C. Brown and Henry C. Blackwell were budding pilots learning their skills in the Civilian Pilot Training Program. Their buddy Sgt. Warren D. Rasmussen was along for the ride. After being on an alert assignment, they returned to the regiment's base at camp where they had the weekend off and time for their next lesson. Starting their trip along the coast, the bright yellow, defenseless pleasure planes encountered Japanese Zeroes escorting the attack formation. They were both promptly shot down over the ocean. These men are believed to be the first service members killed during the attack. (5)

The 98ᵗʰ Coast Artillery had been established at Schofield Barracks in the summer of 1941. It was still organizing, training, and getting familiar with its weapons and assignments at the time of Pearl Harbor. Its 1200 men and two battalions were at the station, in an encampment area known as Upper Schofield. The twelve 3-inch M3 guns of the 1ˢᵗ Battalion (combat station of nearby Wheeler Field) and twelve older 3-inch M1918 guns of the 2ⁿᵈ Battalion (combat station at Kaneohe Bay/Bellows Field) were all parked at Schofield. Lt. Stephen G. Salzman was the regimental Signal Officer. Saltzman and another signalman of the 98ᵗʰ's 1ˢᵗ Battalion, Sergeant Lowell Klatt, left an interesting account of their encounter with aircraft on the morning of the attack.

Saltzman had still been in bed in his post quarters on the 7ᵗʰ when he heard a large formation fly over. Going outside and seeing Japanese markings on the planes, he wrapped a towel around himself and went to the switchboard to have his C.O. alerted. Dressing and grabbing a pistol, he headed to the regiment's command

During the attack light AA guns, predominantly 0.30 and 0.50-caliber machine guns, were deployed by available units—as here on the tarmac at Wheeler Field. *NARA.*

In what appears to be a staged photograph, this 0.30-cal AA pit at Wheeler appears to be more "permanent" than the expedient positions used during the attack. *LOC.*

post just east of Wheeler Field. He remembered that an alert had sounded at 8:20 a.m. He helped dispatch units going to field positions and dropped guards at several terminals. At the C.P. he and Klatt heard the approach of two planes coming across the Kam Highway towards his position. He grabbed an '03 Springfield and the staff sergeant had a Browning Automatic Rifle. One of the planes headed towards them and they dropped to their knees and opened up and brought it down. It crashed out of sight around a building and immediately began to burn. (6)

The two soldiers were credited with a "kill", both were awarded the Silver Star. The lieutenant testified in the Pearl Harbor Investigation that there were two dead pilots in the wreck and that it had a Pratt-Whitney engine. Klatt added more detail in reporting that it had an American-made engine and prop, and that a parachute they saw was American. That worried Saltzman, he thought maybe he had made a mistake, but an inspection for Air Corps Intelligence assured him that it was a Japanese plane. The assessment that the plane was Japanese was reported in a number of accounts and books after the war, including in Gordon Prange's *December 7, 1941* published in 1988. But they were wrong, this was an American P-36.

For one thing the description the soldiers gave must be mistaken. The Japanese planes used in the attack did not have American Pratt & Whitney radial engines, or American props, and most, if not all, didn't carry any parachutes. The strafing A6M "Zeroes" were all single-place aircraft. One plane that *was* lost while attempting to land at Wheeler was a P-36 fighter flown by Lt. John Dains. Dains had made two flights from Haleiwa Field as one of the few Americans to get airborne this morning, though he is not credited with any combat success. On his third flight he had to switch to a P-36 from his previous P-40s. On returning to Wheeler, he was shot down. The P-36 did have a Pratt & Whitney engine, American prop, and would have had an American parachute, but still was just a single-place airplane. Recent histories like *The Air Force Story* by Leatrice R. Arakaki and John R. Kuborn do describe the real events, and it is even fictionalized in James Jones' *From Here to Eternity*. Unfortunately, it seems that one of very few claims for an aircraft by the Coast Artillery during the Pearl Harbor attack was a victim of a friendly-fire incident.

Being immediately adjacent to Hickam Airfield and the Pearl Harbor Navy Base, the anti-aircraft batteries of Fort Kamehameha were heavily engaged during both waves of the attack. While there were no direct attacks on coastal defense facilities, the closeness of the air and navy targets made the involvement of the weapons of

Army Air Force pilot Lt. John Dains of the 47th Pursuit Squadron was killed when his P-36 fighter was shot down by friendly fire approaching Wheeler Field for landing on December 7th. *NARA.*

Kamehameha inevitable. AA defense here was normally provided by Battery F of the 97th (3-in AA). The fixed battery near Battery Closson wasn't ready to fire until 8:35 a.m. after the first wave had passed. Besides the AA units, most mobile seacoast batteries had sections of AA machineguns. Batteries A, B and C of the 55th CA almost immediately set up their 0.30-cal AA machineguns in front of their barracks and fired away at Japanese aircraft lining up for their runs on the ships at Pearl Harbor. In the case of Battery C, one soldier had to use his 0.45 to "unlock" a locked storeroom and locker to get to their machineguns. (7)

When battery B of the 55th was alerted, Captain Frank W. Ebey, the commanding officer, began the process of moving his 155 mm guns from their assigned quarters at Fort Kamehameha out to their ready position at Barbers Point. He ordered the unit's machineguns unlimbered and set-up in the tennis courts behind the quarters. The captain noted the time as 8:13 a.m. Even after pulling everything together, he still had to make numerous trips by barge across the Pearl Harbor channel to get the unit to the western side. A motley collection of soldiers, sailors, and civilians helped Quartermaster Jack Barros operate his barge. Around 8:15 a.m. Japanese Zero pilot Petty Officer 1st class Takashi Hirano turned over the Pearl Harbor channel. He attempted to strafe the barge but was taken under fire simultaneously by American destroyer USS *Helm* and Captain Ebey's gunners. *Helm* was just about the only ship in Pearl Harbor actually underway when the attack started, and she immediately started to depart the harbor with urgency. Hirano's plane was hit and went down, post-battle credit being shared between the destroyer and the gunners. The plane crashed at Fort Kamehameha against a building, photographs of the wreckage were prominently reproduced subsequently in newspapers throughout the country. (8)

The Japanese targeted the Marine Corps airfield at Ewa during both waves of the attack. Antiaircraft protection was mostly provided by the army's Camp Malakole that was adjacent and just west of the field. This was the home of the 251st Coast Artillery (AA). Like so many other posts, the jarring awakening from the peace of Sunday morning breakfast occurred. Private Herbert J. Elfring had just finished his at the camp and was scanning upcoming assignments on the bulletin board when the sounds of bombing at Pearl Harbor intruded. Within minutes the post was being strafed, and he observed: "This line of bullets only missed me by 15 feet. When I saw this red ball on the airplane, I thought, My God that's a Japanese plane!" Elfring scrambled to get to his post with the radar squad. From there he watched his post and the airbase at Ewa get roughly handled. (9)

After he was unable to contact the 53rd CA (AA) Brigade HQ at Fort Shafter, Capt. Melbourne H. West, officer of the day of the 251st CA, ordered the call to arms sounded at 8:10 a.m. Captain West's Battery E, the 2nd Bn's searchlight battery, remained at Camp Malakole temporarily, while the remainder of the battalion (AAMG Batteries F, G, and H) quickly moved out to positions at the navy yard, the navy recreation area, and the tank farm. The 1st Bn also began moving to its assigned positions south and west of Pearl Harbor at West Loch and near Ewa. All three gun batteries (B, C, and D) of the 1st Bn were in position and ready to fire at 11:45. By late afternoon Battery E had moved its 15 mobile 60-inch searchlights into its field position on the Ewa plain west of Pearl Harbor.

Sergeant Eugene Camp was with the regiment's Battery B, one of the mobile 3-inch AA units. It took him an hour to organize a small convoy to move to their preassigned deployment position at the West Loch navy ammunition depot. He reported that the trucks were repeatedly strafed. He was taken aback with the presence of an M.P. at each intersection. They did get into position and found ammunition immediately available at their emplacements—not a common occurrence this particular morning. The artillerymen stood by, assisted by about 20 willing volunteer survivors from sunk minelayer USS *Ogalala*. But also like many others, once ready the attack was already over, and no targets appeared. (10)

Likewise, Lt. Willis T. Lyman of the same unit was on his way to church at Camp Malakole when the attack started. Receiving the battalion alert, he got his machine gunners busy. They quickly nicked a strafing plane (presumably an A6M Zero) at just an estimated 75-foot distance. Hit near the wing root it banked away to turn and then wheeled out and kept descending until it hit the water some distance away. The Japanese records confirm the loss of one Zero during the attack on Malakole. (11)

The anti-aircraft guns at Fort Weaver were assigned to Battery G of the 97th CA which was on post. They were able to get firing within about 30 minutes from the start of the attack, at 8:30 a.m. Being just on the southwest shore of Pearl Harbor, they were hotly engaged for a significant part of the action. Many of the losses, particularly during the second wave over Pearl Harbor, were not specifically credited to individual units. With the Pacific Fleet, the guns at Fort Kamehameha and Fort Weaver all readily in range, there was a hail of bullets and explosions over the harbor. (12)

The final morning friendly fire loss is unfortunately shared with the gunners at Fort Weaver. The flight of 18 SBD dive bombers dispatched from USS *Enterprise* at 6:00 a.m. to Hickam and Ford Island arrived just at the height of the first wave. Several were shot down by Japanese fighters. The final loss was the plane of Ensign Edward T. Deacon, shot down by friendly fire from the artilleryman at Weaver. Fort Weaver was also the home of the Fleet Machine Gun School. These men were alerted by 8:00 that morning, and within ten minutes had their 0.50-cal and 20 mm automatic guns ready to fire. Like their army counterparts, the marine 3-inch AA ammunition was in storage at the navy's Lualualei Ammunition Depot, and they did not receive ammunition until 11 a.m. As a consequence, only the machine guns and the 20 mm automatic weapons were engaged. They were joined by members of the 97th Coast Artillery AA Regiment with its AA machineguns and 3-inch guns. When Ensign Deacon and his radioman/gunner RM3c Audrey Coslett in their SBD tried to fly to Hickam along the coast to avoid the smoke and fire at Ewa they flew directly over Fort Weaver. The navy trainees and professional artilleryman let go with all they had. The plane was heavily hit and both occupants wounded. Deacon had to ditch with wheels up about 200-yards from Hickam's western shoreline in the channel. He inflated a raft, loaded his more seriously wounded radioman and managed to reach an army boat sent out to investigate. Both men survived but the plane was lost. (13, 14)

The first warning the Kaneohe NAS received of the Japanese attack came at 7:53 a.m. with the sound of several low-flying Japanese A6M2 "Zero" fighter aircraft from the aircraft carriers *Zuikaku* and *Shokaku* that came streaking in over the air station from the west. These fighter aircraft had escorted Nakajima Kate high-level bombers. The Zeros escorted the Kates as far as the middle of Oahu when they detached from the main flight of attacking aircraft to make strafing attacks on Kaneohe. This strafing attack started several fires on the station and set several of the PBY-5 amphibians parked on the ramp afire. This initial attack lasted some ten to 15 minutes according to Commander Martin, the station's commanding officer, and all available antiaircraft defense measures with the limited resources available were taken. Most of the navy personnel just watched as the attackers strafed the PBYs moored in the bay and parked on the ramps. During the lull that followed the initial attack marines were assigned to assist in the evacuation of dependents from the air station to a large storage bunker at the base of Puu Hawaiiloa.

The second wave attack came about 40 minutes after the first attack, 18 more "Kates" from the Japanese aircraft carrier *Shokaku* escorted by some 17 "Zeros" from the carriers *Hiryu* and *Soryu* composed this wave. During the second raid, a combination of incendiary as well as high explosive 100-pound bombs were dropped causing still more fires. Hanger Number 1 was hit and set afire destroying the four aircraft inside. The air station sailors and Marines had little in the way of antiaircraft matériel to return fire to the Japanese. What AA defenses there were was limited to machineguns, many taken out of the PBYs, rifles and other small arms. The antiaircraft machine gun fire from various locations on the air station that had been sporadic during the first attack, increased considerably during the second wave attack and the station complement gave a good account in responding to the second attack. At the end of the raid there were only nine PBYs at Kaneohe that had not been destroyed, and of that nine, only three, those that had been on patrol at the time of the attack, were undamaged. It was also during the second raid that most of the casualties at the air station were caused. Seventeen were killed, including two civilian employees of the Navy, and 67 wounded.

The sole armament available to the army's small guard detachment at the new cantonment at Camp Ulupau was their small arms. There was little Lieutenant Dunlop and his men could do as they watched the Japanese planes plaster the nearby air station except to return fire with their '03 rifles and watch the ocean approaches to Oahu from Station "J." Because Camp Ulupau had been scarcely begun, there was little in the way of targets on the eastern half of the Mokapu Peninsula to attract the attention of the attacking Japanese. There were no deployed Army coast artillerymen on station at the base. Batteries F and G of the 97th CA with mobile 3-inch guns had an assigned station at Kaneohe Bay, but they normally (and so on this morning) were quartered at Schofield Barrack until deployed. They moved out during the attack but weren't reported as ready for fire until 1:15 p.m. in the afternoon.

The Navy and Marine Corps personnel made desperate attempts to jury-rig machine guns and deploy with small arms to take on the attackers. Navy Chief Ordnanceman John William Finn found both a 0.30 and 0.50-cal machine gun in his ordnance shop and set them up on the parking ramp. Even though wounded, the Chief kept up his fire and won the Medal of Honor for his bravery. In this attack Lt. Fusata Iida's plane was hit, it began leaking fuel rapidly. The airman knew he wouldn't be able to return to his carrier and crashed the fighter into the base instead. While not an army score, it could be credited to defensive ground fire. (15)

The Japanese carefully chose their targets for their attack. The ships of the Pacific Fleet and the combat aircraft assets of the U.S. military were their primary objective. Coast artillery defenses were not. No attempt was made to bomb or strafe any of the gun emplacements. Except for the attempt to send midget submarines into Pearl Harbor, no enemy ships approached Oahu. Actions of the various coast artillery posts was limited to antiaircraft fire if adjacent to an airfield or to firing at stray or imagined enemy airplanes. At the time concern was pervasive for what follow-on enemy action would occur, and preparations were made to engage surface targets, fire on paratroopers, defend assets from saboteurs or partisans. Fort Weaver and Barrette in southwest Oahu were hosts to the most important seaward defenses—the 16-inch gun batteries. Weaver also provided quarters for the newly formed 97th Coast Artillery (AA). The single battalion of this unit was still housed in tents at the time of the attack. The duty stations for the 3-inch mobile batteries were at the defenses of Pearl Harbor; one each for Fort Kamehameha, Weaver, and Barrette. Crew members were quickly dispatched the morning of the 7th, and the guns at Kamehameha and Weaver fired at Japanese airplanes approaching or leaving their harbor attacks. At all three posts both antiaircraft and seacoast personnel manned their positions either on their principal armament or in perimeter and security details. Battery H, 97th CA, was also posted at Fort Weaver, although its battle station was at Fort Barrette. The battery, alerted at 7:55 a.m., departed Fort Weaver by truck at 8:30 for Fort Barrette. The battery arrived at 9:10, having engaged some low-flying Japanese planes enroute with small arms. The detachment guarding the Barrette AA battery engaged one Japanese plane with small arms while the main body of the battery was enroute. (16)

Battery G of the 97th, also quartered in the tents at Fort Weaver, was assigned to the fixed 3-inch AA battery in the rear of 16-inch Battery Williston. Battery G rapidly broke out the boxed ammunition and was ready to fire by 8:30 a.m. Battery G fired 30 rounds of 3-inch shrapnel. Its commanding officer, Lieutenant King, reported that the battery "definitely turned back one formation of 15 planes." First Lieutenant William Glover Sylvester of Battery G was the first coast artillery officer killed in action in World War II. Sylvester was driving his automobile across Hickam Field enroute to the ferry mooring at Fort Kamehameha to cross to his battery at Fort Weaver when Japanese fire struck his vehicle, killing him. (17)

The M1917 3-inch AA guns of the Sand Island AA Battery were manned by a detachment of Battery F, 55th CA Regiment, on the morning of December 7th. The detachment was credited with downing two attacking Japanese planes, though Japanese records do not confirm any specific losses to this battery so they should be considered just claims, not confirmed kills.

The 3-inch fixed AA guns at Sand Island, manned by a detachment of Battery F, 55th CA were fully engaged during the raid, and credited with two "probable" kills. *USAMH.*

On this morning the Hawaiian Coast Artillery Command at Fort DeRussy was awakened with the word of the Japanese air raid at Pearl Harbor nearly 25-miles to the west of the post. The HCAC took up its battle position in a prepared command post in a bombproof room located on the lower level of Battery Randolph's left flank traverse. By 8:20 a.m. the post was fully on alert and all armament was manned and ready. Battery A had Randolph's 14-inch guns as well as the 6-inch guns of Battery Dudley manned and ready to fire on the Japanese invasion fleet that was expected to appear imminently. The AA detachment was manning the 3-inch

M1917M1 AA guns at the fort in anticipation of additional air raids. Those personnel not actually at the guns were busily engaged at the battery position strengthening the local defenses with additional field fortifications. Although the batteries were all manned and ready for action, the Japanese planes did not come within range of either the seacoast or AA batteries at the Waikiki Post. Although Fort DeRussy was not attacked by the Japanese it was hit several times by 5-inch naval AA shells that were fired by the AA guns on board ships in the harbor. Designed to burst in mid-air, many of the AA shells did not detonate until they returned to earth because of hastily set or faulty fuse settings.

Although the HD of Honolulu was alerted within minutes of the commencement of the Japanese air raid, there was little the Fort Ruger garrison could do to. Fort Ruger was not a target; the Japanese were well aware of the antiaircraft guns at Diamond Head. They skirted that part of the island, giving the fort a wide berth, but antiaircraft shells from 5-inch naval AA guns aboard ships in Pearl Harbor caused some damage when they landed on Diamond Head. One improperly fused shell damaged one of the four 12-inch mortars in Battery Birkhimer. The 16th CA at Fort Ruger and Fort DeRussy remained on 24-hour alert for several days following the attack, with manning details standing by both the antiaircraft and seacoast batteries. (18)

On the morning of December 7, 1941, the 64th CA was in its quarters at Fort Shafter. Like nearly every other army unit, it too was looking forward to a quiet Sunday that would for the first time in several weeks be alert free. Most of the batteries of the four AA regiments on the island were alerted to the Japanese attack by 8:15 a.m., soon after the first explosions in Pearl Harbor. The batteries of the 64th began preparing to deploy to their assigned positions in the Pearl Harbor and Honolulu Harbor area. Battery A of the 64th, one of the two regimental searchlight batteries, manned their 0.30 cal. AAMGs at Fort Shafter. At 09:00 and 10:00 a.m. they engaged enemy planes with machine gun fire, expending 1000 rounds. In the aftermath of the attack, a detachment of Battery A was deployed to Windward Oahu, where it set up a number of 60-inch searchlights between the town of Kaneohe and the Koolau Range. During the first night of the war, this battery's positions had the dubious distinction of being fired upon by inexperienced troops of the 98th CA Regiment's 2nd Bn at the new Kaneohe Naval Air Station.

Battery B lost little time in hooking up their 3-inch M3 guns and departing for their battle position at Aiea, just north of Pearl Harbor. The vehicles of the ammunition train moved out to nearby Aliamanu Crater, where they were issued an initial allowance of 1200 rounds of AA ammunition. By 10:00 a.m. they were in position and ready to fire. The guns of Battery C followed and arrived at the battery's battle position at Aliamanu, south of Fort Shafter. After drawing ammunition, it too was ready by 10:30. Battery D took up position just south of Aliamanu's south rim about 11:00 a.m. While Fort Shafter was not a target for the Japanese, the post also sustained some damage during the air raid, not from the Japanese, but from the U.S. Navy. At least one round from a navy or marine 5-inch/25 cal. AA gun, the fuse of which had not been set for air burst, landed on the barracks of Battery E, killing one enlisted man and wounding three others. The units at Fort Shafter fired some 3000 rounds of 0.30 cal. ammunition. Later in the day Battery E went to the Honolulu docks, where it set up its mobile searchlights for service that night. (19)

A Japanese D3 Val (note the presence of fixed wheels) hit by antiaircraft fire during the attack and heading down. *NARA.*

An A6M2 Zero after being shot down and crashing on the grounds of Fort Kamehameha. *NARA*.

The remaining gun batteries of the 2nd Bn, 64th CA, deployed to the west of Pearl Harbor. By 11:05 Battery F had set up its guns in defense of the naval station. Battery G headed for Fort Kamehameha as quickly as it could hook up its guns to the prime movers. By 10:30 a.m. the battery was in place at Ahua Point, ready to begin firing. Battery H arrived at its battle station at Fort Weaver, and was ready to fire at 11:45. While the prime movers of the gun batteries were towing their guns to the battle positions, the ammunition trucks were converging on the Departmental Ammunition Storage Facility at Aliamanu Crater to draw the 1200 rounds of 3-inch HE shells allocated to each battery. The automatic weapons batteries of the 3rd Battalion were also sent out to various pre-selected positions late in the morning. (20)

Reacting to what proved to be a false report of enemy aircraft carriers south of Barbers Point, USS *Enterprise* launched a strike force at 4:42 p.m. of 19 TBDs and 6 SBDs escorted by 6 F4Fs led by Lt. (j.g.) Francis Hebel. They didn't find any targets, and it was dark by time to return and the carrier was in black-out condition. Most of the bombers were allowed to land on the carrier deck, but the six fighters were instructed to land in Oahu. Approaching the island at about 8:45 p.m. and seeing the lights at Ford Island, the flight of six fighters approached to land. The arriving flight was misinterpreted as a new attack and a tremendous amount of defensive antiaircraft fire came from the ships and bases surrounding Pearl Harbor. Five of the six planes were shot down. Hebel himself abandoned the attempt for Ford Island and headed for Wheeler Field. That decision did not save him, as army anti-aircraft gunners still shot him down. The pilot was forced to make a rough landing in a gulch near Kunia Road. Hebel was mortally wounded in the landing, dying later at Tripler hospital. (21)

The exact number of American casualties from the December 7th attack is still not known, even today. Close to 2388 American deaths were recorded. According to the Pearl Harbor Investigation Report, the army's Hawaiian Department is reported to have suffered 600 casualties, 215 died in action or died of wounds received in action, two died in non-battle situations. Most of the Army's losses were with the Army Air Force during the Japanese air attacks on their bases. There were only 16 army ground troop deaths, 11 were members of Coast Artillery units. Three deaths were the student pilots of the 251st CA (AA) from Camp Malakole, one soldier of the 15th CA was killed at Fort Barrette, five of the 41st CA at Fort Kamehameha, one of the 64th CA (AA) at Fort Shafter, and one of the 97th CA (AA) at Fort Weaver. No weapons or gun emplacements were destroyed or even damaged. (22)

The Marine Corps Defense Battalions were in training near the Pearl Harbor Navy Yard but were unable to get anything except light machineguns in action during the attack. *NARA.*

The Army thought, or at least claimed they were responsible for, shooting down 11 enemy airplanes (not including those that fell to pursuit aircraft). It was very common, in confused actions such as this with excited, inexperienced troops, for the claims to largely exceed the actual count. In the Philippine campaign that would last considerably longer, actual Japanese losses were only one-third or one-quarter of the losses claimed by the defenders. Twenty-nine Japanese aircraft were loss in the raid or its return flight, more were seriously damaged and thrown overboard once they returned to their aircraft carriers. Eleven Japanese airplanes were lost in aerial combat. Thus about 18 were either shot down by ground fire, or perhaps damaged to such an extent (either by ground or air attack) that they ditched on the return journey. Due to the variety and intensity of antiaircraft fire from navy ships and army guns it is now not possible to precisely quantify the number shot down by the coast artillery antiaircraft units. Only the one fighter hit over Ewa/Malakole and the shared credit for the plane shot down and crashing at Fort Kamehameha seem to be likely to be credited to the Army AA units.

Rumors of sinister enemy activity began almost immediately. The tremendous shock of the attack and the feelings of helplessness made for imagined dangers. Also it seemed the Japanese were full of deceit for staging a surprise attack and giving no warning, or maybe that it was executed so well that the enemy seemed capable of anything. In any event the hours and days immediately following the attack generated a large number of false and even fanciful rumors and claims. And remember that about 40% of the Hawaiian population was of Japanese heritage and as a whole never really trusted by the local government and the U.S. military administrations.

Many rumors were about Japanese attempts to generate panic and confusion. One good example was by Lieutenant Gray of the 17th Air Base Group: (23)

> Almost everyone had a rumor to tell, some of them (entirely false of course) initiated by the Japanese themselves aboard their nearby ships. Broadcasting in a Hawaiian frequency, their rumors were designed to confuse both the military and civilians, The announcer spoke flawless English to make it appear that the program originated in the islands. One admonition at about 10 a.m. on December 7 was to drink no water because the Honolulu water reservoir had been poisoned. Other rumors later in the day were that San Francisco was under bombardment; that the Panama Canal had fallen to Japanese forces, and that Kansas City was the target for enemy planes.

The only Japanese naval craft to approach the Hawaiian seacoast defenses were the midget submarines released off the entrance to Pearl Harbor, and they were spotted and engaged by naval forces, not the Coast Artillery. *NARA.*

Enemy downed flyers were said to have American college class rings on their fingers. Airplane parts and other equipment was purported to be American. Local newspaper boys, barbers, housekeepers were supposedly knowledgeable about the attack in advance, or had secret radios, or cut giant arrows in cane fields to direct attacking planes or were seen in enemy uniforms scurrying about on secretive missions. Radios were said to be hidden in homes and shrines for use in coordinating the air attack or arranging sabotage assignments.

To be clear, there never was any confirmed local or "fifth-column" activity in Hawaii by the Japanese or any part of the population. These were all rumors, some of which persist to this day. Also, false reports, while not really rumors, abounded. A host of observations of suspicious air, sea, and land activity were reported to local defenders, including the coast artillery seacoast and antiaircraft command. They had to be checked out, though with decreasing enthusiasm as the extent of false reports accumulated. Here is the log page of the War Diary of the Hawaiian Coast Artillery Command for December 8th that easily conveys the plethora of incoming reports: (24)

0145  From Harbor Control. Enemy attempting landing at Waiamea.

0428  All units notified to be on alert for landing attack at dawn

0438  Flash. 30 Enemy planes approaching from Kauai

0507  Enemy planes dive bombing Wheeler field (Some firing by Btys in Wheeler Field took place)

0525  Schofield AA group reports barrage fire against planes, later reported as friendly.

0608  53rd Brigade reports small arms firing on friendly aircraft definitely established as from Marines or Navy.

0912  From HD it OH. P-40 fired on Sampans; four are dispersing one is afire (Crew later taken off by friendly destroyer)

0940  From HD of PH. Submarine has been fired on by shore batteries (Later established as friendly SS).

1150  From CO, HD of Hon. Enemy submarine bombed by Navy is sinking off Kaneohe

1507  Troops ordered into olive drab cotton or woold uniforms, with all cloth insignia except chevrons removed, because of reported presence of saboteurs wearing blue uniform with red shoulder patches.

2020  Landing in Wailua Bay

2257  From S-3, HD of PH, Glider troops reported as landing on salt beds near Fort Weaver (Some small arms firing occurred).

It would seem that many of these imagined threats would have easily been resolved with a small dose of logical thinking such as that performed by Vice Admiral Halsey. After he returned to Pearl Harbor on USS *Enterprise* on the afternoon of December 8th, he had dinner with Admiral Kimmel. During the meal the admiral received a report of gliders and parachute troops landing near Fort Weaver. Halsey just laughed. He is reported to have said: "I have heard many damn fool reports in my life, but without exception that is the damnest fool one I have ever heard. The Japs are not using their precious carrier decks to take off paratroopers and gliders and it is not possible for them to tow gliders from the nearest base to Oahu because of the distance…" (25)

The day after the attack, Department commander General Walter Short sent a radiogram to Washington reporting losses and requesting reinforcements. He requested 60 heavy bombers, 36 dive bombers, and 200 fighters. To restock the air force's ammunition supplies, Short requested ten million rounds of 0.50-caliber ammunition with links and almost 20,000 bombs of various sizes. For the AA guns the most urgently needed were 115 37 mm weapons the automatic AA types which were severely lacking in the antiaircraft regiments. Also needed were more 3-inch AA, releasing the smaller guns for use on the outlying islands. For troops Short wanted 3300 harbor defense (coast artillery) soldiers to make up the deficiencies in existing units. Likewise, 1500 more troops and forty-eight 105 mm howitzers were needed for the field artillery deficiencies in the two Hawaiian divisions. The only new combat unit desired at this point was for a small tank force. The manpower requests then grew quickly. On December 20, just two weeks after the attack, the reinforcement recommendation from Hawaii was for 3648 men for the seacoast artillery, and 2642 for the antiaircraft brigade. (26)

A high need was perceived to enhance the food store for the resident civilian population. Throughout this early period the lack of diversity in the agricultural production of the islands was deplored—sugar cane and pineapples were great for the peacetime export economy but not to feed a potentially besieged population. Food supplies were requested for resident civilian population in case of the isolation of the islands. And the evacuation of military dependents to the mainland was recommended. The arrest of perceived agents and foreign citizens of the axis nations was already underway. (27)

By December 11, at 6:00 p.m. the Hawaiian Department reported that they had expended almost ten million rounds of ammunition. Most of this (over nine million) was from the 0.30-caliber machinegun ammunition liberally sprayed from the department's antiaircraft mounts during the raid. The reported breakdown by type was: (28)

| | |
|---|---|
| 3-inch AA | 10,676 rounds of 136,924 on hand |
| 37 mm AA | 473 rounds of 11,527 on hand |
| 0.50-caliber | 278,000 rounds of 2,676,000 on hand |
| 0.30-caliber | 9,645,100 rounds of 52,388.000 on hand |

While certainly not depleted, in the days following the attack no one was sure of what was coming next. Ammunition resupply, and indeed significant enhancement, was vigorously pursued.

The Quartermaster General authorized immediate action to supply 71,000 new rounds of 3-inch AA, 20,000 rounds of 37 mm, and 5,000,000 of 0.50-caliber. The first convoy of units and equipment from the West Coast to Hawaii left San Francisco on December 16. It carried 2,800,000 rounds of 0.50-caliber ammunition. These rounds were also shared with the Army Air Force. The next convoy of seven freighters sailing on December 27th had 50,000 rounds of 3-inch AA. (29)

Immediate steps were taken to reinforce the Hawaiian Islands following the raid. Real concerns persisted that the attack was just preliminary to an invasion attempt to occupy the islands by the Japanese. The Coast Artillery was not the only beneficiary. The Air Corps was rapidly supplied not only replacement aircraft, but also new units of both bombers and fighters. A new infantry division and a whole array of special support units were sent to Oahu. As previously noted, the troop strength of the Hawaiian Coast Artillery Command had been chronically understrength before the start of war. To some extent this was intentional. It was thought that if given a fixed number of troops, it was better to have them organized into understrength units that could be filled out while under a wartime setting with trainees or transferees, rather than try to organize new troops into entirely new units. One way the local command dealt with this situation was to have troops assigned to multiple responsibilities. Units manning fixed guns might have an assigned secondary or a tertiary role to man antiaircraft guns or mobile batteries. Some emplacements were maintained in reserve status—fully functional but not regularly expected to be manned unless a given situation warranted it. In any event the Hawaiian Command requested over 3000 soldiers which was eventually satisfied by sending over either "fillers" or casual artilleryman not currently assigned to a unit, or by sending entire units without armament to be assigned as needed or broken-up to fill other more critical needs.

Fortunately, the Army (and Army Air Force) had a convenient reserve of organized, trained units to pull from. In the fall of 1941 San Francisco filled up with units awaiting shipment to reinforce General MacArthur's forces in the Philippines. The appointed units to be transferred to the department there exceeded the capacity of the military transport system. Upwards of 25,000 men in a variety of regiments and battalions were quartered in northern California awaiting their turn to ship out—much of which would have been done starting in December. While not a perfect match for the needs of Hawaii, much of it was. There were two infantry regiments waiting—just what was needed to complete the two Infantry Divisions in Oahu. Field artillery, signal, ordnance, quartermaster, chemical detachments all proved useful. There were no new regular seacoast artillery units, but there were new anti-aircraft and mobile gun units. (30)

As early as November 6, 1941 the War Department had alerted the 57th CA (TD) Regiment to prepare one of its battalions for deployment to Oahu. The Army had selected them as the unit that would provide the manning detachments for the 155 mm batteries. The regiment was then participating in the First Army maneuvers in North Carolina. The 2nd Battalion, 57th CA was quickly released from First Army and returned to its home station at Camp Pendleton, Virginia. The battalion's eight 155 mm GPF guns were loaded aboard a train at the Camp Pendleton railhead, and accompanied by a guard detail, had left for San Francisco on November 10. (31)

The War Department hoped to have the battalion in San Francisco by November 27, but the maneuvers delayed the departure of the 2nd Battalion's personnel until two days after Pearl Harbor. Upon return to Camp Pendleton, additional men were transferred from other elements of the regiment to the battalion to bring it up to wartime strength. The battalion, consisting of its headquarters and headquarters battery and two firing batteries, C and D, began boarding troop trains at the camp's railhead during the night of December 10. The ensuing trip across the continent covered 3419 miles and six days. The first of the two troop trains arrived at the Fort Mason Port of Embarkation in San Francisco at 9:30 a.m. on December 16. They were to be sent with the first relief convoy to Hawaii since the attack on the 7th. The men remained aboard the trains at Fort Mason until 3 o'clock the following morning when the battalion left the trains and filed on board SS *President Garfield*, recently a passenger ship of the American President Line. The guns and heavy equipment that had preceded the men of the battalion were loaded aboard. Although now operating as a troop transport, *Garfield* was still manned by its civilian crew. Just after dawn on December 17, *President Garfield* passed through the Golden Gate and set course for Hawaii beginning the eight-day voyage to Honolulu. (32)

Although the 2nd Battalion was traveling under sealed orders to Oahu, under the code name COPPER, the 21 officers and 508 enlisted personnel knew that they were enroute to Oahu. A submarine, identified as presumably Japanese, was spotted on the second day out of San Francisco, but no other incidents occurred. On the afternoon of Christmas Eve, 1941, *President Garfield* entered Honolulu Harbor. Little time was lost in disembarking the battalion and they were soon moving by truck convoy to Camp Ulupau arriving there at 4 p.m. The battalion's guns and ordnance equipment off loaded from *President Garfield* and moved to Fort Ruger. Christmas dinner for the men of the 57th and the other troops on Mokapu consisted of field rations. The battalion spent the next two days getting settled and organized at their new station and on December 27, took over the defenses. (33)

On that same day the balance of the 57th followed its 2nd Battalion. The Regiment HQ and the 1st Battalion loaded on to USAT *President Johnson*. This included 61 officers and 1115 men. Convoy No. 2007, escorted by USS *Detroit, Cummings*, and *Clark* departed San Francisco on December 27th and arrived in Honolulu on January 7th.

The advent of war in December 1941 upset the War Department's carefully prepared timetable for an orderly and timely reinforcement of the Hawaiian Islands. In its prewar planning, the 96th CA (AA) Regiment was to arrive on Oahu in March 1942. Upon its arrival, one of its gun battalions and its automatic weapons battalion were to be allocated to the air defense of Kaneohe Bay while its other gun battalion was projected for use elsewhere. Within days following the Japanese attack on Hawaii, the War Department directed that the 95th CA (AA) Regiment, which had already completed its training at Camp Davis, North Carolina be allotted to the Hawaiian Islands in lieu of the 96th CA. The latter unit was still in the process of completing its training, having just received a draft of 500 selectees in October 1941.

Within a week following the Pearl Harbor attack, the 95th CA was preparing for transit to the west coast from its station at Camp Davis. At 4 p.m. on December 15, 1941, the regiment entrained at Camp Davis, arriving at San Francisco on December 21. It boarded the US Army Transport *Maui* late that afternoon, joining the 57th on its journey with Convoy No. 2007 destined for the defenders of the new Kaneohe Bay base. It had 78 officers and 1520 enlisted men, but was less its 1st Battalion, and sailed without that organization and its Batteries A and D. Arriving in Honolulu on January 7th, it was immediately trucked to its new station at Camp Ulupau, Kaneohe Bay. The heavy gun batteries were armed with twelve 3-inch AA guns and the 3rd battalion (Batteries I, K, L, and M) armed with 0.50 caliber AA machine guns were posted at various locations in the Kaneohe Bay-Bellows Field Area from January 7 until February 5, 1942. The batteries assigned to Bellows Field relieved the AA machine gun battery of the 251st CA. In February, plans were implemented to transfer three batteries to the air defenses near Pearl Harbor. Battery E, the searchlight battery of the 2nd Battalion, 95th CA, remained in the Kaneohe Bay area where it took over the antiaircraft searchlight responsibilities on the windward side of the island.

The next coast artillery unit dispatched after Pearl Harbor attack was the 52nd CA (Ry). Like its sister the 41st, it was one of the few railway artillery regiments in the U.S. Army. Within days following the Japanese attack, orders were sent to the regiment, to immediately send a battalion to reinforce the 41st CA on Oahu. At 2 p.m. on December 16, the 1st Battalion began its journey. They were to ship out from San Francisco on the same December 27th that carried the 57th, and 95th CA. The battalion was assigned transport on the ex-American President Line passenger ship USAT *President Johnson* now converted to a troop transport.

When the convoy bearing the men of the 52nd reached Honolulu, the 1st Battalion HHB and Battery A were sent to Fort Kamehameha, while Battery B's personnel were immediately trucked out to Camp Ulupau. The unit had been sent without any primary armament specifically to help address the coast artillery manpower shortage. There the battery's five officers and 179 enlisted personnel took charge of the four 8-inch railway guns and began the process of camouflaging the position and preparing the guns for active service. The men found the platforms holding the rails of the carriages unstable and began the work of erecting the guns on new platforms. On January 24, 1942, a settling round was fired from each gun. Five days later four more rounds were expended in the battery's first practice shoot. Late in January 1942, the personnel of the 52nd Coast Artillery were reassigned to the reorganized Battery C, 41st CA. They would man Battery Sylvester through early 1944.

Also on this same convoy of late December 1941 was additional help for the Coast Artillery Corps. On Matson liner SS *Maui* were loaded 7 officers and 459 unassigned coast artillery "fillers". After arrival in early January, they were gratefully accepted and distributed as needed to several active units.

The Pearl Harbor attack restarted discussions about how to defend the outer islands. In particular there were concerns about the three largest—Hawaii, Kauai, and Maui. While none had vital industrial or military establishments, they were positioned close enough to Oahu to potentially be steppingstones for invasion. Enemy forces could easily land and conquer one of these islands and use it as a base for air and short-ranged landing operations against Oahu. They had large populations of foreign agricultural workers that might perform acts of sabotage or conduct support of Japanese submarines or raiders. Prior to the attack battalions of infantry were distributed to these islands. The Hawaiian National Guard 299th Infantry, after Federalization, sent one battalion each to Hawaii, Kauai and Maui in May 1941. At the time of the attack, there were no anti-aircraft or seacoast capability outside of what the organic artillery and machineguns of these battalions could offer.

On December 19 orders were issued to dispatch at least a token of weapons and personnel from the Coast Artillery Command to the three islands. Six 155 mm GPF guns were taken from batteries that were lacking personnel, as well as 12 old 3-inch Model 1918 mobile AA guns taken from reserve storage, along with three searchlights. The detachments were: (34)

> For Hawaii, departing Oahu December 23, 1941. From Battery B, 55th CA were two 155 mm guns with one officer and 50 men. From the 53rd Anti-aircraft Brigade were four 3-inch AA guns and one officer and 26 men. Also included were one searchlight, one tractor, five trucks, two 0.30-caliber AA machineguns, a coincidence rangefinder from the Ordnance Storage Depot, two units of fire for each weapon.

> For Kauai, departing Oahu December 26, 1941. From Battery D, 55th CA were four 155 mm guns with two officers and 100 men. From 53rd Anti-aircraft Brigade were four 3-inch AA guns and one officer and 32 men. Also included were two portable searchlights, two tractors, five trucks, two 0.30-caliber AA machineguns, a 9-foot coincidence rangefinder contributed from either seacoast Battery Hawkins or Tiernon, two units of fire for each weapon.

> For Maui, departing Oahu December 26, 1941. Maui did not receive any 155 mm guns. But it did get from the 53rd Anti-aircraft Brigade four 3-inch AA guns and one officer and 20 men. Also, three trucks, two 0.30-caliber AA machineguns, a 9-foot coincidence rangefinder contributed from either seacoast Battery Hawkins or Tiernon, two units of fire for each weapon.

Within the jurisdiction of the Hawaiian Department was Christmas Island, a large though almost uninhabited atoll south of Hawaii. It was being developed as an air ferry route stop at the time of the war's start. The Hawaiian Department sent a small garrison to the island, an advance party of one lieutenant and 20 men of the 53rd Anti-aircraft Brigade was sent on transport *Halekole* on December 12th. A second contingent joined late in the month pulled from the 64th Coast Artillery Regiment. Two 155 mm GPF guns and four 3-inch Model 1918 AA guns were subsequently sent from Oahu. (35)

The 96th Coast Artillery Regiment was selected to be sent to the outer islands as a permanent garrison. This was a new unit, it had been constituted on March 15, 1941 at the AA Training Center at Camp Davis, North Carolina. When ready a couple of months later, it was sent with its two battalions and 2589-man complement on the convoy of February 27th. It was armed with modern 90 mm heavy M1 AA guns, making its first appearance in the Hawaiian Islands, and had its full complement of 37 mm automatic AA guns, a weapon that had been sorely lacking in the defenses up to this point. Upon arrival, the 1st Battalion and two batteries (K and L) of automatic weapons went to the Island of Hawaii. Maui received the 3rd Battalion (less batteries K and L sent to Hawaii). Kauai received the 2nd Battalion and automatic battery M.

During the December 7 attack, the mobile seacoast batteries were deployed to their wartime stations. Alerted shortly after 8 a.m. Batteries D and E of the 55th CA began preparations for their rapid move to the Camp Ulupau battle positions. By 9:15 a.m. Battery E's heavy column had limbered up the battery's three serviceable 155 mm guns and was moving out of the battalion gun park at Fort Ruger. Battery D followed shortly after. Peacetime speed regulations for the powerful diesel tractors fell by the wayside this morning. The governors on the D-7 Caterpillar tractor throttles were soon burned out and the huge Cats roared along the coast highway at speeds that at times approached 22 miles per hour; a speed four times the normal pace for the gun towing tractors. The heavy column of Battery E arrived at Camp Ulupau only 85 minutes after leaving Fort Ruger. Surprisingly, neither battery experienced equipment failure nor casualties in the high-speed movement.

The new Naval Air Station at Kaneohe Bay was heavily hit during the December 7th attack, with most all of its patrol aircraft destroyed. *NARA.*

Upon arrival at Camp Ulupau the guns were winched into their previously prepared field positions amidst the dunes of North Beach to cover the seaward approaches to Kaneohe Bay. Battery D emplaced its four guns in the gun-pit type field emplacements prepared the previous summer on East Beach at the south end of Ulupau crater. (36)

When attacked there were no deployed Army coast artillerymen on station at the Kaneohe base. The 97th CA with 3-inch guns were assigned stations at Kaneohe Bay, but they were quartered at Schofield Barrack until needed. *Gaines Collection.*

Kaneohe Station seaplane ramps and hangars two days after the attack—the extent of the devastation is obvious. *NARA*.

At the time of the attack the 8-inch railway guns were at Fort Kamehameha with their assigned batteries of the 41st CA (Ry). Four guns intended for Camp Ulupau were also there pending construction of special rail sidings being built at the Kaneohe Bay post. In the course of the raid, the Japanese managed to damage the military railroad at Hickam Field that connected Fort Kamehameha with the OR&L main line. The day after the attack, Battery A was able to move to its battle position at Browns Camp by using OR&L locomotives. On December 9, Battery B's train finally moved out for its battle position on the North Shore.

For the remote sites, the detachments for fire control stations and searchlights were dispatched and supplies transferred to make the small stations habitable. For these detachments there were long hours of gazing out onto the sea with binoculars and optical azimuth and depression finders trying to locate the enemy fleets that everyone thought were certainly coming. The alert period for the first few days after the attack proved exhausting, to the point that the command had to issue directives to commanders to allow plenty of time for rest of the men to assure the longevity of the assignments. (37)

During the days that followed the entry into World War II, the seacoast fort buildings were prepared for blackouts at night, and overhead camouflage of the batteries was improved. Three days after the attack engineers arranged visits of all batteries to assist in arranging and approving camouflage practices. Early in 1942 a specialized engineer camouflage company was sent from the mainland to help supervise the camouflaging of military structures on Oahu. Extensive use of native plants was emphasized in their approach.

Existing field fortifications were also improved and new field works were constructed for local defense. When not standing by at the guns the men of the batteries were kept busy providing fatigue details and to string barbed wire along the beaches, dig slit trenches as a caution against future air raids, build sandbagged machine gun emplacements, and erect camouflage nets and other devices that would obscure or otherwise protect the gun emplacements or add to the local beach defenses. Although elements of the infantry divisions were assigned to the beach defenses along the shoreline in the Honolulu Harbor defense area, the coast artillery units were expected to provide their own close-in defense for the most part, and only occasionally were they supported by a small infantry detachment manning a machine gun or a 37 mm anti-tank gun. (38)

More ambitious engineering efforts were simultaneously undertaken. The department's engineering office saw an immediate expansion of projects, including those long-delayed for want of funding. The construction of the new defenses of Kaneohe Bay were accelerated. Many new airfields were started, obstacles were placed to thwart paratroopers and gliders at all fields. Small steps, like immediately painting Geneva red crosses on the roofs of hospitals to large steps, like the construction of massive new underground aircraft maintenance, cold storage, ammunition and command facilities were undertaken. (39)

Guards were maintained on all the defensive positions on a 24-hour basis. While passes had been permitted to a small percentage of the battery personnel in late December and early January all passes were canceled for Sundays and holidays. After a series of air raid alerts in early March including one on the night of March 3, 1942, when several bombs were dropped on the Tantalus Hills north of Waikiki by a Japanese Kawanishi flying boat, all antiaircraft machine guns were manned one hour prior to sunrise and for an hour afterward. This heighten state of alert was required because the Japanese threat of invasion was considered highly probable through the first six months of the war. Japanese submarines continued to reconnoiter Oahu and the other islands in the Hawaiian chain through much of 1942.

Another early engineering task was the creation of a confinement center for civilian detainees. The Honolulu immigration station and adjacent Territorial Quarantine Hospital grounds were used for this purpose immediately following the establishment of martial law late on Sunday December 7. Using the FBI's detention list, individuals were rounded up. By December 10 almost 500 civilians, including several American citizens, priests, teachers, and Japanese consular staff were moved to and held in this camp. Initial conditions were poor, but they improved somewhat in succeeding months and by late 1943 most of the retained individuals were relocated. (40)

The HCAC moved its headquarters from Fort DeRussy to Fort Shafter in the spring of 1942. A small headquarters detachment that remained at Fort DeRussy briefly retained the title of Headquarters, Hawaiian Separate Coast Artillery Brigade. (41)

Large gun tubes generate huge pressure and erode rapidly with extensive firing. Guns over about 6-inch bore size have a barrel life (defined with a view towards accuracy, not really safety) of just a few hundred rounds or even less. Typically guns need to be replaced with new guns or relined tubes after prolonged or intensive use. For an island fortress facing the possibility of extended isolation or even siege, sending out new tubes from the mainland was a problem. Consequently Hawaii (and the Philippines) was allotted a spare tube for each of the modern, long-ranged batteries. In March 1926 a spare 12-inch tube specifically intended for Battery Closson was shipped and stored near that battery. However, the 16-inch tubes (different models for the two batteries concerned) were much less common and expensive to use just as replacements. Still, before the start of the war spare tubes were supplied. The Army Model 1919 tube for Battery Williston at Fort Weaver was approved for transfer in late 1940, though reconditioning the second tube took until April 1942. The spare navy Mk II tubes for Battery Hatch at Fort Barrette were shipped during August 1940, and special concrete pedestals to hold the tubes were built to the rear of the two casemated gun emplacements. (42)

# CHAPTER 9
## THE EMERGENCY PROMPTS NEW GUN BATTERIES

The attack of December 7th created a new sense of urgency to enhance the seacoast defenses of Oahu. The U.S. Navy almost immediately offered to assist the Army to augment the defenses of the local bases. Spare guns stored by the yard's ordnance department were released for loan to the Army. More guns were available from the ships heavily damaged or even constructively lost. This included a number of anti-aircraft guns that might be very useful if subsequent air attacks ensued. Available were a relatively large number of the Navy's standard 5-in/25 AA gun (and a few 5-in/38s). Sixteen guns were immediately prepared for emplacement.

Eight guns, none requiring repair, were removed from critically damaged USS *California*. The Mk 10 Mod 2 5-inch/25 with their Mk19 directors were allocated to new four-gun batteries at Hickam Field and a site the navy referred to as Puuloa (later the army called this site Ewa Beach). Another eight guns (the similar 5-inch/25 Mk 11) were taken off the heavily damaged USS *West Virginia*. Four guns were needed for ship installations, but the other four were released for use in a new naval battery emplaced at West Loch. Initially the latter location was referred to in correspondence as Honouliuli. It was found that the guns on USS *Arizona* were too badly burned or damaged, and USS *Oklahoma*'s guns were still underwater as the ship had capsized.

The final four guns of this first batch were the more modern 5-inch/38 open destroyer mounts. USS *Downes* and USS *Shaw* had both been severely damaged while docked. One gun and mount from *Downes* and three from *Shaw* were found usable for mounting in one of the new batteries. These guns were located near the old airship mooring mast at Ewa Field, and subsequently the site was known as Ewa Mooring Mast. It appears that this was the only battery with this type gun utilized on Oahu, and probably the only battery of this type the U.S. emplaced anywhere.

These first four sites were selected by the Navy, all deployed for coverage of the navy base. The Navy did the design and construction work with their own personnel. Work was pressed, and a letter of February 8 1942 reports all sites were completed and armed. A follow-on fifth site at Ahua Point, Fort Kamehameha had been started, using another four 5-inch/25s released for use by the Navy on January 7. Eventually guns were found for four more emplacements. One was never begun, resulting in eight individual batteries totaling 32 guns of navy origin being emplaced in early 1942.

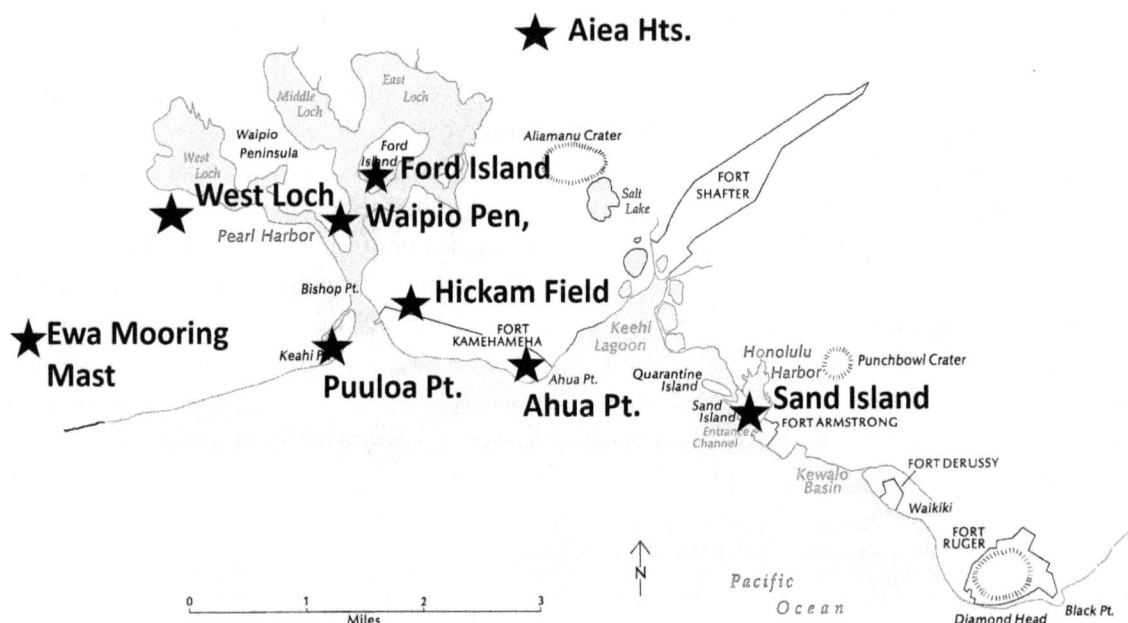

Map showing locations of emergency 5-inch AA guns donated by the Navy for antiaircraft defense surrounding the Navy Yard in early 1942. *Williford Collection.*

The status of these sites, as of a report of March 16, 1942 was: (1)

| Battery Number | Location | Armament | Status |
|---|---|---|---|
| 1 | Hickam Field | 4 x 5-in/25 | Complete, Navy built |
| 2 | West Loch | 4 x 5-in/25 | Complete, Navy built |
| 3 | Ewa Beach | 4 x 5-in/25 | Complete, Navy built |
| 4 | Mooring Mast | 4 x 5-in/38 | Complete, Navy built |
| 5 | Kamehameha | 4 x 5-in/25 | Under construction, transferred to Army Engineers |
| 6 | Waipio | 4 x 5-in/25 | Site tentatively selected |
| 7 | Ford Island | 4 x 5-in/25 | Site selected, Navy to construct |
| 8 | Aiea | 4 x 5-in/25 | Subject to Approval |
| 9 | Sand Island | 4 x 5-in/25 | Not yet scheduled |

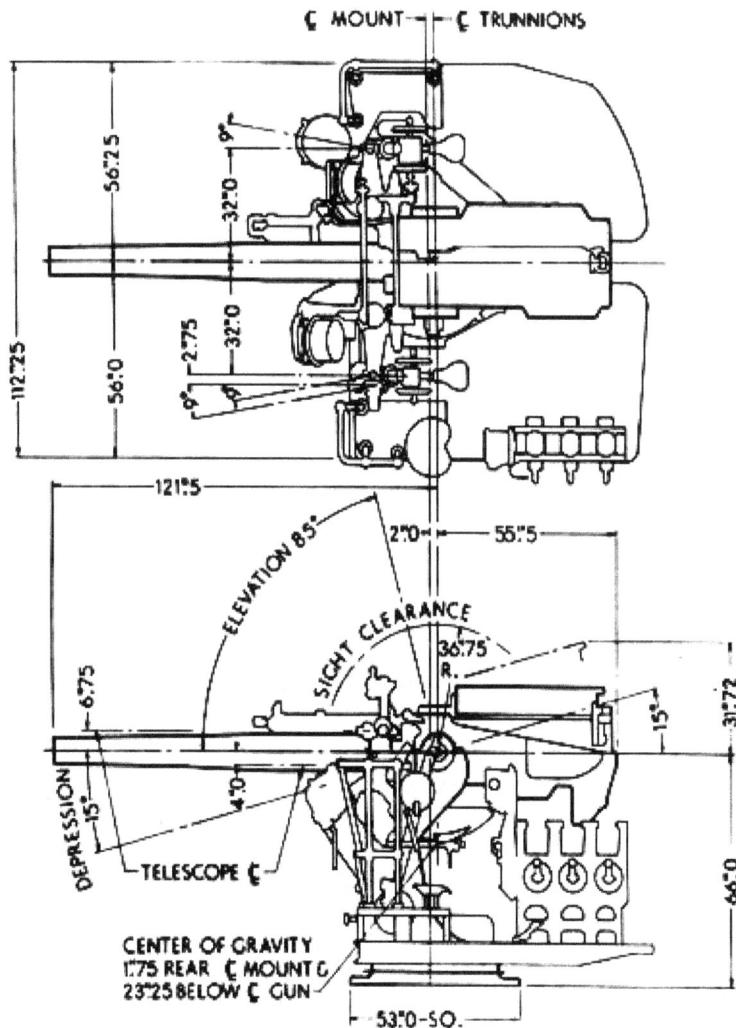

Navy Ordnance Pamphlet No. 1112 sketch of the 5-inch/25 caliber ship mount, all but one battery of the emergency fixed AA emplacements was of this gun type. *NARA.*

At this stage of the war, there was excellent cooperation between the Army and Navy. The Navy freely made available personnel, shops, and supervisors to assist the army's takeover of both the construction projects and instruction and maintenance of the equipment itself.

The emplacement design was kept simple. The four guns of the battery were arranged at the corners of a square of about 120-feet on a side. Each individual gun mounting was secured by hold-down bolts to a massive gun block of reinforced concrete. This square block was 15-feet on a side and 4.5-feet deep; it was estimated to weigh 76 tons. Circling the gun pit was a thin retaining wall of either concrete to cement-covered sandbags that included the ready ammunition boxes. Between the guns, in the center of the square was another pit and block to hold the naval director. Other rooms, all of splinter-proof light concrete construction, were built for the commander's post, switchboard, and one or two rooms for backup generators and compressors. These guns required 150 p.s.i. compressed air for power ramming of shells, supplied either directly by the compressor or a pressurized vessel. Finally, a primary magazine was required. Initially any required crew buildings were of just light wooden construction or even tentage.

Unfortunately, by May 1943 the hastily constructed emplacements were beginning to deteriorate. In particular the surrounding retaining walls of sandbags covered with thin layer of concrete was cracking and breaking up in a number of emplacements. A request to replace the walls with more substantial concrete or concrete block construction was refused, but temporary stabilizing repairs were performed by battery troop labor. Plans were in place to replace the navy guns with standard army-issue 90 mm antiaircraft guns. By year end of 1943 it was decided that to place new, heavy guns in these existing emplacements would be impractical. They needed a new gun block, a wider platform, and an opening in the retaining wall to allow the mobile guns easy entrance and exit. It was cheaper and faster to just build new emplacements nearby. (2)

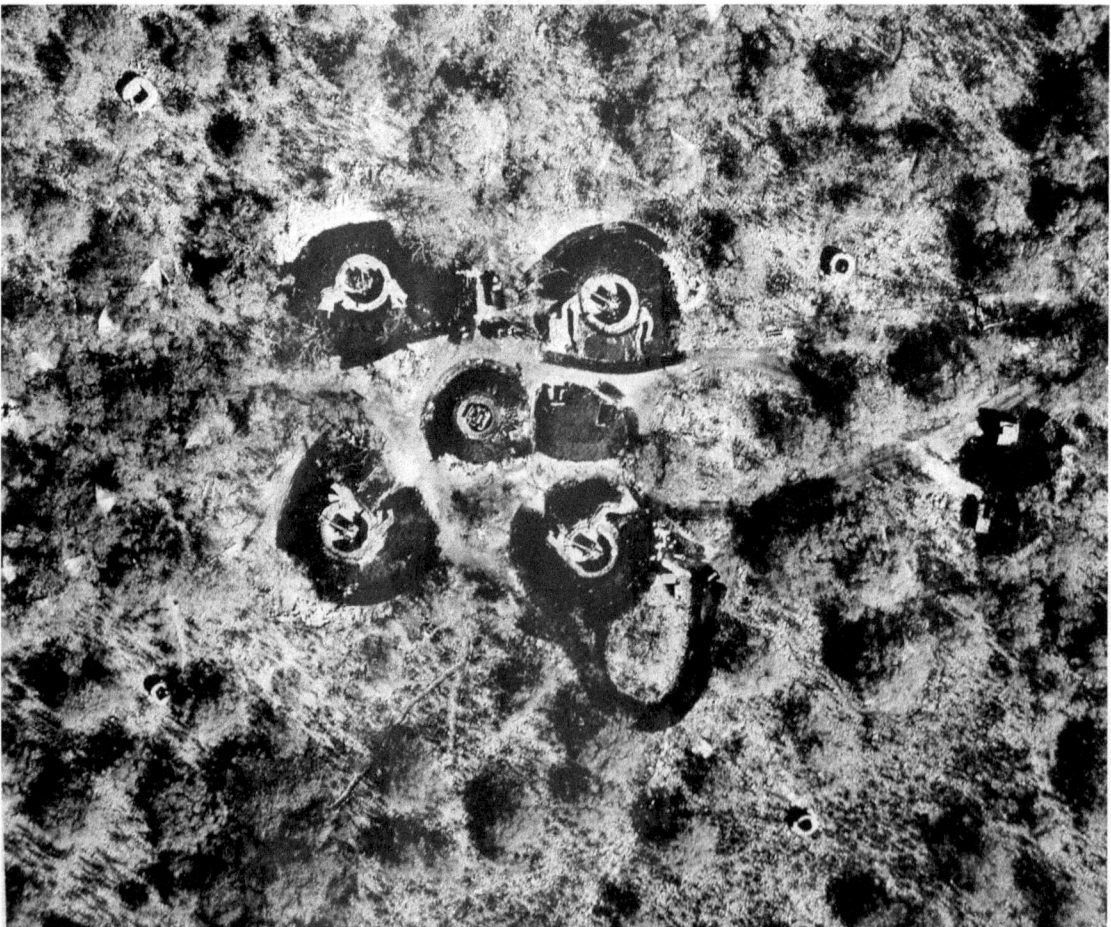

Early 1942 aerial photograph showing the layout of the four-gun 5-inch AA battery at Ewa. The smaller pit in the center is for the director instrument. *NARA.*

One of the guns of the Ewa Battery as emplaced. Photograph dated to January 20, 1942. *NARA.*

In September of 1943 directions were issued to return to the Navy its 5-inch AA guns, which after all had just been loaned and not permanently transferred. After determining that the Navy was in no rush to get them back and there was no other Army use for the weapons, it was agreed to leave them in place until new batteries of either 90 mm or 120 mm guns were emplaced. By mid-November at least four batteries were disarmed in compliance with these conditions. At the same time the older 3-inch fixed Model 1917 guns were also being phased out of service permanently. They too were initially authorized for returning to storage from their fixed emplacements on September 19, 1943. Eventually they too were left in their old emplacements. At first there was concern about just leaving them unattended in place, but after removal of their breechblocks and all attached sights and equipment, the risk was considered acceptable. They remained in their old emplacements until scrapped under contract after the war. (3)

A 5-inch gun at the Puuloa Battery on February 11, 1942. *NARA.*

The ex-shipboard AA Director of West Loch Battery on January 8, 1942 while battery construction was clearly still underway. *NARA.*

Underground switchboard room for the same battery at West Loch. *NARA.*

In the months following the outbreak of World War II in the Pacific there was a massive buildup of the Army's antiaircraft defenses on Oahu. A large number of filler troops already on the island in December 1941 were used to bring the already organized coast artillery batteries up to near full wartime strength. In May 1942, several provisional CA (AA) gun batteries were authorized on Oahu to bolster the air defenses of the island with particular emphasis on the Pearl Harbor Naval Base. The personnel for these provisional batteries were obtained by using a portion of the nearly 2,000 raw coast artillery replacement troops that had arrived on Oahu during the summer of 1941 and during the early months of the war. These troops were assembled at Camp Malakole where they were organized on May 18, 1942, into the 710th 711th, 712th, 713th, 714th, and 715th Coast Artillery (AA) Batteries (Separate). After training a few weeks, the newly formed units were deployed to locations generally encircling Pearl Harbor. Three of these 700 Series batteries were assigned to man new batteries of 3-inch semi-mobile AA guns that were shipped to Oahu early in the war. Most of these 3-inch batteries were established in the vicinity of Pearl Harbor at Pearl City, the Puuloa Dumps, and at Aiea. The remaining three 700 series batteries were eventually assigned to man three batteries of 5-inch/25 cal. emergency naval AA guns that were positioned in close proximity to the naval station. (4)

Details of the War Emergency Navy 5-inch AA battery sites can be seen in Appendix VII.

A few ex-navy 3-inch guns were available and loaned to the army command in Oahu. Several unidentified surface contact incidents in the summer of 1942 in Kaneohe Bay contributed towards the decision to establish additional improvised batteries in that harbor defense. In July 1942, the local command obtained two 3-inch guns from navy stock. A pair of the navy 3-inch/50-caliber Mark VI Mod 6 deck guns on Navy Mark VII Mod 4 pedestal mounts was emplaced on Kii Point at the south end of Ulupau Crater's east rim to cover the waters of Kailua Bay. These older guns had been used on small ships and could fire a 13-lb. projectile about 7000-yards. This battery was named Battery Kii (also known as the South Ulupau Battery). By October 1942, three more 3-inch/50-caliber navy guns were made available and emplaced on the north rim of Ulupau at Pukaulua Point. Named Battery Puku, these three guns increased the coverage of the northeastern approaches to the Mokapu Peninsula and uncovered a dead zone behind Moku Manu Island. The guns remained in place and manned into late summer of 1944. (5)

Similarly a few excess army 3-inch seacoast guns were available for emplacement for the emergency. Two ex-navy 5-inch low-angle guns had been mounted right after Pearl Harbor in a position covering Waimanalo Bay on a ridge at Wailea Point. Within a few months the Hawaiian Coast Artillery Command (HCAC) replaced these guns with two 3-inch Model 1903 seacoast guns once mounted at Fort Kamehameha's Battery Chandler and still in army storage until made available for use at this location. The battery site selected was not far from Bellows Field along the coast southeast of Kaneohe Bay. A detachment of the 34th Engineer Regiment assisted by coast artillery personnel built a battery commander's station, plotting room, magazines, and gun blocks for two 3-inch seacoast guns. The two searchlights already in this position were removed from the ridge and placed atop towers further to the south of the ridge near the mouth of Waimanalo Stream. The work on the battery was apparently poorly planned and executed as neither the District Engineer Office nor the HSAC (Hawaiian Seacoast Artillery Command) was consulted in the planning or in the construction. As a result, much of the work had to be redone. The pair of 3-inch, rapid fire guns were retained at Wailea Point until early 1944, when they were replaced briefly by two 90 mm dual purpose guns. Towards the end of the war the 155 mm Panama Mount Battery Pyramid at Fort Hase were scheduled to be relocated to the Wailea site when the armament of Kaneohe Bay was completed. (6)

Details of the war emergency 3-inch batteries can be viewed in Appendix VII.

Brig. Gen. Robert C. Garrett, Commanding General, HSAC, wrote General Emmons on April 10, 1942, that twenty-five 4-inch 50-caliber navy guns were being delivered to the Hawaiian Department Ordnance Officer. The navy had agreed earlier in the year to release these excess guns from the Pearl Harbor Navy Yard ordnance shop to the army for use during the war emergency. Garrett advocated emplacing the 4-inch guns at the earliest practical date, and "gladly offered troop labor" to mount the guns. He noted that the 14th Naval District (Pearl Harbor) had designed a wooden firing platform for mounting the 4-inch guns. On April 20, 1942, Emmons directed the Honolulu District Engineer to fabricate 24 timber firing platforms for the guns, following the navy specifications. (7)

Sixteen of the guns, supplied in batteries of four each, quickly were allocated to neighboring islands otherwise almost devoid of gun defense. Four 4-inch guns were shipped to Canton Island ("Holly" in Army location code), to defend the new Air Ferry Route base on the atoll. The other three 4-inch gun sets were distributed one battery each to the islands of Kauai, Maui, and Hawaii. They were emplaced in simple emplacements manned by special detachments. (8)

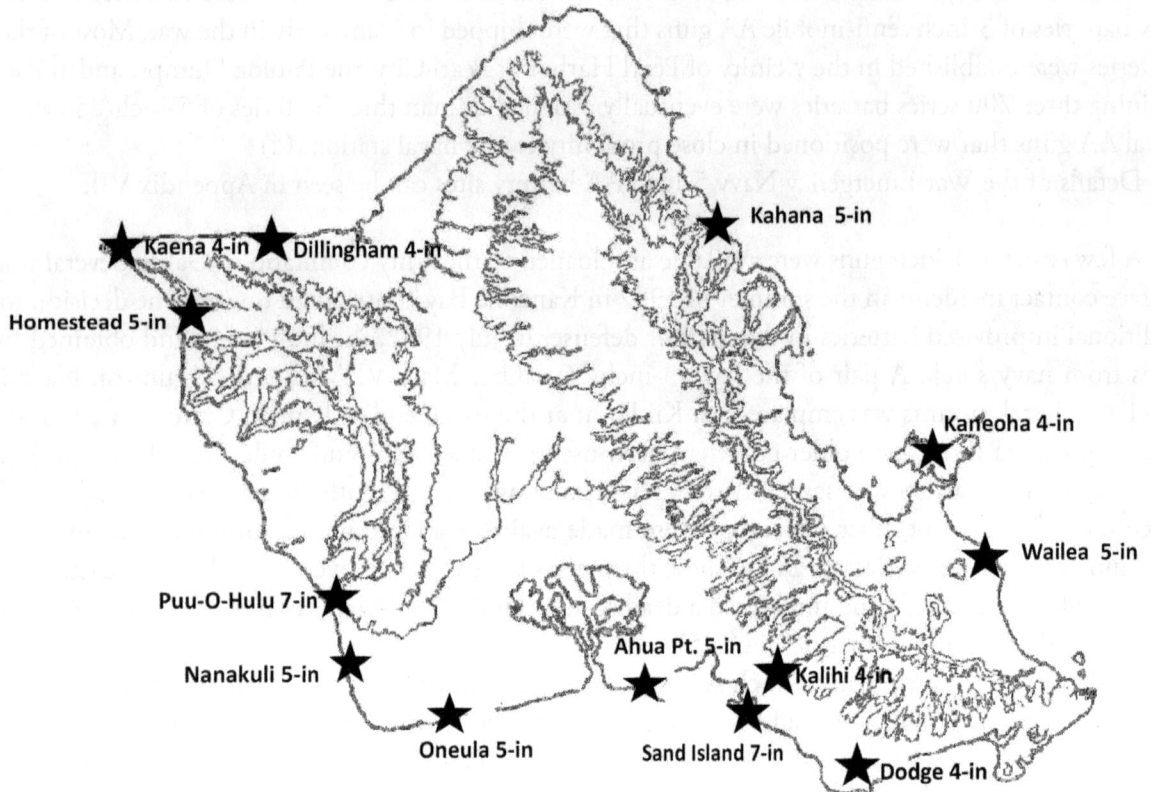

Map showing locations for the ex-navy 4, 5, and 7-inch gun batteries built in the emergency immediately following the Pearl Harbor attack. *Williford Collection.*

For use on Oahu, the Army retained nine 4-inch navy guns. All of the guns used were standard navy 4-inch/50 Mk IX guns on Mk XII pedestals. These were low-angle guns (20° maximum elevation) used since the late 1910s primarily as destroyer and auxiliary main armament. They fired a 35-lb. projectile at 2700-f.p.s. to a range of 15,920-yards. New emplacements were constructed at the north shore near Kaena Point (two guns) and Dillingham (two guns), Battery Dodge at Diamond Head (two guns), and Battery Kalihi (three guns) near Keehi Lagoon between Honolulu Harbor and Fort Kamehameha. In addition, the Marine Corps received and manned a battery of 4-inch guns at the Kaneohe Bay base (eventually named Fort Hase).

Less than a year later the Navy informed the Hawaiian Department on February 25, 1943 that all the 4-inch pedestal mounted guns loaned to the department were being recalled. They were wanted for a program to arm merchant vessels. As some of the locations slated to lose guns were still a priority for defending, the Army had to adjust some of its other armament. (9)

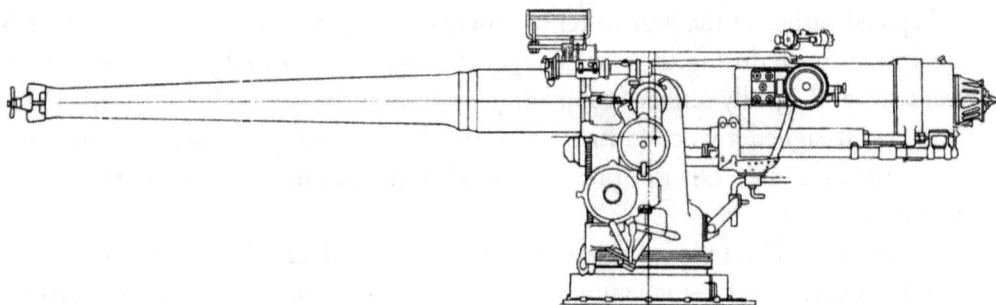

Navy drawing of the 4-inch Mk IX gun and its pedestal mount. Due to the emergency nature of the emplacements and wartime security restrictions, photographs of the guns and emplacements of the emergency batteries are either very rare or non-existent. *NARA.*

At Battery Kalihi in April 1943 four M1917 155 mm GPF guns (part of a shipment of 25 such guns sent to Oahu from the mainland), were earmarked for the Battery School/Kalihi position to replace the 4-inch guns. Similarly on May 21,1943, two 155 mm GPF guns on Panama mounts replaced the 4-inch at Battery Kaena, and were manned by HQ Battery, 3rd Bn, 57th CA. Construction work of the Panama Mounts was supervised by the Hawaiian Department Engineers. The 155s in turn were withdrawn on January 13, 1944, and this position was inactivated. A second pair of 155 mm Panama Mounts was constructed as replacement for Battery Dillingham at its location in May 1943. (10)

Details of the war emergency 4-inch gun battery sites can be seen in Appendix VII.

The sense of surprise and inadequacy that confronted military commanders in Hawaii following the Pearl Harbor raid cannot be overstated. In the first weeks Lieutenant General Delos C. Emmons, succeeding Lieutenant General Walter C. Short as commander of the Hawaiian Department, explored his options. As the probability seemed high that the Japanese forces might well mount additional attacks and even attempt an invasion of the islands, Emmons lost little time in making efforts to bolster Oahu's defenses. By early January 1942, in addition to the 4-inch guns mentioned above, arrangements were made to obtain several 5-inch low-angle guns. On January 7, 1942, the Navy offered, and the Army accepted an offer of eleven 5-inch naval guns for the emergency. Three general locations on Oahu's south coast and one on the island's northeast coast had been selected by General Henry T. Burgin and by January 22 construction by the district engineer began.

These were 5-inch/51-caliber low-angle broadside guns, formerly utilized as casemated guns in battleships. The Mk VIII guns (also designated Mk XV if relined with a uniform twist) on Mk XIII pedestals were fairly easy to emplace, requiring just a stable platform with hold-down bolts. The gun could be elevated to 20° and fired a 50-lb. projectile to a range of 16,000 yards. While using separate, bagged ammunition loading, they had a useful range and their pedestal-type mount allowed sufficient traversing rate to be used against ship targets. The Navy agreed to supply 150-200 rounds per gun. Sights were modified to allow greater deflection. As battleships had been decreasing the number of these gun carried for several years, a relatively abundant stock had accumulated in the navy, enough for use as the primary anti-ship armament of Marine Corps defense battalions just forming. Almost 20 were removed during the salvage work on USS *California* and *West Virginia* after they were sunk in shallow water in the Pearl Harbor attack. At least one from the former and six from the latter found their way into the transfer lot to the army.

The eleven 5-inch guns of this first allotment were initially allocated to four batteries; with emplacements for 2-3 guns each. Sites were selected where small boat landings could be made on beaches connecting to roadways. There were early changes to sites and gun numbers. Originally there were four main locations: one on the eastern coast (Kahana), two on the southern coast (Ahua and Oneula), and one on the western coast (Makua). However, two guns were mounted at a temporary position at Wailea Point covering Waimanalo Bay and these were soon replaced with 3-inch seacoast guns. These guns and one more were transferred to the delayed battery at Kahana Bay. A position for 5-inch guns was also established at Nanakuli on the southwest coast. The Marines used it for defense battalion training, initially manned with Marine Corps personnel from one of the resident defense battalions.

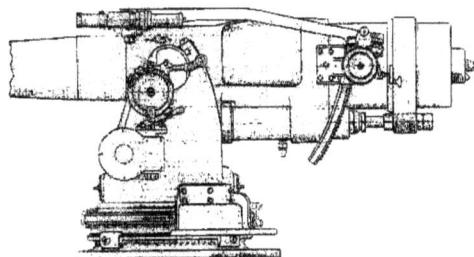

Navy Ordnance drawing of the breech end of a low-angle 5-inch gun, frequently used in battleship broadside mounts. In early January 1942 eleven of these guns were released by the Navy to the Army for emplacement in Oahu emergency batteries. *NARA*.

While in poor focus, this is the only known photograph of a wartime, emergency battery 5-inch gun in its emplacement. This is Battery Homestead in 1942. *Bennett Collection.*

For most of their service period the 5-inch guns were deployed as five separate gun batteries. The Navy had already worked out specifications for fixed, ground mountings for the 5-inch/51. They used a 2-inch steel plate on a 2-inch white oak wooden platform with steel hold-down bolts and nuts. Emplacement requirements were for two 15 x 8 x 6-foot splinter-proof magazines (shells stored separate from powder), a gas-proofed plotting room, a power generator, and a battery commander's station. Depending on the topography of the site, some of the BC stations were simple instrument stands in lightly splinter-proofed structures, some were placed on 15-foot wooden towers for better sighting.

Work began with contractors under the supervision of the 34th Engineer Regiment. Orders to proceed on construction were issued on May 28, 1942. The batteries were completed with armament and crew by October 1942. Later in 1942 Hawaii engineers recommended that the batteries at Oneula, Ahua, and Homestead be casemated with overhead protection. The first two sites had this request denied, the last one was tentatively approved, but (permanently) postponed due to higher priority projects in the Department. In mid-1943 General Burgin surveyed the sites, determining that the 5-inch were not equal to their expected tasks, and should be taken out of service. At places that still required an active defense, new 155 mm Panama mount batteries were substituted. (11)

Detailed summary of the war emergency 5-inch gun battery sites can be seen in Appendix VII.

The navy's 7-inch/45-caliber Mk II gun had been used as a broadside secondary gun mounted in eight pre-Dreadnought battleships from around 1906-1910. It was the heaviest "intermediate" gun, supposedly capable of manual loading and operations. The Mk II gun had a muzzle velocity of 2700-f.p.s. and could elevate to 15°. Maximum range is listed in navy manuals at something like 16,500-yards. As it had been developed for use in ship broadside casemates, it did not have an independent shield, and for land use was mounted in the open on a simple sunken concrete platform.

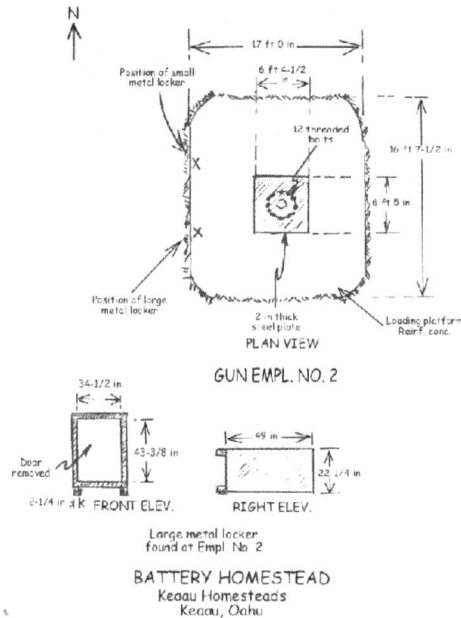

2004 field note drawing by John Bennett of his inspection of the remains of 5-inch Battery Homestead. *Bennett Collection.*

A few guns removed from ships early during the First World War were used in Marine Corps coastal batteries in the Portuguese Azores Islands and the newly acquired American Virgin Islands in the Caribbean. While two of the battleships were sold to Greece in 1914, the other ships all survived the war but not the scrapping demanded by the Washington Naval Treaty. These relatively heavy intermediate guns, many still with good bore life, became available for alternate uses. Some were placed on tracked mobile mounts also used by the marines, and some were transferred to the Army and placed on railway carriages. At the start of the Second World War, there were still dozens of intact guns and mount available for employment.

After analyzing its requirements at Pearl Harbor, twelve (thirteen also mentioned in some correspondence) 7-inch guns were offered to the coast artillery command in Oahu. Two emplacements were selected on Oahu. A battery for four guns was to be emplaced on the Sand Island Military Reservation near Honolulu, and two others to a new site on the slope of Puu O Hulu on Oahu's western coast. Army inspections revealed only four guns in condition for immediate mounting, so Sand Island only got two of the guns. Eventually all would be reconditioned for issue. As there was also an urgent need to get guns to the other major Hawaiian Islands, four others were sent to Kauai. They were soon emplaced at batteries at Ahukini and Monument (near Nawiliwili Bay). In total just ten were emplaced, with two or three spares. (12)

By mid-February the gun locations for the two Oahu batteries were being prepared. Besides the fact the Puu O Hulu was only to have two guns, and Sand Island four, there was one other major difference between the emplacements. Sand Island had open emplacements; the gun blocks were surrounded by a concrete retaining wall embanked on the exterior with sand. Ready ammunition niches were built into the walls. The two emplacements of Puu O Hulu had what was referred to as "tunnel casemates". The guns were placed into an enclosed concrete structure with a front opening for the gun. Separate powder and shell magazines were included to the rear of the gun. The rationale for the casemate use at Puu O Hulu was to protect the guns from rocks rolling down from the hill behind. Incidentally, both Kauai emplacements were also casemated. (13)

The layout of the battery structures followed the design described for the 4 and 5-inch navy gun emergency batteries. The guns were placed on concrete blocks, with a cushioning layer of wood, and then a steel plate with the hold-down bolts. Separate magazines for the shells and the powder (bagged powder stored in canisters) were either in the gun casemate structure or otherwise in protected magazines. The emplacement had splin-

Navy Ordnance drawing of a 7-inch/45-caliber Mk II broadside gun. Twelve were supplied to the Army command in Oahu after the December 7th attack and emplaced at two locations on Oahu and others eventually went to Canton Island and Kauai. *NARA.*

ter-proofed plotting room, a battery commander's station with observation slot at an elevated position, a small power room for a reserve generator, and in 1943 the locations added an aid room and facilities for SCR-296 radar and its associated operating room. The plotting room and commander's station received gas airlocks.

The guns were emplaced in the two Oahu batteries relatively quickly and were in service by mid-summer 1942. On October 21 the names of Battery Hulu for the Puu O Hulu site and Battery Harbor for the Sand Island site were confirmed. However, within not too long a time the batteries became obsolete in terms of armament and the new generation of permanent defenses were replacing their function. On October 30, 1943, the HQ of the Army Forces in the Central Pacific recommended both Oahu 7-inch batteries be replaced by modern 6-inch batteries, and two months later the local command was given authorization to abandon the batteries at a time of their own selection. On December 14, 1943, a letter to the army's adjutant general contained the following description of the 7-inch/45: (14)

October 1942 sketch of the plan for the two-gun 7-inch battery at Puu-O-Hulu. There are no known photographs of the 7-inch guns emplaced during wartime in the Hawaiian Islands. *Smith Collection.*

"The 7" Navy guns, Hulu and Harbor, are a discard from the Navy and are wholly obsolete. The guns in these two batteries have been fired so many times that they are inaccurate – some of them being practically smooth bore. No 7" ammunition is being manufactured and when the small supply on hand is used, these guns will be of no further use".

Two more 7-inch guns, probably the pair originally kept as spares, were released, and sent for emplacement on Canton Island. This station was one of the aircraft ferry bases on the leg south from Hawaii. Elsewhere in the Pacific the Navy supplied eight 7-inch guns for use by the army garrison at the new refueling depot at Bora Bora, and another navy-supplied set of four guns were emplaced with the Marine Corps defenders of Midway Island early in 1942. It's not clear precisely when the 7-inch guns on Oahu were deactivated or removed. A note from March 1946 confirms that Battery Harbor was abandoned with guns and equipment removed. Two of the guns and carriages from this emplacement are now display pieces at the U.S. Army Museum of the Pacific at Fort DeRussy, Honolulu.

Details of the war emergency 7-inch gun battery sites can be seen in Appendix VII.

The two sites selected for the emergency 7-inch batteries were already existing military reservations. While not large enough to be granted fort status, they were more than just expropriated commercial property intended for temporary use.

The Puu O Hulu Military Reservation was established April 27, 1923, by Executive Order of the Territorial Governor. It consisted of approximately 0.15 of an acre atop Puu O Hulu Kai. It was originally acquired for use as a fire control site for the newly emplaced 16-inch guns of Battery Williston at Fort Weaver. Over the next twenty years, an additional 67.76 acres were obtained from various owners. The reservation was first used as fire control station "U," and later became the command post for the Hulu Group, a gun group under the Pearl Harbor Defenses. During the period between the two world wars the reservation was under the administrative control of Schofield Barracks then in 1939 it was under the jurisdiction of HSCAB. During World War II two batteries were established at Puu O Hulu. The first was the casemated emergency battery for two 7-inch guns, named Battery Hulu. The second was for the new modernization program 6-inch barbette battery of Battery Construction Number 303. (15)

Other fire control stations were later built on the heights of Puu O Hulu. A cable car right of way was obtained in April 1931, to improve access to the fire control stations that were sited some 650-feet above sea level. The cable tramway was powered by a six- cylinder GMC cable engine and pulley. The site was further enlarged in 1934, when a fourth fire control station was added for the newly emplaced 16-inch guns of Battery Hatch. This station was adjacent to, and slightly below that of Battery Williston's fire control station but was somewhat larger to accommodate the pedestals for two DPF instruments. A fifth single story fire control station with two pedestals for DPF Instruments, and space in the rear of the observing area of the station for a plotting room and a fire control switchboard was partially dug into the crest of the puu about 1940. During World War II, the Hulu Group provided tactical command and control of the 7-inch guns of Battery Hulu, the 8-inch railway guns at Browns Camp, and its alternate position at Waianae. It also controlled Battery Awanui's 155 mm guns at Browns Camp, the 5-inch and 155 mm guns at Nanakuli, the 155 mm guns at Kahe Point and the 5-inch and later 155 mm guns of Battery Homestead. In 1945 one of Battery Arizona's base end station was established in the structure formerly assigned to casemated Battery Hatch. The 1920s station formerly reserved for the railway batteries along the Waianae coast was reassigned as a base end station for Battery Salt Lake's two 8-inch turrets.

In addition to the fire control stations, the group command post, and the landing for the tramway system, there were two timber dormitory buildings and mess hall as well as a fire control cable terminal hut perched on the steep slopes of the puu. The installation was abandoned by the Army in 1948 and jurisdiction passed to the territorial government. The five fire control stations and the fire control cable terminal hut on the rugged summit of Puu O Hulu Kai are still extant. (16)

The searchlight installation at Puu O Hulu Kai, consisted of two 60-inch portable lights, Numbers 57 and 58. These lights were on concrete emplacements on the northwest slopes of the puu at an elevation of about 100-feet. The manning details for the lights were provided by headquarters elements of Headquarters Battery, 1st Battalion, 41st CA until Battery G, 15th CA was activated in August 1942. Detachment continued to man the lights until the end of the war.

The Kaholaloa Reef formed the base for what would eventually become Sand Island. Dredging of the Honolulu harbor channel placed spoil upon the reef forming the small islands of Quarantine Island and Sand Island. The Quarantine Station was built by the federal government on the former, while the Territorial Magazine and a Department of Commerce Light House Reservation was established on Sand Island. Additional widening and deepening of the harbor eventually resulted in the joining of the two isles into today's Sand Island with an area of 641 acres.

The secondary mine station for Honolulu defenses was built in 1913 and was the first element of the fortifications built on the Sand Island. It was a small building some nine feet above the ground level supported on an iron framework. The station was transferred to the Fort Armstrong garrison on December 29, 1913. The structure was not retained in service after the mid-1920s when submarine mines were deleted from the defenses. The increased military use of the island came when the War Department exchanged the Queen Emma Wharf area at Fort Armstrong for the Territorial Government's Immigration and Quarantine Station. The island's principal military function in the years prior to World War I was as the location of a submarine cable that traversed the length of the island and its cable hut. In 1917 it was selected for construction of two emplacements for 3-inch antiaircraft guns. A second pair of antiaircraft gun emplacements was built in 1928 adjacent to the initial pair near the island's southeast end. After March 1921, the Sand Island reservation was included within Honolulu's Harbor defenses. A new communications cable terminal hut was built of timber in the latter part of 1931.

In 1933, Sand Island was designated as the site for one of the twelve concrete machine gun shelters, or "pillboxes" to be built on Oahu. Two others were built on nearby Mokauea Island, and two more on Mokuoeo Island just to the west of Sand Island. All five formed a part of the beach defenses of Oahu, and also covered the shallow channel leading into Honolulu Harbor from Keehi Lagoon on the west side of Sand Island. All were of reinforced concrete construction and measured 13-feet square. When elements of the 25th Infantry Division deployed to Sand Island on the afternoon of December 7, 1941, troops of the 2nd Battalion, 27th Infantry Regiment occupied these and other beach defense positions along the south shore of Oahu in the Honolulu and Waikiki areas. (17)

During the 1930s, the growth and development of the Waikiki and Diamond Head areas of Honolulu encroached upon Forts DeRussy and Ruger to such an extent that the coast artillery could not conduct effective live fire exercises. As a consequence, the Sand Island tract was adopted as a firing point for 155 mm guns. Two pairs of Panama mounts had been constructed for 155 mm mobile seacoast guns by 1937. The Panama mounts were located about 300-feet to seaward from the antiaircraft gun battery. The 155 mm battery was soon provided with an emergency power generator housed in a small concrete splinter-proof building. Fire direction and control was provided by a 15-foot CRF installed atop a 35-foot-high steel tower. The 20-foot square instrument house was provided with a splinter-proof sides and roof. From this height a ship could be tracked out to a distance of 20,000-yards.

A pair of 60-inch seacoast searchlights on fixed mounts atop two six-legged rigid towers 26-feet, six inches tall constructed of welded steel were erected on Sand Island in 1941. A corrugated galvanized iron shed measuring 12-feet square with 12-feet, six-inches of headroom atop the tower housed the light. The lights were controlled from a 50-foot-tall tower acquired in 1941 that was erected 597-feet to the northwest of the searchlight towers. Sand Island was also the final site for the combined Harbor Entrance Control Post Observation Post (HECP-OP) for the Honolulu Harbor defenses as well as a signal station and Battery Commander's Station for Battery Harbor.

Sand Island became a major Quartermaster Corps installation in the last two years of the war and as the coast artillery presence on Oahu was being reduced, many of the coast artillery units awaiting inactivation or disbandment were assigned to quartermaster duties here. Here also was located one of the primary installations of the Army Port and Service Command in Hawaii. The Sand Island Military Reservation was inactivated at the end of the war. (18)

Periodic reviews of the island's defenses during the period between the two world wars continued to consider Oahu's north and northeastern coastline between Kaena Point and Makapuu Head to be adequately guarded from large scale invasion by the numerous offshore reefs. There were few sheltered bays that could be used to land and support an invasion force, and the rugged Koolau Range that commanded Windward Oahu was still considered an easily defended natural rampart.

During the period between the two world wars, the Army and Navy conducted joint maneuvers at various locations in the United States. Because the Hawaiian Department was one of the principal training commands in the U.S. Army during that period, Hawaii was from time to time selected as the location for the joint maneuvers. Those carried out in 1932 were designed to test the coastal defenses of Oahu against invasion. In addition to the participation of the forces on Oahu, the battle fleet of the U.S. Navy based at San Diego, the 30th Infantry Regiment, and Battery D of the 76th FA, both stationed at the Presidio of San Francisco were selected to provide the "invasion force." (19)

After a period of training disembarking from a ship by way of rope ladders into small boats, the 30 officers and 610 men of the 30th Infantry and Battery D boarded the U.S. Army Transport *St. Mihiel* at the Fort Mason on January 31, 1932, and joined the fleet at sea off Southern California. The fleet arrived off Waimea Bay on Oahu's North Shore at dawn on February 11. After making a feint against that side of the island, the fleet withdrew and after dropping over the horizon, turned and made their approach to Oahu's southwest coast under cover of darkness. The landing was made from launches manned by the Navy. The surprise landing at Makua was accomplished at dawn on February 12, and no opposition was encountered on this isolated part of Oahu's southwest coast until the advance elements were entering the small town of Waianae some eight miles to the south. Here the "invaders" were briefly opposed by only a single platoon of the defending force. At that point a force of Marines began their landing over the beach at Waianae in a flanking attack. The maneuver ended by late afternoon and the "invasion" force moved by rail on the O.R.&L. to Schofield Barracks where they went into bivouac. (20)

Among other lessons learned from these exercises was the relative ease that landings, even of considerable size, could be made on Oahu beaches. One of the deficiencies noted on the 1932 Joint Exercises was the lack of strong points to serve as "anchors" of the beach defenses. Remedies were soon proposed.

In November 1933, the Hawaiian Department authorized the construction of twelve concrete machine-gun shelters, or "pillboxes," at various locations around the island of Oahu. While most of these were constructed along the island's south and southwest shorelines, two locations were selected on the North Shore. One of these reinforced concrete structures was built at Puuiki to cover the beaches of Waialua Bay, and two more on the high ground at Waimea Bay. All three of these pillboxes, as well as those constructed at other locations on Oahu at this time, were identical in design and construction. The structures were 13-feet square in area and had a height, from floor to the top of the concrete roof, of 7-feet, 3-inches and were dug into the ground so that the embrasures for the machine guns were just above the surface of the surrounding terrain. The firing doors of the embrasures were composed of one and a half-inch thick boilerplate. Upon completion of the shelters in June 1934, the small amount of exposed concrete was painted in camouflage colors and a thick layer of sod was placed on their roofs. Fittings were included to hold the Model 1917A1 0.30-caliber Browning machinegun. All 12 were finished soon, most being turned over to troops for a calculated construction cost of $1229 apiece. (21)

Key ingredients to a successful defense were the lines of communication and supply. It is worthwhile to note that during World War I, there were just 171 miles of roadway on the island, of which only about 18 miles were actually hard-surfaced. On Oahu's windward side there were several dirt roads. A coastwise road, designated Highway 1, ran from Honolulu to the junction with Pali Road via Makapuu Point. At the time of World War I, it was unpaved from the south side of the island around the east end of the Koolau Range through the hamlet of Waimanalo to Kaneohe. This part of the road was especially susceptible to washouts after even moderate rainstorms, a prevalent characteristic of the island's windward side. The roads in and around Kaneohe, the island's second largest community, were little more than dirt tracks.

The (Nuuanu) Pali Road, later designated Highway 13, was the only road from Honolulu over the Koolau Range. The Pali section of the road was provided with a short section of concrete slab pavement about 1917. From the Pali junction northward to Kahuku, Highway 1 hugged the shore for much of the distance, and at several places traveled between nearly sheer cliffs on one hand, and a rock-strewn beach on the other. On the windward side of Oahu, it was commanded by the towering heights of the Koolau Range. Up in the vicinity of Laie, a few dirt plantation tracks ran inland through the cane fields for a few miles before petering out near the foot of the Koolau Range or in a few cases being reduced to a pathway trail leading up into the Koolau. At Kahuku, the highway passed around the north end of the Koolau Range.

During the 1930s considerable attention was given to improving the road and trail network on the island of Oahu, but nowhere was it more evident than on the North Shore. Extensive examinations of the island were undertaken between 1932 and 1935 to determine the exact location and characteristics of the trails on the North Shore. The mountains and their many spurs were explored to determine what work would be required to make them suitable for the movement of the Hawaiian Division's infantry and machine gun carts, and to locate the best observation points along the trails. Similar surveys were conducted by the Hawaiian Separate Coast Artillery Brigade between June 1932 and December 1936 for the purpose of determining positions for the 155 mm guns and railway artillery as well as for observation posts for the coast artillery batteries. After completion, the findings were used to develop the Hawaiian Defense Projects and the data acquired was used to enhance the road net on the island.

An alternate northern route was developed by the 3rd Engineer Regiment of the Hawaiian Division. Funds from the depression era's Emergency Recovery Act were appropriated for the new road's construction. This alternate route, originally known as the Wahiawa-Pupukea Trail was begun in early 1936, at a point about two miles north of Wahiawa where it intersected with Route 135. It then followed a generally northerly route that roughly conformed to the foot of the southern slopes of the spurs of the Koolau Range winding through the rugged terrain of Oahu's interior to the eastern terminus of the Pupukea Road. There the road turned westward to an eventual junction with Highway Number 1. When it was completed in June 1937, this new, but still primitive track, enabled traffic to move between the Pupukea area and the interior of the island near Schofield Barracks while avoiding the coastal area. Although it received the designation of Route 135, and its southern branch Route 124, upon completion, the military road was named Drum Road in honor of Major General Hugh A. Drum, who commanded the Hawaiian Department from March 19, 1935, to July 30, 1937.

# CHAPTER 10
## THE NAVY CONTRIBUTES TURRET BATTERIES

Breakthrough new weapon systems have frequently been accompanied by tactical usage errors. When the aircraft carrier was "new," just how it would best be employed was a less than obvious answer. For a period in the 1920s and 1930s there was a debate about whether it was a scout ship incidentally equipped with some airplanes, or a fleet airplane carrier incidentally needing at least a modicum of self-defense. The defense role was further defined with the arms limitation agreement of the Washington Treaty that put an upper limit of allowed guns on carriers of an 8-inch bore. In any event the large carriers converted from capital ships of both the Japanese and American navies carried an 8-inch battery of guns.

In the late 1920s the battle cruisers USS *Lexington* and *Saratoga* were completed as aircraft carriers with eight modern 8-inch/55 caliber Mk 9 guns. These guns were mounted in four dual, lightly armored gun houses, two forward and two aft of the starboard island structure. Though not technically turrets, most later army correspondence utilized that term, and it will be used similarly here. The housing had only 3/16-inch armor thickness, barely enough for protection from bombardment splinters. The low-angle guns were suitable only for surface action. Their 2800-fps muzzle velocity could deliver a 260-lb. armored projectile at 41° elevation to an impressive range of 31,800-yards. Control was either local at the turret, or more generally by the General Electric Mk 18 director from the control tops. The mounts were fed ammunition by hand from handling rooms immediately below the turret. (1)

There was a persistent debate within the Navy concerning the wisdom of devoting the space and weight for this armament on these ships. Even before the design was finalized the Navy Department Bureau of Aeronautics was opposed to the heavy surface armament. Finally, the General Board approved its installation with the observation that if proven undesirable, it would be relatively easy to remove it later. As war approached the clamor to remove the turrets and replace the guns with heavier antiaircraft batteries increased. It appears that the Navy had already made this decision at the time of Pearl Harbor, but it wasn't to be accomplished until a convenient period of planned ship modernization. Obviously, the start of the war had a serious impact on the frequency of routine maintenance availabilities at the naval dockyards.

While operating some 500 miles west of Hawaii at 19:15 in the evening of January 11, 1942 USS *Saratoga* was torpedoed by Commander Michimune Inaba's submarine *I-6*. A single torpedo hit low on the port side at about frame 105. Three boiler rooms were flooded and a 6° list developed, but it was corrected using counterflooding and fuel transfer. There were six casualties. The ship was not endangered but would have to suspend operations and did require major repairs. The carrier made Pearl Harbor at 17 knots and tied to Pier 9 on the 13th. Two days later she went into drydock No. 2. The intent was to make temporary repairs at Oahu and then proceed to Puget Sound where complete repairs could be carried out. These repairs prompted the navy to address the changeout of main armament. The four deck mounts would go, being replaced by a like number of dual 5-inch/38 twin dual-purpose mounts. (2)

The earlier opinions about removing the 8-inch armament of the carriers would not have foreseen the dire situation and perceived threat of invasion to Hawaii since the Pearl Harbor raid. The Navy was preparing to loan the Army its grab bag of excess deck guns of 3, 4, 5, and 7-inch size. They were even planning on using salvaged anti-aircraft batteries in southern Oahu on their own accord and with crews drawn from disabled battleships. A perfect solution for this now excess 8-inch armament was to loan it to the army for use in defending Oahu.

It is not known whose idea this precisely was, or even which service came up with it, but things seem to begin moving quickly after the torpedoing. On January 17th (just six days following the damage) the Commander of the Pacific Fleet sent an inquiry to the Army about any interest they might have in the mounts. Army documents recognize the 18th as the official authorization date for the site projects. On the 24th of January; just 9 days after going into the dock, and 16 days before *Saratoga* left for Bremerton, the Hawaiian Engineer Office in a document entitled "Coast Artillery Projects" issued a priority for emplacing the four guns at two sites. (3)

Coordinated plans for the removal came quickly. On the 27th the local navy yard authorized the removal work. For the first step the Army was requested to immediately take away the ship's ammunition for the guns. 870 of the 260-lb. projectiles (and 16 drill rounds with 40 drill charges) were emptied from *Saratoga's* magazines as she sat in Dockyard No. 2. Already removed and at the navy's Lualualei Ammunition Depot were 870 full powder charges. By the 30th both the projectiles and powder were relocated into the army's hands at their Aliamanu Ammunition facility. The Army would design emplacement magazines to hold 250 rounds for each turret, so this quantity was almost a complete allotment for this first set of turrets. (4)

The following day the turrets were cast loose, detaching the forward two mounts first while in dock, and then moving the ship so the aft pair could come off under the yard's hammerhead crane. All four mounts were to be removed, and moved by the Army no later than February 5. As any dock or yard space was at a premium (remember all the urgent work that must have been going on since the December 7 attack), the Army needed to remove any part of the mounts or ammunition as quick as possible.

The Coast Artillery Command made a final decision on sites on February 5. Two turrets were to form a battery at sites already staked out at Brodie Camp, and Aliamanu Crater (later referred to as Salt Lake). Probably aiding this selection was the fact that both were already army land and negotiations or condemnation action wouldn't be necessary. Engineers were authorized to start work on the 8th, based on design plans. Already alert to the probability of getting the four turrets from the sistership USS *Lexington*, a second set of battery locations were selected on February 15. The third set would go to the north shore to augment those at Brodie Camp—Opaeula. Then the final battery would go to the south shore to reinforce the Salt Lake position—but further east at Wiliwilinui Ridge. (5)

It was also realized at this stage that these guns and their extensive coverage of virtually all the beaches on Oahu, could be valuable in opposing landing operations and forces. In addition to the ex-navy 260-lb. heavy armored projectiles, a reasonable number of high-explosive rounds and appropriate fuses were made available.

The two 8-inch gun houses aft of the superstructure of USS *Saratoga* on March 28, 1928, not long after commissioning. *NARA*.

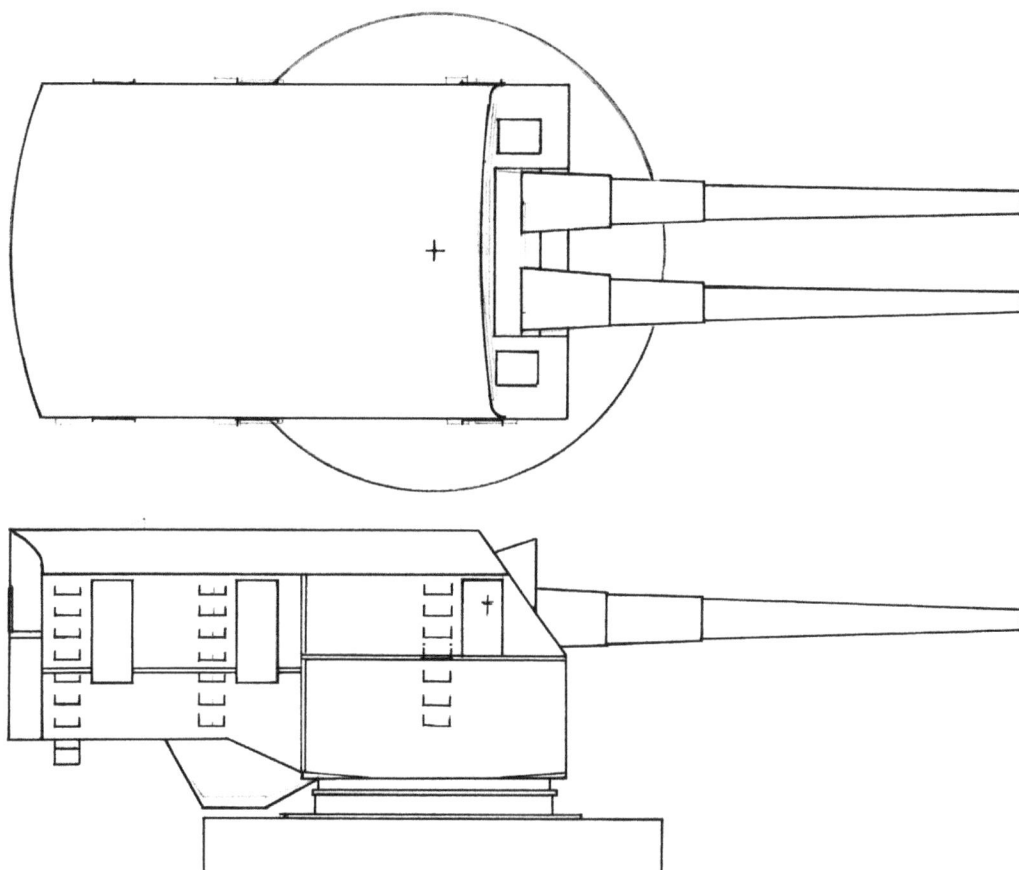

Diagram of the 8-inch aircraft carrier gun house such as the type used on USS *Lexington* and *Saratoga. Rowbottom*

Progressive pictures showing the removal of the 8-inch mounts from USS *Lexington* in March 1942 at the Pearl Harbor Navy Yard. *NARA.*

8-inch turret landed adjacent to the aircraft carrier. *NARA.*

Two of the mounts from USS *Lexington* now sitting dockside. *NARA.*

Battery emplacement plans were modified for all four sites to add an above-ground storage rack for these 200-lb. shells.

On March 19th, just a month after the start of work with *Saratoga* turrets, *Lexington* was scheduled for a period of yard availability at Pearl Harbor for four major modifications. She was to have her 8-inch turrets removed, have eight 1.1-inch automatic AA mounts installed (one already on board, and seven shipping from Mare Island to Hawaii for this), mount eighteen 20 mm light AA guns, and get an overhaul of wing void space flood valves. While the intention was to mount 5-inch DP guns in place of the 8-inch turrets, they were only available on the West Coast so just the 8-inch turrets would be removed at Pearl Harbor. Interestingly the navy was insistent that the ship be ready for combat operations on a 24-hour notice if necessary. It was estimated that it would take two days to prep, and then six days to affect the removal. Prior to the ship's arrival dockside two 500-ton barges were shored and cribbed each to carry one complete "unit" (the three lift loads of housing, guns, rotating base), Then the ship would be moved under the 250-ton hammerhead crane for lifting the other two units and placing them on pre-set cribbing. All the units could then be removed by rail from the yard. (6)

The Army was in receipt of the four turrets by March 30. This time a total of 950 projectiles and charges were available. On April 9th orders were issued to start work on the first site selected for these batteries, that at Opaeula. There were second thoughts about the location of the final site, and its start and eventual completion was slightly delayed. By the end of May all four sites were well underway with just about the highest priority on the island for fortification projects. The Navy helped with technical assistance when needed. All of the equipment needed was immediately available—guns, mounts, handling room equipment, ammunition, directors, even navy sound-powered telephone sets. Later the Army would add SCR-296A radar and contribute their type of plotting, reserve electrical power, and switchboard equipment. This was a major accomplishment in a short timeframe; the installation of 16 heavy guns in protected turrets sited to give coverage of the entire island in less than a year. While there were plenty of delays and frustrations with some of the other heavy gun projects in Oahu, this was not one of them. (7)

One of the 8-inch turrets being moved by rail to its new seacoast emplacement site. *Berhow Collection.*

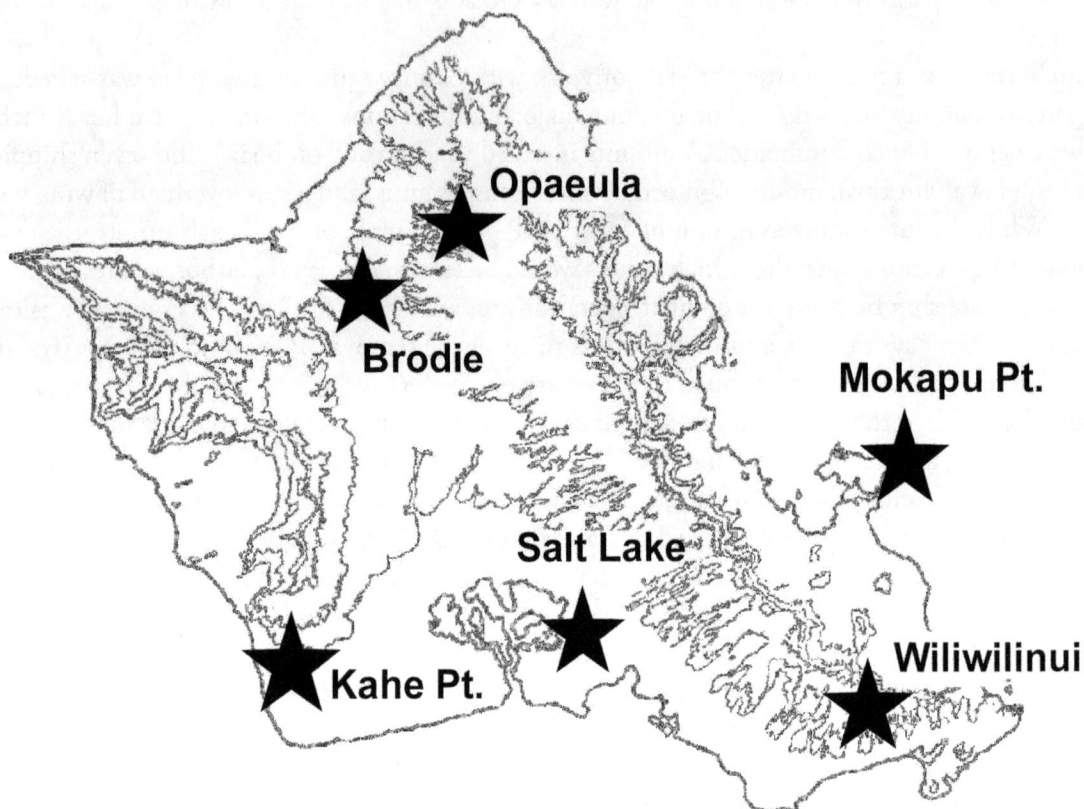

Map showing general locations for the six "turret batteries" emplaced on Oahu during the Second World War.
*Williford Collection.*

The two North Shore batteries (those at Brodie and Opaeula) followed a simple layout plan. That plan had to be adjusted because of topography for the two South Shore batteries (Salt Lake and Wiliwilinui). The preferred method of construction employed was that of the cut and cover type. Pits 20-feet deep were excavated, and timber forms built for the steel reinforced concrete gun barbettes and magazines, the power plant, plotting and fire control switchboard rooms. When the concrete work was finished the structures were buried and covered with between five and ten feet of earth. This minimal covering of concrete and earth was necessitated by the decision to copy the gun mount layout used aboard the carriers.

The two turrets as they were termed by the Army, were installed atop large cylindrical concrete barbettes approximately 450-feet apart. The barbettes descended some ten feet below the surface of the ground. At the base of the barbettes were naval type watertight doors opening into the magazines. One opening was for the service of the powder charges, and the other for the projectiles. The separate magazines formed a "V" with the gun barbette being located at the apex. The floor of the powder magazine was about 15-feet below the surface of the ground and was 13-feet wide and 40-feet long. The floor of the projectile magazine, also some 15-feet below the surface, was 46-feet in length and ten-feet wide. Along the right side of the shell room a projectile storage table extended nearly the full length of the room, just 30-feet. At the rear of each of the magazine spaces there was a tool room and a gallery to the exterior of the structure.

Each of the batteries had a battery commander's station centered at about the mid-point between the two turrets. This structure was one of the batteries few above ground elements. When the batteries were being built, it had been contemplated that the Mk XVIII Optical Gun Directors used with the guns aboard the carriers would be also used with the shore-based armament. However, eventually the navy fire control equipment was supplanted by more traditional Coast Artillery range finding methods utilizing DPFs and horizontal baselines and base end stations, and later, radar. The battery commander's stations had pedestals for the naval fire control director as well as a DPF and an azimuth instrument or spotting telescope. Later, in the war the batteries

## 8-inch Turret Layout
## August 29, 1942

GENERATOR BUILDING

To Power Room

DIRECTOR BUILDING

The spacing of the conduit should be reduced within the last length in order to allow cables to enter hole in the roof

GUN POSITION NO. 1

GUN POSITION NO. 2

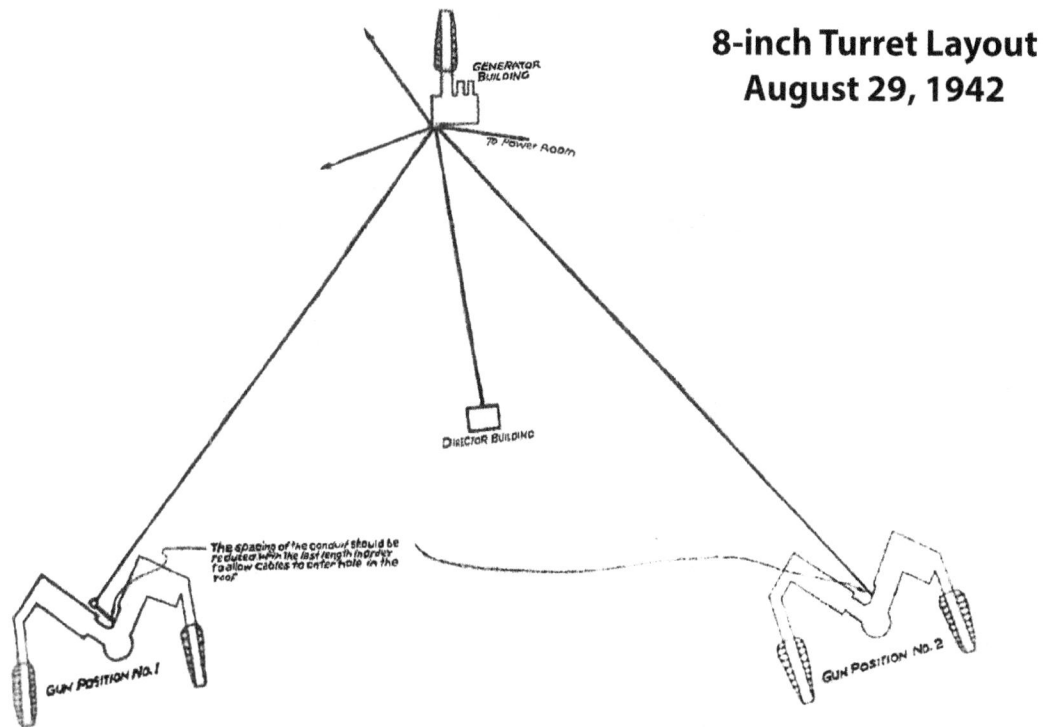

Generalized August 1942 plan for the 8-inch turret mount battery. Note the two separate emplacements each for one turret and adjacent underground magazine and power generator room between the gun blocks. *NARA.*

were equipped with the SCR-296A type fire control radar. This required a separate operating room structure, usually added adjacent aside the plotting room or the battery commander's station.

When the extra HE rounds were allotted, a new building was partially dug into the ground to the rear of the turrets. This building had an arched roof of heavy corrugated steel sometimes referred to as "Elephant Iron." The building was entered by way of a concrete staircase from the ground surface. An additional 600 projectiles were to be stored in the open outside the magazine.

Underground plotting rooms were also provided. Initially, only one plotting room was contemplated for each pair of turrets. However, as the batteries approached completion in the late summer of 1942, the decision was made to add a second plotting room so that each turret could engage its own target. The plotting rooms contained M1 Plotting and Relocating Boards. Attached to the walls were the appropriate fire control maps, meteorological data boards, and other fire control data boards. Telephone boxes were ranged along one wall of the plotting room and the telephone lines connected with the headset connections. The plotting room was gas proofed and equipped with an escape tunnel and shaft for use in the event that the main entrance became blocked. A horizontal escape corridor led to a vertical shaft containing a staple type ladder up to the ground surface that terminated in a small concrete house with a pyramid-shaped roof.

150 feet directly to the rear of the battery commander's stations, and some 15-feet below the surface of the ground, was the battery power generator room. This structure was about 30-feet long and 21-feet wide. Although all four of the turret batteries operated on commercial electric power, this structure and its power generators provided an emergency supply of electricity to the two turrets and the battery camps. Each battery location had a camp for the operating crew. The barracks, mess halls, latrines, and other structures were of the Theater of Operation-type construction: light wooden frame buildings with an exterior finish of 15-pound tar coated felt paper. Most of these structures were generally located to the rear of the turrets in the vicinity of the power generator room. Concrete pillboxes and sandbagged emplacements armed with 0.30 caliber machine guns for local defense, were placed at various points all around the battery perimeters to cover their approaches. (8)

Gun Unit No. 1

Storehouse

Battery Commander's Station

Barracks

Barracks

Barracks

Gun unit No. 2    Barracks

Generator Bldg.

Barracks

Barracks

Barracks

Barracks

**BATTERY BRODIE**

**BRODIE CAMP No. 4**

Barracks

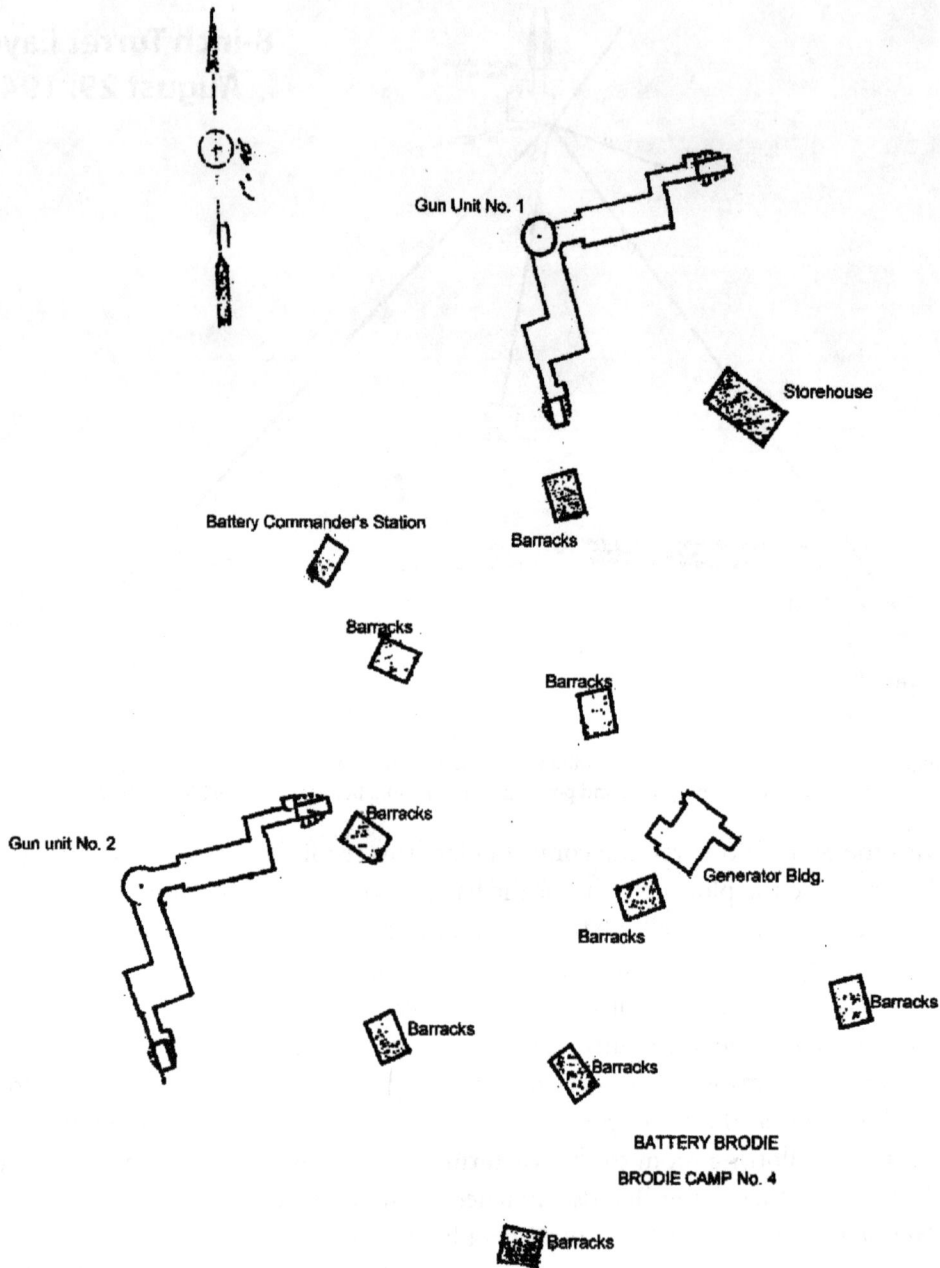

Layout for the 8-inch turret reservation of Brodie Camp, later the battery was named Battery George Ricker. *Bennett Collection.*

There is an interesting postwar remembrance by an enlisted man that was stationed at the Battery Brodie turret position: (9)

Battery Brodie was adjacent to Brodie Camp No. 4, on the north side of the pineapple-workers' village in the pineapple fields several miles above the village of Haleiwa. Both 8-inch "turrets" were emplaced at the north end of the camp, facing Waialua Bay and the ocean. The mounts were painted their original navy gray and retained this scheme as long as I was with the battery. Each "turret" was supervised by a "three striper" (sergeant) gun captain, and a pointer and setter were also assigned to the turret, all equipped with earphones, and in communication with the command center and officer in charge. The pointer traversed the turret; his seat was on the left side as you faced forward. The setter on the right elevated the guns. The door to the turret was on the left, and two peepholes on either side of the guns were for the pointer and setter; a window was at the rear of the turret. The other members of the crew assisted with loading the weapons. The entire turret was covered with a timber latticework affair, with a corrugated iron roof that resembled the half-sized rooftops of the native huts of the nearby village; the whole affair was removed before the guns were fired. When loading the guns, the projectile came up first and was rammed into the gun by a power rammer; then by the same action each powder bag was put into place. The breechblock was then closed and locked; a primer cartridge was implanted into the outside of the breechblock, a wire placed on the primer, and the gun was fired, either from inside turret or from the command post. I was glad I was not assigned to one of the gun crews; they had to train repeatedly, day after day, in the high temperatures found inside the turrets during the daytime.

Side view of one of the turrets (and the second beyond) installed and in service at Brodie Camp. Structure on top of the turret is an attempt at camouflage from aerial recognition. *Smith Collection.*

Appearance of one of the Brodie Camp turrets head-on. *Smith Collection.*

Two turrets from USS *Saratoga* were projected for emplacement on Oahu's south shore at a site at the foot of the south rim of Aliamanu Crater. It consisted of guns and mounts for two twin 8-inch/55 Caliber Mark IX Mod 1 naval guns that were received in a somewhat disassembled condition. In the meantime, construction of the battery emplacements, magazines and fire control structures had begun on February 10, 1942. Construction was given a high priority, second only to the construction of 8-inch gun emplacements on the Mokapu Peninsula and at Kahuku Mill on the island's windward side. Even use of lights at night was authorized to accelerate progress. Unlike the level ground plan described above, the presence of the crater rim immediately behind the battery allowed for the use of some tunneling. The battery's magazines, plotting rooms, and power plant were a combination of tunnels (to a depth of about 50-feet) and cut and cover type construction. When construction was complete, these structures were given a covering of earth and rock a few feet thick to render them splinter-proof.

The two barbettes were about 350-feet apart and extended some eight-feet into the ground. Extending out from the base of each barbette were two concrete lined cut and cover tunnel structures, each about 50-feet in overall length, which formed a V with the barbette at its apex. By April 9, 1942, the concrete floors of the magazine galleries and the bases of the barbettes had been poured and forms and reinforcing had been placed for the gallery sidewalls. Bridges and roads had also been strengthened and repaired in preparation for hauling the barbettes to the battery site. The battery commander's station was a reinforced concrete structure. Because of a construction error in calculating the height of the naval director's optical system, a hole was cut in the floor to lower the entire unit for mounting on a sunken metal table. This unit proved to be unsuitable for shore-based fire control and was soon replaced with standard Army fire control equipment. In August 1943, a SCR-296A fire control radar set was installed. Its cylindrical antenna housing measuring 12-feet in diameter was camouflaged to resemble a water tank atop a 25-foot high. Because the battery was built into the slope of a crater, the battery did not have all around fire. The turrets were equipped with both mechanical limit stops, and electrical cutoff stops.

Brodie Camp's Battery Ricker photograph of the 8-inch gun breeches inside the turret. *Smith Collection.*

Brodie Camp's Battery Ricker underground shell magazine loading tables with projectiles. *Smith Collection.*

The plotting room for the two turrets when completed was also dug into the south slope of Aliamanu Crater. This gas proof structure housed the 18-foot by 25-foot room was accessed by a ramp. The plotting room itself had an escape shaft that exited some 20-feet farther up the slope. A power generator building 21-feet wide, and 30-feet long was located to the rear of the battery. Like most of the other structures on the reservation, it too was of the cut and cover type reinforced concrete construction. An extra magazine for high explosive rounds was also built. This partially sunk reinforced concrete structure measured 20-feet long and 13-feet wide.

On May 21, 1942, the 805th CA (HD) Battery (Separate) was activated with a full wartime strength of three officers and 156 enlisted men providing the battery's manning detachment. The same day the 805th was activated, the guns were proof fired with six rounds being fired from each tube. During the remaining days of May the unit was brought up to its full wartime strength. Ten prefabricated 16-man huts were assembled for the battery personnel, along with theater of operations-style mess hall, latrines and various supply and truck sheds during the summer of 1942. The battery's fortification elements were generally complete by October 1942. Battery Salt Lake and its associated camp was formally transferred to troops on September 15, 1943.

During a proof firing of the battery conducted in October 1942, loose rock around the entries to the magazines became dislodged. Repeated firing threatened to plug up the magazine entrances with falling rock and earth, and requests were made to apply a coating of gunite or construction of revetments for the walls at the magazine entrances. The 805th manned Battery Salt Lake until May 24, 1943, when it was redesignated Battery F, 15th CA Regiment, and then again as Battery A, 54th CA who manned the battery through the end of December 1945, after which the crew was reduced to a caretaking detachment of some 18 enlisted men.

With implementation of a combination of vertical and horizontal fire control systems and the installation of fire control radar, the need for the navy fire direction system became superfluous. With the recommendation of the Hawaiian Seacoast Artillery Command in September 1944, the Navy requested the return of its original equipment. During the remaining months of 1944 the process of uninstalling the navy director and its related equipment was carried out and by the end of February 1945, the work had been completed and returned to the Navy. In March 1946, the Salt Lake Battery was named in honor of Colonel Louis R. Burgess. (10)

One of the turrets of Battery Riggs at the Opaeula Reservation. *Gaines Collection.*

The second pair of North Shore *Saratoga* mounts were positioned about halfway between Kahuku and Kaena Points. On February 5, 1942, the site for the first of what would be two of such batteries on the North Shore was approved by the Hawaiian Department. This site was on Dole Pineapple plantation land at Brodie Camp Number 4, some 7,900-yards southeast of Waialua Bay on Oahu's north shore. The site was already on a junction point in the army's communications and fire control cable system. Construction commenced on March 10, 1942, and by April 9, 1942, the concrete had been poured to form the floors and sidewalls of the galleries at both turrets and at the base of the Number 1 turret's barbette. In spite of the progress made it was not found to be proceeding fast enough and a new supervisor, Captain Bullock, of the Corps of Engineers was placed in charge of the construction in mid-May 1942.

With the guns installed, four rounds were fired from turret No. 1 on May 25, 1942. Work continued on the powder magazines, chiefly with installation of the electrical service. The trenches had been excavated for the laying of the underground power cables from the power plant to the battery commander's station and the turrets. The cut and cover power plant building's construction had begun, and its concrete floor was being laid. With the completion of the electrical work in the magazines, the battery was placed in service using standby electrical generators that enabled the guns to be fired. At the end of July 1942, the tactical elements of Battery Brodie were practically complete and by September, the battery was nearly ready to be fired using the naval director.

The camp and a fresh water and sewage system for the battery personnel was completed by about September 15. The gun emplacements were largely buried except for the turrets, a battery commander's station, and a base end station atop a tower (that was built later in 1942). The gun turrets were the specified 450-feet apart. When the group command post for the Saratoga Group (consisting of Batteries Brodie, Opaeula, and Haleiwa) was established at Brodie Camp, an underground structure that had originally served as a storeroom was adapted for use as a gas proof combination plotting room and fire control switchboard room for the group's command personnel.

Fire-control data for the two new turrets were obtained from a series of fire-control stations along the shore between Kepuhi New on Keaau Ridge on Oahu's southwest coast, and Station "M" on the island's windward coast. In all but two cases, the base end stations for the 8-inch turrets used existing coast artillery fire control station locations. An SCR-296A fire control radar sets was provided and installed by early August 1943. A radar operating room measuring 16 by 17-feet was built onto the left side of the BCS. This construction required sealing up of the observation slot on the observation room's left side. The radar antenna was housed atop a 25-foot tower erected on the roof of the antenna house.

Battery Brodie was initially manned by the 809th CA (HD) Battery (Separate) that was activated by Captain L. A. Simon on May 21, 1942. Target practice generally paralleled that of Battery Opaeula as the two batteries operated together as a general rule. The guns of the battery were exercised on a regular basis. The 809th Battery provided the manning details for the turrets until May 22, 1943, when the unit was redesignated Battery D, 41st CA. The battery continued to be manned until early 1945, when the unit was transferred to the Sand Island Quartermaster Depot leaving a small caretaking detail at Battery Brodie. Battery Brodie was formally named Battery Ricker in honor of Lieutenant Colonel George W. Ricker, CAC in 1946. (11)

The turrets from USS *Lexington* were obtained from the navy at the end of March 1942. By April 4, 1942, the four twin gun mounts were removed from *Lexington* intact, rather than being disassembled like the mounts from the USS *Saratoga*. The mounts were stored at Waipio Point in Pearl Harbor under army guard until provisions could be made to move them to their respective battery sites. One pair of twin turrets was sent to the North Shore Groupment. Early in 1942, the HCAC had selected the Anahulu Flats, an upland region that lay between the sprawling Schofield Barracks Military Reservation and Haleiwa for this emplacement. The actual battery site was located in the midst of the sugar cane fields on the north side of Opaeula Gulch, some 9,000-yards back from the north shore at an elevation of 1,100-feet above sea level. By April 9, the battery position had been staked out and construction plans and estimates as well as drawings were being prepared. Construction of Battery Opaeula, as it was initially termed, began in May 1942, and was pushed forward urgently. By June 30, the battery was considered about 30% finished.

The two twin turrets were armed with the 8-inch/55-caliber Mark IX Mod 2 naval guns. On August 6, 1942, the four guns of Battery Opaeula were proof fired using a temporary electrical connection. Three reduced charge rounds were fired from each gun followed by three service rounds. The proof firing was successful, and the guns were turned over to the HSAC and the manning detachment of the 810th CA Battery. By September 1942, the BCS and the plotting room were still awaiting construction and a fire control system had yet to be established. The additional ammunition storage magazine of Elephant Iron was still to be built. Sanitation facilities were about 50% completed, while the mess hall was about 60% complete. (12)

Upon completion later in 1942, Battery Opaeula followed the standardized layout plan. Its fortification elements were largely buried except for the two turrets and the BCS. Most of the battery's structures were of the cut and cover type constructed of reinforced concrete placed 15 to 40-feet underground. The gun turret barbettes were installed the standard 450-feet apart. The Battery Commander Stations (BCS) had a pedestal for a DPF and an azimuth instrument or spotting telescope. The lower level of the BCS at Battery Opaeula was below grade and contained a room that measured about ten-feet by ten-feet. This room was initially used to house the electro-mechanical equipment for the navy gun director. Later, when the batteries were equipped with the SCR-296A type fire control radar, a separate structure was constructed to serve as the radar operating room. This was a 40-foot-long building partially dug into the ground. This building was originally built as a magazine had an arched roof of "Elephant Iron" heavy corrugated steel. The building, located to the right rear of the Number 1 turret, near the Opaeula Road, was entered by way of a concrete staircase from the ground surface. At Battery Opaeula, two underground plotting rooms were provided. Both were located about 200-feet to the rear of the Number 2 turret in Opaeula Gulch at the rear of the battery and about 100-feet from each other.

Directly to the rear of the battery commander's station by nearly 150-feet, and some 15-feet below the surface of the ground, was the battery power generator room, which served as an emergency supply of electricity to the two turrets and the post at Opaeula. The camp's barracks, mess halls, latrines, and other structures at the battery were of the Theater of Operation-type construction. Most of these structures were generally located to the rear of the two turrets on the reverse slope of the ridge and between the power generator room and Opaeula

Road. Concrete pillboxes for local defense, armed with machine guns, were placed at various points around the perimeter of the battery to cover the approaches to the battery position.

Battery Opaeula was initially manned by the 810th CA (HD) Battery (Separate) until May 1943, when it was inactivated, and its personnel reassigned to Battery E of the reactivated and reorganized 41st CA. As one of the primary gun batteries on the north shore, the guns of Opaeula Battery conducted regular target practices, usually once a week in which one round was fired from each gun at a hypothetical target in the waters off the north shore. In addition, about three times a year, service practices in which about two-dozen rounds of armor piercing (AP) 8-inch shells were fired at a high-speed target towed by a destroyer or destroyer-minesweeper. In July 1943, two such practices were fired. Nearly 50 rounds of armor piercing projectiles were fired out some 20,000-yards into the Pacific Ocean. On December 2, 1943, ten rounds were fired from each turret in a battle practice. The last practice by Battery Opaeula was on November 9, 1944. On December 26, 1944, the battery was taken out of service. Belatedly, on August 27, 1946, Battery Opaeula was named Battery Carroll G. Riggs in honor of Colonel Carroll G. Riggs, CAC.

In the latter part of 1942, the gun batteries on the North Shore were each provided with a SCR-296A fire control radar sets by the Signal Corps. Set Number 2 was installed at Battery Opaeula. Although the coast artillery commanders considered these turret batteries to be among the best on the island, the Saratoga gun group commander, Major Albert W. Sparrow Jr., battalion commander of the 3rd Battalion, 57th Coast Artillery, observed that Battery Brodie's four gun tubes from USS *Saratoga* were so worn that they were "hopelessly inaccurate."

The Wiliwilinui reservation was at a location east of Fort Ruger between Diamond Head and Koko Head about one and three quarters miles inland from Maunalua Bay on Wiliwilinui Ridge, one of many southern spurs of the Koolau Range. The military reservation was established at an elevation of about 1,050-feet above the waters of Maunalua Bay; it consisted of two principal tracts, as well as easements for utilities and roadways; all of that totaled nearly 86 acres. The property was secured in 1942, by condemnation for defense purposes from its owners, the Bishop Estate, and Robert Hind, Ltd. When work began to wind down on the construction of the turreted 8-inch Battery Wilridge, new projects at the reservation took over.

Wiliwilinui was the fourth and final site for a pair of twin 8-inch gun mounts that were removed from USS *Lexington* for use as an Army shore battery early in 1942. Initial planning for the south coast battery called for it to be built at Wilhelmina Rise to the north of Diamond Head. During April and May, several staff discussions and studies by the Hawaiian Seacoast Artillery Command and the Hawaiian Artillery Command considered relocation of the battery site to Makapuu Head on Oahu's easternmost point, Ulupau Head, or Kaaawa Point, on the island's windward coast. Finally, on May 27, the battery site was located at a compromise site on Wiliwilinui Ridge to the east of Diamond Head. The site for this battery, initially named Battery Wilridge, was located on the newly established Wiliwilinui Military Reservation, on the upper reaches of the ridge east of Fort Ruger. From this position the battery could provide direct fire at middle or long-range targets along Oahu's shoreline from Honolulu Harbor on the south shore, eastward to Makapuu Head on the eastern end of the island. In addition, the battery could provide indirect fire on Waimanalo Bay area to the northeast and along the south shore between Honolulu and Pearl Harbor. (13)

Construction was given a high priority with work commencing on June 15, 1942. By Mid-September, the magazines and their galleries were finished, the guns were emplaced in their barbettes and ready to fire using temporary electric power, although the plotting room, BCS, power plant building and auxiliary ammunition shelter were still under construction. Work continued on the battery until December 26, 1942, by that time, the battery's various elements and support systems had been brought to completion. The battery was formally turned over to the HSAC on September 20, 1943.

The two mounts each carried the familiar 8-inch/55 caliber Mk IX MOD 1 navy guns. The construction was of the cut and cover type. Battery Wilridge was the last of the four 8-inch naval turret batteries to be completed and as a result, it incorporated improved design features when compared to the other turret batteries.

Two 8-inch tubes peeking out of the camouflage netting at the battery on Wiliwilinui Ridge (later to be named Battery Kirkpatrick). *NARA*.

The mounts were provided with better protection with bunkers surrounding them and their barbettes. These splinter-proof bunkers were constructed of parallel timber walls with earth fill between them that protected the mounts from all but direct hits.

The most obvious difference was in the size and layout of the battery commander's station. That at Battery Wilridge was far more spacious than those built on the north shore. The battery commander's station was a two story "L" shaped structure of reinforced concrete about midway between the two turrets. The building contained a combination fire control switchboard room and radar operating room measuring 15-feet by 11-feet, and a two-tiered observation station. Changes to accommodate an army fire control method and operations resulted in a much larger battery commander's station than those provided at the other batteries of this type. After the SCR-296A fire control radar set was installed, its antenna was placed in a cylindrical container camouflaged to resemble a water tank atop a 25-foot-high tower at the rear of the battery commander's station adjacent to the radar operating room. Inside the BCS, a short staircase some two feet, six inches wide on the left side of the station connected the upper and lower observing levels.

A second departure from the standardized plans used elsewhere on the island was that the battery on Wiliwilinui Ridge had only one plotting room for the two turrets and that the fire control switchboard room was incorporated into the battery commander's station. The single battery plotting room was an underground building dug into the west slope of Wiliwilinui Ridge about 30-feet below its crest. The plotting room had an escape shaft that exited some 20-feet further up the slope of the ridge. The two concrete barbettes of Battery Wilridge were about 350-feet apart and extended about ten feet into the ground. Underground they had the same separate projectile and powder rooms forming a V with the barbette at the apex. (14)

A power generator building was located some 100-feet directly behind the battery commander's station. 100-feet west of the power building a concrete stairway led down the west slope of the ridge about another 100-feet to a small plateau area on the slope. Here the barracks camp for the battery personnel in the form of 29 six-man huts were located, along with a mess hall, and various supply and truck sheds. These structures were constructed on a timber frame and provided with walls of heavy-duty black tarpaper sheeting. Situated 800-feet south of the turrets and about 100-feet lower down the ridge, was an outdoor motor pool and a Quonset hut-like structure that was dug partially into the ground. This structure was used as an auxiliary storage magazine for about 600 additional rounds. Battery Wilridge was only about half completed when it was assigned a provisional crew comprised of detail personnel from a variety of units. This provisional manning detachment was designated the 806th CA (HD) Battery. The 806th manned the two turrets from October 1942, until May 26, 1943, when it was redesignated Battery D, 16th CA. On May 30, 1944, Battery D fired an eight round target service practice. Finally, a caretaking detachment of some 34 men from the 608th CA (HD) Battery (Separate), provided a maintenance detail at the battery until early 1946, when the 608th was inactivated. In March 1946, the War Department formally designated Battery Wilridge as Battery Lewis S. Kirkpatrick.

A new battery position for 155 mm GPF guns was also planned for this reservation. In early December 1942, fatigue parties from Battery E, 55th CA Regiment began developing a 1.5-acre site for a new battery on Panama mounts between Battery Wilridge and Maunalua Bay. By December 21, the initial phase of the project was complete. When complete, Battery Wili, as it was termed, consisted of four Panama mounts for 155 mm GPF guns that were constructed on the spine of Wiliwilinui Ridge some 3,000-feet south of Battery Wilridge at an elevation of about 500-feet above sea level. (15)

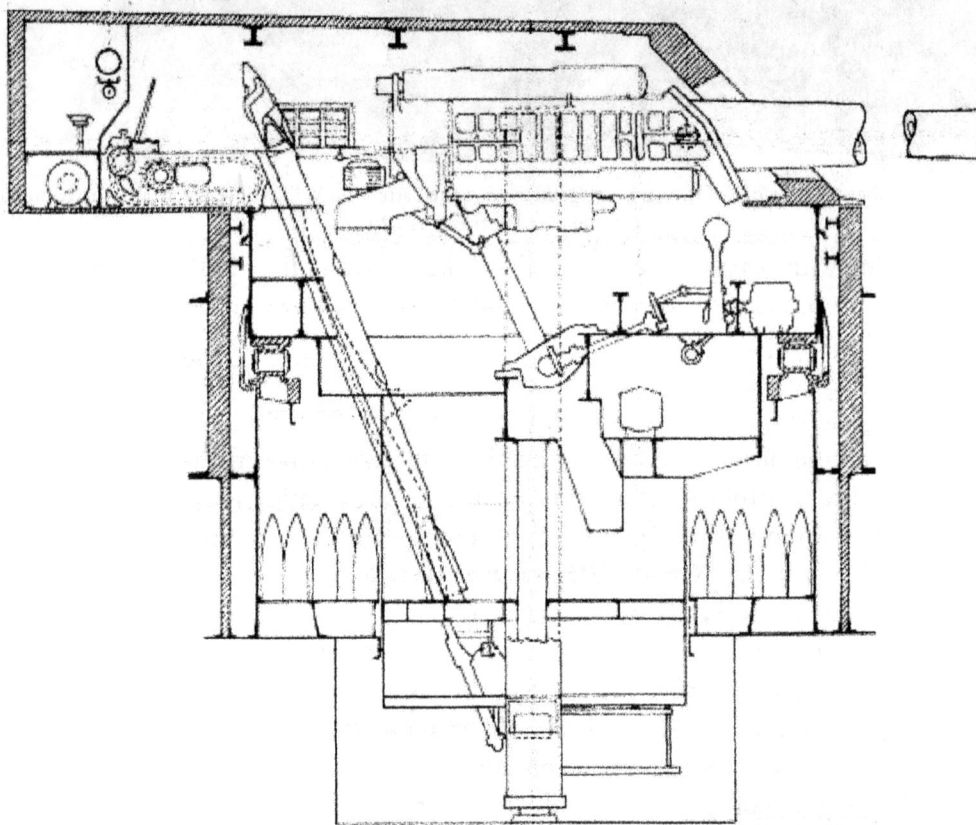

Navy general side-view plan of a 14-inch turret similar to that mounted on battleship USS *Arizona. Smith Collection.*

It was obvious immediately after the attack on Pearl Harbor that USS *Arizona* and possibly the *Oklahoma* were total losses. The damage to the *Arizona* from the explosion of the forward magazines had destroyed far too much of the forward hull; no amount of patching or use of bulkheads would restore flotation and any repair. However, the aft of the ship, and the two triple 14-inch gun turrets there, were mostly intact. Six weeks after the loss conversations began with the Commander in Chief of the Pacific Fleet about the use of turrets by the Army. Initially the commander held the opinion that the turrets were too heavy to permit them to be mounted on land. Or rather that a stable, protected platform able to withstand the shock of firing was not possible for a land mount. The Army, however, preferred to study the engineering first. Additionally, the initial discussion thought that a third turret, the one least damage from the No. 2 position might also be salvaged, but that idea was later abandoned.

The *Arizona* was armed with twelve 14-inch/45-caliber guns Mark VIII arranged in four triple turrets, one superfiring over the other fore and aft. This gun and its related marks had been used on five classes of American battleships and was a very powerful and proven weapon. It fired a 1500-lb. armored-piercing shell at 2600 f.p.s. at 30° elevation to a distance of 34,500-yards. This performance would be second only to the two 16-inch gun batteries installed at Forts Barrette and Weaver. The guns were housed in heavily armored steel turrets, turret faces were armored to 18-inches, sides to 10-inches, and the roof 5-inches. Once emplaced, they would not need casemates of concrete thus providing a 360° field of fire. However, the turret weight was massive—over 700 tons. And they needed an intricately balanced, precision roller path supported by a very strong barbette capable of both carrying the weight and withstanding the shock of firing. (16)

USS *Arizona* resting on the bottom at Pearl Harbor after her loss in the December 7th Attack. The two aft turrets that were eventually salvaged are to the left. *NHC.*

Work on salvaging the two battleship turrets is well underway in the photograph of February 25, 1942. *NHC.*

The army's Hawaiian ordnance office visited the Navy and the ship on January 14th. Their feasibility study showed that they could design an emplacement to hold and fire guns in mounts of this weight. A key consideration was the knowledge that conventional sources of new heavy guns—such as 16-inch domestically produced—would take at least 2-3 years to procure, if even that. These available naval turrets, even with the challenges of reconditioning the mounts and designing a unique emplacement, could be in place in as little as nine months. (17)

While discussions continued, location sites were surveyed for the turrets pending authority to proceed. The Navy suggested heavy gun coverage for the new Kaneohe Bay air station on the east coast and the naval ammunition depot at Lualualei on the west coast. The Army agreed and the first turret was tentatively assigned to Kaneohe Bay. After some discussion about placing it on the northern side at Lae O Kaoio Point or at Puu Papaa Ridge, the advantages of placing it on the crater lip at Ulupau, despite construction challenges, prevailed. The second turret was slated for the west side—the initial choice was on the summit at Puu Mailiilii Ridge. However, soon preference shifted to a better protected spot on the elevation at Kahe Point. (18)

Approval for a project of this scale was understandably staged as information was accumulated. The Army in Washington tentatively gave the go-ahead in December 1942. In May of 1943 the plan for the emplacement layout was completed and reviewed. Formal approval and full funding and priority finally came on August 13, 1943. (19)

Meanwhile the huge job of removing and reconditioning the turrets themselves was underway. The gun tubes were found to be relatively clean and intact, just requiring relatively easy routine cleaning and inspection, which could be done at the Pearl Harbor Navy Yard. Unfortunately, the initial Navy removal of tubes and ammunition had resulted in the turrets being cut open without regard to any possible future usage. It was found that the turret shells were cut into pieces at an extremely delicate point, requiring them to be later welded

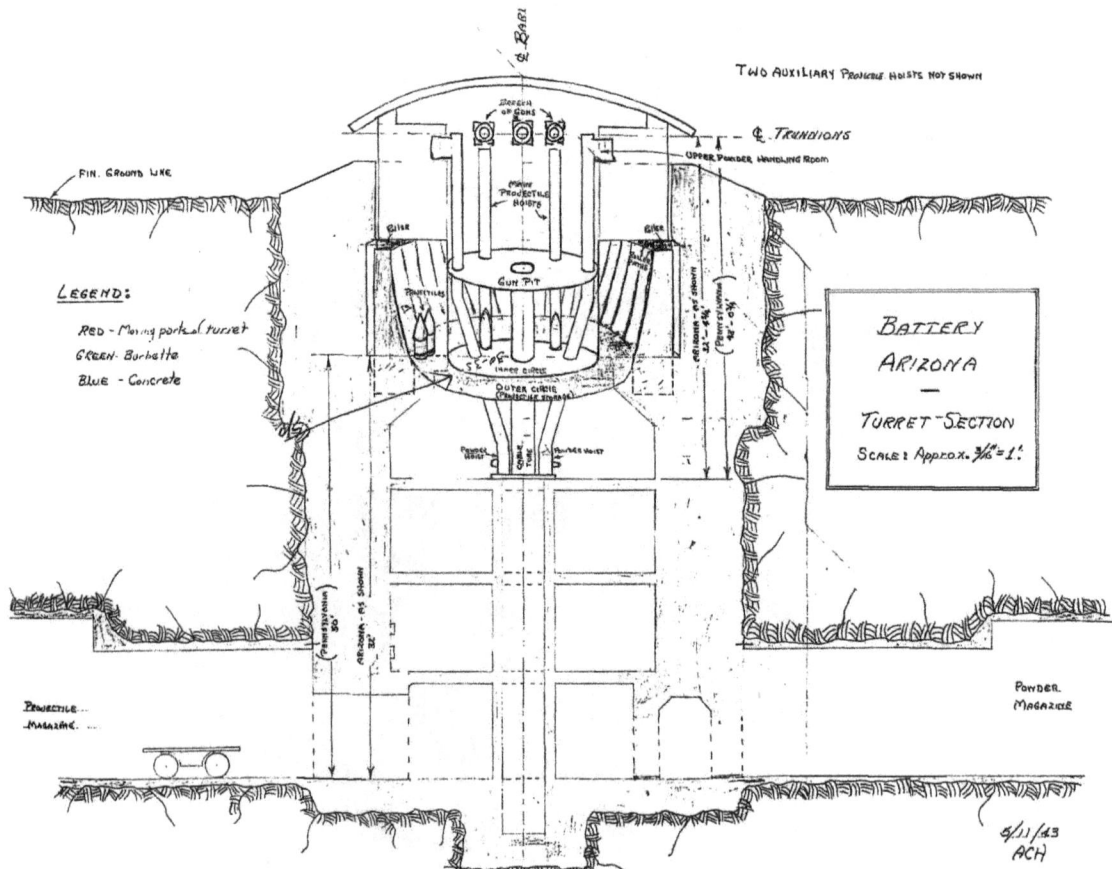

May 1943 sketch for the major structural elements of the foundation for one of the 14-inch turrets from *Arizona*. NARA.

back together with great accuracy. Some parts were simply missing: anything electrical had to be completely rewound or replaced due to water damage. Much of the turret rotational mechanism and the lower ammunition handling apparatus had been submerged in salt water for many months.

Even detailed ship plans to help in reassembly were hard to come by. Drawings were located in Washington D.C., but 25 years of service life had led to so many modifications and repairs that they were outdated. The Army assembled a special "Project Installation Committee" with full-time representatives from Signal, Ordnance, Engineers, and Coast Artillery along with a Navy representative to oversee the entire project in March 1943. Later that month the Navy reported that the first turret was going to be set ashore on April 30, the second following about a month later. A warehouse was set up in Pearl City to store parts for the project, and part repairs and manufacture occupied both army and navy ordnance workshops. Some heavy parts had to be fabricated in the U.S., which proved frustrating with all the wartime priorities. At one point the assembly of a roller path awaited a turning gear worm and pinion from the Washington Navy Yard for eleven months. (20)

Securing the necessary ammunition was not a problem. There were still nine battleships in service which used the 14-inch gun and its ammunition. Inquiries conducted in April 1943 revealed that the navy was planning to supply 600 complete rounds and charges. For each gun there would be 100 rounds: 85 were 1500-lb. AP, and 15 were 1275-lb. HC. The emplacement magazines were built to accommodate 840 complete rounds—280 per gun tube. Additional reserves were stored at the navy's Lualualei Ammunition Depot on Oahu and could be released in an emergency. The army did need complete ballistic information on the shells so it could properly design and program the gun data computers necessary for the army range-finding and firing apparatus. (21)

The designs for the two emplacements were circulated for approval in May 1943. With minor questions and modifications, they were approved, and work orders issued. Due to the local terrain topography the final design for the two emplacements were different though the basic design elements and requirements were the same. Each battery had a single barbette well where the turret would sit on its heavy roller path. The well was

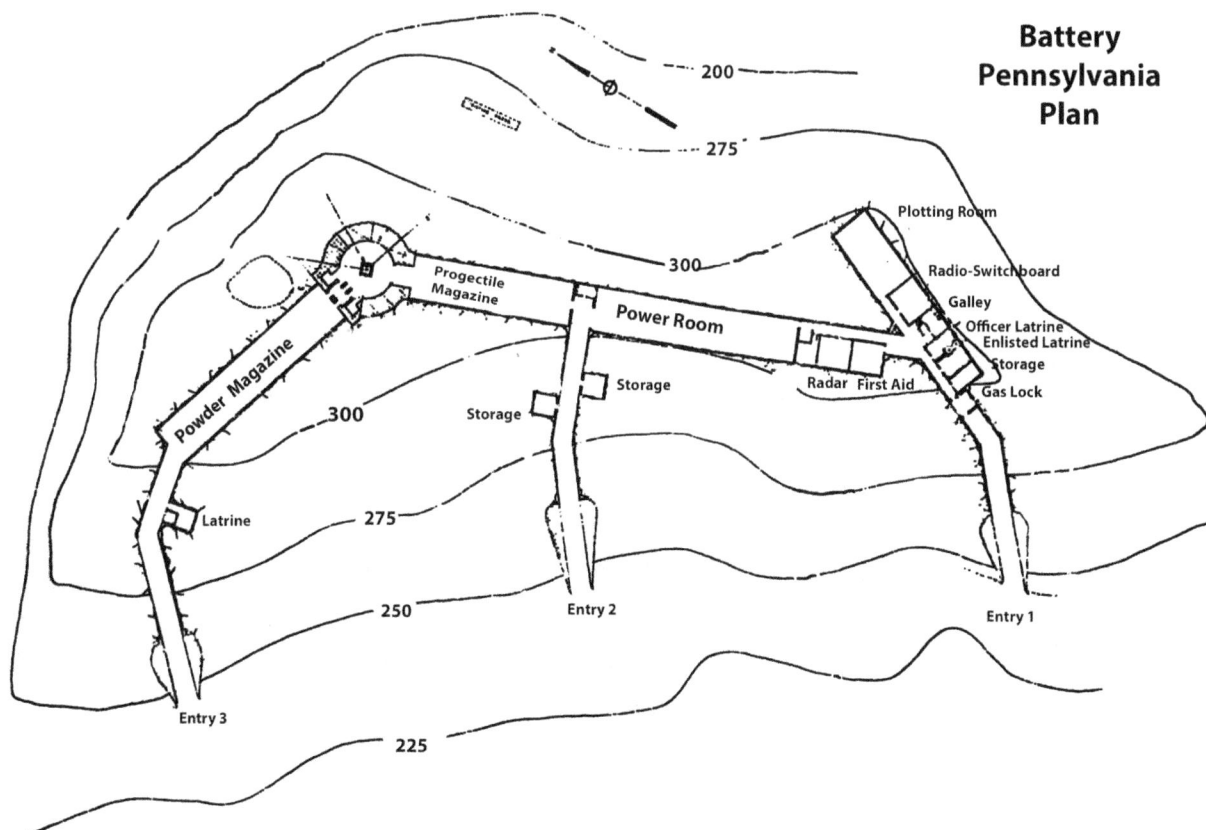

Layout plan for Battery Pennsylvania at Ulupau Head, Fort Hase. *NARA.*

of 3-ft. concrete thickness set in rock. It had a depth of 60-feet and had a 24-foot inside diameter. There were three levels below ground. Because the protection requirements for the magazines worked out to a 40-foot overhead earth or rock coverage, the depth of the magazines (and thus well) was greater than the 25-foot used on ship. Servicing the turret with projectiles and charges required a new apparatus; the old ship motor-powered chain shell lifts could not be used. There was still a handling room immediately below the turret access. The weight of the turret assembly and the expectation of heavy shock during firing required a heavily strengthened structure—the turret supported ring required 600 1.25-inch square reinforcing bars welded to the shell and web. (22)

General recommendations called for separate powder and projectile magazines underground adjacent to the turret well. Many of the other support facilities were along separate access tunnels ending at the munitions and shell lift area. This tunnel was to include a power room for three 125 kW diesel generators, a fuel tank, and a 10,000-gallon water tank. In a section of the tunnel protected by a gas-proof airlock were the operational rooms: radio-switchboard room, plotting room, separate men's and officers' latrines, first aid room, small galley, office, and storerooms. As both batteries were isolated from any support camp, an operations or emergency manning crew would need to live in the battery. Routine operation of each battery was calculated for a crew of four officers and 157 enlisted men. Hooks for hanging bunks in galleries and stores for longer occupation were provided in the design.

Aerial photograph showing general area of Ulupau crater and the rest of Fort Hase and the Kaneohe Bay Naval Air Station beyond in January, 1945. *NARA.*

Work on site re-assembling the turret for Battery Pennsylvania in late February 1945. *NARA*.

Further work on the final turret assembly at Battery Pennsylvania in March 1945. *NARA*.

The battery commander's station had specific siting needs. It needed to have a wide surface view. In both locations the BCS was located on a peak or rise adjacent to the turret. The underground gallery had to extend to directly underneath the BCS to allow for a shaft for access, and to run the coaxial cable needed for communications and radar. Radar structures supporting the SCR-296A radar had been incorporated in the designs from the beginning. The tower was to be emplaced on top or next to the BCS, with the operating room immediately below. (23)

The battery sites were named even during their planning and construction phases. The western, Kahe Point location was called Battery Arizona, the eastern Mokapu location Battery Pennsylvania. These were the names of the two battleships in the *Pennsylvania* class of which the USS *Arizona* was a member. The names were never confirmed by the official General Orders process, but still seemed to have been widely adopted and used by the Army. After approval of the finalized design in May 1943 (and the successful removal of the barbettes from the battleship wreck in the same month), serious construction began. While the Mokapu site already had some camp facilities, these were still needed at Kahe Point. For several reasons work on the eastern Ulupau site for Battery Pennsylvania progressed faster, being maybe 4-6 months ahead of the western Kahe Point Battery Arizona. Both sites made extensive use of the tunnel technique already practiced on previous engineering projects for the military on Oahu. Most of the underground tunneling and squaring out of magazine and support rooms was done in the last half of 1943.

On August 15, 1942, the navy advised the Hawaiian Department that transfer of the two USS *Arizona* turrets had been approved and the next day General Emmons requested authority from the War Department to accept the turrets from the Navy. Site surveys in July did not initially favor a Ulupau site. The site was considered to be in "difficult and inaccessible terrain...requiring considerable rock excavation." They expected that it would require seven and a half months to construct the battery to the point where its guns could be test fired.

Despite these reservations, the final decision regarding the Kaneohe turret was made and General Emmons finally approved the site on Ulupau crater's east rim on Mokapu Point on October 10, 1942. On November 21, 1942, a work order was issued for the construction of this emplacement on Ulupau Head. An extensive amount of tunneling was applied to the Mokapu site. It was found that the volcanic rock here was good material, most of the tunnels only needed occasional caulking to make waterproof, applying widespread gunite to tunnel surfaces was unnecessary. The department engineers were delighted to find something break their way for this project.

The crest of the crater rim was a hogback ridge that rose more than 145-feet above the sea. The steep ridge had two promontories, but limited space to place as big a structure as needed for a 14-inch turret emplacement. It would require extensive excavation, and more than 20-feet were cut away to create a wide enough space to accommodate the barbette of the turret. The complex had a central barbette well of reinforced concrete set in the volcanic rock that rose some 66-feet to the new crest of the east rim. Below, tunnels were bored into the crater's rim to serve as magazines, and to house various service and operational rooms. Three separate access tunnels of ten-foot bore were drilled into the interior slope of the crater's east rim. One provided entry to the powder magazine tunnel, a second to the projectile magazine and power room tunnels, and a third into the operations tunnel. The sections of the battery's tunnel complex were all interconnected. (24)

After the various parts of the guns and the disassembled turret had been refurbished, they were moved by barge around the island to East Beach on Kailua Bay where they were unloaded and moved to the assembly point on Ulupau crater's floor behind Kii Point. There the turret was partially reassembled and prepared for hoisting up onto the barbette that had been prepared atop the east rim.

By the end of December 1944, the installation of Battery Pennsylvania's turret was about 50% complete. The lower powder hoist assembly, the lower and upper turret sections had been installed and all sections welded together. Templates of all necessary structural steel work and special powder passing tray assemblies for the upper turret section had been completed so that fabrication could be started on those items. The mechanical equipment for the operation of the turret, including hydraulic gears, motors, blowers, etc., had been installed

Battery Pennsylvania underground projectile room on March 22, 1945. *NARA.*

Battery Pennsylvania underground powder magazine room on March 22, 1945. *NARA.*

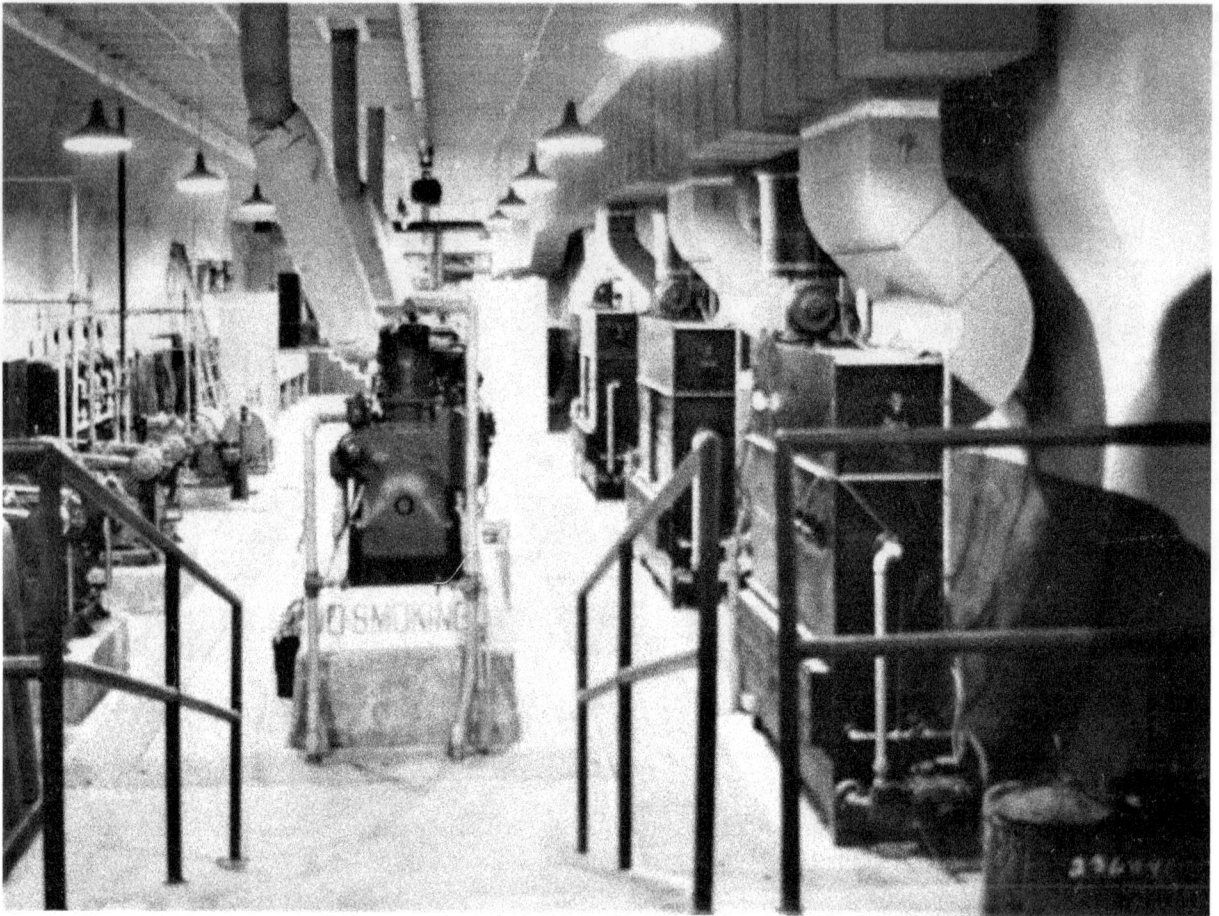

Battery Pennsylvania underground power generator room on March 22, 1945. *NARA.*

Battery Pennsylvania underground plotting room on March 22, 1945. *NARA.*

before the closing of various points of access to sections within the turret. The deck lugs, side armor plate, main ventilating blowers, fire sprinkling system, and water storage tanks had also been installed. More than three years would have elapsed before the turret and its three 14-inch naval rifles were ready to be placed in service atop Ulupau's east rim on Mokapu Point. (25)

The northern most of the three access tunnels provided access to the powder magazine. It was about 60-feet long and ten-feet wide. It had a dogleg turn just past its mid-point. Here also was a short ten-feet deep and ten-feet wide branch tunnel that served as a storage room. Farther into the crater rim, the access tunnel took another dogleg of about 45° and entered a 20-foot-wide powder magazine tunnel. This powder storage room had a length of 65-feet and a crown of 14-feet. The powder magazine had a capacity of about 800 powder charges stored four high in racks on each side of the tunnel. At the far end of the magazine, doors opened onto the two powder handling rooms. Each handling room was adjacent to the massive circular barbette of concrete masonry. The three lower levels were connected by a spiral staircase set against the barbette's interior wall while a vertical steel ladder afforded access to the uppermost level of the concrete barbette and the gun turret's lower level. At a 45° angle across from the barbette end of the powder magazine was the doorway to the projectile magazine tunnel. This room was 68-feet long and 20-feet wide. Its capacity was approximately 300 rounds stored three high on either side of the projectile room. At the opposite end of this magazine another doorway opened on to the central entry tunnel. This tunnel also had a dogleg turn some 16-feet inside its entrance and then a 70-foot-long section that terminated at the water tank tunnel This location was also the juncture of the projectile magazine tunnel on the left, and the power room tunnel on the right. (26)

Directly across the entry tunnel from the projectile magazine entrance was the entry to the 105-foot-long power room tunnel. Its concrete floor was three feet lower than the other tunnels, to accommodate large three-foot high concrete footings supporting the powerful generators. The metal flooring was floated above the concrete footings leaving a void beneath the metal floor for the large electrical conduits and fuel lines for the generators. Exhaust pipes for the three 125 kW power generator sets were extended above the generators

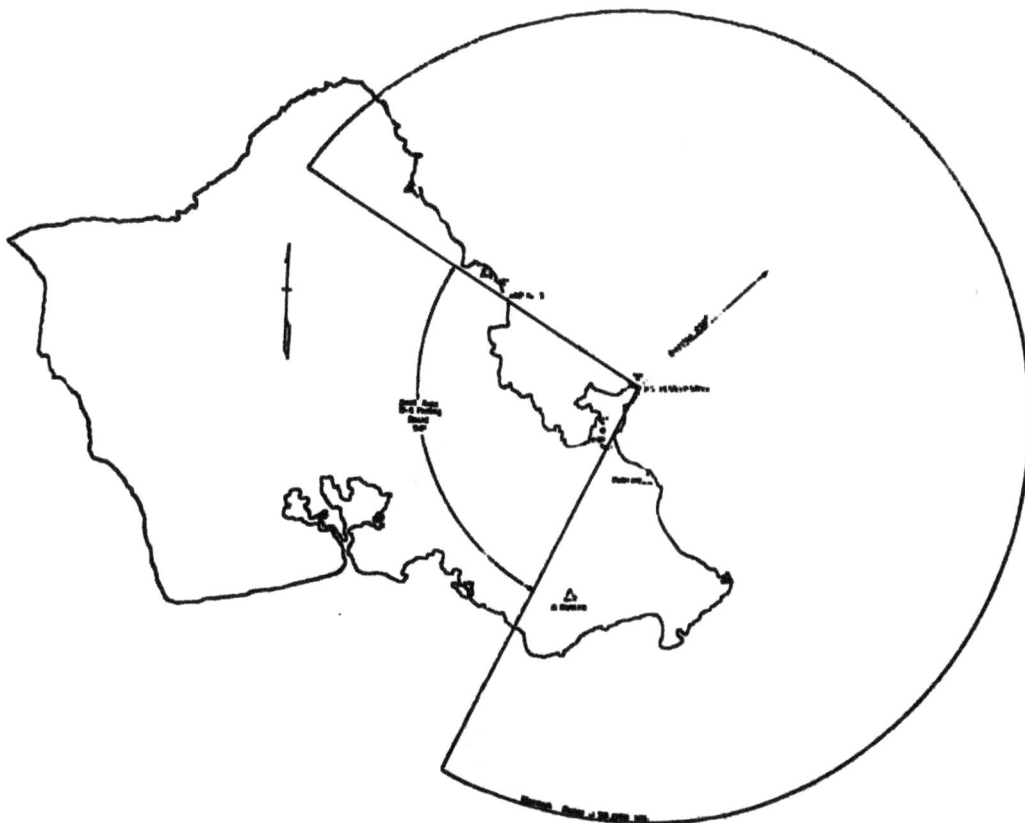

Map showing field of fire for Battery Pennsylvania, Fort Hase. *NARA.*

into the ceiling, and thence through a concrete shaft out into the open air. The power room also contained the air conditioning and ventilation system machinery as well as the necessary electrical switchboard panels and monitoring equipment for the battery's mechanical systems.

The operations section of the battery occupied a 30-foot-long extension of the power room tunnel and an 88-foot long main operations tunnel that intersected the power room tunnel extension at an angle of about 45°. The operations tunnels had a five-foot wide passageway down one side that provided space for a series of rooms on the east side of the tunnel. A room 15-feet by 14-feet, originally planned as the radar operating room, opened into this corridor. Even before the battery was nearing completion, the proximity of the radar antenna and its operating room to the turret was too close subjecting the radar instruments to severe concussion and blast effect from the 14-inch guns. The decision was then made to relocate the BCS as well as the radar antenna and radar operating room across the crater floor to the west rim of the crater, the former radar operating room in the operations tunnel was adapted for use as the battery office. The battery's first aid room and dispensary also occupied a 15-foot by 14-foot space and opened on to the corridor where it intersected the passageway of the main operations tunnel. At the east end of the main tunnel corridor was the battery plotting room. Just to its west was the radio and fire control switchboard room The battery's small galley and mess area was directly opposite the power plant extension corridor. (27)

The original battery commander's station built atop the east rim of the crater on its second promontory was located some 200-feet south of the turret. This station was redesignated as the B3S3 base end station for Battery Pennsylvania. A replacement new combination battery commander's station and radar operating room was constructed of reinforced concrete and was located somewhat to the north of the old coast artillery observation station "J" on Ulupau's west rim. In addition to moving the sensitive radar and optical instrumentation further away from the concussion and blast effect of the guns, the new location gave the battery commander an improved field of vision of the seaward approaches to Oahu from the north. Electric power supply for the new station and the radar set was provided from a dedicated small power plant dug into the interior slope of the crater rim a few hundred feet below the station.

Modern aerial photo showing location of the Battery Pennsylvania Battery Commander's Station. *Williford Collection.*

Proof firing of Battery Pennsylvania on August 10, 1945, almost simultaneous with the war's end. *NARA.*

The three 14-inch rifles of Battery Pennsylvania were proof-fired on August 10, 1945, this was the same day that the Empire of Japan announced its acceptance of the terms of surrender offered them. Earlier concerns that when the 14-inch guns were fired, the concussion would cause the crater rim to fracture causing the battery to fall into the waters below proved to be totally unfounded. But the blast effect and concussion created by the firing of the guns confirmed that the original site for the battery commander's station and radar antenna would have been located too close to the guns. Even though proofed, the battery was not completed in all respects. It appears that some of the signal and communications wiring and equipment were missing. No unit was ever assigned to man the battery, nor was it ever officially turned over to the coast artillery.

With the Navy's input, the Hawaiian command determined that one of the turrets should be emplaced to protect the U.S. Navy's Lualualei Ammunition Depot. Consideration of several sites eventually ended in the selection of Kahe Point for the 14-inch turret on the Waianae (western) coast. On October 13, 1942, the HSAC ordered an expedited project at this western location for a battery of three 14-inch MkVIII Mod 4 guns mounted in a naval turret. Physcial construction of the large underground emplacement was begun early in 1943.

This battery was located on a site of about 325-foot elevation on a hilly area to the east of the western coast. It was enough inland to place it behind the final infantry defense line in this sector, giving it some protection by ground forces. The position up the Kahe Point Ridge gave it coverage from Kaena Point to the northwest all around to Honolulu. A higher ridge about 400-feet to the east, northeast was selected as the site for the BC station and radar. Engineers admitted that connecting the turret to the BCS underground created a necessarily long gallery, but that it was the best location, and any relevant cost increase was justified. (28)

Two main portals entered the underground battery structure, lying about 170-feet apart east to west. The western entry led to a "Y", with dogleg to the north with the support gallery, and one to the east which fed two wide galleries. The closest was the southern gallery which first led through the power room for the three generators and fuel storage, and then the second to the projectile magazine. The second corridor or gallery paralleled the first but was shorter. It contained the powder magazines with entry from the east and exit to the west directly into the gallery at the base of the turret well. Likewise, the southern gallery projectile room also fed into lower gallery of the well. The custom fabricated power lifts would raise projectiles into the turret for power ramming and powder to the handling room immediately below the turret for passage through scuttles to the breech. In the southern gallery at the mid-point between the magazine and power room at 90° branched the corridor which led to the second, western entry portal. It passed a couple of storerooms and carried the exhaust piping from the power room to a vent in the exit portal. (29)

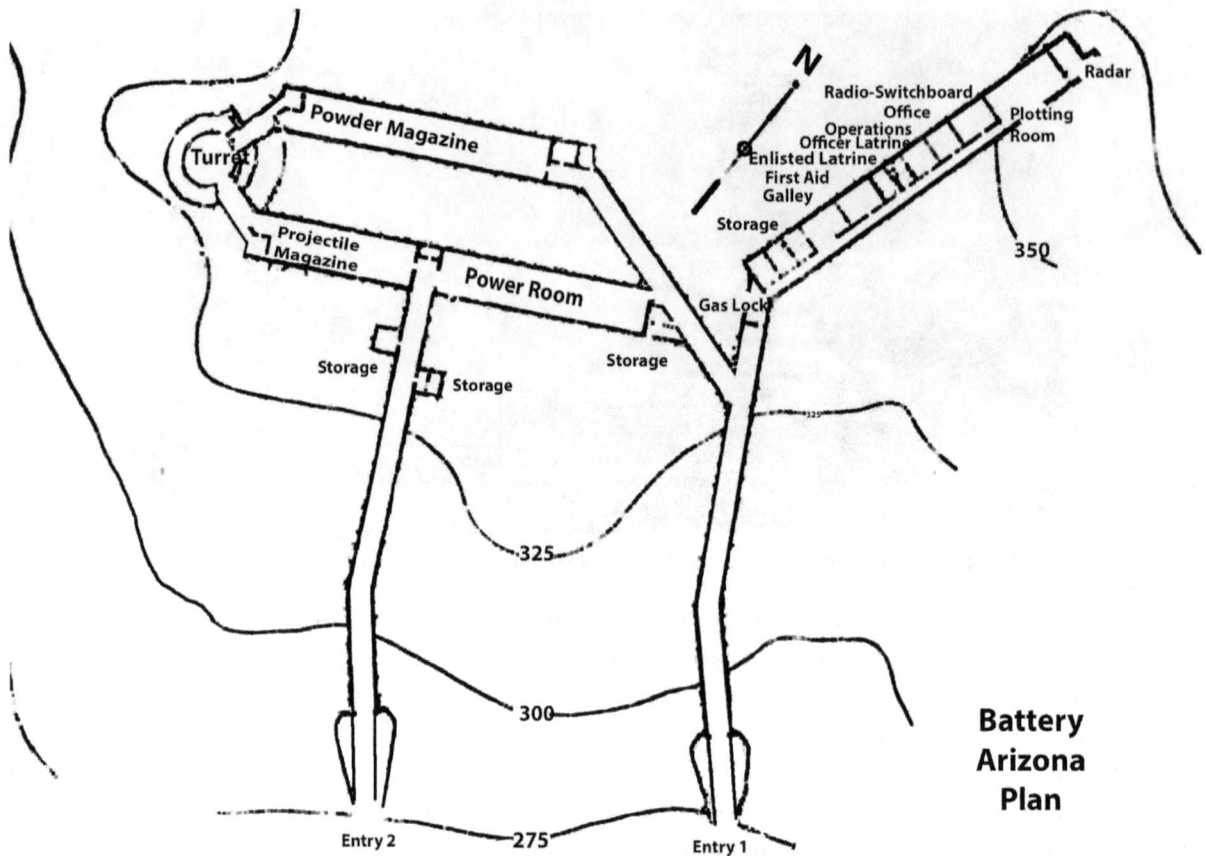

Layout plan for Battery Arizona at Kahe Point Military Reservation. *NARA.*

The support gallery branched off from Entry 1 (eastern-most) with a dogleg. A long gallery it made a dead-end directly under the Battery Commander's Station. Along the western side of the gallery were the various support and operations room. In order from the southern entry of the gallery were storage rooms (three), galley, first aid station, enlisted latrine, officer latrine, operations room, office, radio & switchboard room and finally plotting room. The latter was accessible with an 80-foot vertical shaft. The shaft would carry the communications wire and radar cable into the plotting room below. However, the stations did have normal exterior entry portals. Above ground this was a double station, one for the battery commander, the other for

Very early gathering of equipment and forms to start excavation work on what would become Battery Arizona. *NARA.*

14-inch turret of Battery Arizona just after the war in January 1946, though never quite completed. *NARA*.

the groupment. Immediately behind the upper station was the fire control SCR-296A operating room, the radar set, disguised in the usual manner as a water tank, directly on top of the room.

The battery was served by a series of base end stations: B1S1 at Station "S" at Kaena Point, B2S2 at Station "S' " at Kepuhi New, B3S3 at Station "U" at Puu O Hulu, B4S4 at the battery commander's station at the battery, B5S5 on the crest of Makakilo, and B6S6 at Station "C" on the crest of the north rim of Aliamanu Crater. While Battery Arizona had its own SCR-296A fire control radar set that was established adjacent to the battery commander's station, target ranges and azimuths could also be obtained from other SCR-296A radars along the south and west coasts of Oahu at Batteries Hulu, Hatch, Williston and Salt Lake. The battery was still lacking some of its fire control radar equipment, when the war ended in August 1945.

In 1944 and early 1945 construction progress slowed on both sites. The guns and major turret parts were delivered to their field locations. The actual installation of the barbette rotating parts and turret assembly with the insertion of the gun tubes was found to be the most complex and time consuming. In early 1945, the project came down to having a couple dozen of technical specialists at work rather than a large crew of laborers and heavy construction equipment. The war had moved to the western Pacific and the threat of an attack, particularly a serious attack by surface ships or an attempted landing seemed more and more remote. The priority for use of the technicians and for key shop assignments slipped. The nine-month expected completion time had now increased from two to three years.

A review of the entire project in December 1944 noted that the battery facilities were practically complete. Both batteries still needed full wiring by the Signal Corps. Ceilings were uncompleted for a specific reason. The large M4 plotting table (a very big circular board with mechanical arms used to plot range settings using observations from base stations) had to be moved and installed as a single piece. It was so large that the ceiling would prevent its movement to the plotting room. So, until these highly precise boards were built and delivered, the ceiling couldn't be installed. The turret for Battery Pennsylvania was estimated at 50% installed, presumably that for Battery Arizona somewhat less. (30)

Work progressed in 1945, but very slowly. In June the priority for these projects was lowered, with the possibility of work being suspended altogether. However, at this point Battery Pennsylvania was so close that it was allowed to be "finished", at least to the point of allowing a proof firing in August 1945. (31)

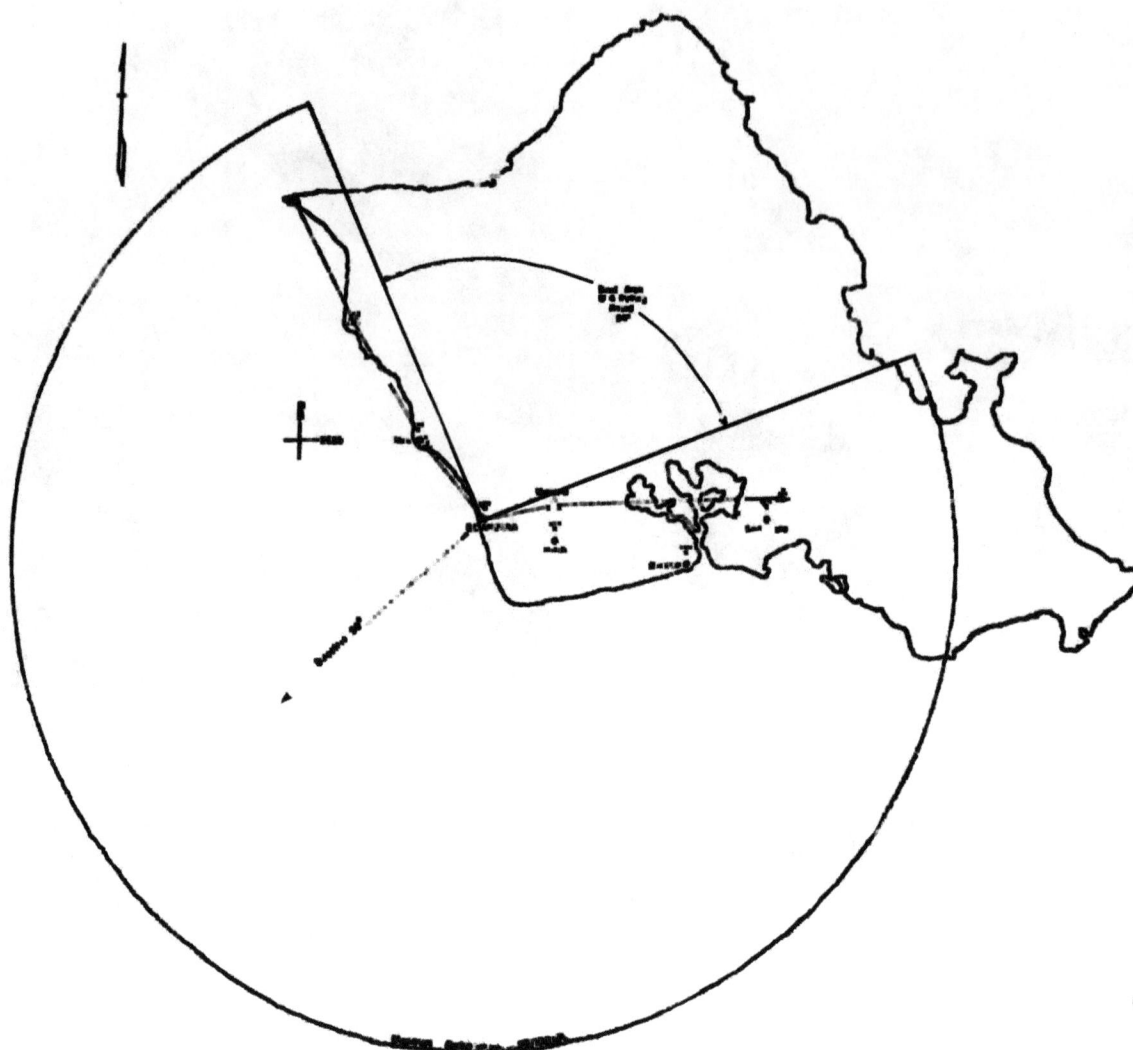

Map showing field of fire for Battery Arizona, Kahe Point. *NARA*.

In October 1942 the Navy Department indicated that additional battleship turrets might be made available. The Navy suggested that the four 14-inch turrets from the capsized battleship USS *Oklahoma* might be renovated and made available to the Army for use as shore batteries. These were 14-inch/45 caliber guns similar to those mounted in USS *Arizona*, but as just two triple mounts and two double mounts. In light of this development the coast artillery command quickly selected several sites on the island of Oahu for these turrets. One, mounting a pair of 14-inch guns, was projected for emplacement on the high ground behind Kahuku Point at Paumalu from which location it could command a field of fire from Windward Oahu around the shoreline towards Kaena Point at the island's northwest end. Another battery, consisting of a triple 14-inch turret from Oklahoma was proposed for emplacement atop Kaena Point itself. From this location its powerful guns could reach eastward along the south shore of Oahu or northward along the North Shore in the direction of Waialua Bay. Two other locations along the shores of Oahu were projected as sites for the other battleship turrets. One was to be emplaced on Makapuu Head near Station "H" at the east end of the island while the other was to be emplaced atop the south rim of Diamond Head.

Unfortunately, *Oklahoma* had capsized on December 7th and remained overtured for over a year after her sinking. While still mounted in place, the four turrets were totally submerged in salt water at the time of the Navy's offer; as it turns out the ship wouldn't even be righted to begin any removal until early June 1943. It was discovered that *Oklahoma*'s turrets had been ruined beyond repair and their ordnance rendered unusable after being submerged in the mud and salt water of Pearl Harbor for all this time. Although the plans to mount battleship turrets on the heights above the North Shore failed to materialize, providing additional heavy caliber gun batteries on this coast was not abandoned by the HAC, and the decade-old proposal of emplacing of 16-inch guns on the North Shore was resurrected for a brief period. Eventually the decision to pursue emplacing 8-inch permanent system batteries at Paumalu, Diamond Head. and Kaena Point probably were made considering the disappointment in not getting the *Oklahoma*'s turrets.

Righting USS *Oklahoma* in 1943 after it capsized from damage during the Pearl Harbor attack. While some discussion about possible coast artillery emplacement of its four turrets ensued, ultimately it was found that the damage to the mounts was greater than could be reasonably repaired. *NARA*.

Lieutenant Lewis Kirkpatrick, for whom the 8-inch turret battery at Wiliwilinui Ridge was eventually named. *Bennett Collection*

# CHAPTER 11
# THE SECOND WORLD WAR PROGRAM

In 1940 the United States initiated a new program to replace its antiquated seacoast artillery with new modern weapons. With Europe already at war, the future status of the French and British naval fleets problematic and a hostile Japanese fleet in the Pacific, it was certainly possible that enemy fleets could approach U.S. territories or even the continental coasts. Studies and limited construction of new types of emplacements for long range weapons had continued throughout the 1930s. As Europe plunged into war the expansion of American military forces began in earnest. A program to thoroughly update coast defense with replacement of obsolete armament and emplacements was recommended by a War Department Special Board and approved by the Administration and Congress. (1)

Initially the new program was restricted to the defenses of the Continental U.S. The list of harbors to be defended was considerably reduced from those prominent 50 years earlier, to 18 locations with significant commercial or military importance. Within a year of the program's start special addendums for selected "overseas" locations were added. Four harbors in Alaska and defenses for San Juan and Roosevelt Roads Puerto Rico were approved. Then a few of the new naval bases obtained in the 1940 "Destroyers for Bases" deal with the United Kingdom were allocated new American fortifications; Argentia base in Newfoundland, Bermuda, Trinidad, and Jamaica. Existing plans to build batteries to defend the new Naval Air Station in Kaneohe, Oahu benefited from the designs and weapons being used in the domestic program. Finally a small number of new intermediate range gun batteries envisioned as "permanent" elements to replace the plethora of expedient obsolete or primarily mobile guns spread around the island of Oahu in 1941-1942 was approved.

For the domestic program, a set of four types of new guns and batteries was approved for emplacement. One notable difference in design for this new generation was the provision for heavy overhead cover. Since the 1915 Board of Review airplanes had become a potent offensive weapon, and all the new generation emplacements were to receive overhead protection. The largest standard battery design was for two 16-inch guns in a heavily protected, fully casemated emplacement. Providentially, the expensive and hard to produce guns for the most part already existed. They were the battleship and battlecruiser guns never used when the Washington Treaty cancelled their intended recipients. A few had already been emplaced in prototype designs, including the two guns at Oahu's Fort Barrette. New barbette carriages had already been designed and put into limited production. Next most of the 12-inch long-range guns from the 1915 Program would keep their guns, but also be given heavy new overhead concrete protection against bombing and plunging gunfire. All but two of the few 16-inch batteries from this same period would be similarly protected.

Then a smaller emplacement was designed for 8-inch intermediate guns. A few of these tubes were available; part of a batch of 66 ex-navy guns transferred to the Army in 1930 for use in both railway and fixed emplacements. The gun was an ex-naval Mk VI 45-caliber weapon, originally mounted as secondary armament on pre-Dreadnaught battleships and armored cruisers. They were relined and slight changes made in the powder chamber. Some were placed on new railway carriages and offered a significantly improved performance over the old Model 1888 seacoast gun used on the 8-inch railway carriages employed in Hawaii. When used as a fixed seacoast gun they were placed on a specifically developed barbette carriage. With an elevation of 0 to 45°, they could range out to 33,000-yards. Firing either a 260-lb. armored-piercing or 240-lb. high explosive round to this distance was an important performance. Renovation of 20 guns for modern system batteries was approved in June 1941, 24 new barbettes were produced, eight of each were supplied to Hawaii. Unlike some of the other weapons scheduled for modernizing the defenses, the 8-inch weapons seemed to have shown up on schedule. (2)

Unfortunately, the emplacements designed for this weapon had a complete lack of protection of the actual gun and mount. As deployed, they were completely exposed in open circular emplacements and susceptible to bombing, shelling, and even splinter damage from close misses. While they could, and were, camouflaged they were in need of better splinter protection. The plan had been to produce a steel wrap-around shield that could be installed after emplacement. However, this plan never was realized and it appears no such shields were ever produced or supplied. The guns served their mission under nothing heavier than camouflage netting.

While there was an adopted standard plan for 8-inch batteries, only a prototype in San Diego and a couple of batteries in Alaska were built following the recommended design. In Oahu one battery was built for the Kaneohe Bay defenses (Battery No. 405, later named Battery DeMerritt), but its magazines and operational rooms were tunneled into the volcanic rock immediately behind it. The gun emplacements were separated to the standard design specifications. The same would apply to the three new batteries ultimately approved for elsewhere on Oahu. It was found to be much cheaper to tunnel into volcanic rock to form galleries and rooms for magazines and operations than to use the conventional cut and cover technique. It seems it may have also been quicker and required less labor. One result of using the commonly available Conway Hucking Machine for boring large-diameter tunnels was that they typically branched off or joined at a 45° angle rather than a perpendicular 90°degree angle. (3)

Despite the difference in construction technique, the standard service rooms were always included. Common to whether the emplacement was conventional cut and cover or tunnels, as many of the support rooms as possible considering the local topographical conditions were included in the main protected structure: the separate powder and projectile magazines (one of each for each gun), a compressor room for the air scavenger, a generator room and fuel tank for primary or emergency power, a plotting room, a radio and switchboard room, a ventilator and air conditioning room (or at least space), latrines, first aid room, an appropriate store or tool room, tanks for emergency water supply, and a radar control room (if radar was co-located with the battery structure). While there were never crew accommodations at the battery, sometimes a small galley was included in the design if the battery was far from its supporting encampment. Due to the topography of the site and need for wide field of fire, the tunneled batteries varied considerably, none of the plans used were ever precisely replicated elsewhere. The approved emplacement was to have the guns on circular blocks 240-feet apart entirely in the open, protected only with camouflage. Between and behind the guns was the protected magazines and support rooms. (4)

More numerous would be 6-inch batteries. The new 6-inch batteries used either Model 1903 or 1905 6-inch tubes. These had been mounted early in the century on disappearing carriages at coast defense forts but removed in 1917 for use of expedient wheeled carriages for use with the American Expeditionary Force in France. The war ended before they could be issued or even dispatched to Europe. After a period in storage, and with still a useful barrel life, they were selected for use as the primary armament of the new intermediate 6-inch battery of the 1940 Program. Like the 8-inch, they were to be used with a thoroughly modern long range barbette carriage capable of all-around fire. As configured the guns could elevate to 47° and had a range of 27,150-yards, longer range than the 20,000-yards of the 155 mm Panama mount guns they would replace. The gun was hand loaded with separate bagged charges, but with a trained crew could achieve 4 rounds/minute. Eventually the early model tubes were all used, and the army ordnance department produced a number of new tubes with similar ballistics, known as the T2 6-inch seacoast gun. One of the few places the new tubes were allocated were to several of the Oahu emplacements. (5)

Protection guidelines for batteries of this generation recommended they be given gunfire protection from cannon of equal caliber. A 6-inch battery should thus be protected from fire of guns of 6-inch caliber. In the last variant of the standard design (1941-42) that meant magazine front wall of 7-feet reinforced concrete and overhead protection of 6-feet. However, the three 6-inch batteries planned for the permanent defenses in Oahu were to be built as modified, tunnel type structures. The guns were emplaced in open circular concrete platforms, but these were supplied with a cast steel (4-6 inch thick) wrap-around shield protected the operating crew from flying splinters.

Besides the different space requirements needed for 8-inch versus 6-inch shells and powder, there were few differences in the internal space layout between the two types of batteries. One was for electrical power. The power requirements for the electrically-driven carriage were greater. The Hawaiian 6-inch batteries were all to be connected to commercial power, but still needed room for fuel and three back up 125 kW generator sets. The 8-inch batteries had a single set of motor driven generators primarily used for lighting as the gun and carriage were all manually maneuvered.

A new approach to battery naming was applied to this generation of guns. During planning and then construction the emplacements were given three-digit numerical designations. The 6-inch batteries assigned to Hawaii were designated battery construction numbers 301, 302, 303, 304, and 305. The 8-inch batteries were designated battery construction numbers 405, 407, 408 and 409. The intention was for these batteries to be given an official name via the usual general orders process after completion and transfer. However, many batteries, even when completely finished and in service, served with just their number designation. Some, including those in Hawaii, weren't named until after the war, at times barely before they were taken out of service. Technological advances had also occurred in position finding. Sophisticated new plotting boards and communications equipment were available for this new generation of gun emplacements along with the use of radar for both detecting enemy vessels and in directing the fire. Other innovations included auto-ramming of large guns, and compressed air scavenging of powder chambers after firing. Emplacements were built with positive pressure airlocks and air filters to protect critical spaces from gas attack.

Soon after the first elements of the U.S. Army arrived on the Island of Oahu in 1898, the process of mapping the island and preparation of defense plans began. While initially the survey work was focused on the south shore of Oahu, by 1906 other parts of the island were also receiving the Army's attention. In 1908, companies of the Engineer Battalion began a comprehensive military survey of the island that continued until 1912, when the land defense survey of the island was completed. (6)

The island's windward, or northeastern coast lies in the shadow of the 2,000-foot high Koolau Range. These mountains parallel the shore northwesterly from Makapuu Head, Oahu's easternmost point, to Kahuku Point, the island's northernmost promontory. For most of its length, the windward side of the island is bounded by offshore reefs. In the early years of the 20th century, only a few of its beaches were considered expansive enough to support landings by hostile forces in numbers greater than small raiding parties. Only at Laie Bay, Kahana Bay, Kaneohe Bay, Kailua Bay and Waimanalo Bay were there beaches capable of landing even moderate sized forces from small boats.

The Kaneohe Bay Naval Air Station as it appeared from the air on October 1, 1941. *USAF.*

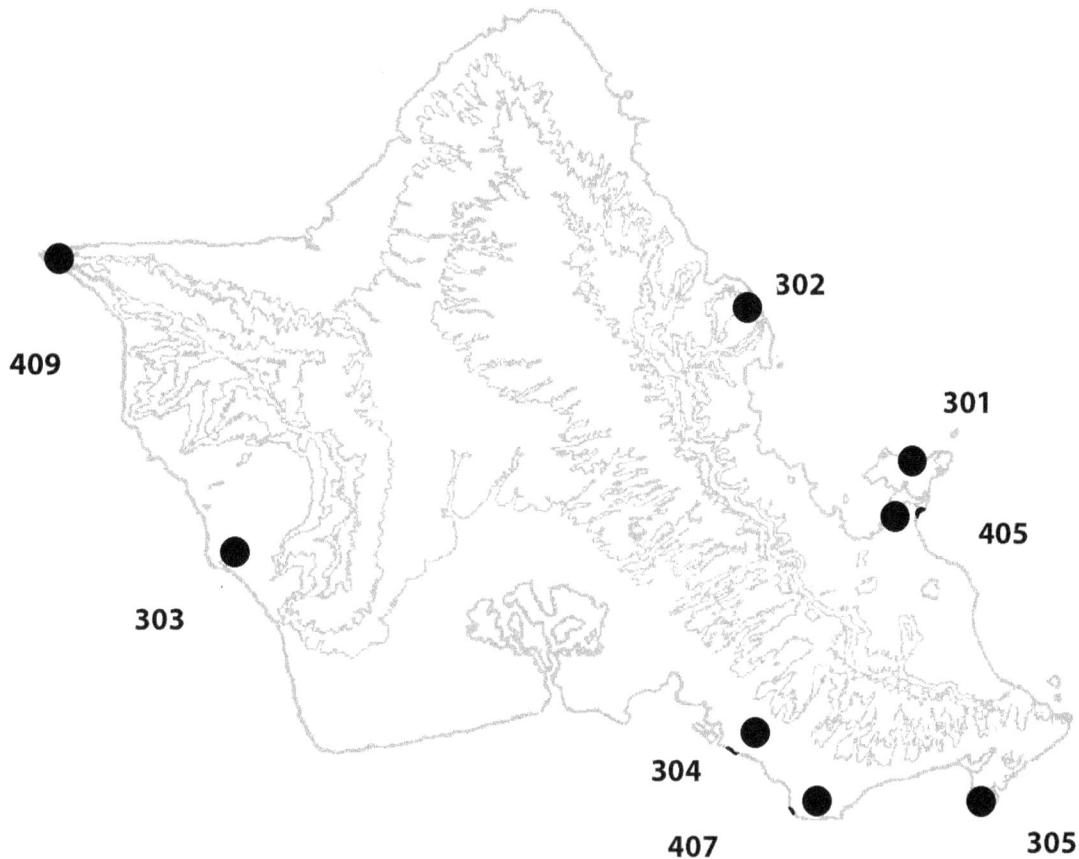

Map showing locations of new proposed standard 1940 program batteries for Oahu. *Williford Collection.*

Kaneohe Bay, expansive at least in surface area if not depth, was well sheltered from the ocean by the Moka-pu Peninsula on its eastern shore, and it had a large area of land between the beach and the Koolau Range suitable for military operations. As originally surveyed, it was strewn with coral patches, shoals, and rock out-croppings; some of that were covered by only a scant foot or two of water. In the early years of the 20th century only shallow draft fishing craft could operate safely in the two narrow channels and constricted anchorages of Kaneohe Bay. Prior to the years immediately preceding World War II, dredging of the bay had been minimal and was chiefly associated with accommodating tuna boats operating out of Kaneohe's small harbor. (7)

Two small, but relatively unprotected bays existed on the east side of the Mokapu Peninsula. Kailua Bay and Waimanalo Bay were also strewn with rocks and were generally unprotected except for the offshore reefs that broke the waves to some extent. The Mokapu Peninsula constitutes perhaps the most prominent feature on Oahu's windward shore. This anvil shaped projection forms the eastern limits of Kaneohe Bay and the northern limits of Kailua Bay. It is about four miles long and varies in width from about one mile near its base to nearly two and a half miles on its seaward side. The peninsula's base is formed by the Papaa Ridge, a series of low hills of which Puu Papaa is the highest rising to 542-feet above sea level. Then around the northeastern extreme of the Mokapu Peninsula is volcanic Ulupau crater.

The Kuwaaohe Military Reservation was established during World War I by executive order. Situated near the center of the peninsula's head, this 322-acre tract consisted of a strip of land about one-half mile wide extending about two-miles in a northeasterly-southwesterly line encompassing the west and north rims of Ulupau crater. Until acquisition of the Waimanalo Military Reservation east of Mokapu, Kuwaaohe was the largest army tract on Windward Oahu. A strip of beach front along the northern end of the peninsula was known as Mokapu Beach, in the years prior to World War II. After the nation entered World War II, the beach was known as North Beach.

The U.S. Navy's decision to build a major aviation facility on the Mokapu Peninsula led to substantial increases in the Army's defense scheme for Windward Oahu. A report by the Secretary of Navy to Congress in late 1938 pressed the need for critical new naval bases. Known as the Hepburn Board, it recommended a naval air station at Kaneohe Bay as an adjunct to the Ford Island Naval Air Station at Pearl Harbor. The Navy's plan to develop a seaplane base at Kaneohe Bay called for the blasting and dredging of a seadrome in the reef-strewn bay, as well as opening a passage in the offshore reef for deep draft vessels. Although this dredging of channels increased its military value to the United States, it also made the bay more accessible to an enemy's landing attempt. It was also between the town of Kaneohe and the towering Koolau Range that hostile ground forces would find the greatest amount of maneuvering room on Oahu's windward coast. (8)

On August 5, 1939, contracts were let to the Pacific Naval Air Base Contractors and construction of the Kaneohe Bay Naval Air Station began the following month. Even before the initial plans were complete, the air station scheme was expanded to include a major runway for land-based aircraft, and the process began to acquire an additional 464 acres at the base of the peninsula for housing, gasoline storage, and shop buildings. By September 1939, the construction engineers had begun a dike on the western fringe of Mokapu that would be back-filled with dredged spoil from the bay. When completed some four years later, this dredging and filling had added another 280 acres to the air station. (9)

The land on the west side of the Mokapu Peninsula was projected for the seaplane ramps, aprons, and hangers and a runway was laid out across the northwest part of the peninsula. Initially 1830 acres on the western half of the Mokapu Peninsula were acquired for the Navy. Construction of the hangers, and quarters of the air station, along with the pouring of the concrete runways and taxiways for land-based planes and the ramps and aprons for the seaplanes was rapidly advanced in 1940, and by November, a small detachment of Marines arrived from Pearl Harbor to guard the budding installation. The initial station complement moved in December 7, 1940 and by early 1941, the air station had been developed into what at that time was the U.S. Navy's largest combined land and seaplane base. (10)

The Army's responsibility for the defense of the new naval air station at Kaneohe Bay had a major effect on the wartime allocation of coast artillery resources on the island. A new harbor defense command was authorized for Kaneohe Bay; its army garrison, initially a reinforced battalion of tractor-drawn coast artillery from the mainland United States, was to be quartered on the eastern part of the Mokapu Peninsula in a cantonment to be built on the existing Kuwaahoe Military Reservation. This cantonment was renamed Camp Ulupau in the spring of 1941. (11)

Soon after assuming command in Hawaii, General Walter Short recommended substantial improvements in the island's defenses. In mid-February 1941 that included a separate harbor defense project for the Kaneohe Bay area. On April 8, 1941, the first Kaneohe Bay Project was approved and published. An important First Revision was issued at the end of July 1941. The ensuing defense projects for Kaneohe Bay were independent of the national 1940 Program. It did, however, adopt the same standard and generalized plans for battery emplacements, but the project and approval of expense to construct stood apart from the general U.S. initiative. (12)

The Hawaiian command initially had recommended a primary armament of 12-inch guns on long-range barbette carriages, of the type emplaced in Battery Closson at Fort Kamehameha. The Hawaiian Department suggested that Battery Closson's two spare M1895M1 gun tubes be used for this new battery at Kaneohe Bay as an immediate stopgap measure. In the July revision the 12-inch project was replaced with a standard 1940 Program casemated battery for two 16-inch guns and additionally two batteries of 6-inch guns. Two, later increased to three, batteries of 155 mm guns on Panama mounts were also recommended as part of a temporary project pending the completion of the permanent emplacements. Soon the 16-inch gun battery idea was deleted (the immediate needs were incompatible with the two-year wait anticipated), but the War Department did authorize a battery of four 8-inch M1888 guns on railway carriages. This was ordnance that was already available on Oahu. (13)

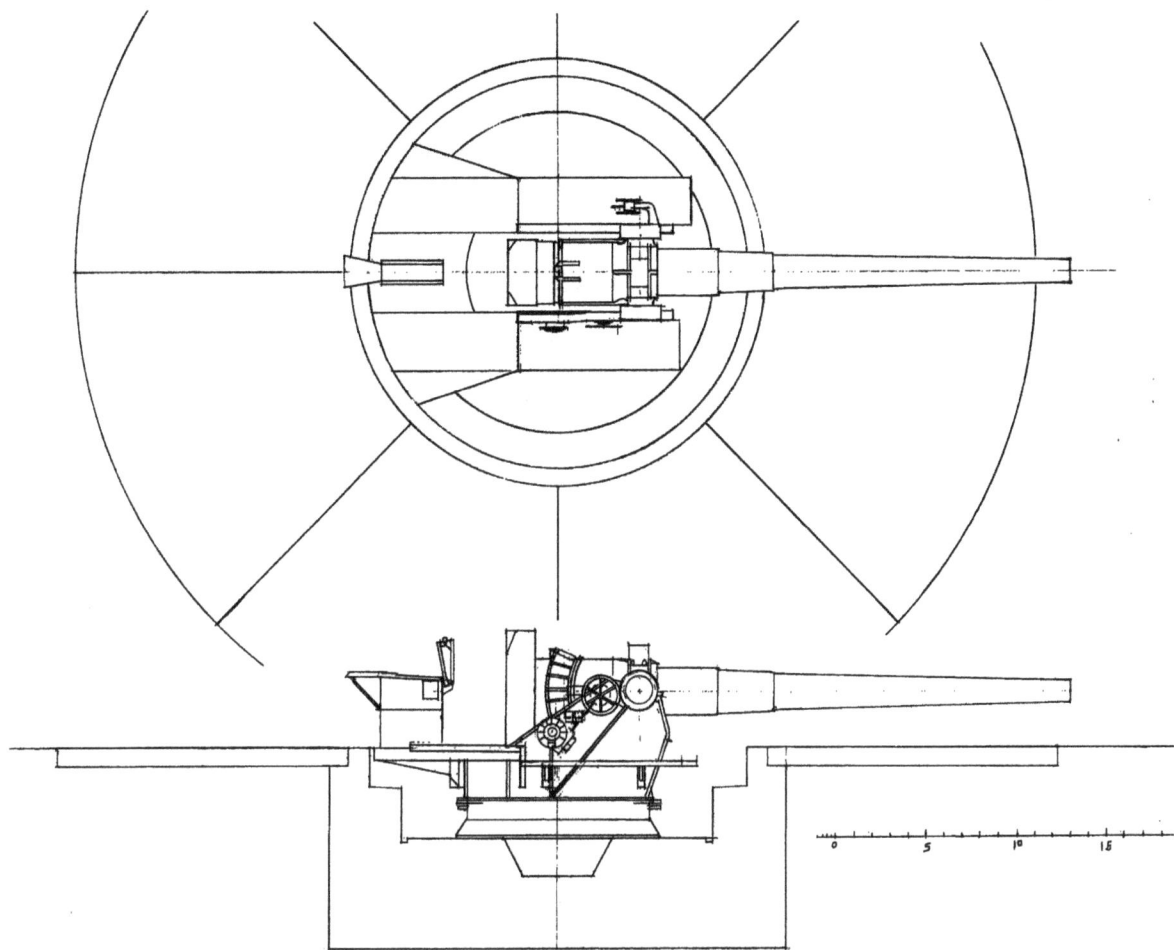

Drawing of the mount for the Navy-type 8-inch gun on barbette carriage intended for the 1940 Program batteries. *Rowbottom.*

Soon after the Army assumed defense responsibilities for the Kaneohe base, various units began conducting regular exercises and maneuvers at Camp Ulupau. During one of these practice alerts in late March 1941, Battery D, 55th CA spent three days conducting surveys for potential gun and searchlight positions on the Mokapu Peninsula and its environs, and generally familiarizing itself with the terrain. Gun positions were selected on North Beach, as it was now termed, and on East Beach that faced Kailua Bay, while future sites for the projected Ulupau Searchlight Group's positions were located near Pyramid Rock at the naval air station, and at the south end of Ulupau's east rim on Kii Point. Additional positions for powerful portable searchlights were selected at the foot of the cliffs at Ka Lae O Kaoio Point, on the high ground above Wailea Point, at Makapuu Point, and on Punaluu Beach near Kahana Bay to the north.

For the remainder of 1941, firing batteries of the 2nd Bn, 55th CA conducted several deployment exercises to the Mokapu Peninsula. Usually, the movement of a battery of 155 mm GPF guns from Fort Ruger to Camp Ulupau involved two columns. The trucks of one column carried the battery personnel, equipment and ammunition and required about an hour and a half for a daytime movement and an additional 15 minutes of travel time at night. The second column, consisting of the D7 Caterpillar diesel tractors towing the M1918 limbers, caissons, and 155 mm guns constituted the advance element. In peacetime the tractors, traveling at speeds varying between two and a half and ten miles per hour, required about four hours to travel the 33 miles to the battery positions on the Mokapu Peninsula.

Batteries D and E of the 55th CA spent the entire month of July 1941, constructing extensive field fortifications and beach defenses on the Mokapu Peninsula. When they departed in August, they left a small guard detachment of 17 enlisted men to look after the material at Camp Ulupau. Fortunately, the Marines provided

Ordnance Department sketch of an 8-inch gun on barbette carriage of the type made available for emplacing in one of the batteries for Kaneohe Bay. *Army Technical Manual.*

bivouacking privileges to their army friends at the air station's Marine Barracks. Battery E's small guard detachment was still posted on the peninsula when the Japanese attacked.

Also included in the approved Harbor Defense Project for Kaneohe Bay were recommendations for garrison size. It would consist of one regiment of coast artillery (antiaircraft), less one gun battalion and its regimental band; one reinforced coast artillery battalion armed with tractor drawn 155 mm guns; and one separate coast artillery battery manning the railway 8-inch guns. However, no new garrison units would be sent until "the situation develops a need therefore and provided such additional troops can be made available in connection with other requirements." Thus, the coast artillerymen of the 55th would have to provide not only Kaneohe's seaward defenses but provide for their own local and beach defenses on the Mokapu Peninsula. The Army quickly moved to complete the installation of its temporary defenses—the 155 mm batteries and the 8-inch railway carriage battery. (14)

Copy of a page from Army Ordnance Manual of the 8-inch barbette gun on its mounting. *Army Technical Manual.*

While the permanent harbor defense construction project for Kaneohe Bay had been approved by the War Department in October 1941, little initial progress was made. Then the attack of December and the danger of a possible subsequent invasion attempt created new urgency. That helped push the construction priority forward rapidly. Detailed planning of the permanent project was rapidly advanced by the spring of 1942. Construction projects were begun for the following batteries that would compose the permanent defenses for the HDKB:

> Two 8-inch guns on barbette carriages with tunnel shields with a range of 38,800 yards were to be emplaced on the north slope of Puu Papaa, in the hills near the base of the Mokapu Peninsula. This project was initially designated Battery Construction No. 405, and in 1946, it was named Battery Robert E. DeMerritt.

> Two 6-inch shielded guns on barbette carriages were to be located atop the bluffs overlooking North Beach on the seaward side of the naval air station. The battery plan was an adaptation of the standardized design used in the 200 Series batteries then under construction on the mainland. The project's designation was Battery Construction No. 301. Following the war it was named Battery Forest J. French.

> Two 6-inch shielded guns on barbette carriages were to be emplaced in casemates at the foot of the cliffs above Kualoa Army Airfield at Ka Lae O Kaoio Point. Designated Battery Construction No. 302, its design departed from that of Battery No, 301, or any other 6-inch gun batteries as its magazines, power generators and operations rooms were tunneled into the face of the cliffs. After the war this battery was named Battery Avery J. Cooper.

Although these plans would undergo considerable modification during 1942, and 1943, by the end of 1943 most of the design work for the 6-inch and 8-inch batteries would be completed. (15)

Battery Construction No. 301 (Battery Forrest J. French) was one of the first of the modern gun battery projects undertaken. Planning of this battery was well advanced by November 1942. It was to be located on the grounds of the Kaneohe NAS atop the low bluff north of Puu Hawaiiloa about 100-yards inland from the northern shoreline. From this location, the 6-inch guns could cover the seaward approaches to Kaneohe Bay some 15 miles out. This battery used what was probably the closest approximation to mainland standard plan of any Hawaiian Second World War battery. Because the battery was to be constructed on government property, there were no land acquisition costs. Construction costs were estimated at $322,086. (16)

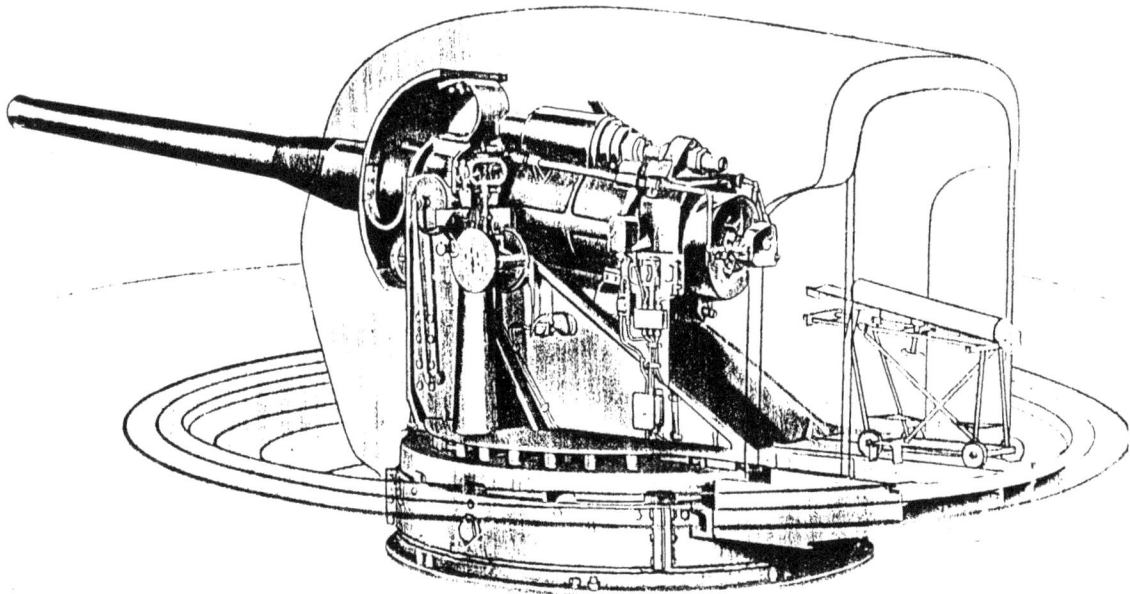

Army Ordnance Department drawing of the gun and mounting for the standard type of 6-inch barbette and mounting. *Army Technical Manual.*

When completed, Battery No. 301 was armed with a pair of two 6-inch M1903A2 guns on M1 shielded barbette carriages. It was equipped with the new cast steel armor wrap-around shields. In an attempt to camouflage the pair of shielded 6-inch guns on their open barbette emplacements, a lightweight roof of lath-like metal slats was installed over each gun so that it resembled the roof of a small house or shed from the air. This "roof" was attached to the shield and carriage so that it rotated with the gun when it was traversed. Tone-down paint and artificial bushes mounted on rollers were placed on the loading platform that added to the camouflage of the battery.

Between the battery's two gun emplacements, a structure of steel reinforced concrete (6-feet thick) covered with several feet of earth was partially built with a cut and cover technique into the crest of the bluff. This building that formed a traverse between the gun emplacements housed the powder and projectile magazines, storerooms, power plant, an air conditioning and ventilation system, fire control switchboard, radio room, plotting room, and the battery's radar operating room. A 25-foot-wide concrete apron connected each of the gun platforms with the battery structure. A short internal stairwell in the rear of the structure provided access down to the central corridor of the battery traverse while another staircase led up to the rear of the battery commander's station atop the bombproof structure. (17)

The battery commander's station and the magazine/service traverse were covered by a minimum of three additional feet of earth contoured to the general slope of the hill on which it stood. An SCR-296A fire control radar antenna was positioned atop the battery commander's station and camouflaged as a water tank. As the battery neared completion in 1943, arrangements were made with the Navy to erect a small camp for the battery personnel at the foot of the bluff in front of the gun battery. There, a barracks, mess hall, latrine, and recreation buildings, all of the Theater of Operations type, were constructed for the coast artillery manning detachment. When completed Battery No. 301 was manned by Battery C of the 41st CA Regiment, and then in late 1944 by Battery C of the 55th CA Battalion.

A standard 6-inch gun in its wrap-around shield on barbette carriage of the 1940 Program, this one was in the Defenses of Southern New York. *NPS.*

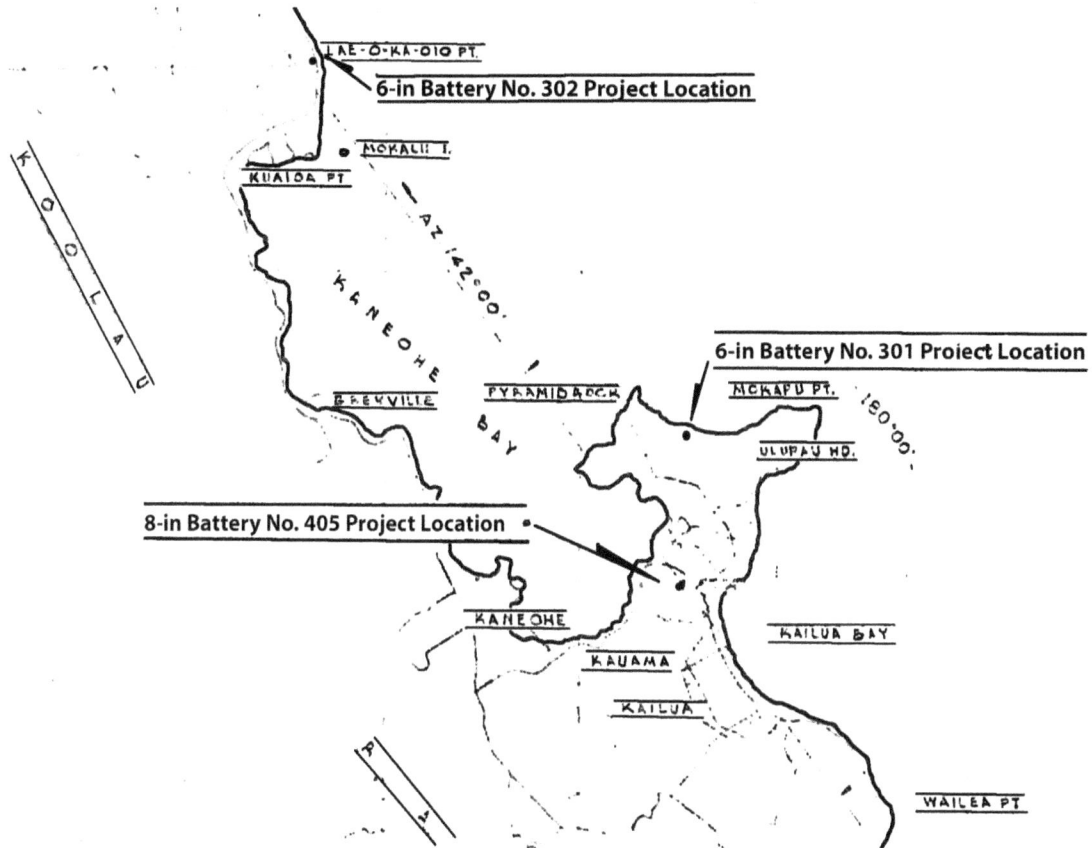

Army Project map showing the locations for the three permanent batteries intended for the Defenses of Kaneohe Bay. *Williford Collection.*

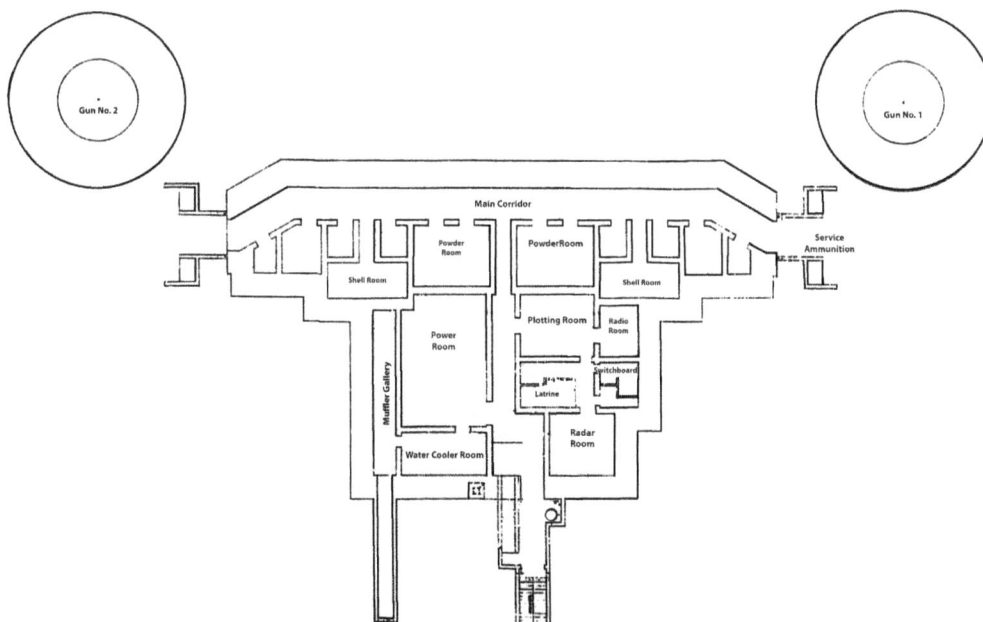

Plan of 6-inch Battery 301 (later named Battery French) on the Mokapuu Peninsula of Fort Hase. *NARA.*

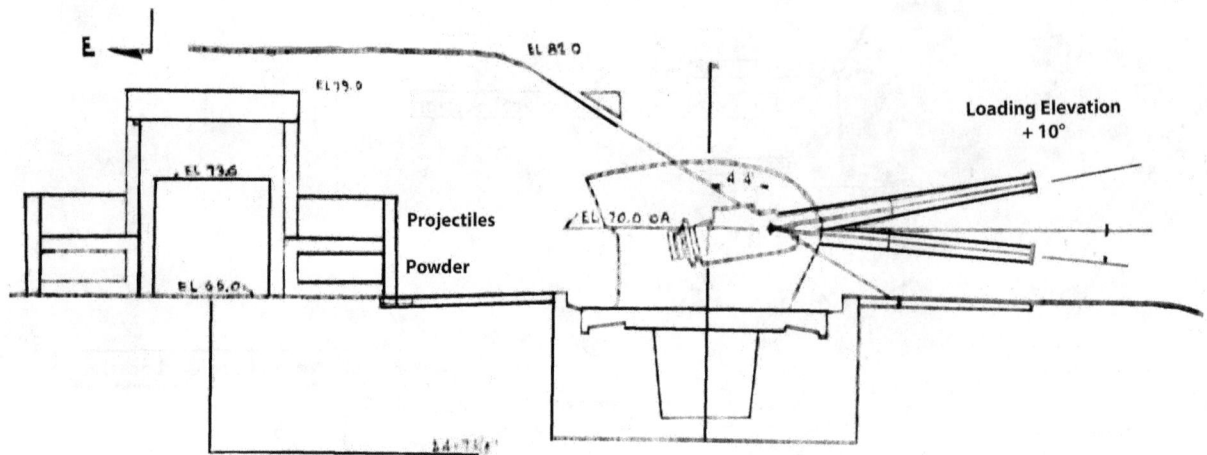

Side view elevation of 6-inch Battery 301 (later named Battery French) of Fort Hase. *NARA.*

Battery Construction No. 302, named Battery Avery J. Cooper in 1946, was located at Ka Lae O Kaoio Point north of Kaneohe Bay. Design work got underway in 1942, but it was not until well into 1943 that the final plans for the battery were approved and its construction start authorized. Cost estimates for land acquisition was $113,227 and battery construction $182,534. Note that this was $140,000 less than Battery Construction No. 301 carrying the same exact armament. This really points out the substantial cost savings of tunneled designs versus cut and cover construction and probably a primary reason for extensive use of the former technique on Oahu.

The battery site was built into the face of the cliffs of Puu Kanehoalani above Ka Lae O Kaoio Point at the northwest end of the Kualoa Army Airfield. It was still to be armed with a pair of M1903A2 guns mounted on M1 Barbette Carriages equipped with cast steel wrap-around gun shields. Unlike Battery No. 301, the shielded guns of Battery No. 302 were emplaced in casemates at the mouths of tunnels excavated into the side of the cliff. The two reinforced concrete casemates were built at the entrances to tunnels to provide additional protection to the guns and their manning detachments. The casemates were of a somewhat less massive design and thickness than those that were being provided for the 12-inch and 16-inch guns on Oahu's south coast and were intended more for the purpose of deflecting falling rock from the cliffs above than for protection from enemy projectiles. (18)

There is a good series of photographs of Battery French from March 22, 1945. Here is one of the M1903A2 guns on barbette carriage M1 with its camouflage overhead screen. *NARA.*

1945 Battery French projectile room. *NARA.*

1945 Battery French powder room (powder for these guns was stored in cylindrical cans). *NARA.*

At the rear of each casemate a concrete lined tunnel ten-feet wide and 150-feet in length was bored into the cliff side. At intervals, other tunnels of varying lengths branched off at 45° angles. These branch tunnels were used for powder and projectile magazines, motor generator rooms and as storerooms. At the rear of the two tunnels a gas-proofing airlock provided access to a transverse operations tunnel that linked the casemate tunnels. The operations tunnel itself was 106-feet long, 20-feet wide and had a 14-foot crown. This tunnel contained the battery's plotting room, the fire control switchboard and radio rooms, an officer's bunk room, a first aid room, latrines for officers and enlisted men, as well as a small galley. Airlocks and filtration equipment at the entrances to the casemate and power plant tunnels rendered the operations tunnel gas proof.

Midway between the pair of casemates another tunnel ten-feet wide was bored into the cliff face to a length of 50-feet. There it opened into the battery's power generator room. The power plant consisted of three electrical generators that powered and controlled the 440-volt three phase 60-cycle system that operated the guns and provided electrical service. At the rear of the power room another tunnel 20-feet long and ten-feet wide led farther into the mountain ending at an airlock that afforded access to the center point of the transverse operations tunnel.

1945 Battery French engine room. *NARA*.

1945 Battery French plotting room, with its semi-circular plotting board. *NARA.*

The battery commander's station was built into the face of the cliff some 75-feet above the entrance to the power plant tunnel. In the rear of the battery commander's observation room, a passageway eight-feet wide and ten-feet high drilled into the mountainside gave access to the operating room of the battery's SCR-296A fire control radar. Even deeper into the mountain was another compartment housing a 10,000-gallon water tank The battery's SCR-296A antenna was mounted on top of the observation room of the battery commander's station. It was contained within a cylindrical housing that resembled a water tank. This upper-level tunnel complex was accessed from within the battery by a series of vertical steel ladders in a 75-foot shaft that extended up through the roof of the power generator room and opened into the right rear corner of the radar room. (19)

When first fired, a problem was experienced with a vacuum created within the magazine access tunnels that required building additional air baffles and strengthening some fixtures. Once in service the guns were manned continuously until the postwar period. The initial manning detachment was provided by Battery A of the 41st CA Regiment, replaced in late 1944 with Battery D of the 55th Coast Artillery Battalion.

The early months of the war were times of tension marked with high levels of activity for the defenders at Kaneohe Bay. The 704 acres occupied by Puu Papaa and Papaa Ridge controlled access to the Mokapu Peninsula from the rest of the island. The Army secured this important reservation. For most of the war, a battery of field artillery supported by infantry was posted in these hills, and later when the permanent defense system was implemented, Puu Papaa was selected as the site for a battery of 8-inch ex-naval guns. These hills at the base of the Mokapu Peninsula were of great importance to the security of the installations on the peninsula. If an enemy force were to land on the shores of Kailua or Kaneohe Bays and move swiftly onto Papaa Ridge,

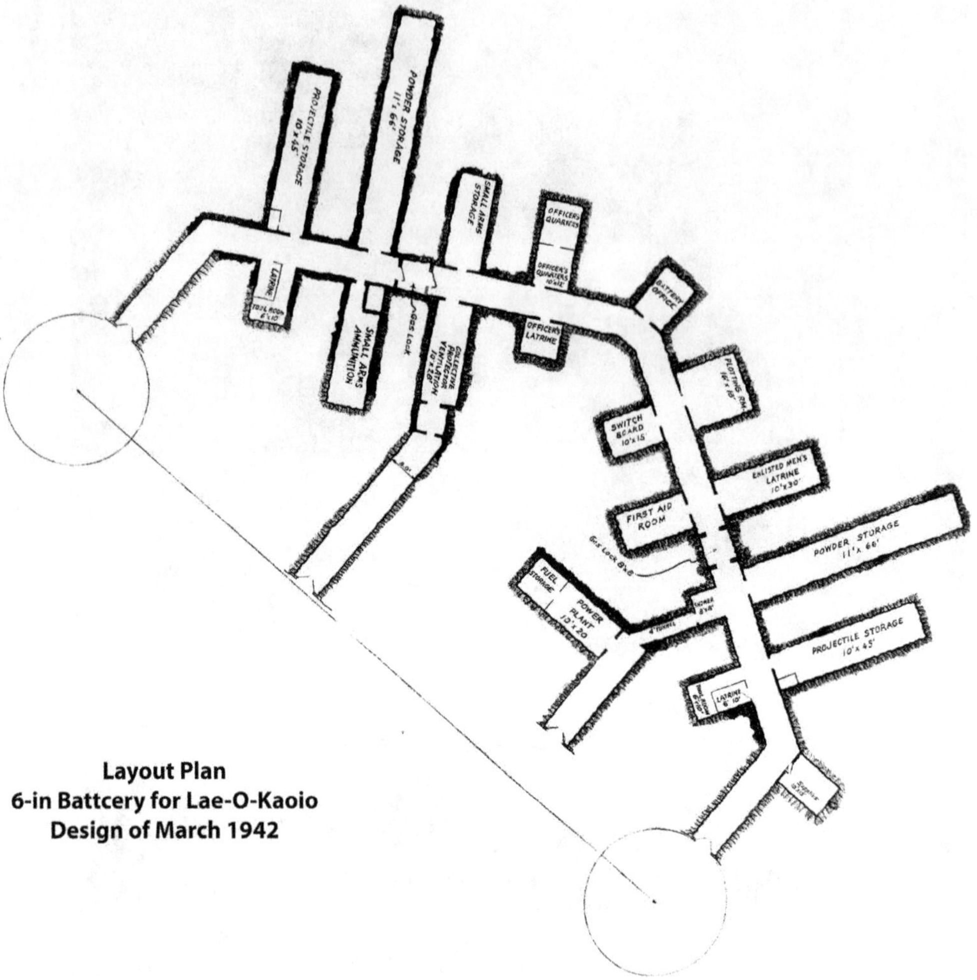

**Layout Plan
6-in Battcery for Lae-O-Kaoio
Design of March 1942**

Early plan for the 6-inch battery at Lae-O-Kaoio in March 1942. Note how different it was from the next plan which shows the final version. Many of the 1940 Program batteries here and elsewhere on Oahu went through fairly extensive plan revisions. *NARA*.

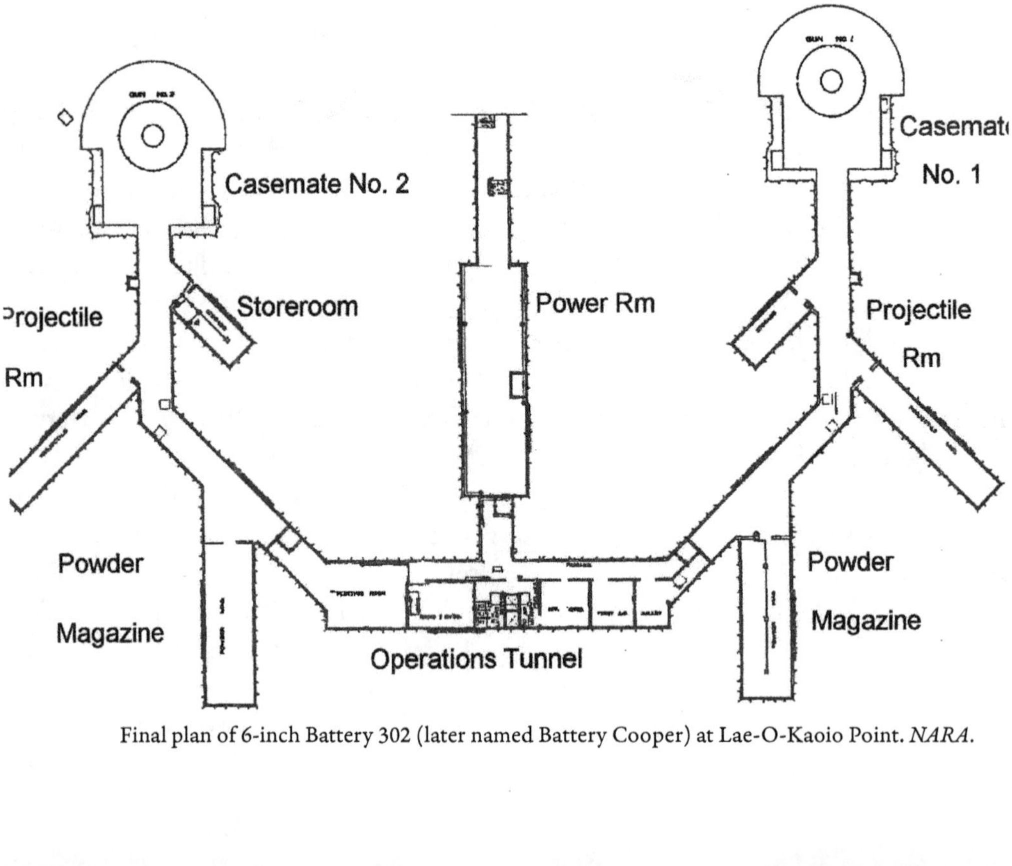

Final plan of 6-inch Battery 302 (later named Battery Cooper) at Lae-O-Kaoio Point. *NARA*.

The Lae-O-Kaoio Point location of Battery 302; it would be built into the cliff immediately adjacent to the road. *NARA.*

they could cut off the forces on the peninsula from the rest of the island and take the defenses under fire from the rear. (20)

The limited number of troops available in the early months of the war prompted the elements of the 57th CA to regularly practice retrograde movements from the peninsula back into the hills overlooking Kaneohe and Kailua Bays; movements that would be necessary if overwhelming enemy forces were to land in the vicinity of Kaneohe Bay. Alternate field positions for the 155 mm guns were selected on the slopes of Papaa Ridge overlooking the bays. Although these alternate battle positions were to be provided with the requisite magazines, plotting rooms, and bunker emplacements, the pair of Panama mount positions on Papaa Ridge were never fully developed.

Battery Construction No. 405 (Battery Robert E. DeMerritt) was the third of the batteries projected for the permanent harbor defense project at Kaneohe Bay. The battery site was on the northern slope of Puu Papaa at the base of the Mokapu Peninsula. From this location, the battery's two 8-inch MKVIM3A2 naval guns mounted on army M1 barbette carriages had a 180° field of fire that covered a 20-mile radius along the eastern shore of Oahu from Laie in the north to beyond Makapuu Point to the southeast.

Planning began early in 1942; the preliminary designs that were developed called for a tunnel complex running roughly east and west. The various support functions and magazines of the battery were located in shorter tunnels branching off at right angles from the main east-west transverse tunnel. Two short tunnels entered the hillside at the rear of the gun emplacements that provided access to the main transverse tunnel. This initial plan was, however, rejected because its rooms were considered excessive in both number and size for a battery of that type. A completely different design was developed in the summer of 1942 that eliminated the small arms ammunition rooms, quarters, offices, and a paints and materials storeroom thus resulting in considerable savings. The eliminated structures were to be provided in Theater of Operations type buildings outside the battery tunnels. Projected cost as of July 1941 for land acquisition ($152,160 for leasing from the Kaneohe Ranch) and construction of the battery was estimated at $436, 066. (21)

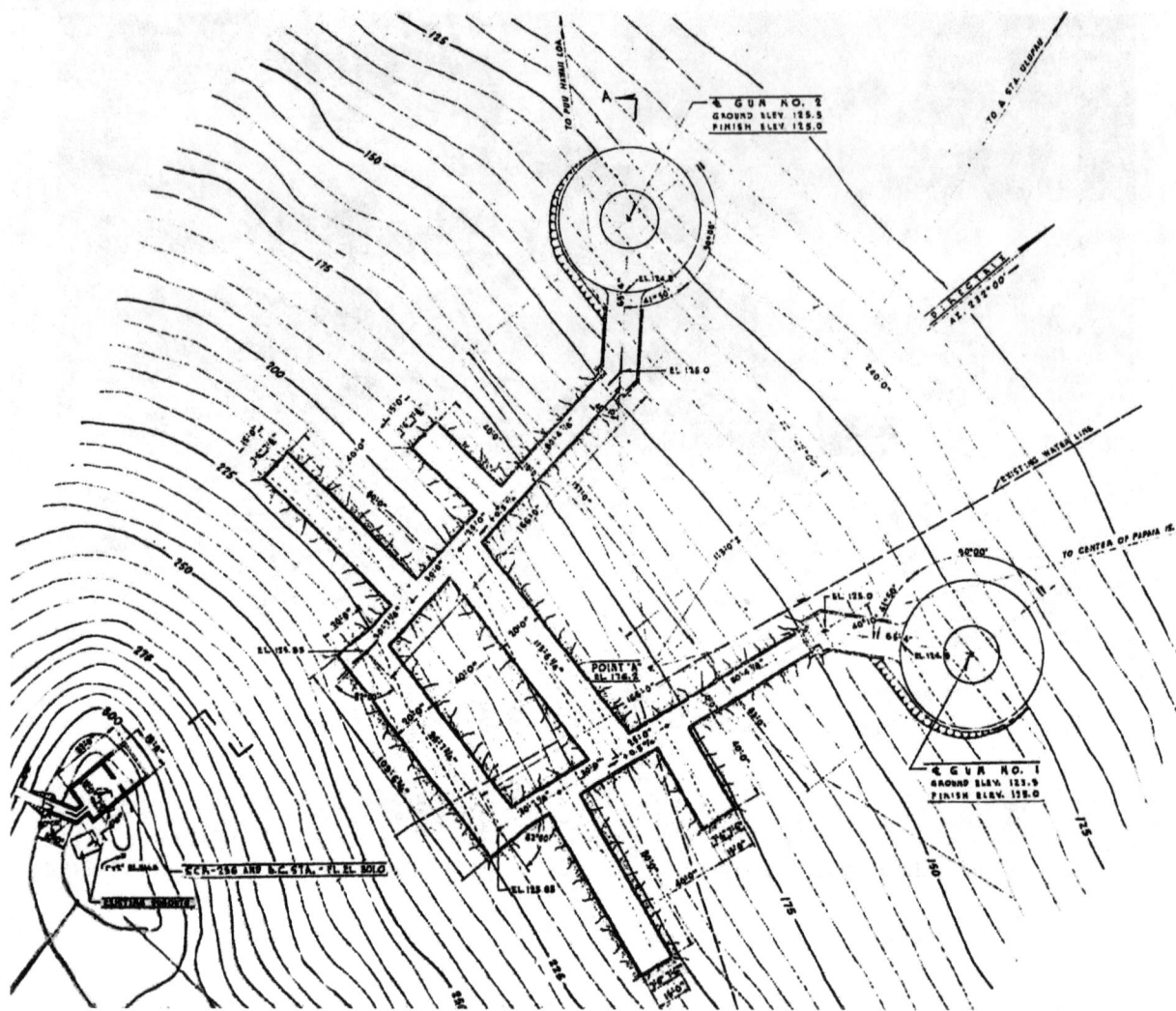

Plan adopted for 8-inch gun on barbette carriage of Battery 405 (later named Battery DeMerritt). *NARA*.

Construction at the Puu Papaa site finally began early in 1943. Certain aspects of the revised design reflected the design that had been adopted for Battery No. 302. A tunnel network drilled into the hillside about 125-feet above sea level, housed its magazines, service, and operation rooms. The battery's two circular gun emplacements were placed significantly outside their ammunition tunnels, such that no casemate or overhang was needed or even practical. The guns were protected only by camouflage composed of light weight metal lathes assembled as to resemble the roofs of farm buildings built on the slopes of the hill. These lath structures were anchored to the unshielded gun carriages. Battery No. 405's two guns were 240-feet apart as measured from the centers of the emplacements, the concrete loading platforms and blast aprons had a diameter of 65-feet.

At the rear of each gun platform a concrete apron 14-feet wide and about 35-feet long led to a tunnel bored into the hillside. The two tunnel adits were also camouflaged by additional metal frame "buildings". These tunnels were about 155-feet long, ten-feet wide and had a crown of 12-feet. Some 65-feet into each of the entry tunnels, another shorter tunnel branched off at about 45°. Nearly 55-feet long and 15-feet wide with a 12-foot crown, these two branch tunnels were used as projectile magazines. Farther into the hillside at a point about 100-feet from the outside entrances, was the first of two transverse tunnels. This first transverse tunnel measured 20-feet in width and was about 95-feet in length with a 14-foot crown. This tunnel housed the fan and air conditioning equipment rooms, four storerooms, and a small power generator room housing a small emergency electrical generator. (22)

The battery's primary source for electricity was Oahu's commercial power plants. However, the reserve 5 kW generator in the power room could provide sufficient electric power to meet the battery illumination

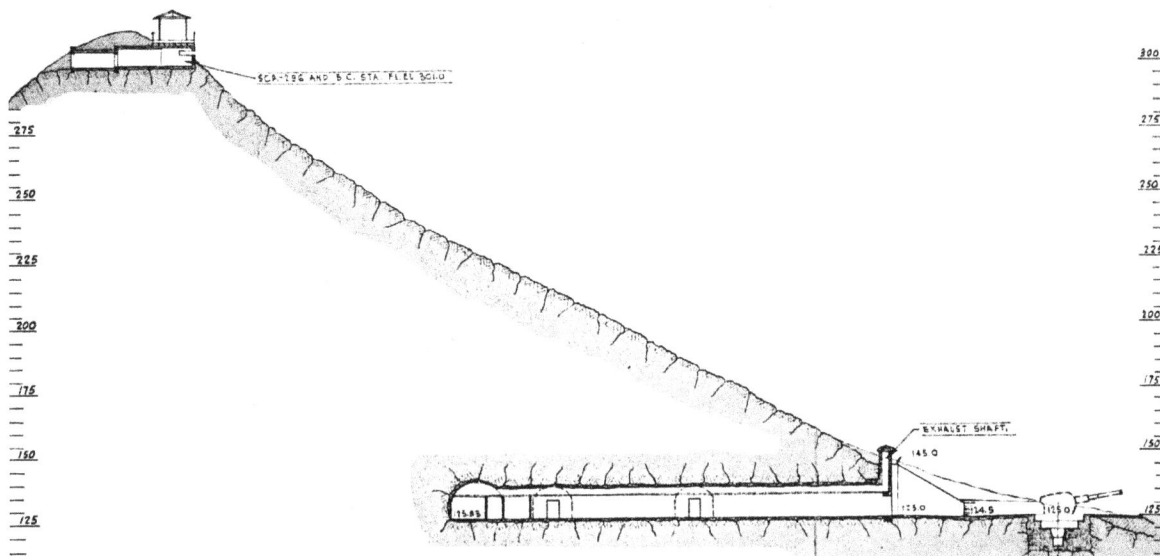

Side-view of Battery 405 with its Battery Commander and radar station high on the hill behind the guns. *NARA.*

needs, operation of the radios and radar, the CWS protective collectors, and ventilation system in the event of an emergency, or loss of commercial power. The guns were operated manually and required no electric power. Some 30-feet further up the emplacement tunnels a second tunnel branched off each of the main tunnels. These tunnels, nearly 95-feet long and 15-feet wide, also with 12-foot crowns, served as powder magazines. After passing through airlocks, one entered the operations tunnel. This 110-foot-long tunnel was 20-feet wide with a 14-foot crown. It contained a large plotting room, the radio and fire control switchboard room, a bunkroom for the battery officers, latrines and showers for officers and enlisted men, a first aid room and a small galley and mess area.

A battery commander's station constructed of reinforced concrete was built on the site of two temporary dugout type bunkers on the summit of Puu Papaa at an elevation of 300-feet above sea level and about 175-feet above the battery's gun emplacements. The battery's radar operating room was in the rear of the battery commander's observation station. Access to the structure was gained from the southwest or reverse side of the hill just below its crest, through a 20-foot long, four-foot wide, dog-legged passage that led into the partially

One of the 8-inch Mk VI ex-navy guns on its M1 barbette carriage at Battery 405 on March 22, 1945. *NARA.*

1945 Battery 405 projectile room. *NARA.*

1945 Battery 405 powder room. *NARA.*

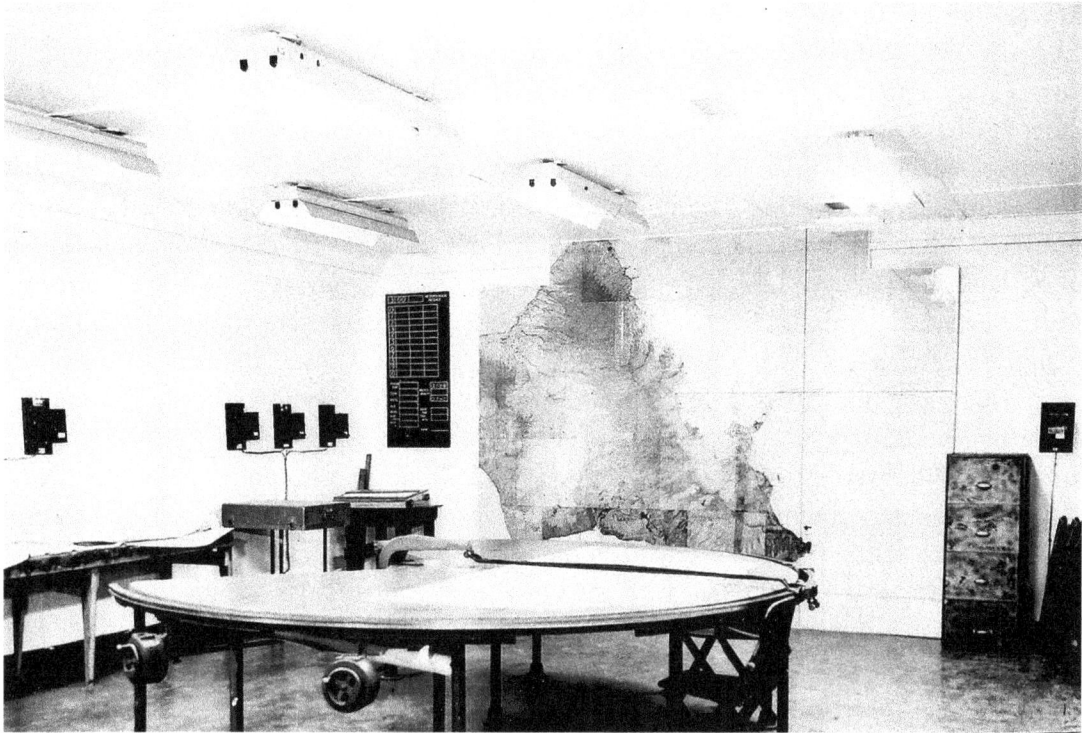

1945 Battery 405 plotting room. *NARA.*

Stations above Battery 405 seen in an overflight of 2009. *Williford Collection.*

buried radar room. The battery commander's station was located at the front of the radar room and was accessed through a doorway in the radar room's right front. A gravity fed emergency water tank for the battery was dug into the hillside 70-feet southeast of the BC station.

Initially, plans had called for Battery No. 405 to receive heavy tunnel type shields for its guns, but by the time the battery was completed, the procurement of the shields had been canceled.

When the new batteries of the Kaneohe Bay permanent project were completed, they utilized seven fire control stations; of which only three had been built prior to the war. Most of the stations projected to serve Battery Pennsylvania, and all of those serving Batteries No. 301, 302, and 405, were located along the coastline of Windward Oahu. These stations had been preceded in most cases by a series of temporary fire control or base end stations that had been built early in 1942 for the 155 mm gun batteries and for Battery Sylvester. While the batteries of the permanent system were being built, new permanent stations were built adjacent to the sites occupied by the temporary stations. Typically, these permanent installations consisted of one or more splinter-proof structures built of reinforced concrete, constructed at elevations that were frequently 400 to 600-feet above sea level. (23)

The batteries on Windward Oahu were served by several SCR-296A fire control radar sets. One was located at the 8-inch gun Battery Kahuku in the North Sector near the island's northernmost point. Another was located at Kaneohe Bay Battle Position Number 3, the site of Battery No. 302 above Ka Lae O Kaoio Point. Battery No. 301 at Fort Hase was supplied with the third set. A fourth radar set was located at Kaneohe Bay Battle Position Number 1, the battery commander's station for Battery No. 405 on the summit of Puu Papaa at the base of the Mokapu Peninsula. A fifth radar site was located at Battery Wilridge in the Harbor Defenses of Honolulu on the southeast side of the island. Battery Wilridge although in the island's southern harbor defenses, could provide indirect fire on targets off the eastern shore of Oahu. A sixth SCR-296A set was provided for Battery Pennsylvania near the end of the war. Its antenna was situated atop Battery Pennsylvania's combination battery commander's station and radar operating room on Ulupau's west rim across the crater floor from the battery. (24)

By early 1944, Battery No. 301 with its two long-range 6-inch guns had been completed and Battery C of the 41st CA took over the new armament. Battery Sylvester became an alternate battery assignment for Battery C and eventually was placed in maintenance status. Battery H, 41st CA was the fourth of the regiment's firing batteries to serve in the Kaneohe Bay area. Upon its activation in 1943, it was assigned to the battery position

Plotting room with M1 Board in the Harbor Defense Command Post of Fort Hase in 1945. *NARA.*

at Puu Papaa where initially it manned a battery of 155 mm GPF guns and provided a manning detachment for the two 3-inch guns at Wailea Point. By early 1944, it was preparing to take over the pair of 8-inch guns of Battery No. 405 as construction of this powerful battery neared completion. Battery H manned Battery No. 405 until May 29, 1944. (25)

It is not known when the program for new permanent batteries for the defenses of Honolulu, Pearl Harbor, and the North Shore were authorized. The separate and slightly earlier Defenses of Kaneohe Bay were proposed and approved in July 1941. However, the 1940 Hawaiian Defense Project version of December 1, 1940 did not foresee or propose any of these batteries. As some of the initial siting surveys were underway in the summer of 1942, it is reasonable to imagine that the project, whether collectively or by individual battery project, was brought forward and scheduled in the six months following the Pearl Harbor attack during the first half of 1942. The first reference to specific site plans dates to October 1943.

Six new batteries (not including the previous project for Kaneohe Bay) were suggested in the planning program. Three were for 6-inch, shielded barbette batteries and the other three for 8-inch, unshielded barbette batteries. Four were structurally begun, mostly with initial tunneling of their magazine spaces; one of each type had work suspended or not undertaken. None were armed and all sites permanently suspended in the immediate postwar period.

A battery for Fort Ruger was designated Battery Construction No. 407. This battery for the new permanent system was to consist of two 8-inch casemated guns to extend and improve coverage of the waters east and south of Diamond Head. Originally slated for emplacement at Fort Kamehameha, the site was moved to Diamond Head in January 1944. In May of 1944 tunneling was reported underway, even to the point of applying gunite. Compared to some of the other projects of this program, it was relatively advanced in construction when work was suspended in mid-1945. In May of that year it was assigned a deferred status after being assessed at 75% completed. Battery No. 407 was to be armed with two 8-inch MKVI Mod 3A2 naval guns on army M1 barbette carriages. The barbette gun carriages were shipped to Hawaii from Aberdeen Proving Ground on May 8, 1944. While the armament was apparently brought to Diamond Head, the guns were never mounted.

The layout of the emplacement adopted a different gun center separation distance because the gun casemates were placed on adjacent ridges. It was built with a tunnel complex consisting of two access tunnels 10-feet wide with a crown of 14-feet, driven through Diamond Head's south rim several hundred feet apart. On the exterior slope of the crater at the end of each tunnel, a casemate of reinforced concrete was constructed for the gun on barbette carriage. About 100-feet inside the rear entry of the two access tunnels a transverse tunnel 20-feet wide with a 20-foot crown connected the two access tunnels. Between the transverse tunnel and the gun casemates, two magazine tunnels and a storeroom branched off from each access tunnel. One magazine in each tunnel was for powder and the other for projectiles. The connecting transverse tunnel would house the operations rooms; a battery plotting room, fire control switchboard room, first aid station, latrines, and a small galley and mess area. Another tunnel at a right angle from the operations tunnel near its center point housed an emergency generator room and the radiator and air conditioning rooms. Airlocks at the intersections with the access tunnels kept the operations tunnel gasproof. The battery also still lacked a portion of its radar and fire control equipment when the war ended in 1945. (26)

A permanent battery emplacement was planned in 1943 for Waihaiwe as Construction No. 408. It was advanced forward on January 8, 1944, when the decision was made to not proceed with emplacing the 14-inch *Oklahoma* turrets. The site planned was on the west side of Kahuku Point at Paumalu. Construction of this battery, which was designated Battery Construction No. 408, was deferred, however, before work could begin. Later, the project was dropped from the defense plans. The design was for a reinforced concrete cut and cover battery similar in some was to Battery No. 301 (Battery French) built at Fort Hase. Most of the structure would be placed below ground, having the guns moved forward to an elevation of approximately 3-feet high-

**N**

6'x6'
Shaft up
Approx. 50' to
Surface

Water Tank

**8-inch Nattery No. 407**
**Diamond Head Crater, Fort Ruger**

**Preliminary Layout Sketch**
**By Lee Guidry, 1994**

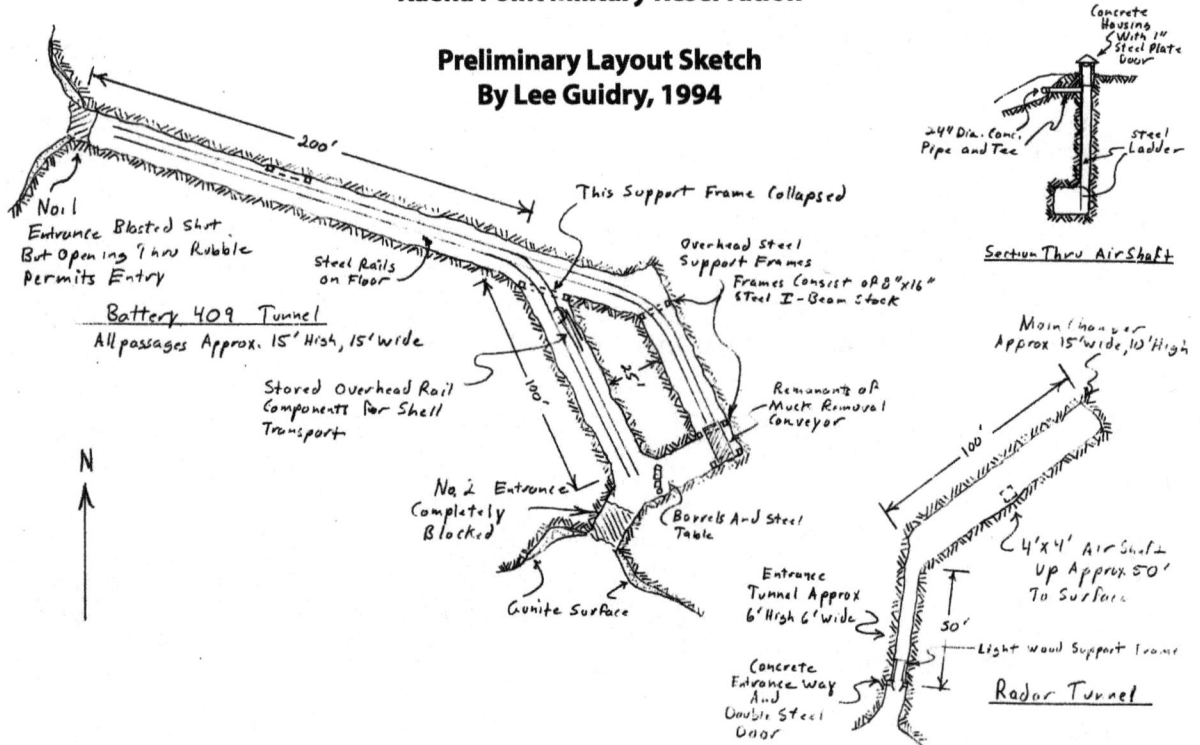

Gun No. 1

Gun No. 2

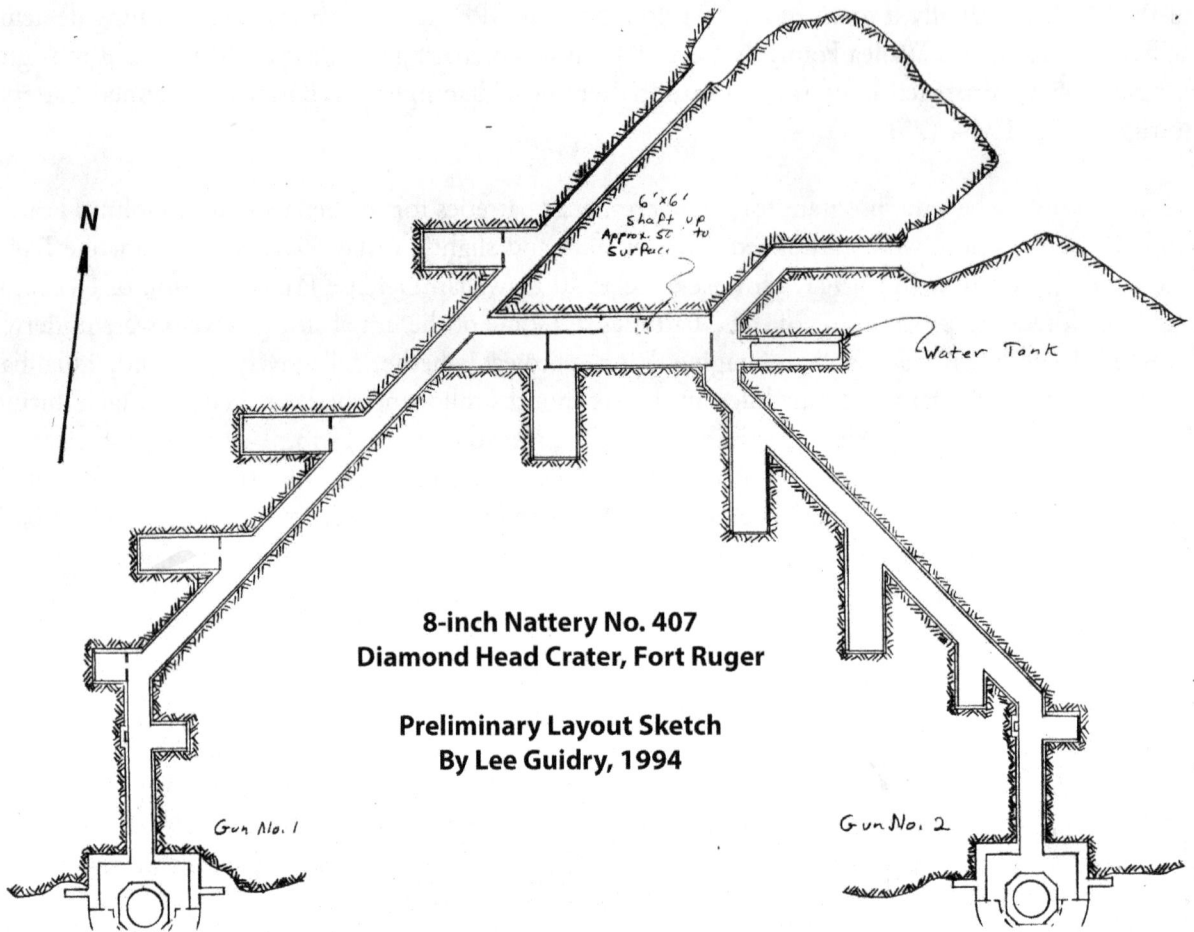

Preliminary sketch from field inspection of the layout for Battery No. 407 at Fort Ruger by Lee Guidry, 1994. *Guidry.*

**8-inch Battery No. 409**
**Kaena Point Military Reservation**

**Preliminary Layout Sketch**
**By Lee Guidry, 1994**

Concrete
Housing
With 1"
Steel Plate
Door

24" Dia. Conc.
Pipe and Tee

Steel
Ladder

*Section Thru Air Shaft*

200'

This Support Frame Collapsed

Overhead Steel
Support Frames
Frames Consist of 8"x16"
Steel I-Beam Stock

No. 1
Entrance Blasted Shut
But Opening Thru Rubble
Permits Entry

Steel Rails
on Floor

Main Chamber
Approx 15' Wide, 10' High

Battery 409 Tunnel
All passages Approx. 15' High, 15' wide

Stored Overhead Rail
Components for Shell
Transport

Remnants of
Muck Removal
Conveyor

100'

**N**

No. 2 Entrance
Completely
Blocked

Barrels And Steel
Table

4'x4' Air Shaft
Up Approx 50'
To Surface

Gunite Surface

Entrance
Tunnel Approx
6' High 6' Wide

50'

Light Wood Support Frame

Concrete
Entrance Way
And
Double Steel
Door

*Radar Tunnel*

Preliminary sketch from field inspection of the layout for Battery No. 409 at Kaena Point by Lee Guidry, 1994. *Guidry.*

er than the roof. Ammunition for the guns was to be carried by continuous hoists from the main structure's magazines to ready boxes adjacent to the open gun platforms.

Kaena Point was the site for Construction No. 409. This was the final 8-inch emplacement planned, also developed as a substitute when usage of USS *Oklahoma*'s 14-inch gun turrets for a position at this point was dropped. The HSAC then proposed that a pair of 16-inch gun batteries be provided at the North Shore sites selected for the 14-inch turrets. This proposal was not viewed favorably by the War Department, but two batteries of 8-inch naval guns were approved January 8, 1944. On January 26, 1944, construction numbers were assigned to these projected 8-inch batteries.

An access road was built from the Farrington Highway to the battery site and the tunnels that would be bored into the cliffs of Puu Pueo. In May 1944 all survey work had been completed, and tunnel excavation was to start at once. By May 1945, only 19% of the project had been completed, although about 60 % of the tunnels had been excavated. Similar to Battery No. 405 at Puu Papaa, its two 8-inch emplacements were located out in the open without any protection from bombardment. Each of the guns was to be provided with a protected magazine with a capacity of 100 rounds adjacent to each gun. Fifteen-foot-wide access tunnels, with crowns of nearly 15-feet afforded access to the tunnel complex. The primary entry tunnel into the hillside from Entrance Number 1 was some 200-feet long. About 40-feet from the end of the tunnel another tunnel branched off at about a 45° angle. The main tunnel continued on another 40-feet before curving to the right for another 75-feet before making a right-angled turn. These unfinished tunnels upon completion would have contained storage spaces for the battery's additional powder and projectiles as well as provide gas proofed spaces for a first aid station, plotting room, latrines, radio room, power generator room and a small galley and mess area, and a freshwater tank.

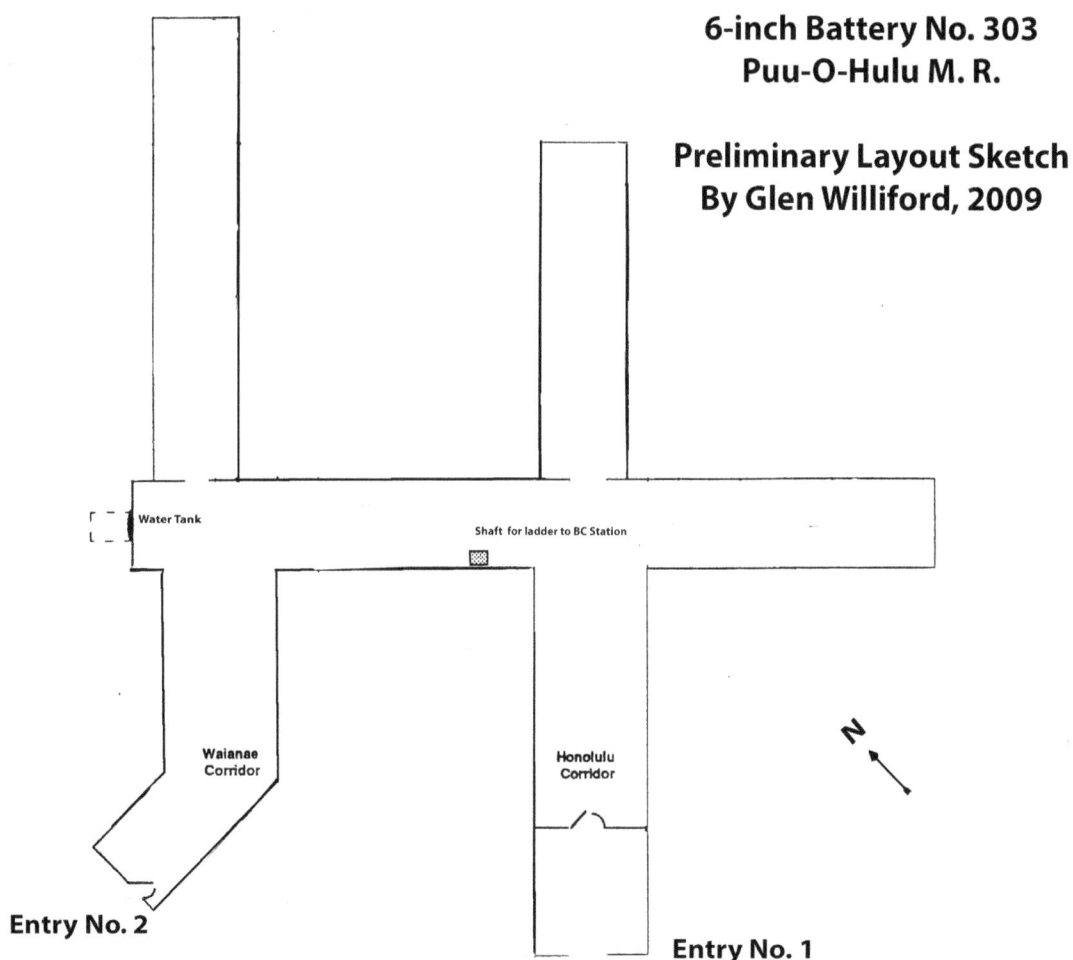

Preliminary sketch from field inspection of the layout for Battery 303 at Puu-O-Hulu by Glen Williford, 1994.
*Williford Collection.*

Some distance up the slopes of Puu Pueo a battery commander's station was begun that was to have been equipped with a DPF. Only the concrete floor and a portion of the walls had been constructed with no lining yet when the project was suspended in June 1945. The battery was also to be provided with SCR-296A fire control radar set assigned to it. Even before the battery was well advanced toward completion, the type of radar to be supplied was changed to the newer AN/FPG –1 Radar with a tracking range of 28,000-yards. The entrance to the radar operating room was located in a tunnel just below the BCS and slightly to its east. When the project was suspended in 1945 the concrete aprons of the gun emplacements were already laid. No armament was ever provided, and the project was abandoned when the war ended. (27)

Kaena Point had previously been used for several other military purposes. A pair of portable 60-inch Seacoast Searchlights numbered 49 and 50 (later changed to Numbers 55 and 56), along with their controller were placed in prepared concrete emplacements east of the Kaena Point Lighthouse at an elevation of about 100-feet. The searchlights were manned by Battery G, 57th CA from January 1942 until July 9, 1944. In January 1945 they were taken out of service and returned to base camp. On the eve of World War II, a system of AAIS observation stations was authorized for construction around the perimeter of Oahu. Initially these were visual stations, but gradually as the war progressed radar was authorized for some of the stations. Antiaircraft Artillery Intelligence Service Station No. 7 at Puu Pueo was built in the spring of 1942 at an elevation of 860-feet and was manned by a detachment from the 53rd CA (AA) Intelligence Battery. When the decision was made to build Battery No. 409 the site of the AAIS station was designated as the site for the battery commander's station, and the two functions were to be combined into a single structure. The BCS was never completed, and it was not until after the war that SCR-271A radar was installed. (28)

Three new 6-inch permanent battery emplacements were approved during the war. One was on the west coast as part of the Pearl Harbor defenses, replacing the 7-inch battery at Puu O Hulu. The other two were for the south coast Honolulu defenses; one at the Punchbowl and the final one at Koko Head east of Fort Ruger. All were placed in existing army reservations.

Puu O Hulu on the west coast of Oahu was selected for a 6-inch battery of the permanent system as Construction No. 303 and was commenced in 1942. The battery was to consist of two 6-inch guns T2 on M4 barbette carriages that were to replace the 7-inch ex-naval guns of Battery Hulu at the same location. A tunnel complex was drilled and lined for the magazines and support rooms directly behind and into the mountainous rise of Puu O Hulu. Apparently, the ultimate design was to have the two shielded 6-inch guns emplaced into the concrete casemates or embrasures built originally for the 7-inch guns—with of course a new carriage rotating ring. While the casemate shell racks could still be used for ready ammunition, most of the munitions would be stored in new magazines of the tunnel complex. Recommendation for use of the casemates was for protection from rock dislodged about the position by enemy fire. A new BC station was tunneled out of the rock above the entry level, between the two tunnel exits. This two-tiered station contained observation rooms and a radar operating room that was accessed via a vertical shaft from within the cross passageway connecting the operations tunnel and the power plant tunnels.

The tunnels afforded access to three branch tunnels: The power plant branch was 75-feet long and 20-feet wide; the magazine tunnel was 60-feet long and 20-feet wide. A 60-foot-long cross passageway connected the magazine, power plant and operations tunnels. Located at the northwest end of this connecting tunnel was a 10,000-gallon water tank that provided an emergency source of fresh water. The operations tunnel was about 120-feet long and 20-feet wide. A five and a half-foot wide corridor extended along the right side of the tunnel's length providing access to the battery plotting room, a fire control switchboard room, officers', and enlisted men's latrines, a first aid room, an officers' bunk room and a small galley and mess area.

Site selection was confirmed in October 1943, and some tunneling apparently started in mid-May 1944. A year later in May 1945 the battery was assigned to a deferred status, with work assessed at 53% done when suspended. While the ordnance was shipped from the mainland United States to Oahu in 1945, it is not believed that the guns and carriages were ever emplaced, and probably not delivered to the construction site.

The second new 6-inch battery was designated Battery Construction No. 304. The original site selected (October 1943) for this permanent battery of the Honolulu defenses was to be at the Sand Island Military Reservation. It was another example of replacing obsolete ex-navy 7-inch guns were more modern weapons. As late in the war the supply of Model 1903-05 gun tubes was exhausted, the battery would have been armed with the ballistically identical 6-inch T2 on M4 barbette carriage. On July 15, 1944, the decision was made to relocate the battery to the Punchbowl, as the Sand Island site was too close to the shoreline and subject to commando attacks. The project began at the Punchbowl in 1944 and was assessed at nearly 36% complete in May 1945. Work on the battery was suspended soon thereafter and never resumed. At the time the project

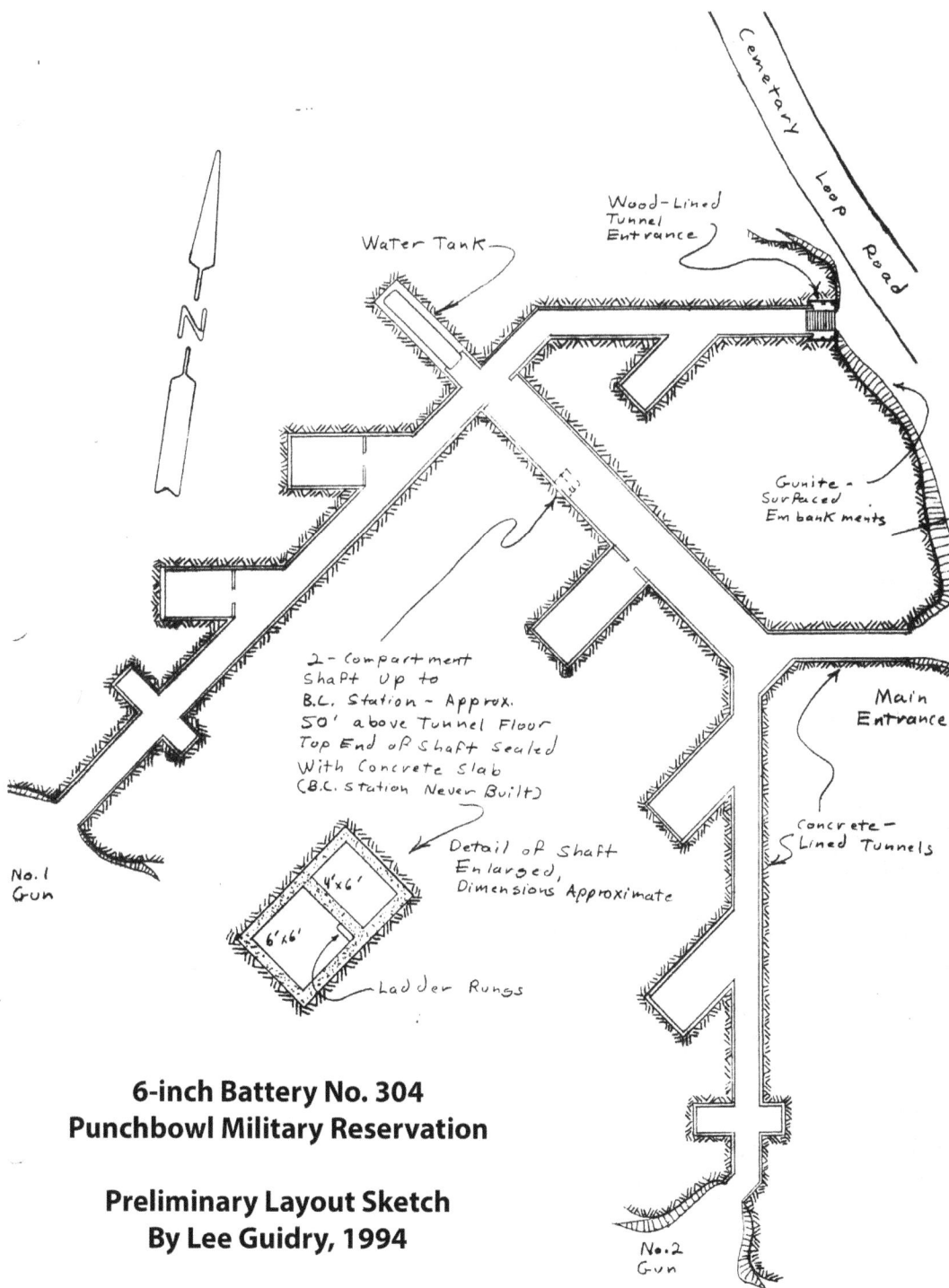

**6-inch Battery No. 304**
**Punchbowl Military Reservation**

**Preliminary Layout Sketch**
**By Lee Guidry, 1994**

Preliminary sketch from field inspection of the layout for Battery 304 at Punchbowl by Lee Guidry, 1994. *Guidry.*

Aerial photograph of the Punchbowl Crater, with the two openings that would have been completed for the 6-inch guns of Battery 304 if it had ever been completed. *NARA*.

was interrupted, the battery consisted of two tunnels dug 300-feet through the crater rim about 60-feet below the almost 500-foot-high crest of the rim. The tunnels were partially lined with concrete to prevent cave-ins.

At a point about 200-feet inside the rim from the exterior slope openings, a transverse tunnel connected the two entry tunnels. This tunnel, 193-feet long and 20-feet wide, was planned to house the operations rooms and the electric power generator room. In this transverse tunnel at a point some 66-feet from tunnel Number 2, a second 20-foot-wide tunnel 56-feet long branched off at a 90° angle. This branch was intended for the battery plotting room. A steel enclosed shaft provided vertical access between the plotting room and the battery commander's station. An SCR-296A fire control radar operating room was to be built atop the crater rim. This vertical shaft extending up some 66-feet also served as an escape shaft for the personnel in the tunnels and opened into a corner of the battery radar operating room. Opposite the junction of the transverse Operations Tunnel and Tunnel Number 1 was another branch tunnel 30-feet long that contained an emergency water tank with a capacity of 10,000 gallons. (29)

When the war ended, the tunnels had been lined with concrete and part of their floors poured, but the casemates at the south, or seaward, end of the access tunnels had not been begun, nor had the BCS, its fire control radar installation, or the internal shaft connecting the operations tunnel with the projected BCS. A pair of guns and their M4 shielded barbette carriages were shipped to Oahu but were never emplaced as the battery project was suspended before it could be completed.

Battery Construction No. 305 was the last of the three 6-inch barbette batteries planned for Oahu as part of the permanent seacoast defense program, located in the Defenses of Honolulu. In October 1943 the initial site selected for the battery was Fort DeRussy, and it was slated to replace the old Taft-generation disappearing Battery Dudley. In April 1944 the site was changed to the Koko Head reservation (geographically a spot midway between Koko Head and Koko Crater). A month later it was reported that the design study was in progress. Apparently, that was all that was accomplished. In May 1945 the battery was described as 0% completed, never having been started and subsequently suspended. (30)

Page of a 1945 document showing the fields of fire and coverage of the 1940 Program batteries when completed. *NARA.*

Many nations saw the need for lighter, rapid-fire/rapid traversing guns in harbor defense. The evolution of small, fast motor torpedo boats produced a dangerous adversary during the Second World War. Small ships could slip into a harbor, torpedo vessels or land special operatives and escape quickly. Conventional large-bore guns with slow rates of fire and cumbersome traverse couldn't cope with this threat. The British effectively developed a new 57 mm (6-pdr) gun and mount for this purpose, and it used it to effect in the Mediterranean Sea against Italian adversaries. It might seem that Hawaii (or even the mainland American ports) was too far from any MTB base, however these guns would work just as well against probing submarines and any light commercial vessel trying to sneak into a harbor for nefarious purposes. The 90 mm anti-aircraft gun was married to a new mount for this purpose in 1942, and in 1943 an entire new network of emplacements was authorized and built throughout U.S. ports, domestic and in overseas locations.

It was found that the excellent new M1 90 mm antiaircraft gun could provide an already existing platform on which to build the required new AMTB weapon system. The standard M1 gun was used with a new 360° traversing M3 pedestal mount. For crew protection in this role the gun was encased in a six-sided, covered steel shield open only to the rear. The gun could be used in anti-ship role, as well as still anti-aircraft and even for ground fire. It was handloaded. Fire was usually directed by a remote system, and capable of delivering 25-30 rounds per minute to a maximum range of 19,500-yards. (31)

The emplacement was extremely simple. It had a concrete block set flush with the ground holding the necessary hold-down bolts and electrical connections. Ready ammunition lockers were nearby, fed by splinter-proof magazines. The usual plotting room and battery commander station were conveniently placed nearby. For most batteries the complete armament of an anti-motor torpedo boat battery was two fixed 90 mm with M3 mounts, flanked on either side by a mobile 90 mm M1 gun (often in a position protected with a sandbag berm, but still with an entry way to allow redeployment if necessary). A section of two automatic 37 mm, replaced later in the war with 40 mm AA guns was also assigned to the battery emplacement.

On October 24, 1942, the Hawaiian Department directed the Chief of Hawaiian Artillery to establish Anti-Motor Torpedo Boat (AMTB) batteries. Their purpose was to provide defense against enemy torpedo boats and submarines with a secondary mission of supplementing the antiaircraft defenses. On November 11, 1942, Hawaii forwarded a study regarding AMTB batteries to the Adjutant General's Office, calling for eight batteries. Washington approved construction of six of the sites, confirmed the availability of the new weapons and instructed construction to begin in May of 1943. Physical work on the new emplacements started in July—Two batteries each for Pearl Harbor; Honolulu; and Kaneohe Bay. (32)

They were erected and manned fairly quickly, after all the physical construction work was minimal. Several served as examination batteries, manned batteries supporting actions to stop and approve vessels and cargoes entering a harbor. In 1944 the primary assignment of AMTB batteries switched to an anti-aircraft role, but they retained their AMTB capabilities. All six of the Oahu batteries continued to be manned and in service to the end of the war, but they were deactivated, and the armament removed from all by the end of 1946.

Details of the Oahu AMTB Batteries can be seen in Appendix VIII.

Ordnance Department sketch of the 90 mm M3 anti-motortorpedoboat gun and barbette with shield. This type of gun was the primary armament of the AMTB Program. No photographs of the gun in service in Oahu have been found. *Army Technical Manual.*

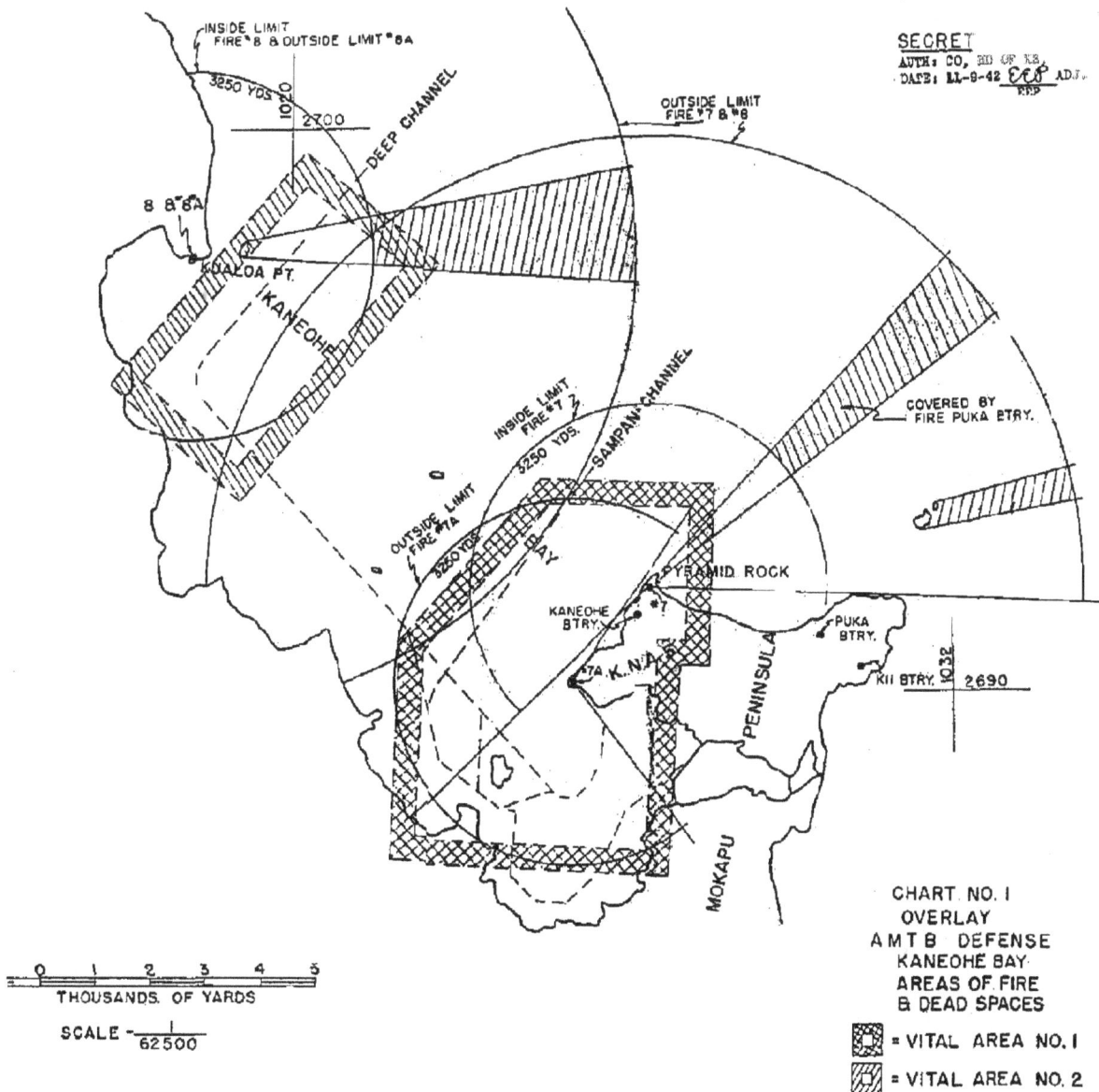

Map of the fields of fire for the various AMTB batteries of Kaneohe Bay in November 1942. *NARA.*

# CHAPTER 12
# DURING THE SECOND WORLD WAR

The original Taft Generation forts in Oahu continued active service during the war. While most of their heavy gun batteries were technically obsolete, they weren't abandoned until suitable replacements were operational. All of the posts had tactical support elements—batteries of antiaircraft guns, searchlights, fire control stations and command and communication facilities. Their extensive infrastructure of garrison housing and logistical support were still an essential part of the defenses.

Fort Kamehameha bustled with activity in the weeks following the raid. The seacoast defense modernization program of 1940 recommended the provision of heavy overhead protection to most of the 12 and 16-inch barbette batteries built since the First World War. Standard plans provided for reinforced concrete and earth covered casemates around each gun. The new standard was protection of guns and magazines from bombs or shells up to 1000 pounds. On July 11, 1941, the War Department authorized the Hawaiian Department to undertake the planning for the overhead protection of Battery Closson's two 12-inch guns. On December 7, 1941, this project was still in the planning stages, but five days after the attack an interim project was started to provide a 210-foot-long splinter-proof concrete wall along the open back of the battery. Providing only splinter-proof protection from near misses, it did not address protection against direct hits. However, the full casemating project was still considered urgent, and by early January 1942, the plans had been approved and the work commenced.

During modernization troops continued to actively man the battery. Just one gun emplacement was modified at a time, allowing the other mount to be fully serviceable in case of an attack. Gun Number 2 was taken out of service first to facilitate the construction of its concrete casemate in February 1942 and by April about 20% of the work was already completed. Initial estimates for completion of the project by October 1942 proved to be overly optimistic. When the Number 2 casemate was completed in the fall of 1942, construction on the Number 1 emplacement was begun on November 8, 1942. The casemate construction project continued until June 1, 1944, when the job was declared 99% complete. (1)

Army Engineer plan for a fully-casemated, protected 12-inch BCLR battery. Battery Closson's modifications closely adhered to this plan. *NARA.*

There were concerns that the soft ground at the site would not be able to hold the increased weight of the massive casemate. Pilings were avoided by placing the new casemates on "rafts" on the soil. Fortunately, while the first casemate did sink six inches within a month of erection, it then stabilized. The new casemates were 52-feet by 37.5-feet and had 10-foot-thick walls to the firing front and sides, and a roof thickness of 8-feet. Twenty feet of earth on top of the gun houses provided additional overhead cover. A compressor room for the breech air scavengers was provided in each casemate. From the casemated gunhouse a corridor 10-feet wide by about 160-feet long exited to the rear to facilitate tube changes. Off this corridor a 12-foot by 185-foot gallery connected the existing traverse magazine. This gallery covered the railway path to the ammunition. To the rear of the ammunition magazines a new powerplant room for three 125 kW generators with evaporative coolers was added. A new BC station with 100-foot tower for an SCR-296A radar and its operation room was also built during this modernization. The guns were to be provided with new steel shields, but these were delayed during manufacture until almost the end of the war. (2)

The personnel at Battery Closson were reassigned to Battery B, 53rd CA (HD) Battalion on August 13, 1944, and continued to provide Battery Closson's manning detachment until early 1945. Service practice firings were conducted regularly throughout the war years. However, by April 1945 the battery had just a small detachment serving as a caretaking unit. The battery was disarmed, and the armament scrapped postwar in 1948.

As mentioned above, on December 12, 1941, the Department Engineer requested that the Honolulu District construct splinter-proof protection in the rear of the various fixed seacoast batteries. An 80-foot-long concrete barrier two-feet thick was authorized for the rear of 12-inch disappearing Battery Selfridge. This splinter-proof protection for the battery was simply a straight concrete wall to help protect battery personnel using the rear corridors from aerial and naval bombardment shrapnel. Construction of the splinter-proof wall was completed by June 1942. When Battery Closson was placed in a reduced manning level while being casemated in the spring of 1942, detachments from Battery Selfridge managed to man both batteries. On August 7, 1942, Battery E, 15th CA was activated and relieved Battery B of its manning responsibility for Battery Selfridge. Battery E manned the two disappearing guns until they were again placed in a reduced manning status in August 1943.

The 1943 survey of the seacoast artillery recommended that Battery Selfridge be eliminated from the Hawaiian Department Defense Project. While General Burgin considered the battery to be slow and cumbersome in action and comparatively short in range, he felt that the battery, because of its enormous striking power, should be retained until replaced by more modern armament. The recommendation was approved by the War Department. The battery was kept in firing condition; 22 rounds were fired by its guns between October 23 and November 29, 1944. As more powerful modern armament was installed only slowly, Battery Selfridge was retained in semi-caretaking status through the end of the war.

Fort Kamehameha's mortar battery was also authorized to receive new splinter-proof protection on December 12th. A 290-foot-long concrete wall two-feet thick was authorized for the rear of the battery to help protect battery personnel. Construction of this wall was completed by June 1942. Battery Hasbrouck continued in use as the Harbor Defense Command Post for the Pearl Harbor defenses from December 1941 until June 1942, when a new HDCP was completed at Salt Lake. Soon afterward, the emplacement was placed in service for a brief period as a Category B or standby battery, but by the end of 1942 the mortars were again back in Category C or maintenance status. Battery Hasbrouck was not manned during the remainder of the war. In March 1945 a study of obsolescent armament by the Artillery Officer, Pacific Ocean recommended that the mortars be removed. This report was submitted to the War Department on March 27 and soon afterward the War Department authorized the scrapping of Battery Hasbrouck's armament.

Battery Hawkins on the morning of December 7, 1941, was assigned to a detachment from Headquarters Battery, 15th CA. Battery Hawkins and Battery Jackson alternated as the Alert Battery during much of 1942. It is probable that a small manning detail from the 15th CA was assigned to Battery Hawkins until August 1942, when Battery D, 15th CA was activated and a detachment from that unit assumed responsibility for manning the battery. Battery Hawkins was manned until July 1943, when the new AMTB Battery was ready. By the end of 1943 the armament was deemed to be "wholly obsolete." The battery had been fired so much that it was no longer accurate. The guns remained mounted, however, and during the latter months of the war were used on a regular basis as a training battery for the Officers Gunnery School at Fort Kamehameha. Following the war, the 3-inch guns of were finally removed. The fort's 6-inch disappearing Battery Jackson appears to have been little used during the war, though it also got a protective splinter wall behind its central traverse. (3)

In 1941 jurisdiction of the Bishop Point site was passed back to the U.S. Navy as property for the expansion of the Pearl Harbor Naval Base. The battery emplacement there (Battery Chandler along with Battery Barri, which had been disarmed for some years) was removed during World War II when the U.S. Navy demolished it to allow new construction at the site as its Bishop Point Section Base.

During the Second World War the population at Fort Kamehameha increased exponentially. Large additions had to be made to base housing, though most were of the conventional utilitarian theatre type buildings of the period. The post had a very large tent encampment between Batteries Selfridge and Closson serving as temporary quarters for units transiting through Oahu to or from the war front further west. In May of 1943 about 18,000 troops were on this reservation, and by March 1945 the population had grown to about 37,000. There was a new exchange, theatre, and even a beer garden constructed on post.

The units posted at Fort Kamehameha underwent several redesignations and reorganizations during the war. In early 1942, several provisional batteries organized in the spring were reorganized as the 803rd CA (HD) Battery (Separate). In the summer of 1942, additional elements of the 15th CA were activated. This regiment eventually was responsible for all the Pearl Harbor defenses until the latter part of World War II when it was finally inactivated.

An interesting unit to be stationed at Fort Kamehameha was the only barrage balloon unit to serve in the island. The 305th Barrage Balloon Battalion had been activated on November 1, 1941, at Camp Davis, North Carolina. Immediately after the Pearl Harbor attack the Hawaiian Coast Artillery Command had been offered a barrage balloon detachment, but initially turned it down with concerns about the crowded airspace on Oahu and potential interference problems. Still, with Washington encouragement a unit was sent. The 305th left San Francisco on July 8 and arrived in Honolulu on July 16, 1942. The battalion had an HQ and four lettered companies, A, B, C and D. Each was equipped with eight balloons for a total of 32. These were hydrogen-filled immobile balloons, tethered by cable to a ground winch. The unit operated at adjacent Hickam Field the last half of 1942 into 1943. Balloons, usually at an altitude of 5000-feet, were deployed during slow times (mostly night) and taken down for landings and takeoff of large formations. Later in 1943 the use of barrage balloons in Hawaii was discontinued and the unit eventually converted to an automatic weapons battalion and redeployed. (4)

During the war several battery positions were established at Fort Kamehameha for mobile antiaircraft guns. Battery No. 15 for four 3-inch M3 mobile guns was deployed to Ahua Point immediately after the Pearl Harbor attack. Permanent blocks were not constructed, but it probably had sandbag revetment protection. By the end of 1943 the 3-inch guns here were replaced with 90 mm M1 AA guns.

When the Japanese attack came in December 1941, Battery Tiernon at old Fort Armstrong was unmanned and in caretaking status. A provisional manning party from Headquarters Battery, 16th CA Regiment, reactivated the battery. Battery Tiernon's 3-inch guns served as the Examination Battery for Honolulu Harbor in 1942 and 1943. Disarmed (meaning the guns, carriages, implements, and ammunition removed for storage, use elsewhere, or sold for scrap) at the end of the war. The Tiernon emplacement was demolished in 1973.

At Fort DeRussy the 14-inch guns of Battery Randolph were manned on the morning of December 7, 1941, but not fired during the raid. The large guns were practice fired on December 13, 1941 (the first time since the 1920s), with the guns and carriages functioning very well. The December 12 construction order for battery splinter-proof protection included Battery Randolph. A 580-foot-long concrete wall two-feet thick

One of the barrage balloons of the 305th Battalion ascends at Fort Kamehameha during the war. *USAMHI.*

and 10-feet tall was authorized for the rear of the battery, which was finished in the fall of 1942. The October 1943 survey of the seacoast artillery, subsequently approved by the War Department, recommended that the battery be eliminated as obsolete. Similar to the assessment of Battery Selfridge at Fort Kamehameha, Battery Randolph, though slow and cumbersome in action and comparatively short in range, was retained until replaced by more modern armament. As such armament was not quickly forthcoming, Battery Randolph was retained in semi-caretaking status ready for action through the end of the war. The powerful 14-inch guns remained in place for only a few years following the war before they were dismounted and scrapped. (5)

Besides the splinter walls and extensive efforts at camouflaging, little else could be done to modernize the defenses at DeRussy. Some of the usual shuffling of units continued. The 16th Coast Artillery began activating those firing batteries that had been inactive since 1924. One of the first batteries activated was Battery B. On August 6, First Lieutenant August L. Boyd assumed command of the battery and its 157 enlisted men. It was organized at Fort DeRussy and assigned to man Battery Dudley's 6-inch disappearing guns relieving the 2nd Battalion, 55th Coast Artillery Regiment that had manned the battery since January.

Not until the end of September 1942, did the coast artillerymen at Fort DeRussy see a noticeable drop in the frequency of the alerts and gun drills. Although both night and day gun drills remained a regular part of the weekly schedule, physical training and field exercises began to partially displace the near constant manning of the seacoast batteries in the daily routines. Twelve-mile conditioning hikes by about half of the battery personnel also became a part of the weekly training schedule. In time the troops would be sent through the various jungle combat training programs that were implemented on the island. The usual post duties of inspections and routine training again became the norm. Men spent more time on the rifle range as well. By the last quarter of 1942, Sundays had once again become a day of rest and recreation.

The Army used Fort DeRussy as a staging area for troops, erecting numerous Quonset huts and other temporary structures for billeting, service clubs, and recreation facilities. The Fort DeRussy Recreation Center, initially established to accommodate about 700 men at the east side of the reservation, was expanded in the summer of 1942. The pavilion and its adjoining swimming pool built for the use of the officer personnel in the 1930s was also turned over to the Recreation Center for the use of enlisted men. Personnel from other army units on the island were rotated to Fort DeRussy to spend a few days of rest and recreation throughout most of the war. (6)

Each of the lettered batteries stationed at Fort DeRussy numbered between five and seven officers and from 300 to 400 enlisted men by late 1942. On January 25, 1943, Battery A conducted a night target practice with the 14-inch guns of Battery Randolph to fire seven salvos. On January 27, the newly reactivated Battery B fired Battery Dudley's 6-inch guns for the first time in many years. Two rounds of service ammunition were fired from the Number 1 gun.

On October 24, 1942, the commanding general, Hawaiian Department, had directed the Chief of Hawaiian Artillery to establish Anti-Motor Torpedo Boat (AMTB) batteries with the newly developed 90-mm M1 AA guns. Work on one of these new batteries at Fort DeRussy began on June 16, 1943. This battery was also intended to replace the fixed battery of M1917M1 AA guns at the fort. By July 6, these mounts were in place and although the battery position was far from complete, the troops began conducting drills with the new armament. Work on the 90mm AA gun battery continued through the month of July and into August. (7)

Battery Dudley was placed in caretaking status and the personnel transferred to Sand Island. Battery Randolph was also taken out of service and placed in caretaking status before the end of 1943. Both Batteries Randolph and Dudley were among the many batteries earmarked for elimination in the fall of 1943. By 1944 the number of seacoast batteries required for the defense of Oahu had declined markedly. The newer generation of gun batteries that had been constructed since the beginning of the war were coming on line. Consequently, the number of coast artillerymen required to man the remaining batteries also declined. (8)

Late in the war nearby Sand Island and even the old 3-inch battery at Fort Armstrong were still being manned. Battery Tiernon was the examination battery for Honolulu Harbor, and the newly emplaced AMTB Battery on Sand Island was the Antiaircraft Alert Battery for Honolulu Harbor. As a result of these crucial roles, they were fully manned around the clock. However, at Fort DeRussy the old disappearing guns were preserved, but unmanned. Fort DeRussy finished out the war as mostly an administrative and recreation post rather than an active artillery post.

During the Pearl Harbor attack all the active elements at Fort Ruger went on immediate alert. New bomb shelters were built and the construction of new field fortifications at the Tennis Court battery position enhanced the local defense. During the spring of 1942, a wider two-lane tunnel project through the northeast rim of the crater was authorized and built. With no further attacks by air or sea full alert gradually turned to routine. (9)

On January 18, 1942, a new service command enabled the Harbor Defenses of Honolulu to separate administrative and tactical duties. This new unit assumed many general post duties such as guards, mess halls, police, etc., leaving the batteries to concentrate on manning the guns. A new building construction program was also undertaken, and new mobilization-type barracks were built in the northwest and southeast parts of the post. In early summer of 1942 sufficient replacement personnel had arrived in Hawaii, and many were assigned to Fort Ruger, and new coast artillery organizations were being formed. Long-inactive battalion headquarters and firing batteries of the 16th CA were finally activated and brought up to full wartime strength. As new units were being organized, their cadres were formed by reassigning men from existing outfits and filling out their ranks with newly arrived men.

On December 22, 1941, the Hawaiian Seacoast Artillery Command (HSAC) was constituted as a subordinate command of the Hawaiian Coast Artillery Command (HCAC). On March 16, 1942, with the arrival of Brig. Gen. Robert C. Garrett, the HSAC was formally activated at Fort Ruger. On August 10, 1943, the Hawaiian Coast Artillery Command was abolished and replaced by the Hawaiian Artillery Command (HAC). Maj. Gen. Henry Burgin commanded the HCAC and its successor, the HAC, as chief of Hawaiian artillery until the HAC was disbanded and replaced by the Central Pacific Base Command (CPBC) at Fort Shafter. CPBC Headquarters remained at Fort Shafter until it was moved to Fort Ruger later in 1945. Early in 1944, the HD of Honolulu and the HD of Pearl Harbor were combined into a single harbor defense command as part of a major reorganization of the coast artillery on Oahu. When the CPBC was inactivated and replaced by U.S. Army Forces Middle Pacific (USAFMIDPAC) in 1945, the old Hawaiian Artillery Command was resurrected. The Hawaiian Seacoast Artillery Command was also reorganized and redesignated the 2274th Hawaiian Seacoast Artillery Command. HAC established its headquarters at Fort Ruger alongside that of the 2274th HSAC.

Battery Granger Adams ended its "silent sentinel" status early in 1942, when its guns were test fired for the first time in many years. It also benefited from the December order to construct splinter wall protection behind the batteries, the battery's plotting room getting an 80-foot long concrete wall. The 8-inch guns were exercised thereafter on an infrequent basis using full-service charges, and sub-caliber practice was held more often. The 16th CA, continued to man Battery Granger Adams until May 30, 1944, when it was reassigned to the Harbor Defenses of Kaneohe Bay and North Shore. Little was done at Battery Granger Adams in the way of subsequent modernization. The July 1941 policy letter for increased emplacement protection included the recommendation for use of protective shields at the battery. However, it seems these were never developed or produced, and no installation ever made. In 1943 the plotting room was given gas-proofing protection. It appears the guns went into reduced maintenance status in late 1944, and subsequently scrapped right after the end of the war.

As part of the July 1941 program for increasing the protection of Hawaiian seacoast batteries, studies were conducted for casemating the 16-inch guns at Fort Weaver. One complicating consideration was that the additional weight of casemates would require complete reconstruction of the barbette emplacements, likely using pilings. Probably more than that was the desire, particularly understanding that both Battery Closson and Hatch were being casemated, to retain at least one battery with full around-Oahu coverage. If all three long-range batteries were casemated and obviously kept for firing to the most important southern approach, no super-heavy guns would be available to protect the west, north, and east shore. Finally, there were concerns about Williston's closeness to the beach, and periodically the recommendation came forward to relocate the guns to a more secure site further inland.

In an effort to provide some degree of protection for the guns and their crews, two tunnel shields of armor plate were authorized for Battery Williston. Orders were given to manufacture and then ship them to Oahu upon completion for installation. After considerable delay two-inch thick steel shields were eventually provided. The shields were to be attached directly to the M1919 Barbette Gun Carriages and their racers. When installed, these splinter-proof shields would provide a modicum of overhead protection for the gun carriages from shrapnel, while still enabling maintenance of an all-around fire capability. The two shields had a cylindrical design not unlike the shape of a 19th century "Conestoga" covered wagon, or as the term implied, a "tunnel". The shields did arrive during the war but missing some parts that delayed installation. When the parts did arrive the arguments about relocating the battery to near Aliamanu Crater were being waged again. It does not appear the shields were ever actually installed. (10)

A final wartime project to help modernize Battery Williston was for the provision of a fully bombproof, 350-round ammunition magazine. Authorization for such was granted in February 1941. The Hawaiian Department recommended that this powder magazine be constructed midway between the two guns of the emplacements. This project was approved and authorized later that year, but construction was never implemented. In May 1945 only the old splinter-proof magazines were in use. (11)

Battery Williston was fully manned during the war by Battery A, 15th CA, and ready to engage the Japanese Fleet. However, the Japanese surface warships never came anywhere near to the Oahu coast. Battery Williston was never required to counter an invasion attempt by the enemy. The guns were, however, practice fired on a regular basis. Normally one round from each gun was fired during these exercises. During the last full year

Ordnance Department model of the tunnel shield developed to envelope the guns of Battery Williston at Fort Weaver. *NARA*.

of its active service, 1944, the powerful weapons were fired 79 times in target practices. These last practices consisted of 18 rounds fired in a radar-controlled target practice on November 25, 1944. The guns of Battery Williston were manned by the 15th (HD) Regiment until August 13, 1944, when the 15th CA was inactivated, and the soldiers reassigned. During the early months of the war, the searchlights at Fort Weaver were manned by a detail from Headquarters Battery or later Battery G, 15th CA. These lights were still being used through and to the end of the war.

At Fort Barrette, need for further camouflaging of Battery Hatch received the attention of the Hawaiian Department early in 1941. The effort to reduce 16-inch Battery Hatch's visibility from hostile aerial observation began in late February 1942 when measures were taken to enhance the existing rudimentary camouflage. Numerous concealment methods were utilized, including overhead cover of garnished nets, dummy positions, extension of roads past the batteries, tone-down painting, and the planting of bushes and shrubbery. Cost of the project for the battery was estimated at $8500.

Simultaneously, plans to casemate Battery Hatch's two 16-inch guns moved ahead. Approval to start physical construction was given when the departmental commander General Walter Short endorsed the recommendation of Fulton Q.C. Gardner, commander of the HSCAB. However, the crush of the priority projects in the department delayed start until early 1942. In February the project stood second to Battery Closson's casemating in priority. A sum of $900,000 was made available for this work. (12)

The guns were casemated one at a time during the war, once again to have at least one mount immediately available at all times. Construction began on the first of the casemates on May 17, 1942. The sequential process of providing overhead protection to the guns one at a time dragged out the construction until June 1, 1944, when the project was estimated to be about 95% complete and requiring only minor finish work. In a report of May 1945, it was reported complete except for some earth fill work over the bursters. The plan for Battery Hatch called for two large gun casemates built of reinforced concrete incorporating attached projectile and powder magazines. The concrete roofs of the casemates had a minimum thickness of 8-feet to which was

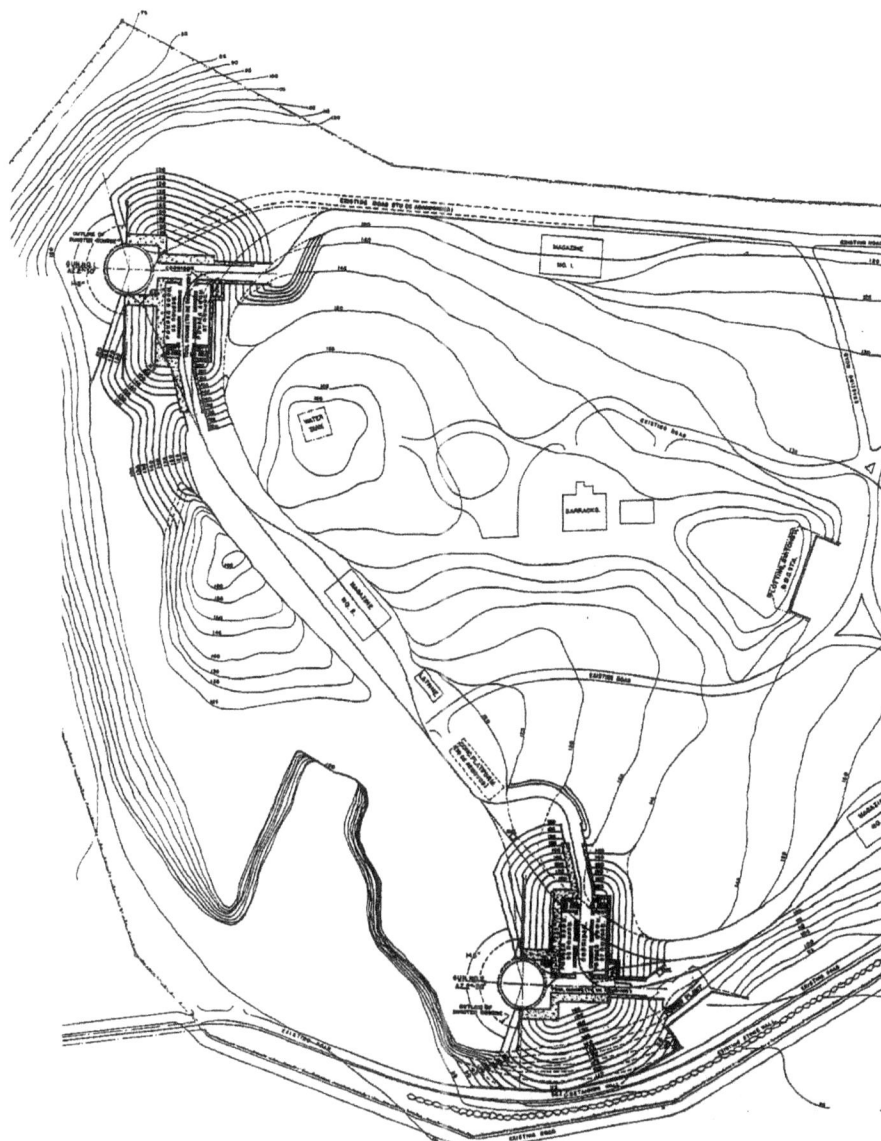

Plan for adding the casemated, overhead protection to the two gun positions of Fort Barrette's Battery Hatch during the Second World War. *NARA*.

provided an additional covering of 20-feet of earth. The soil cover enhanced their camouflage as well to added to the bombproofing of the structure.

Each of the two separated casemates were massive concrete structures. The walls were 12-feet thick around the gunrooms, 10-feet in the rear. Upon completion of construction, each casemate stood as a separate structure containing two powder magazines 73-feet long and 15-feet wide and two projectile rooms 18-feet long and 12-feet wide, as well as a 17-foot long by 10-foot-wide storeroom. The gunrooms of the casemate structures themselves measured 51-feet wide and 34-feet deep. A corridor 16 by 100-feet ran from the casemate directly to the rear and was kept straight in order to directly move a gun tube in or out. A branch corridor 15 by 175-feet ran from this corridor at a right angle past the magazine spaces and out the side, along the still-used railway ammunition line. (13)

A one-foot-thick burster course was placed atop the earthen cover and provided additional protection. The burster course was a commonly used protective scheme often used in pre and early war fortification designs. The theory was that a bomb or shell would detonate on this layer and not the actual structure itself, insulating the emplacement. This was the engineering equivalent of spaced armor. Later experiments showed that the benefits were limited, and in some cases even deleterious, and its use was discontinued in later war construction.

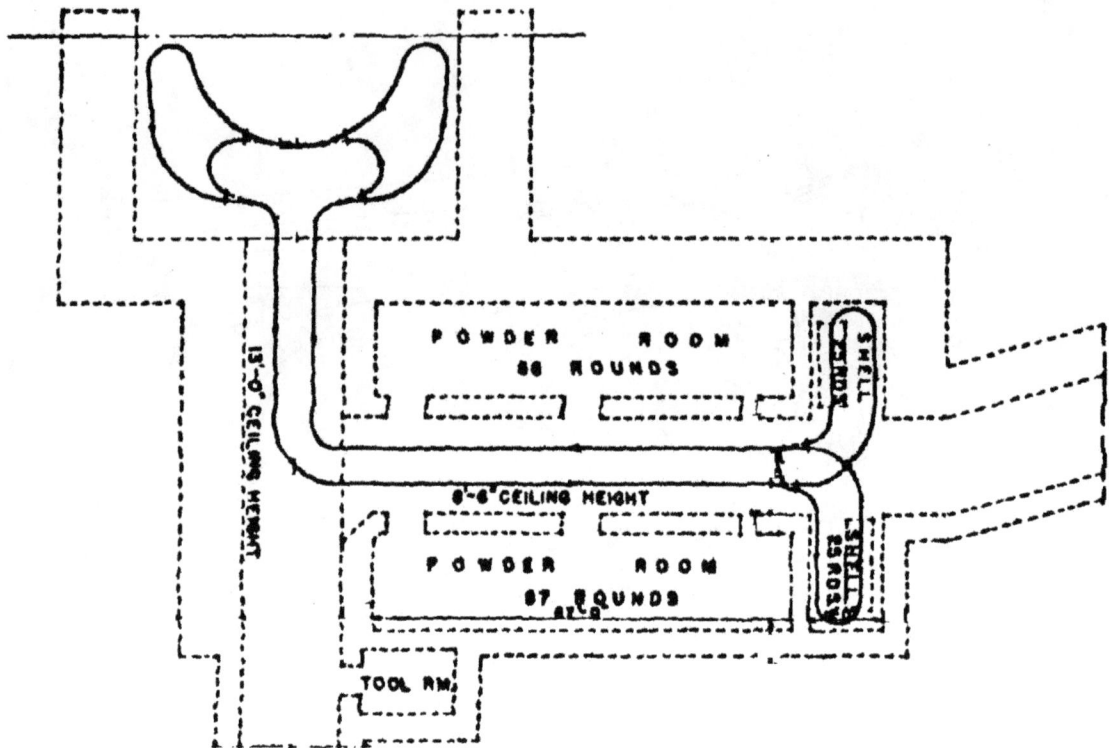

Detailed plan of one of Battery Hatch's gun casemate showing how the new structures include rooms for ammunition. *NARA*.

Completed casemate room for one of Battery Hatch's guns, probably in 1943. *USAMH*.

Business end of a 16-inch emplaced in a casemated gun room and protected by a wrap-around steel shield. Photograph is dated 1946 and already shows a little lack of maintenance. *USAMH.*

Each of the two casemated emplacements had its own two powder and two projectile rooms. Shells were moved on and off shell tables or racks and transported with an overhead trolley system to the central corridor where they could be placed on the rail transport cars and taken to the breech in the gun room. These rooms were built with a capacity of 175 powder charges and 50 projectiles. The only other significant wartime modernization was the installation of SCR-296A radar for fire control of the guns. A tower 25-feet high with the radar, using the usual water tank configuration for the set was added on top of the existing BCS.

Until 1944 target practices were carried out on a regular basis. In the early months of 1944, Battery Hatch was fired in two Target Service Practices each involving the expenditure of four service rounds of 16-inch ammunition. The first practice was on January 13, and the second on April 21. It is not believed the guns were ever fired again. When the 15th CA was inactivated August 15, 1944, Battery C's personnel were reassigned to Battery C of the 53rd CA (HD) Battalion, who took over the manning detail.

After the war began Fort Barrette received new garrison buildings. By 1944 the post was provided with three new sets of family quarters, a garage, cooks' quarters, mess hall, five barracks for enlisted men, a dispensary, fire station, four storage sheds, a gas proof personnel shelter, a shed for six gasoline locomotive 'mules' used to tow the narrow-gauge ammo carts between the magazines and guns. The lower reservation also housed an open-air theater, a post exchange, and even a beer hall. The entire lower reservation was surrounded by 18 concrete machine gun shelters or "pillboxes" as well as rows of barbed wire entanglements to help protect the important guns from an enemy assault or saboteurs.

New wartime priority construction for the seacoast defenses was not limited just to gun batteries. During the war two new major army fuel (mostly aviation gasoline) and ordnance depots were developed. At least partially they were to help with the increased volume of munitions needed in the islands. Furthermore, Oahu had become a significant depot for campaigns and units operating further west on the front as the war approached the Japanese home islands. New depots were constructed in Waikakalua and Kipapa. In both cases the technique of boring out tunnel galleries along a cliff line that was used in Aliamanu Crater was adopted; though in these cases the sites were along the steep sides of deep gulches rather than in a volcanic crater.

The Kipapa Gulch Ammunition and Gasoline Storage Site was on the list of department engineering priorities of January 24, 1942, and it appears construction work was urgently pushed. It consisted of four 62,000-gallon vertical fuel tanks and three sets of horizontal storage galleries or tunnels cut into the sides of steep Kipapa Gulch. The lowest set was on the south side of the Roosevelt Bridge over the Kamehameha Highway, to the southeast of the Wheeler Field reservation. It immediately adjoined the corresponding naval ammunition storage area. Two other sections of similar type constructions were to the east of Miliani Town.

The engineer priority list of 24 January 1942, also included the Waikakalua Ammunition and War Reserve Gasoline Storage Site. It was built primarily as a storage facility for fuel, bombs and other ordnance supplied for large (4-engined) bomber aircraft, including those at the Kipapa Army Airfield. Located to the south of Wheeler Field, the land had formerly been sugar cane fields of the Oahu Sugar Company. It had a large number of galleries, 26 in the lower section and 53 in the upper. A large explosion occurred in 1946 in one of the tunnels, causing much damage but fortunately no casualties. Some galleries contained fuel tanks rather than dry munitions. There was a connecting road and a servicing rail spur. (14)

Many other tunnels were constructed in Oahu just before and during the war. Some were small magazines or storage warehouses for ammunition and chemical gas, some were protected switchboard/radiocommunications centers, some were the actual larger gun batteries (like the 14-inch turret sites), and others major command posts. Three other facilities should be mentioned, not because they were directly associated with the coast artillery function but because they stand out in sheer scale.

The main naval ammunition storage for Pearl Harbor was at Lualualei. The Navy's original site had been on the island of Kuahua within the harbor. It served from 1916 until 1934 until the constricted site proved inadequate and the Navy had to construct a new depot. In 1929 they purchased an 8184-acre piece of property from the McCandless estate at Lualualei in west central Oahu. Opening in May 1934, the depot used dispersed, independent concrete magazines with earth cover. A rail line connected the magazines as well as feeding a spur that went directly to the Pearl Harbor naval base. While not servicing army ordnance, the location of the navy facility was certainly a factor in the location for the guns in both Fort Barrette and Kahe Point (Battery Arizona).

The Department's concern for the lack of food reserve in case of isolation led to several unique solutions. One even included the stockpiling of vegetable seeds to be used in place of the monocultures of sugar cane and pineapple. One project that was requested over many years was for a very large refrigeration plant to hold meat supplies for the garrison. Correspondence in 1938 advocated a budget request of $638,000 for a facility to located in a tunnel at Fort Shafter. It was needed to provide a garrison meat supply for 70 days, in accordance with the Joint Board Plan of 1937. The old cold storage plant at this fort was in a vulnerable, unprotected surface building and could contain just 23 days' worth of meat for the Army's peace garrison. Budgets were too tight in the late 1930s, and the item next appears as an authorized department engineer project in mid-1942 for $839,400. It was put into use in early 1943. (15)

The largest Oahu project was for the Kunia underground Aviation Repair Facility. This was to be a major aircraft assembly plant, presumably for aircraft sent disassembled from the mainland and reassembled here. It was technically not a tunnel, but rather a very large, three-story free-standing building built on a flat piece of land but embanked on the sides and with earth cover. It was located on the north side of the Schofield Barracks reservation. It began as an estimated $2,000,000 project in department engineer authorizations of mid-July 1942. Later descriptions state $23 million was the total invested cost. Entry was by a single tunnel-like portal into a one-quarter mile long tunnel with a bend in it. This led to a set of two elevators (one personnel, one freight) that served three levels. At least one large bay with no divisions could service three 4-engined heavy

Many other strategically important military facilities were constructed in tunnels or at least underground during the Second World War. Here is shown the diesel power room for the huge underground Red Hill Fuel Depot on November 3, 1942. *NARA.*

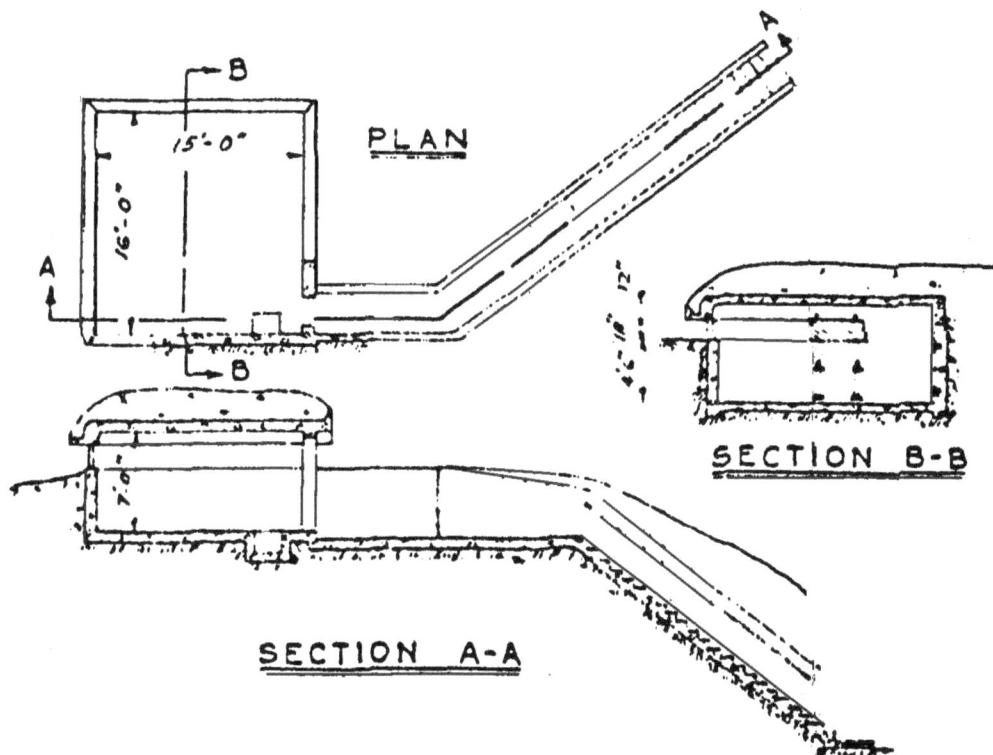

Aliamanu Crater also offered a convenient viewing location for the defenses. Besides a variety of fire control stations, it gained this Forward Echelon Command Post for the Seacoast Defenses during the interwar period. *NARA.*

bombers simultaneously. The facility had its own power, air conditioning, kitchen, bathrooms, and shops. It was subsequently known simply as the "Hole". Basic work was done in 1942-44, but it never served as an aircraft maintenance facility. After the war it was a major site for map and photo reproduction, then served as a signal intelligence processing center. (16)

Construction of the Forward Echelon Command Post at Aliamanu Crater was still underway in December 1941, when the nation entered World War II. It was occupied even though incomplete during construction. General Walter Short, commander of the Hawaiian Department ran operations from here immediately following the December 7th attack. We have an interesting account from a visiting member of the department's legal staff: (17)

> General Short and his staff were installed at the Aliamanu Crater some three miles from Fort Shafter. This headquarters had been constructed by throwing together several ammunition storage bins. It was far underground and unfinished at the time of the attack. There was not at that time any provision for air conditioning and the number of officers who were crowded in there soon exhausted the air supply and staff officers were compelled to go outside frequently to get a breath of fresh air. In company with Colonel Craig, the Provost Marshall, and his Assistant, Colonel Steer, about the third afternoon after the attack I went to Aliamanu Crater to report to General Short and his Chief of Staff. We finished our business and stopped for a free meal, and then hurried to depart for Honolulu.

The U.S. Navy was offered space in the complex in early 1942, but beyond establishing an information center in August the Navy continued to pursue a separate site for its operations center on Oahu. In the fall of 1942, CINCPAC directed the Fourteenth Naval District to participate in a joint operations center. In November, the Navy began operating in a section of five emptied, adjacent laterals to the Army's command post. These tunnels however were found to be too constricted for the Navy's needs and they requested that a much larger Joint Army-Navy Operating Center be built. (18)

The Joint Operating Center (JOC) at Aliamanu Crater was built adjacent to the Alternate Command Center. The need was for an enlarged combined operating center to coordinate army, air, and navy operations in Hawaiian waters. Connecting or expanding the existing army center at Aliamanu Crater was found to be the most logical location. Work was funded 50-50 between the two services. Work was started in 1942, though evolving needs continued to delay completion. A remarkable tunnel complex was drilled into the volcanic mass of Aliamanu. Work on the JOC, however, was suspended in February 1944, before it was fully complete and equipped. The facility continued in use through the end of the war by both the Army and by the operations staff of the Navy's Hawaiian Sea Frontier. With the end of the war use of the JOC was greatly reduced, and by September 1945, the Navy's only use of the command center was as an air-sea rescue coordination center and communications office.

The JOC was accessed by a second vehicular tunnel approximately 630-feet long and 14-feet wide that led to the central control and communication complex. The principal room in this complex was 115-feet wide, and 34-feet deep and had a ceiling height of 24-feet. Adjacent to this compartment were tunnels leading to the Army and Navy command tunnels. Each of these had a set of 18- to 20-foot-wide laterals of varied length as well as several connecting passageways, or laterals. These galleries branched off both sides of the main access tunnel at an angle of about 45° at 60-foot intervals. Many of the laterals had metal bunks lining the walls for the use of personnel assigned to the underground command center. These bunks could be "triced" up when not in use to facilitate traffic through the tunnels.

The Navy laterals contained a radio transmitter room, the admiral's office and quarters, dormitories, the surface plot room, the air plotting room, and the intelligence lateral. The Army also had a radio transmitter room, barracks spaces and latrines and teletype communications rooms, as well as various operations rooms, offices, switchboard rooms and the usual emergency power generator rooms, ventilation and air conditioning plant rooms and an incinerator. In addition, the Seventh Army Air Force and the Seventh Bomber Command were provided with command and communication spaces in the complex. These underground facilities had

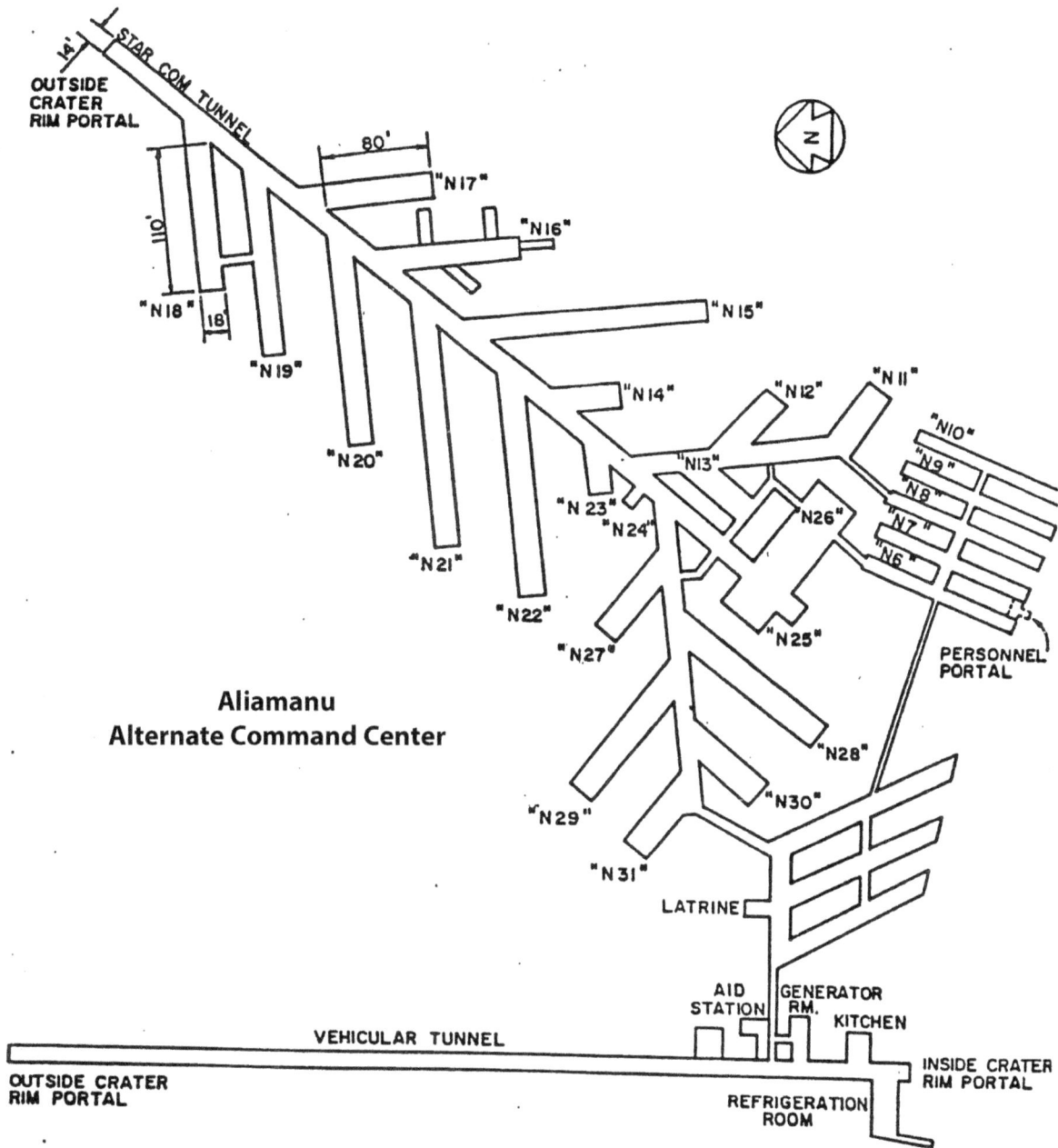

**Aliamanu**
**Alternate Command Center**

Command "tunnels" in Aliamanu evolved in name and structure between 1930 and 1945. Here is the complex known as the Aliamanu Alternate Command Post about 1941. *USAMH.*

seven vertical communication shafts extending up from the tunnel complex to the crest of the rim to accommodate radio antennae and ventilation ducts. (19)

Fort Shafter between the wars had become the headquarters of both the Hawaiian Department and the Anti-aircraft wing of the Harbor Defenses. For effective air defense warfare, a coordinating information and command post was created. This command post, code named "Little Robert" during World War II, provided tactical command and control of the antiaircraft defenses of Oahu. It was originally housed in a collection of World War I-era frame buildings adjacent to the cantonment of the 64th Coast Artillery near the gulch that ran through the south side of the Fort Shafter Reservation. This station was manned by the Intelligence Battery of the 53rd CA (AA) Brigade. Soon after the war began the staff was augmented by the assignment of the female personnel of the Women's Air Raid Defense (WARD). The WARD was later redesignated as the 515th Signal Aircraft Warning Regiment.

Oahu's chief air attack warning center was the Antiaircraft Artillery Intelligence Center command post at Fort Shafter during the war. Here workers plot aircraft sightings on a large maps and a plotting board. *NARA*.

A new AAIS Information and Control Center was tunneled into the side of a hill on the east side of Fort Shafter during the early part of 1942. This facility, code named "Lizard", consisted of a concrete lined main tunnel with branching laterals. One of the laterals was somewhat larger than the others and was used as the Air Defense Plotting Center. The original AAIS center was demolished soon after the war, the underground tunnel facility is still used for other purposes by the Army. (20)

Preparations for extensively enhancing the beach defenses with more pillboxes quickly followed the Japanese attack of December 1941. Sections of the 3rd and 65th Engineer Combat Battalions were detailed to construct these pillboxes along the beaches. Hundreds were built, maybe as many as 200 were earmarked for the southern approaches to Oahu alone. The beaches in and adjacent to Kaneohe Bay were also extensively covered, As compared to the previous generation, these were generally smaller, about seven to ten-feet square, and most had an entry door on the side. They were outfitted with firing "steps" for the mounts of the 0.30-caliber machine gun. It took about a day for a platoon of engineers to construct each pillbox using prefabricated forms for the standard design. By the end of March 1942 most had been placed. Every mile of beach was strung with barbed wire and units of infantry soldiers formed lines of resistance near the beaches and occupied the pillboxes.

There were several variants of these wartime pillboxes. Most were rectangular, but some had a rounded front and firing slit. Some even were double ended, with firing positions on both beachfront and rear landward approach. A few were built in place by the engineers, but it appears that most were made of prefabricated slabs made in a special facility near Honolulu. The slabs were fitted with eyebolts to facilitate loading onto vehicles and then placing in a previously prepared location. (21)

The wartime command post for Harbor Defenses of Honolulu was located in the tunnel piercing Diamond Head Crater on the north side. *NARA.*

This and the previous view are photographs of the Fort Ruger HDCP Operations room in March of 1945. *NARA.*

Finally, an armored steel pillbox had been developed. It was a small dome-shaped affair that could accommodate two men. This dome, sometimes termed a turret although it did not rotate, could also be sunk into a prepared pit type position. Entry to these "machine gun domes" was made by crawling through a buried concrete culvert pipe. Several of these dome type pillboxes could be connected by the concrete pipes to form a strong point. Before the war it was reported that the Army in Hawaii was planning on using these pillboxes throughout the island, but far fewer were ultimately placed. The pillboxes were manufactured by the Blaw Knox Company of Pittsburgh, PA and shipped to Oahu. They were about nine-feet high and projected about four-feet above ground level. The gun port was the size for either 0.30 or 0.50-caliber machineguns. The armored pillbox was equipped with a periscope. (22)

On the morning of the Pearl Harbor attack, at 11:30 a.m. on December 7, 1941, Territorial Governor Joseph B, Poindexter declared Martial Law in the territory and invoked special powers under the Hawaiian Defense Act. Within hours the Commanding General, U.S. Army's Hawaiian Department was authorized to exercise the governor's power. A long list of new legal restrictions for residents (most of which applied to both military personnel and the local population) changed their lives for several years. Martial law was lifted on October 27, 1942, though many local restrictions continued to be managed by the military until the end of the war.

The personal experience for the average coast artilleryman was the same as for other army, air force, and navy servicemen. There were no special privileges based on assignment. For officers and some non-commissioned officers one of the biggest initial changes was the order to relocate military dependent wives and children to the mainland. Starting in late December with the return journey of troopships that had brought reinforcements to the islands, families started to leave. A large convoy departing on February 21st took over 3000 women and children. These were not optional but mandatory evacuations. In the military's eyes the islands were already overcrowded and there were obvious safety and then food supply and residential space needs to be considered. By the fall of 1942 the task had been completed with the final relocation of just over 15,000 dependents. (23)

While it was considered and discussed, the Japanese population in the islands was not either evacuated or placed in retention camps like was done on the American West Coast. The sheer size of this population in the islands and their relative importance to the agricultural and industrial labor force were major considerations arguing against deportation.

Some new restrictions did impact the recreational activities of off-duty personnel. There were new curfews which lasted until July 1945, in some situations beginning as early as 9:00 p.m. Perhaps the most unpopular for the average off-duty soldier was the immediate closing of bars and saloons and the banning of liquor sales. This move was intended to reduce any drunkenness and the possibility of leaking confidential information. After a few months pressure from the affected businesses and even some military commands led to changes. Alcohol sales of retail containers were strictly rationed. In bars there was a limit to the time a patron could be present buying drinks, at certain establishments police would clear everyone out each hour to allow others to enter and to lessen the chance of prolonged drinking.

With the virtual cessation of the tourist business, many hotels converted into military recreation or short-stay options. The famous Royal Hawaiian in Waikiki became a Navy recreation site with low-cost meals, nightly dances, and rooms for those that stayed past curfew. Other hotels catered to the army's needs. Travel was still allowed on Oahu, but gasoline and tires were rationed like they were on the mainland, additionally blackouts had to be observed, including masking out most of a vehicle's headlamps and taillights. Food was rationed, but the importance of restaurants for servicemen morale was also recognized and efforts were made to keep the establishments open. Some issues for the Navy were particularly important. Ships might dock after weeks or months away and be expected to sail again after a short stay. Having some food, drink, entertainment options available during a short shore leave were deemed very helpful to morale. (24)

A very large and handsome Mormon Tabernacle was completed in Honolulu just months before the Pearl Harbor raid, and it quickly became a major attraction both for servicemen and tourists. During the war the church provided an open opportunity for rest, reflection, and even free meals and counseling during the war to service personnel. Meanwhile the local prostitution issue continued to perplex the military governance. It was technically illegal under territorial law but had been tolerated for many years as long as the location of brothels was restricted to one geographical district and that the incidence of venereal disease was controlled. The military government preferred to avoid the issue and have it handled by local or city police and continued the policy of tolerance.

Volunteers of the Hawaii Rifles train on an ex-navy 6-pounder gun immediately following the Pearl Harbor attack. *Bennett Collection*

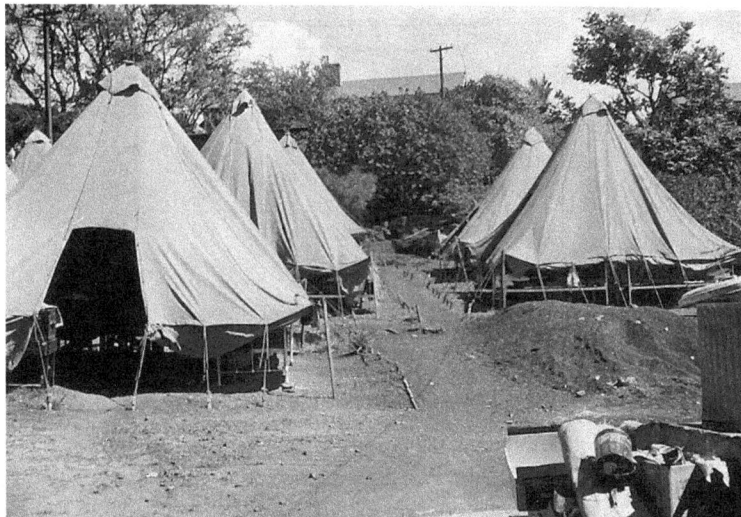

The rapid influx of reinforcing units early in the war easily outstripped housing capacity—numerous units were forced to make camp in tentage. *Berhow Collection*

# CHAPTER 13
## WARTIME ON THE NORTHERN AND EASTERN OAHU SHORES

The project for modern defenses of Kaneohe Bay was approved for construction, but still a long way from completion when the war started. For immediate defense, reliance had to be placed on temporary or emergency gun batteries. The intended emplacement of four 8-inch railway mount guns would prove to be among the easiest of the project's elements to provide. Eight additional 8-inch model M1888Ml guns and an equal number of barbette railway carriages had arrived on Oahu in 1940 and were in storage at Fort Kamehameha for use by the 41ˢᵗ CA (Ry) Regiment. While they were all mounted on their narrow-gauge railway cars, there was no trackage laid to the east end of Oahu and the Kaneohe Bay area. Consequently, in order to move four of the guns out to the Mokapu Peninsula, it was necessary to dismount the guns, truck them, their carriages, and the railway cars from Fort Kamehameha to Camp Ulupau, and remount them on short lengths of railroad track hastily laid by the engineers at the battery site near the Camp Ulupau cantonment area. (1)

By December 1, 1941, a final decision was made as to the precise location of this 8-inch railway battery. It would be located on army land between the cantonment area and the sand dunes of North Beach. Within a few days it became obvious that the stability of the temporary railroad bed and track that had been laid down to accommodate the cars was insufficient as a suitable gun platform. Inasmuch as the four weapons tactically constituted a "fixed battery" and mobility was no longer a factor, the decision was made to remove the upper barbettes from the railway cars and directly emplace them on concrete blocks. The department had experience

Camp Ulupau, immediately adjacent to the Naval Station, was the army post organized chiefly to house the coast artillery units and construction crews for defensive works. Here is the headquarters area in early 1942. *USAMH.*

Also contributing to the early defense of Kaneohe Bay, were several batteries of 155 mm GPF guns on Panama Mounts. Here one of these batteries skirts the Ulupau Crater road in mid-1942. *NARA.*

with this procedure as adopted a few years earlier when emplacing two other such guns at Battery Granger Adams at Fort Ruger. (2)

Soon after emplacement the four 8-inch railway guns were informally named Battery Sylvester. The namesake was 1st Lt. William Sylvester, of the 97th CA killed on the morning of December 7th, 1941. With a range calculated at 24,900 yards, the guns of Battery Sylvester covered the approaches to Kaneohe Bay 14-miles to seaward. Moreover, the guns could be traversed 360° and their projectiles could reach the shore of Oahu from Honolulu Harbor eastward to Makapuu Point. In the event of an attack by Japanese surface vessels these weapons would constitute the primary gun defense for the harbor until late 1943, when the first batteries of the permanent harbor defense project were completed. In the latter part of 1942, new permanent concrete emplacements, service aprons, and magazines were completed for all four guns adjacent to the original emplacements.

SCR-296A Fire Control radar sets were also supplied to Battery Sylvester and the 155 mm gun batteries in the harbor defenses by late 1942. These radar sets had the ability under good conditions to give warning of surface targets at ranges up to 100,000-yards or about 50 miles. But good conditions were often not encountered. More often large warships such as battleships and heavy cruisers could be tracked at ranges up to 40,000-yards, and smaller vessels at 20,000- yards. The SCR-296A had some drawbacks from a fire control standpoint. It was generally unable to discern separate targets unless at close range. The antenna for the radar provided for Battery Sylvester was located on a 100-foot metal framework tower atop Ulupau's west rim near fire control stations. This gave the line of sight radar an antenna height that allowed it to attain ranges up to 30 nautical miles. (3)

While the original antiaircraft provisions of the Kaneohe Bay harbor defense project remained basically unaltered by the War Department, the manpower and matériel resources necessary to implement the plan were still not available. Until the arrival of additional antiaircraft units from the mainland, temporary responsibility for Kaneohe Bay's air defenses fell to the newly organized 2nd Battalion, 98th CA (AA) Regiment stationed at Schofield Barracks. This battalion of about 12 officers and 300 enlisted men exercised its initial deployment to

The first "permanent" battery for the Kaneohe Defenses was an emplacement for four 8-inch railway guns. But as no railway ran to this location, the guns were placed on fixed blocks not unlike Battery Granger Adams. Here the battery is under construction in late 1941. *NARA.*

When completed in early 1942, the 8-inch railway-carriage battery appeared like this. It was later named Battery Sylvester. *Gaines Collection.*

Kaneohe during the first week of December 1941 and set up their eight semi-mobile 3-inch M3 AA guns as a training exercise. On Friday, December 5, after several days at Kaneohe, the men of the 98th limbered up the guns to their Corbett prime movers, packed up their gear, and returned to Schofield Barracks for what would prove to be an eventful weekend. (4)

Upon arrival on the Mokapu Peninsula in December 1941, the headquarters of the 2nd Battalion, 57th Coast Artillery began functioning as the harbor defenses' tactical command and established its first command post in one of the few private residences on the peninsula, a frame stucco construction dwelling located east of the cantonment area. It was promptly dubbed the "Rancho" by the men of the 57th because of its similarity to southern California adobe construction and its red tile roof. This headquarters served until early February 1942, when a larger Theater of Operations structure was completed adjacent to the post headquarters in Camp Ulupau's cantonment area.

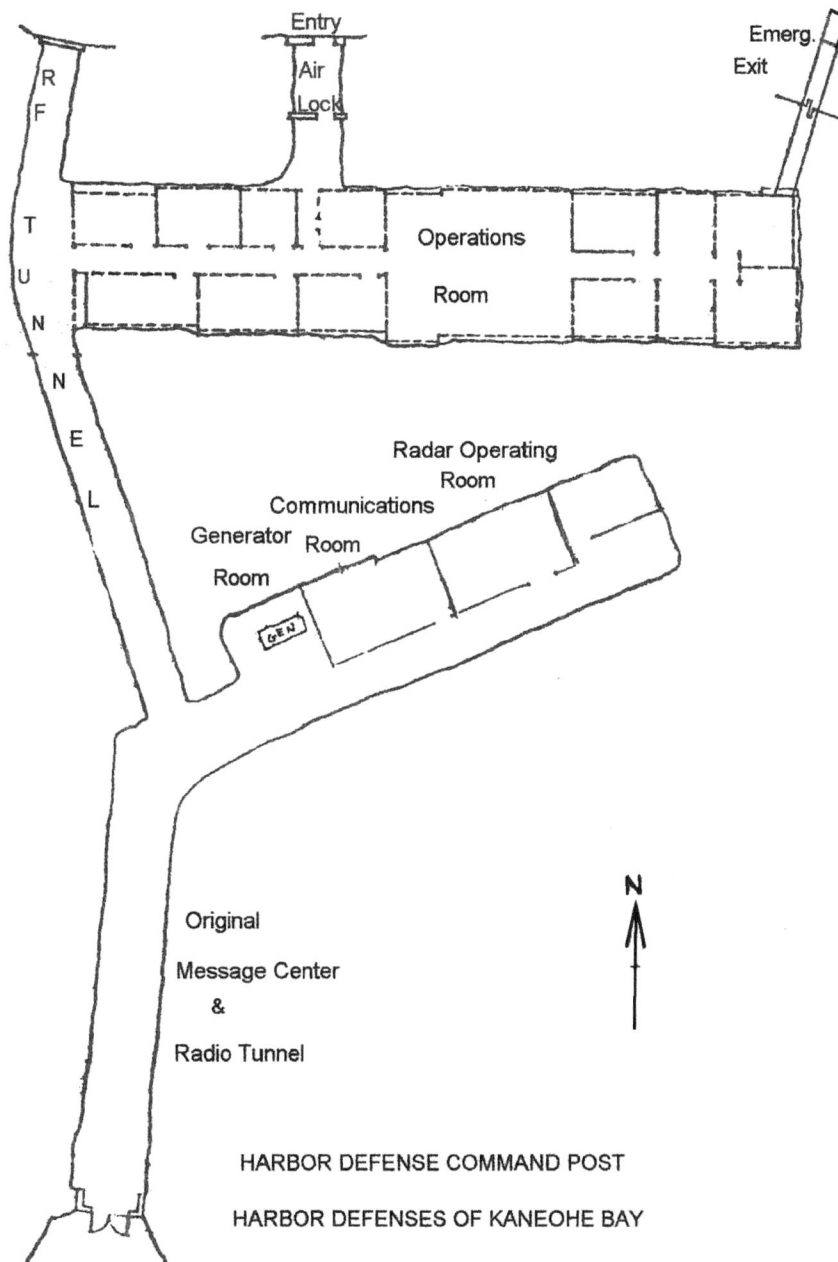

Plan of the wartime command post for Kaneohe Bay located in a tunnel complex at the Ulupau Crater. *Gaines Collection.*

Photograph from 1944 of operations inside the Fort Hase base headquarters. *USAMH.*

Camp Ulupau expanded rapidly across the flatlands adjacent the naval station in 1942. *Gaines Collection.*

The need for a secure bombproof structure for the harbor defense command and communications was recognized early in the war, and preliminary plans were developed for construction of one on the north side of Puu Hawaiiloa. This tunnel complex was to be drilled into the side of the hill and would have two parallel entry tunnels equipped with gas locks. These rooms were to be occupied by the navy in-shore patrol command post, an infantry command post, and a communication center consisting of a message center and a radio room and the harbor defense command post. On one side of the complex an officer's bunk room and latrines for officers and enlisted men were also planned. The Harbor Entrance Control Post (HECP) was to be located on the hill's summit near the navy's Kansas Tower (the naval air station's control tower and the navy signal station). (5)

Little progress was made completing the command post, however, until the harbor defense commander, Lieutenant Colonel Donald L. Dunlop, assumed the initiative in early spring of 1942. He commenced a new protected message center and radio facility for the Kaneohe Defenses. Instead of the site at Puu Hawaiiloa, he selected a location closer to the Camp Ulupau cantonment. Dunlop had a detachment from the 34th Engineer Regiment began excavation of a tunnel into a spur of Ulupau's west rim that was formed by two ravines. Upon completion, this facility consisted of a single unlined tunnel about 9-feet wide driven over 150-feet through the spur roughly south to north. This somewhat narrow underground gallery had no protective doors or air-locks when it was first occupied in mid-1942. Initially, this protected communications facility contained only the harbor defense's message center and a radio room. The radio transmitter lines ran northward through the tunnel exiting at its north end where the antennae were located.

This impromptu tunnel gradually replaced the original HDCP tunnel project. The unlined tunnel became a communication gallery or passageway leading to two new branch tunnels. The small northern exit for the radio transmission lines to the antennae was sealed off and a new branch tunnel for the HDCP was laid out and driven off the original radio transmission tunnel, on an east-west axis within the spur of Ulupau. Later in 1942, an east-west tunnel was bored off the original message center tunnel for use as a Radar Operating Room Tunnel. During construction, the rock was found to be of such good quality that concrete lining was believed unnecessary.

Consequently, the tunnel walls and ceilings were pressure grouted to render them waterproof and the exteriors of the entrances provided with a coating of gunite. Partition walls between rooms in the operation tunnels were constructed of concrete blocks. (6)

Access to the air-conditioned main operations tunnel was gained through a new entrance from the northern ravine. An airlock was established just inside the entry tunnel as part of the CWS gas proofing arrangements. Spaces were also partitioned off for the message center, radio room, fire control switchboard room, latrines, a first aid room, bunkrooms for officers and a small galley with a serving area. In addition, space was provided for the navy's inshore patrol command post and the beach defenses command post. At the east end of the operations tunnel an escape tunnel and shaft fitted with a staple ladder extended 50-feet to the north and the emergency exit opened into the ravine in a thicket of guava trees. The west end of the operations tunnel opened onto the original north-south radio cable tunnel that was enlarged slightly and became a communication gallery that connected the HDCP with the radar operating room tunnel and the south (original) entrance.

The smaller branch tunnel containing the radar operating rooms was 65-feet south of the HDCP Tunnel It contained radar operating rooms and an electric power generator room equipped with a 15-kW generator and the associated switchboard panels and controls. The radar antenna for an SCR-582A surface search radar set was mounted on a metal framework tower also located atop the crater rim above the tunnels. This radar set was one of four seacoast surveillance radar sets sent to the Hawaiian Department in the first year of the war. The SCR-582 set installed at Ulupau was made operational by June 15, 1943. It often proved to be unreliable because of breakdowns and a mobile SCR-270A long range air search radar was set up on the crater rim that provided verifying data for the data supplied by the SCR-582, or in the event of breakdown to act in its stead. (7)

The planning of the Camp Ulupau cantonment began in October 1941 soon after the defense project was approved. At a cost estimate of $2,221,000 facilities were projected that would accommodate an army garrison of 3,000 men. The constructing quartermaster received informal instruction to proceed with the building of a Theater of Operations type camp instead of the more expensive camp of frame buildings originally proposed. He was also advised that the sum of $300,000 would be forthcoming to cover the initial start of construction. In November, the 34th Engineer Regiment began to construct the camp's 186 structures. In addition, the engineers would lay the base course of the roadways in the camp, construct "duck" walks and hard stands, and build the 275 mess tables to be used in the mess halls and 350 tables to be used in the administrative buildings. They were to construct water tanks, provide and install all utilities including all plumbing and electrical work, water and sewage, installation of stoves, refrigerators and provision of chairs used in the camp and apply the running course of the roadways. When the troops occupying the barracks arrived, they were to put in shelving, interior partitions and construct wooden lockers. (8)

Some idea of the scope of the construction can be gained from this list of items built on November 10, 1941. These buildings consisted of the following: (9)

| Number | Structure Description | Size |
|--------|----------------------|------|
| 60 | Enlisted Barracks | 20' x 100' |
| 13 | Mess Halls | 20' x 130' |
| 1 | Mess Hall | 20' x 150' |
| 2 | Mess Halls | 20' x 110' |
| 16 | Co. Administration and Storehouses | 20' x 100' |
| 16 | Co. Bath Houses | 20' x 28' |
| 16 | Latrines | 8' x 16' |
| 16 | Recreation Buildings | 20' x 100' |
| 3 | Battalion Administration Buildings | 20' x 36' |
| 3 | Battalion Storehouses | 20' x 100' |
| 3 | Battalion Guardhouse | 20' x 40' |
| 1 | Regimental Administration Building | 20' x 100' |
| 1 | Regimental Storehouse | 20' x 100' |
| 1 | Regimental Post Exchange | 20' x 100' |
| 2 | Regimental Infirmary Buildings | 20' x 100 |
| 1 | Regimental Recreation Building | 20' x 100' |
| 1 | Regimental Guardhouse | 20' x 60' |
| 7 | Officers Quarters | 20' x 100' |
| 1 | Officers Mess | 20' x 120' |
| 1 | Officers Bath House | 20' x 28' |
| 1 | Officers Latrine | 8' x 15' |
| 1 | Station Complement Administration Building | 20' x 100' |
| 15 | General Warehouses | 20' x 100' |
| 1 | Ordnance Repair Shop & Magazine (Steel) | 20' x 140' |
| 1 | Warehouse Shop | 32' x 48' |
| 2 | Warehouse Shops | 48' x 112' |
| 1 | Gas and Oil Station | 6,000 Gallons |

By November 10, a shortage of tar paper and dynamite and numerous smaller items had developed, and the District Engineer was called upon to provide these from his stocks. These materials and stockpiles of lumber were deposited at the campsite in late November and considered so valuable that they were placed under guard by a detachment from the coast artillery. The sewage plant was designed to be a biofiltration system with a safe effluent discharge low in suspended matter that was to empty about 300-feet out in Kailua Bay. By December 4, 1941, all parties concerned had signed off on the construction program at Camp Ulupau. The events of the following weekend resulted in the 34th Engineers being assigned to more immediate fortification related concerns and the Zone Constructing Quartermaster was assigned the task of building the camp. The entire project was subsequently turned over to a civilian contractor who commenced work later that month. (10)

When initially constructed, Camp Ulupau was planned to accommodate about 3,000 coast artillery troops: an antiaircraft regiment, less one battalion, a reinforced battalion of tractor drawn artillery, and a separate battery. The camp was laid out on a grid composed of four unpaved streets running east and west and four others running north and south. The camp's complex of mess halls, its headquarters and administrative buildings and supply buildings were built along the south side of the cantonment's northern most roadway. Directly south of these structures were the orderly rooms, supply rooms and day rooms for the various batteries and companies of troops posted at the camp. The southern half of the cantonment was occupied by 66 platoon sized barracks buildings and 18 latrines. No hot water was available at the camp for bathing, but cold-water showers were made available to the troops at a shower point erected by the engineers in the northwest corner of the cantonment area near the mess halls.

On February 2, 1942, Camp Ulupau, the headquarters post of the Harbor Defenses of Kaneohe Bay, was renamed Fort Hase in honor of Major General William F. Hase, a former Chief of Coast Artillery and a one-time commander of coast artillery units in Hawaii.

By 1945 Fort Hase had exploded in troop capacity—though by then devoted to units transiting the Pacific for the combat arenas closer to Japan. *USAMH.*

Major General William Hase, namesake for the post when it was renamed in February 1942 to Fort Hase. *Gaines Collection.*

The naval air station occupied the west half of Mokapu and quickly became crowded with personnel, planes, and numerous support facilities. Fort Hase also expanded into a bustling installation occupying the eastern half of the Peninsula and was well on the way to becoming a major military installation. In time, nearly every square foot of flat terrain on the Army's half of the Mokapu Peninsula would be given over to the quartering or support of tens of thousands of troops. Fort Hase would, in addition to the harbor defense elements, eventually become the base for an infantry division.

Kaneohe Bay relied on numerous 60-inch portable searchlights to sweep the waters of the bay and its approaches and illuminate targets for the guns of the defense. These searchlights were designated as the Ulupau Searchlight Group in 1941. They were initially manned by a detachment of Battery D, 16th CA (HD) Regiment from Fort DeRussy. There were four positions each provided with two portable 60-inch lights. Two of

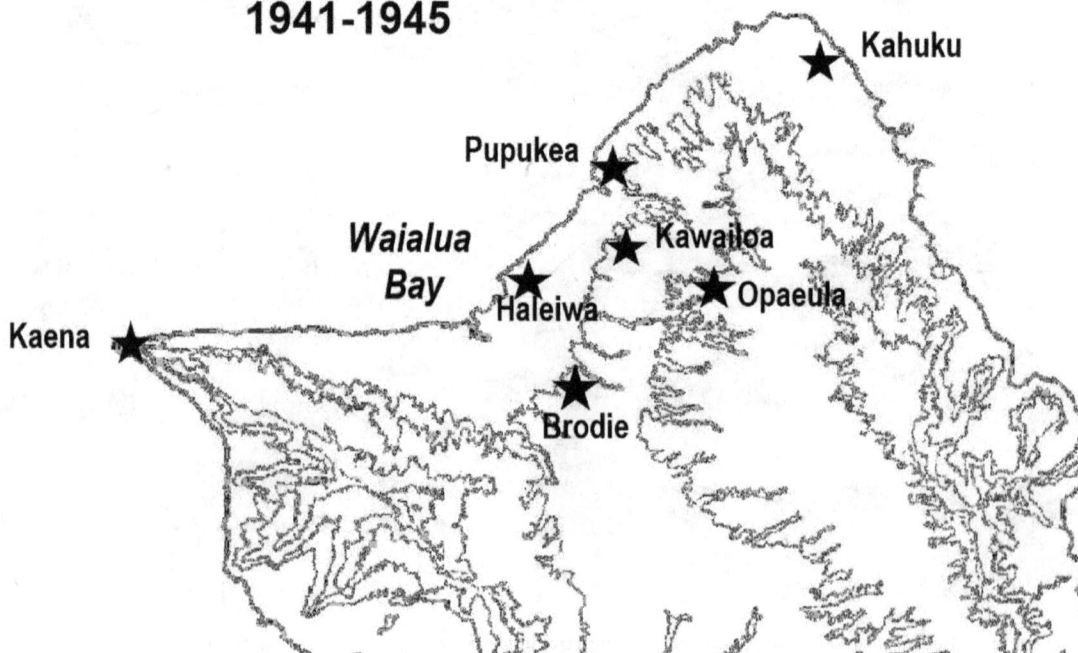

Map of major defensive locations on the North Shore 1941-45. *Williford Collection.*

the lights were located at Makapuu Point, and a second pair were emplaced on concrete pads atop the ridge at Wailea Point and later moved to the towers near the shore of Waimanalo Bay. Two more searchlights were sited on a promontory of land known as Kii Point, 35 to 45-feet above the waters of Kailua Bay at the south end of Ulupau's east rim. A reserve searchlight was later set up near the small boat landing at the south end of the waters of the anchorage that lay off the naval air station. Across the bay at Ka Lae O Kaoio Point two more searchlights were emplaced near the foot of the cliffs above and behind the 155 mm guns of Battery Loko on the edge of the shore north of Kualoa Point.  (11)

Much of the burden of manning active Kaneohe batteries fell to Battery C of the 41st CA. They had manned Battery Sylvester's 8-inch guns throughout 1942 and 1943. Numerous personnel reassignments during 1942, and the early months of 1943, so diminished the strength of Battery C, that by May 1943 the battery had been reduced to only four officers and 60 men, barely enough to man the four 8-inch guns in an emergency. Many of the battery's men had been used to help form the provisional manning detachments of the batteries at Wailea Point and at Kahana Bay. When the 5-inch guns of Battery Homestead near Makua on Oahu's southwest shore were inactivated on May 24, 1943, its three officers and 75 enlisted men were transferred to Fort Hase where they were reassigned to Battery C of the 41st. After being augmented, Battery C continued to man the guns of Battery Sylvester through the end of 1943.

The 1st Battalion, 47th CA (TD) Regiment arrived in Hawaii in February 1944 and was assigned to the Harbor Defense of Kaneohe Bay soon afterward. Upon arrival at Fort Hase, the battalion was reorganized as a two battery 155 mm gun battalion and redesignated as the 32nd CA Battalion (155 mm Guns). The excess firing battery from the 47th, Battery C, was redesignated as the 717th CA (155 mm Gun) Battery and was attached to the 32nd Battalion. On March 5, 1944, the 32nd CA activated the Coral Gun Group in the Kaneohe Bay harbor defenses. This group was composed of 155 mm guns emplaced at Kaaawa and Camp Kahana that was occupied by Battery G (Searchlights) of the 41st Regiment. When the 32nd Battalion arrived at Camp Kahana the 5-inch guns of Battery Kahana, although still in place, were no longer in service and had been placed in maintenance status.

Soon after the arrival of the 32nd Battalion, the 155 mm gun battalions were reduced to one firing unit (two guns) per battery by order of Headquarters, Central Pacific Area. In all probability this reduction was made in order to release the newer model 155 mm M1A1 guns to units in the forward areas. On March 19, the 32nd and its attached 717th CA Battery were released from tactical control of the harbor defenses and assigned to the Jungle Training School in the Kahana Valley although administration and supply responsibility for the units was retained by the harbor defenses.


To the early planners, it seemed the upper, or northern section of Windward Oahu was the least inviting for possible hostile invaders. In the years before World War II, the Army considered the reef-strewn and generally inhospitable landing beaches of Windward Oahu coupled with the massive rampart of the 2,000-foot high Koolau Mountain Range a natural line of defense. The Koolau's near vertical windward slopes formed an especially rugged defense line that was considered impassable. The military felt that the mountain range effectively blocked passage by any large enemy force moving to attack the rear of the principal population centers and the military and naval installations located on the south shore. A small defending force of one or two companies of infantry was considered sufficient to deter any enemy attempt to scale the mountain range.

There were only a few trails that ascended the slopes of Koolau, and these could be easily defended. In the northern sector the Kahuku Trail led from the plantation town of Kahuku up to the crest of the ridge to the Summit Trail that extended along the length of the range. From that junction the Pupukea Trail led down the west side of the range to the North Shore. In addition to the Pupukea Trail, there were three others that led down into the central part of the island. The Kawailoa Trail led to the coast on the North Shore below Pupukea, the Poamoho Trail gave access to the Leilehua Plain north of Wahiawa, and the Waikane Trail led down to the Schofield Barracks Military Reservation.

The 13th Field Artillery Regiment and their 75 mm field guns training for beach defense in the late 1930s. *Gaines Collection.*

In the immediate pre-war period, lack of manpower in the Coast Artillery forced the Department's Field Artillery to assume much of the defense of the North Shore. This is the 11th Field Artillery with 155 mm howitzers in the late 1930s. *Gaines Collection.*

On Oahu's North Shore only two places offered potential landing beaches: Waialua Bay and the somewhat smaller Waimea Bay. Both were poor choices for landings during the winter months when the northerly ocean swells reached their maximum size. In the late spring and summer when the seas are much calmer, the beaches were capable of allowing a limited invasion from the open sea. Amphibious operations in the 1920s and 1930s were in their infancy and generally limited to the landing of troops from ships' small boats or by seizure of a harbor's port facilities that could handle the unloading of troops and equipment from ocean going vessels. Consequently, between the two world wars, the open waters of the bays on both the windward side of Oahu and the North Shore were strong natural obstacles to an invasion force. But then by the late 1930s the Japanese had markedly improved their landing craft and "over the beach" capability. This caused a commensurate increased interest by the U.S. Army in its defensive measures for the North Shore. Unlike the Windward side, should an enemy be successful in landing over the beaches at Waialua, there were few natural barriers capable of blocking an invasion force. An opponent would only have to proceed inland over a terrain that gradually rose to an elevation of some 500-feet to the plateau and then move further inland across the Leilehua Plain to the center of the island and thence down into the rear of Pearl Harbor and Honolulu.

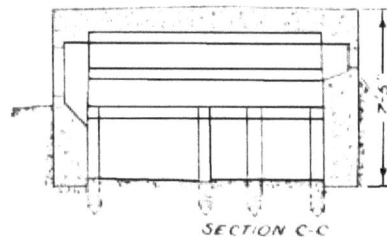

SECTION C-C

Plan for the standardized beach pillboxes of the early 1930s generation. *NARA*.

Photograph captures both one of the beach pillboxes at Oneula Beach, and fire control tower "X" for Battery Williston immediately behind it. *NARA*.

The Army began construction in 1908 of a garrison post 14-miles from Honolulu and about nine-miles northwest of Pearl City. This post was on the Leilehua Plain about midway between the Koolau and Waianae Ranges and about halfway between the North Shore and Honolulu. It was named Schofield Barracks in honor of Lieutenant General John M. Schofield who had commanded a corps during the Civil War and served as Secretary of War under President Andrew Johnson and led the 1873 survey of Oahu for location of a defended naval station. While primarily intended as a spacious location for stationing and exercising a permanent field army garrison, the location strategically also blocked access to the southern cities and naval harbor from a northern landing. Following World War I, Schofield Barracks became the home of the Hawaiian Division. Occupying a land tract of more than 14,000 acres, in central Oahu, the post became the principal base and maneuvering area for the mobile garrison; a force that served as the major blocking force to an invading army landing on the North Shore.

Administratively the defenses for the north shore created an awkward challenge. The north shore had neither a harbor or central fort or reservation. It did have coverage from fixed, long-range batteries elsewhere, from mobile seacoast defense batteries, and eventually during the Second World War a few specific batteries on small individual reservations. The forces dedicated to its defenses were minor, or simply detachments from units primarily assigned elsewhere. Frankly, it did not justify the status of being its own "Harbor Defense". Still, it needed its own fire control stations and command functions. Eventually it was organized into a sort of junior harbor defense described as a "groupment". The North Shore Groupment had been organized soon after the end of World War I as a subordinate element of Hawaiian Coast Artillery District. Later it was reassigned as a tactical sub-element of the Harbor Defenses of Pearl Harbor, and eventually merged with the Harbor Defenses of Kaneohe Bay.

Prior to the war, only one big project was ever seriously proposed for the north shore. Late in the 1930s a third 16-inch battery, armed with the same type of guns as Battery Hatch, was proposed for construction on the North Shore in the vicinity of the Opaeula Gulch on the Anahulu Flats. A battery at this location would improve the field of fire coverage by large caliber artillery of the waters off Kahuku Point. Plans were developed and land acquired for the project in 1936, but it was never approved for construction. (12)

As they were dependent on mobile guns, the defense planning for Oahu's North Shore during the year prior to the nation's entry into war placed a relatively high priority on providing the Groupment with a full allocation of manning detachments. This would relieve the field artillery from its coast artillery missions. The Hawaiian Defense Project of 1940 called for the creation of the 41st CA's 2nd Battalion as a part of the overall augmentation of the island's defenses. This would enable the regiment to man two railway battle positions

While easy to confuse as being a pillbox, this is one of the few surviving searchlight control stations left on the North Shore.
*Gaines Collection.*

on the North Shore, as well as the two on Oahu's southwest coast. In addition, the 55th CA Regiment was to be fully activated and another full regiment of tractor-drawn coast artillery armed with 155 mm guns would reinforce the island's coast artillery garrison and relieve the 11th Field Artillery Brigade of its coast defense mission on the North Shore. If the augmentation of the Hawaiian Islands were carried out during peacetime, the reinforcing regiment would be accompanied by its own armament and equipment. However, if reinforcement occurred after the commencement of hostilities, only the personnel would be sent to the island. There they would take over pre-positioned armament and equipment held in the departmental war reserve and the regiment would be assigned to various battery locations on the island. (13)

The groupment was initially assigned two gun groups: The Kahuku Group and the Haleiwa Group. Each of these groups was composed of railway and tractor drawn mobile artillery batteries that would be deployed to their positions in time of war. Assigned were the 155 mm gun positions at Ashley Station and Kawailoa and a third battle position for 155 mm guns at Kahuku to be manned by one of the batteries from the reinforcing coast artillery regiment. Railway battle positions for 8-inch guns at Puuiki and another at Kahuku, still to be built, were to be manned by the prospective Battery C of the 41st. Overall command and control of the North Shore Groupment was to be provided by Headquarters and Headquarters Battery, 55th CA Regiment.

The need to provide coast artillery units for the new naval air station at Kaneohe Bay pre-empted the above plan and resulted in a reassignment of units to the North Shore. The 11th Field Artillery Brigade of the Hawaiian Division was directed to assume temporarily the manning of 155 mm batteries on the North Shore. On January 3, 1940, battery positions at Ashley Station and at Kawailoa were transferred to the custody of Brigadier General Maxwell Murray's Field Artillery Brigade. The groupment continued to be composed of two gun groups. The Kahuku Group's command post located at Station "K" overlooking Kahuku Point provided tactical command and control of the mobile batteries defending the northern most part of Windward Oahu from the Kahuku Point area down the windward coast as far as the village of Hauula. The Haleiwa Group providing command and control of the batteries along the North Shore from Mokuleia north to Kahuku Point had its command post at Station "T," at Kawailoa.

Station "T" also functioned as the North Shore Groupment's command post and was eventually housed in a 40-foot tall, four-story splinter-proof concrete structure. It had been begun earlier in 1941 about 300-yards north of the Waialua Agricultural Company's plantation camp known as Kawailoa Village. Still unfinished when the war broke out, the structure was quickly brought to completion and transferred to the coast artillery on December 12, 1941. As yet there were no fixed-gun emplacements on the North Shore. (14)

Although in the planning stage for several years, the three battle positions for 155 mm guns for the North Shore mentioned previously were finally built just prior to the outbreak of World War II in Europe. These battery positions, each with four Panama mount emplacements, were sited near the plantation camps at Kawailoa and Ashley Station, and on the land of the Kahuku Ranch atop Kalaeokahipa Ridge northwest of the sugar-milling town of Kahuku. On May 17, 1939, work on the batteries at Ashley Station and Kawailoa Village began, using troop labor and materials supplied by the Corps of Engineers and the Quartermaster Corps. Both were declared complete on October 18, 1939. The battle positions on the North Shore were similarly arranged, with their four concrete 360° Panama Mounts arranged along the trace of a slight arc, with the battery capitals bearing to the northwest. Both positions were provided with battery commander's stations. The third of the 155 mm gun batteries built before the outbreak of the war was commenced atop Kalaeokahipa Ridge on March 11, 1941, and completed about the middle of April, was transferred on August 23, 1941. Named Battery Ranch, for its location on the Kahuku Ranch, the position also consisted of four Panama mounts. (15)

Construction also proceeded on fire control stations, searchlights, and beach defenses. Details from the 3rd Engineer Regiment of the Hawaiian Division were employed in the preparation of beach defense positions on the high ground overlooking potential enemy landing positions on both the island's windward coast and the North Shore. These positions were similar to the strong point pill boxes built in the 1930s but were more hastily built. At one location known locally as the "Crouching Lion," overlooking Kahana Bay, Company A

Early in the war more pillboxes were hastily constructed. Here is one with its water-cooled 0.30-cal machine gun in the spring of 1942. *NARA.*

Pouring concrete in early 1943 for the permanent gun blocks of what would become 8-inch Battery Kahuku. These model 1888 ex-railway mount guns would be fired for settings on March 1, 1943. *Gaines Collection.*

of the 3rd Engineers constructed concrete pill boxes for 37 mm antitank guns, and machine guns as well as splinter-proof observation posts and personnel shelters dug into the slopes. Similar strong points were built on the precipitous slopes of the high ground north of Pupukea that covered the beaches below as well as providing close-in defense for the Pupukea bivouac area and the Opana radar installation.

In October 1941 the Hawaiian Division was dissolved and its units transferred to help create both the 24th and 25th Infantry Divisions. The Japanese air raid was still in progress when the first orders were issued for these divisions to begin deployment to their assigned field and beach defense positions and prepare for an expected Japanese invasion. The forces available for the defense of the North Sector consisted of elements of the 19th and 21st Infantry Regiments. The 21st Infantry's 1st Battalion (Companies A, B, C, and D), was spread thin along the shoreline from Haleiwa northward to Paumalu (today's Sunset Beach area) a front of some eight miles. These units established strong points along the beach around permanent defenses such as the concrete "strong point" pill boxes constructed in the 1930s on the beach at Waialua and on the heights above the mouth of Waimea Stream, as well as more recently dug in positions revetted with sandbags. The 3rd Bn, 21st Infantry was given the task of serving as a trail blocking detachment along the 12 miles of the Koolau Range's ridge line that lay within the North Sector. These troops were equipped to blow up critical sections of the various trails that climbed the north scarp of the Koolau, should an enemy force land and attempt to scale the formidable mountain range. The remainder of the 3rd Bn was posted at about the midpoint of the sector line on Pupukea Heights to function as the North Sector Reserve. The 2nd Bn of the 21st Infantry served as part of the 24th Division's Reserve and was positioned at Schofield Barracks well to the rear of the North Shore beach defense line.

The 2nd Bn, 19th Infantry Regiment was assigned to the western part of the North Shore Groupment. One company was posted along the southwest shore of Oahu from the vicinity of Puu O Hulu to just north of Waianae, while the three remaining companies of the battalion were held back in a sector reserve in the Lualualei area. One company patrolled the section of shoreline from Kaena Point to Waialua while another operated as a trail blocking detachment along the ridge line of the Waianae Range. The 24th Division artillery followed the infantry as they moved out into the wilds of Oahu's North Sector. The two provisional batteries of 240 mm howitzers manned by the 11th FA Battalion brought up the rear of the division's artillery columns. One of the two-gun 240 mm howitzer batteries was emplaced at Quadrupod Peak while the other occupied a position on the Anahulu Flats that lay southeast of the Pupukea area. By the evening of December 7, most of the elements of the 24th Division were in place, occupying defensive strong points from near Puu O Hulu on the island's southwest coast to Paumalu on the North Shore. In addition, motorized patrols were being maintained from Kahuku down the windward side of the island to Kahana Bay.

Upon disembarking on January 7. 1942, the two battalions of the 57th CA were deployed to the defenses of Oahu's North Shore relieving the 24th Infantry Division Artillery of their coast artillery mission. On January 11, an advance party composed of an officer and 30 enlisted men from was sent to the Kahuku area to establish a base camp. That same day another battery moved to Schofield Barracks where it took up duties as an "artillery reserve" for the North Shore Groupment. The following day the regimental headquarters battery departed camp in a truck convoy for Camp Kawailoa and began preparations to assume tactical command and control of the North Shore Groupment. This was the first time the North Groupment received a dedicated unit in its roster. The 1st Bn Headquarters Battery and the main body of Battery B arrived at Kahuku at 4:30 p.m. where the battalion set up a command post and bivouac at Kahuku School, from which the group was initially to be operated. (16)

On December 9, Battery B's gun train finally moved out from Fort Kamehameha for its battle position on the North Shore. Although the prewar plans had called for them to take up their primary position at Puuiki, the Hawaiian Department was by now fully convinced that the locations of all existing battery positions on the island were known to the Japanese and the decision was made to send the men to the alternate battle position near Kawailoa further up the coast. Revetments were hastily built around the guns using scrap lumber, logs, and earth. On Friday, December 12, the battery fired one settling round from each gun. Designated Battery

This is Station "T" built in 1941 at Kawailoa. It also served at the North Shore Groupment Command Post. *Gaines Collection.*

Haleiwa, the four railway guns would constitute the most powerful armament on the north shore until the late summer of 1942, when the two naval turret batteries at Opaeula and Brodie Camp were completed and placed in service. Although the railway guns remained near Kawailoa Station most of the time, they could be moved to their alternate battle position in the North Shore Groupment at Puuiki. The four 8-inch guns of Battery Haleiwa were fired with regularity in 1942 and through 1943, but the frequency of target practices declined during 1944. On December 26, 1944, the railway guns were taken out of service and placed in maintenance status. On March 27, 1945, the artillery officer, Pacific Ocean Area, reported to the War Department on obsolescent armament that should be removed. Shortly afterward the War Department directed that while Browns Camp and Haleiwa Batteries were no longer required in the Hawaiian area, their guns were to be maintained in caretaker status pending the completion of more modern batteries still under construction.

Shortly after assuming command of the Hawaiian Department in February 1941, Lt. Gen. Walter C. Short had again made the case for the construction of the railway link between Wahiawa and Schofield Barracks to the North Shore that had been proposed as early as 1921 and planned in 1935. After much interservice wrangling, funding was forthcoming late in 1941 and the connector was built. When completed the extended railroad was some 14 miles long. Completion of this rail link with the North Shore reduced the movement time and distance between the south coast and the North Shore by some 75% and was less subject to interdiction by naval gunfire than the coastal route. Colonel John Silkman, commanding the 34th Engineers, drove the "Golden Spike" on December 23, 1941, and the new right-of-way was placed in service. (17)

From the time railway artillery had been introduced to Oahu in the 1920s, the coast artillery had relied upon the OR&L for the requisite locomotives and much of the other rolling stock. In the early weeks of the war the War Department finally sent large amounts of rolling stock to Oahu to support the railway artillery operations. The Services of Supply were preparing to send no less than seven railway kitchen cars, eight storage cars, five tank cars, and two 60-ton diesel locomotives to Oahu. While the pair of diesels was welcomed, this considerable array of rolling stock was greatly in excess of the railway artillery's needs, as two of the 8-inch batteries had already, for all intents and purposes, been slated for conversion to fixed batteries, leaving only two mobile batteries. After extensive communications between the War Department and Oahu, it was finally determined that there was sufficient military rolling stock on the island. There was, however, still a need for the locomotives and 20 tank cars. This requirement was finally recognized by mid-summer 1942, and measures taken to begin supply of the equipment. Before shipment began, however, the 60-ton diesels were changed to Whitcomb 44-ton locomotives and two General Electric 25-ton diesel locomotives modified to operate on the 36-inch OR&L gauge tracks, to be supplied by the government, but paid for by the OR&L. Delivery was expected in April 1943. (18)

Plans for a railway artillery battery at Kahuku dated from the 1930s, though undergoing several subsequent revisions. When the Laie railway position was abandoned in favor of a position closer to the village of Kahuku in August 1941, this new site near Kahuku Mills adjacent to the OR&L right-of-way became the third location for firing spurs at Kahuku. When plans were advanced on August 21, 1941, to mount 8-inch guns on fixed emplacements near the Kahuku Golf Course instead of on railway cars, General Burgin recommended constructing temporary firing positions there pending completion of the fixed emplacements. Construction of the temporary firing spurs was not begun until the last weeks of 1941, and the spurs were still incomplete and uncamouflaged when the nation entered World War II. In early February 1942 only two 8-inch guns, on M1918 railway cars were at the partially completed temporary railway battery at Kahuku Mills. The other two guns were still behind on the firing spurs at Fort Kamehameha.

The permanent position at Kahuku, like those at Camp Ulupau, were to be on concrete gun blocks placed in the ground so that the bottom of the gun barrels would be only two feet above the ground at 0° elevation. The battery was to be provided with projectile and powder magazines and splinter-proof plotting rooms. While no bunkers were to be erected, slit trenches with overhead cover were to be provided around each gun emplacement for personnel. The North Shore 8-inch battery projects were of such high importance that the district engineer authorized suspending casemating Batteries Hatch and Closson if necessary to expedite construction of the new batteries. (19)

The Kahuku battery was constructed by civilian contractors from the Honolulu District Engineer's Office. The temporary spur emplacements of Battery Kahuku were retained in service until mid-March 1942, when Battery D began moving to the four newly completed permanent concrete barbette emplacements between Kahuku Mills and the shoreline. By April 1942, the project was about 40% complete. The four concrete emplacements for the 8-inch guns were completed by June 15, 1942. In July modifications to the ammunition service were authorized to permit the use of shot trucks instead of the cranes. Bunkers were also authorized for the ammunition. Finish work and construction of a large bombproof concrete command post and rangefinder building and a cut-and-cover operations tunnel containing a generator room, radar operating room, combination plotting room and fire control switchboard room, and other service facilities were not completed until September 19, 1942. Moving armament to the new barbette emplacements did not begin until October. The two 8-inch railway guns left at Fort Kamehameha in February were brought out to Kahuku, insuring that two guns were in service while the guns were transferred from the railway spurs to the permanent emplacements.

In February 1943 the battery was in service, although it still lacked some of its elements. Only four of the eight projectile magazines had been constructed; only six of the eight powder magazines had been built, and none of the splinter-proof trenches had been commenced. The battery personnel had, however, excavated cut-and-cover trenches and pits for local defense and it was decided that no further work of that type was required. In March of 1943, Battery D had determined that a low coral-concrete wall around each of the emplacements was desirable for both protection and camouflage. The enclosing walls were approved and the 47th Engineers was directed to furnish technical support and materials; the battery personnel furnished the labor. An SCR-296A fire control radar was also located here, set up so that it could also provide fire direction to the other batteries of the North Sector. On the night of March 11, 1943, the battery fired a full target practice of 24 rounds at a high-speed target towed by a destroyer. (20)

During 1944, the guns of Battery Kahuku were being fired regularly. By late April 1944, the battery's four guns had been equipped with locally improvised protective shields. On May 2, in a record service target practice, some 20 rounds were expended and while the results were satisfactory, it was noted that an improved system for blowing the smoke out of the shielded compartment was needed. On June 2, 1944, another ten rounds of target practice ammunition were fired to calibrate the guns and to test a new type of blower that had been installed in the "turrets." While posted at Kahuku Mill, Battery A, fired a special day practice using radar ranges in which 20 rounds were expended. The service of the 8-inch guns was not the only responsibility of Battery A. On October 19, the unit went to Kawailoa where it fired a special day service practice expending 18 rounds from two 155 mm guns. (21)

On December 26, 1944, Battery Kahuku was placed in reserve status. In March 1945 approval was granted by the War Department for the removal of the guns when the commanding general of the Pacific Ocean Area considered them no longer required. However, the 8-inch ordnance was held in reserve until the end of the war, when it was dismounted and scrapped. In the last year of the war the designation of the units assigned to Battery Kahuku changed frequently, finally becoming the 606th CA (HD) Battery (Separate) that continued to man Battery Kahuku through the end of the war. In September 1945, the 606th departed Kahuku for Fort Hase and the installations at Kahuku were left in the hands of caretakers. (22)

The two turret batteries of 8-inch naval guns and the 8-inch railway guns of Battery Haleiwa and Battery Kahuku were formed into a separate tactical gun group designated the Saratoga Group within the North Shore Groupment. They were at this time manned by the 809th and 810th CA Batteries (attached to the 57th CA for administration and supply). The 809th and 810th Batteries usually consisted of about 4 officers and 153 enlisted personnel each. Both batteries continued to man these important emplacements until May 1943, when in a reorganization of the coast artillery on Oahu, the 809th and 810th were redesignated as Batteries D and E, of a reorganized and reactivated 41st CA (HD) Regiment respectively. (23)

The uncertain possibility of a Japanese invasion on Oahu's North Sector prompted the need for a protected command post. It was needed for control of the large number of coast artillery and other U.S. Army units conducting the local defense. The need for such a command post had been proposed even before war came to Oahu. A Forward Echelon Command Post and Message Center for the local defending forces on the north shore was established in mid-1942 at Wahiawa. Soon after construction commenced, however, the function of the structure was changed to that of Headquarters of the North Sector's rear echelon and the building design was altered. Complete bombproofing was no longer considered necessary, and a third story with windows was added to the structure. It finally emerged as a tower backfilled on three sides with earth. It was heavily camouflaged with wingwalls supporting camouflage nets. Gas proofing was provided. It was powered by commercial electrical, but an emergency generator was installed. (24)

The fourth, and final shipment of 25 155 mm GPF guns was sent to the department in the summer of 1943. They were scheduled to be replacements for the 4-inch and 5-inch ex-navy guns that had been loaned to the Army by the Navy. Also, in 1943 some of the new Coast Artillery units arriving in Oahu were armed with the new replacement for the venerable GPF 155 mm—the M1 "Long Tom". This gun was not able to be placed on the older Panama mounts, indeed it had its own, more portable version of fixed mount. In a few cases these guns served in the coast artillery role when those units were posted to former GPF sites for training or temporary duty. Then in October 1943 the Department received 12 of these 155 mm M1 guns directly as intended replacements for deployed M1917/18 guns.

Even before the war ended, it was evident that the need for preventing an enemy surface attack on Oahu was rapidly diminishing. The process of shutting down the defenses began in earnest in September 1944, when authority was received to take the 155 mm GPF batteries out of service. The first two weeks of September saw frequent target practices at Batteries Ranch, Ashley, Kawailoa, and Dillingham as excess 155 mm ammunition was consumed. On September 12, 1944, Batteries Dillingham and Ashley were taken out of service permanently. Two days later Battery Kawailoa and Ranch followed. The next afternoon the North Group Command Post closed and Battery Kahuku, the sole remaining firing battery of the North Group came under the direct command of the North Shore Sector at Station "T".

The 56th CA Battalion continued to operate the three 8-inch batteries of the Saratoga Group until the gun group was inactivated on October 4, 1944. The manning batteries of Batteries Brodie, Opaeula, and Haleiwa then came under the direct tactical control of the 143rd CA Group. Then on December 26, 1944, these batteries were taken out of service permanently. The battalion was transferred to the Honolulu area in the early part of 1945, to help operate the Quartermaster Depot operations at Fort Armstrong and Sand Island. On December 26, 1944, Battery E, 54th Battalion, took Battery Kahuku out of service. The unit was then transferred to Pearl City near Pearl Harbor, leaving a small detachment at the Kahuku Golf Course to care for the guns of the battery. (25)

The reorganization of the harbor defense units during the summer of 1944 completely abolished the coast artillery's regimental system on Oahu that had been in effect since 1924. The regiments were replaced by groups of separate coast artillery battalions. During the transition period in June and July 1944, there was left in place of the old structure a bewildering array of separate batteries and a few still active battery remnants of the partially disbanded or inactivated regiments. The command structure at Fort Hase reflected this gaggle of organizations in a profusion of headquarters units. In August 1944, the coast artillery organizations on Oahu underwent another phase of reorganization and downsizing, and the last of the 15th and 16th CA Regiments were inactivated and disbanded. (26)

During the war the role of Fort Hase expanded far beyond its original function as the headquarters post of the Kaneohe Bay Harbor Defenses. As more troops arrived on Oahu in 1942 and 1943, additional training bases were established all over the island. Oahu's eastern coast was no exception and Fort Hase became the headquarters post for Windward Oahu's training centers: the Waimanalo Amphibious Training Center, the Unit Jungle Training Center (later redesignated as the Unit Combat Training Center and still later the Pacific Combat Training Center) in the Kahana Valley, as well as large training camps at Kahuku, Waiahole, and Heeia. During the war virtually all of the Mokapu Peninsula north of the Nuupia Ponds as well as Papaa Ridge (the area south of the ponds known today as the Oneawa Hills) was leased by the Army and Navy.

At the time of the 1941 attack no less than 21 antiaircraft gun batteries were manned by the 64th, 97th, 98th, and 251st CA AA Regiments, and four more AA batteries were manned by the AA detachments of the 16th and 55th CA regiments. In addition, there were seven mobile AA machine gun batteries and five mobile AA searchlight batteries equipped with antiaircraft machineguns. The old M1918 3-inch mobile guns were still in storage at the Department Ordnance Depot. To this total of AA weapons, the 36 3-inch AA guns of the three Marine Corps AA batteries assigned to marine defense battalions training near Pearl Harbor could also be included.

While these not inconsiderable assets were assigned to the principal military and naval installations, numerous minor outposts such as fire control stations, radar sites, etc. were left completely unprotected, except for camouflage. The SCR-270 radar site at Kaaawa on Oahu's windward coast had only a single 0.45 cal. pistol, and not until after the commencement of hostilities did the site receive any additional small arms. The various emplacements for railway and tractor-drawn artillery around the island depended on their organic 0.30 cal. AAMGs for defense against aerial attack. Still, these assets were considered insufficient for the island's air defense. Hawaii was not alone regarding inadequate antiaircraft defenses. Most of the continental United States was in a similar position and both the Panama Canal Zone and the Philippines suffered similar shortages of both men and materiel. (27)

Plans were being implemented in 1941 to remedy the shortages of men and upgrade the weapons. Production of the new 90 mm M1 AA gun was already underway. Spurred by modern aircraft flying higher and faster, and the need for a larger explosive charge, most nations had guns of about this size. The British developed their 3.7-inch gun, the Russians an 85 mm gun, and the Germans had their famous 88. The Chief of Coast Artillery, the branch responsible for antiaircraft service, advocated the adoption of a gun of 90 mm bore on November 27, 1937. They needed a gun of greater projectile size and shorter time of flight than the previous 3-inch standard. A pilot model was made and evaluated during 1938. It was adopted for production in March 1940 as the 90 mm antiaircraft gun M1. The tube itself was a 50-caliber, monobloc, steel tube with a vertical-sliding drop-block breech. The initial M1 gun used manual loading of a fixed round. On its wheeled mount it could elevate to +80° and fired a new 23.4-lb. projectile. The mount was provided with four outriggers and a single axle. The firing platform could be folded-up and one outrigger leg used as a towing bar for transport.

After a prolonged development period in the late 1930s, the gun was finally put into production in 1940. While available in small quantities in 1940-41, the War Department was reluctant to place the gun in overseas garrisons where they could be isolated in war, or at least difficult to supply for scarce ammunition, parts, and trained crews. Significantly the last regiment sent to the Philippines before the outbreak of war (the Federalized 200th CA) was ordered to turn in its 90 mm gun and take older 3-inch M3s before departure stateside in late 1941. It did turn out to be an excellent gun, particularly when married to the latest generation of fire control and fuses. It was large, but still mobile mount. It could be set up quickly for field deployment, but in static defenses like Oahu it often was placed in prepared positions of a familiar type—an embanked pit with retaining walls, ready ammunition racks, an opening or gate to allow towing it in or out of the emplacement, and separate but nearby magazines, directors, and battery commander bunkers. Eventually many 3-inch AA positions were altered to take the (larger) 90 mm.

The first 24 M1 90 mm guns allotted for Oahu were not scheduled for arrival until 1942. The newly developed 37 mm automatic weapon on its M3 AA mount, capable of 120 rounds per minute, was also in short supply. The defense of Hawaii called for 140 of these weapons to replace the 0.30 and 0.50 cal. AAMGs. Although 20 of these guns were finally delivered in the latter part of 1941, no ammunition was supplied. Just two days before the Japanese attack, 9,600 rounds were received, a mere 500 rounds per gun, enough to enable the guns to fire for a scant five minutes each. Even the 0.50 cal. AAMGs were in short supply on Oahu in 1941. General Burgin noted that only 40% of the 0.50 cal. AAMGs required had been provided, and the defenders still had to use some old 0.30 cal. guns on tripods as substitutes. (28)

To establish an effective AA barrier around the various tactical positions on the island, the AA machinegun batteries were stationed at or near the defense installations. However, it was necessary for positions to be somewhat distant from the installations they were to protect. This required the mobile AA guns to be sited, in many cases, out in sugar cane and pineapple fields. These positions had been selected many months before. While it became standard for the AA regiments to deploy into the field on a regular basis in the last half of 1941, they did so without ammunition and frequently had to remain on roads adjacent to their intended positions in the agricultural fields. Issuance of ammunition to both the field and antiaircraft artillery in peacetime Oahu was made in just one instance during the summer of 1941 when the Hawaiian Department was on full alert. When the alert was concluded and the munitions returned to the magazines of the Ordnance Department, it was found to be dirty and in need of inspection before it could be restocked. Thereafter, requests by both the coast and field artillery for peacetime issuance of ammunition to units in the field were disapproved by the Department's command. All subsequent drills were carried out without ammunition as December of 1941 approached. (29)

Army Ordnance Technical Manual sketch of the M1 90 mm AA gun. *Army Technical Manual*

After the start of the war Oahu began to receive the 90 mm M1 AA gun with new units, though for several years it served alongside the 3-inch rather than being strictly a replacement. *Gaines Collection.*

Following the outbreak of the war the army antiaircraft defense of Oahu was immediately reinforced. Filler troops already on the island in December 1941 were used to bring the organized coast artillery batteries up to full wartime strength. New troops were organized on May 18, 1942, into the 710th, 711th, 712th, 713th, 714th and 715th CA (AA) Batteries (Separate). The new batteries were assigned to either mobile 3-inch AA guns shipped to Oahu early in the war or manned new batteries of the emergency emplaced 5-inch naval AA guns. Early in 1942, entire AA regiments and their equipment arrived from the continental U.S. as reinforcements for Oahu's air defenses. The 95th CA (AA) Regiment with mobile 3-inch AA guns arrived in Honolulu on January 7, 1942, having been rushed to the islands soon after the declaration of war. It was quickly moved to Kaneohe Bay, where it assumed the AA defense of Bellows, Kualoa, and Mokuleia Fields on the North Shore. In March the 96th CA (AA) Regiment arrived and was briefly assigned to the south of Oahu to bolster the air defenses of Pearl Harbor, Schofield Barracks, and Honolulu, replacing the 251st CA that was transferred to the Fiji Islands in May 1942. The 96th was the first regiment to arrive in Oahu with the new 90 mm AA guns. It was however, soon redeployed to positions on the islands of Hawaii, Kauai, and Maui.

The 93rd CA (AA) Regiment of two battalions, with six batteries of mobile 90 mm M1 AA guns and four batteries of 37 mm automatic weapons arrived on May 21, 1942. A third battalion of automatic (37 mm) guns for the 93rd was organized from local resources on June 15. The 369th CA (AA) (Colored) was a former New York National Guard two-battalion regiment armed with 3-inch AA guns, 37 mm automatic weapons, and 0.50 cal. AAMGs. The unit was federalized on January 13, 1941, and arrived on June 21, 1942, and was assigned to the air defenses of the North Shore Groupment. This included the Opana radar station, the Kahuku Army Air Base, and the Haleiwa and Mokuleia Air Fields. (30)

As mentioned above, the 96th CA was the first unit to arrive with the new 90 mm AA gun in early 1942. New guns were also provided to rearm the 64th CA. During 1942 five of its firing batteries were reequipped with 90 mm guns and by the latter part of that year, the 37 mm guns were being replaced by more effective 40 mm automatic weapons. These units were placed under the command and control of the 53rd CA (AA) Brigade of the Hawaiian Antiaircraft Command (HAAC). (31)

Before the war, the new 90 mm AA gun was also scheduled to replace the 3-inch M1917 fixed AA guns at Fort Kamehameha and Fort Weaver that guarded the approaches to Pearl Harbor. After the outbreak of war, this plan was accelerated, and by mid-1942 at least eight of the new guns had been delivered to Oahu. On June 14, 1943, it was reported that 55% of the AA materiel was obsolete and many of the coast artillery AA batteries were still under-strength. It was recommended that the emergency 5-inch naval AA batteries be replaced by 90 mm batteries with remote control and M9 directors. It was requested and that the 18 batteries of fixed 3-inch AA guns at the CA posts along Oahu's south shore be replaced by eight batteries of 120 mm AA guns with M10 directors. All 258 37 mm guns were to be replaced by 40 mm automatic weapons. The War Department approved these requests. (32)

During the first two years of the war, the AA regiments continued a major organizational role in the air defense of the Hawaiian Islands. Most of the 64th CA was deployed in the defense of Honolulu, covering Honolulu Harbor, and Forts DeRussy, Ruger, and Shafter. In the latter weeks of 1943, a complete reorganization of the AA defenses of Oahu was implemented. The venerable 64th CA (AA) Regiment was inactivated and was disbanded later in the year. The redesignated units of the 64th remained on Oahu for a few more months as they completed jungle warfare training and prepared for deployment to more active Pacific campaigns. The six other antiaircraft regiments on Oahu in December 1943 were similarly reorganized. Many remained on Oahu for a time, but most would eventually be forwarded to the central and western Pacific as the war progressed toward Japan.

The largest AA gun to serve the U.S. in the war was the 120 mm. Development of this weapon started in 1938. It had a long 60-caliber tube and fired a 50-pound projective to a maximum altitude of 19,150-yards—though effective range depended more on the abilities of fire control than the ballistics of the gun. It was moved by trailer, carried on two dual-wheeled bogies. On account of its size, it was not found easy to transport,

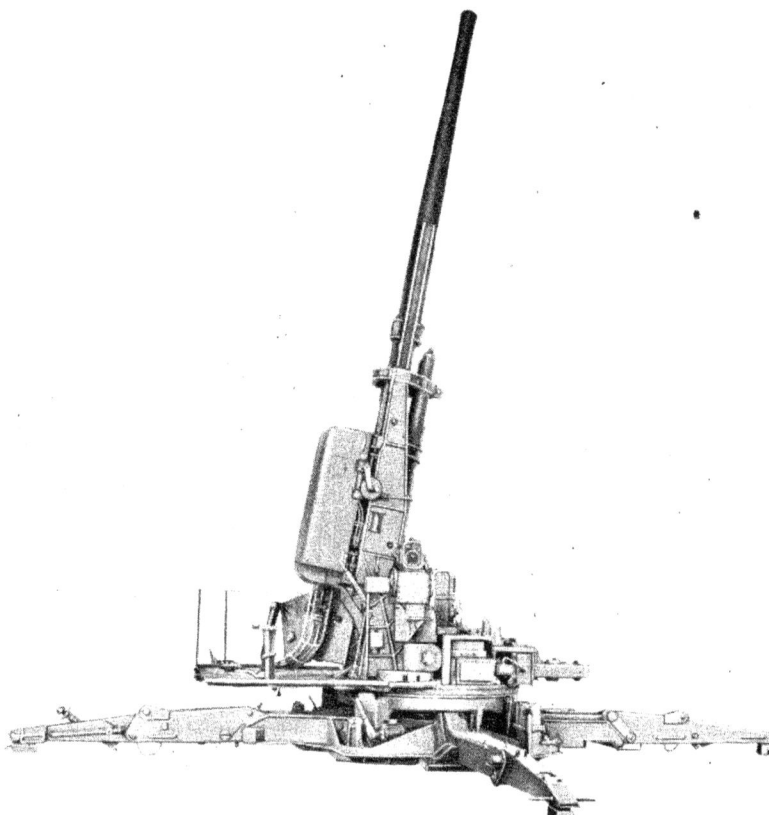

The ultimate American heavy AA gun was the 120 mm M1 type, as shown in this Ordnance Department sketch. *Army Technical Manual.*

Only a few locations outside the continental U.S. received the 120 mm armament, but that did include several batteries on Oahu. Several locations replaced their 90 mm with 120s, and new locations, like this illustration of the battery at Kaena Point were also so armed. *Scrapbook.*

so the gun often served during and after the war in fixed emplacements. Full electric traverse and elevation of -5 to +80° was possible through automatic operation from the remote director M10.

The gun was standardized in October 1942 as the 120 mm M1. Ultimately just 500 guns and carriages were made by February 1, 1944. The 120 mm AA gun was deployed mainly to rear-area strategic defenses in Hawaii, Panama, and Northern Ireland. Building of the new field emplacements and conversion of the permanent AA batteries using the new 120 mm guns began with their arrival in Hawaii early 1944. Most of the new guns were reserved for the permanent forts such as Fort Kamehameha and Fort Weaver. Retaining berms were made of reinforced concrete or cinderblock with a couple feet of fill between courses. The emplacements had 12-foot moveable gates to allow relocation. A few batteries were located at strategic locations such as Camp Malakole and on Kaena Point. At the latter location the guns had experimental sights that allowed firing at surface targets. (33)

By 1945 Oahu bristled with AA guns. No less than seven 90/120 mm, six automatic weapon, and three antiaircraft searchlight battalions were distributed around the island. The AAA gun battalions were by then organized into the 28th and 139th AAA Groups of the 48th AAA Brigade, while the automatic weapons battalions of the 36th and 123rd AAA Groups and the searchlight battalions of the 98th AAA Group formed the 70th AAA Brigade. By August 1945 there were still six AAA gun battalions on Oahu and elements of two others, all armed with 90 mm and 120 mm guns. In addition, there were eight AW battalions and two searchlight battalions. After the end of the war nearly all the AAA units on Oahu were inactivated or transferred to new stations during 1946, leaving only a minimal AAA presence on the island.

Late in the Second World War the seacoast defenses were organized into a total of three full Harbor Defenses and the slightly junior Northern Groupment. Each of these organizations had its own protected command post. By the Second World War these were known as the Harbor Defense Command Post (HDCP) and featured joint Army-Navy commands for the defenses. Additionally, the "ports" of Pearl Harbor, Honolulu, and Kaneohe Bay had Harbor Entrance Control Posts (HECP) to observe and control entry of vessels into the port.

At Fort Kamehameha the first HDCP was established in the magazine spaces of old mortar Battery Hasbrouck. This occurred in 1938 when the battery spaces were gas-proofed. The battery itself was still in reduced maintenance status and the guns mounted, though not actively manned. Battery Hasbrouck continued in use as the HDCP for the Pearl Harbor defenses from December 1941, until June 1942, when a new HDCP was completed at the Salt Lake Reservation. In 1941 a second steel frame tower surmounted by a two-story splinter-proof concrete house was built directly adjacent to the "C" Station at the western group of stations at Fort Kamehameha. It served as the Harbor Entrance Control Post (HECP). This command post was manned by the Headquarters Company at Fort Kamehameha, during the war.

A new HDCP for Pearl Harbor was already in the planning stages by early in World War II. However, it was not until after April 1942 that the plans and site had been decided. Construction of the new command post by a battalion of the 34th Engineer Regiment was expedited and eventually completed in a tunnel complex some 50-feet below the crest on the south rim of Aliamanu Crater. The new HDCP was much larger than its predecessor. The operations tunnel was reached by way of an eight-foot wide, 330-foot long access tunnel. This command post was a separate structure from the Joint Army-Navy Command Post built in the old ammunition tunnels. It had its own power generator room and was connected by a shaft to the harbor entrance observation post (HEOP) on top of the crater rim. The command post contained rooms for the officers assigned to the post, a switchboard room, a message center, small gally and mess area, as well as the operations center. Upon completion the HDCP was operated by the recently created Headquarters and Headquarters Battery, HD of Pearl Harbor. This headquarters battery also continued to man the HECP tower at Fort Kamehameha. (34)

With the formation of the headquarters for the Harbor Defenses of Honolulu at Fort Ruger in January 1927, Diamond Head was an obvious location for the unit's command post. As described previously, in 1934 an expansion of the tunnel through Diamond Head became the new HDCP for the Pearl Harbor Defenses. The Kapahulu Tunnel through the north rim of Diamond Head remained the sole entry to the crater floor until the latter part of 1942, when a wider vehicular tunnel was bored through the crater rim farther east. (35)

The Honolulu Harbor Entrance Control Post was located separately, at the Sand Island Military Reservation. Soon after the U.S. entered World War II, a derrick type steel frame tower 50-feet tall and surmounted by a square two-story house of one-inch steel plate construction was erected on the island. This tower was to be used as a combination HECP Observation Post, Signal Station, and battery commander's station for ex-navy 7-inch gun Battery Harbor. The signal blinker searchlights, flag boxes, and a signal mast with its flag hoist halyards, and yardarm blinker lights were installed on the roof of the tower. The upper of the two enclosed levels served as the observation post for the HECP, while the lower of the two levels was occupied by the Battery Commander's Station. The HECP coordinated the guns of the Examination Battery of Honolulu Harbor.

When the 57th CA arrived at the Mokapu Peninsula in December 1941, it immediately assumed the command functions of the Defenses in Kaneohe Bay. It established its first command post in one of the few private residences on the peninsula, a frame stucco construction dwelling located east of Camp Ulupau. Preliminary plans were cast for construction of a harbor defense command post (HDCP) on the north side of Puu Hawaiiloa. The Harbor Entrance Control Post (HECP) was also to be located on the hill's summit near the navy's Kansas Tower. This plan, however, was never executed. Eventually in the spring of 1942 a protected message center and radio facility for the Kaneohe Groupment was built as a tunnel into a spur of Ulupau's west rim. Subsequently new rooms were added to this tunnel, and it was gas-proofed, emerging as the defenses HDCP. A smaller branch tunnel contained the radar operating room with an antenna for an SCR-582A surface search radar set located atop the crater rim above the tunnels.

While not really possessing a harbor, the growing number of batteries, stations, and units posted to the Oahu North Coast certainly prompted the need for an overall command and a post for its operation. The need for such a command post had been identified and proposed even before war came to Oahu. By mid-spring 1942 a site for a Forward Echelon Command Post and Message Center for the 24th Division had been selected in a secluded location north of Wahiawa, in the Poamoho Gulch. Excavation for the cut-and-cover bombproof structure was finally begun in June 1942. Soon after construction commenced, however, the function of the structure was changed to that of Headquarters of the North Sector's rear echelon and the building design was altered. Complete bombproofing was no longer considered necessary, and a third story with windows was added to the structure.

By the time the project had been completed in January 1943, its design had undergone considerable modification. The enlarged structure with some 9,495 square feet of space was backfilled with earth on three sides with wing walls at each end of the building's front to retain the earthen slopes. A splinter-proof retaining wall was also built adjacent to the command post entrance, and camouflage nets were draped across the front side of the building for concealment purposes. The building was also gas proofed. The command post was supplied with commercial electricity. A separate powerhouse was built to house the standby generators that provided an adequate supply of emergency electric power. (36)

## CHAPTER 14
## SPOTTING THE ENEMY

The increasing range of guns generated the need to locate targets further out at sea and with increasing tracking precision. Predicting the distance to the target generally depended on the use of a known observation—the angle to the target sighting from a defined, "fixed" observation point. With one side and two angles known, the remaining angle and the length of the other two sides of a triangle could be mathematically calculated. The operational choices were between vertical and horizontal triangles. A vertical triangle relied on an elevated observing instrument, or depression position finder (DPF). The angle at which the instrument was depressed to see the target was measured, and from this the range to the target calculated. A horizontal system, on the other hand, used two instruments in separate locations with a carefully measured base line between them, both observing the same target. Each instrument measured the angle between the base line and the target, and this information established the location of the target. (1)

There were advantages and drawbacks with each system. The vertical (DPF) system just required a single station with two soldiers to observe and transmit the information. The instrument could even be located above the plotting room, eliminating the need for lengthy cables or potentially faulty telephones. The disadvantages, however, were considerable. The instrument had to be elevated. DPFs lower than 125 feet above sea level were inaccurate beyond 10,000 yards, which was still a reasonable engagement distance with the early weapons. This height requirement posed little problem where cliffs and mountains abounded on a volcanic island like Oahu, but for the low-lying posts like Fort Kamehameha some other solution would have to be found. DPF observers had to carefully set their telescope crosshairs on the waterline of the target. The further the target, the higher the instrument height should be. The later big guns with a range of 45,000-yards required an instrument about 400-feet above the water level. Mechanical preciseness was required to measure the minute angles accurately, and DPFs were four times as expensive to install as azimuth instruments. A reliable facility to measure the tide level was also required.

The horizontal-baseline system differed in several aspects from the vertical-base system. Unlike the DPF system, horizontal-base stations could be miles apart, and therefore much more accurate, especially at longer

A typical Coast Artillery Corps Depression Position Finding (DPF) instrument observing station, though taken in Portland Maine, this is the same arrangement that would have been used on Oahu. *Smith Collection.*

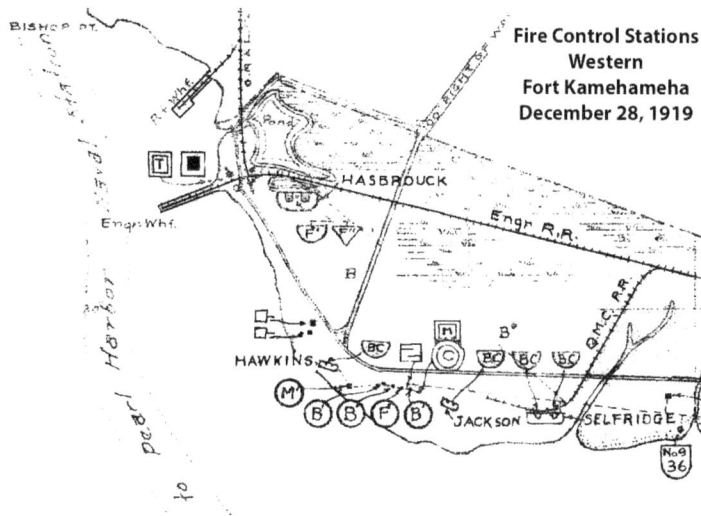

**Fire Control Stations
Western
Fort Kamehameha
December 28, 1919**

Map showing the location of the Fort Kamehameha primary stations west of Battery Jackson as assigned in 1919. *NARA.*

ranges. In addition, the instruments were much simpler, needing only to measure horizontal angles, without allowing for tide. However, horizontal-base systems required reliable long distance communications. Even if the primary base end station was above the plotting room, efficient communications were essential between the second measuring station (conveniently known as the battery's "secondary") and the plotting room. Cable communications were safer and more reliable than radio, but still not always perfect. Secondly, more men and stations were needed. Two stations were needed for every observation, doubling the manpower required. Finally, both stations had to track the same target. Fog (though not common in Hawaii), smoke, haze, rain squalls or any other factor that prevented both stations from seeing the same target would thwart the system. Further, both stations had to read the azimuth to the target at precisely the same instant. Special timing systems were necessary to assure this, adding complexities and costs of the system.

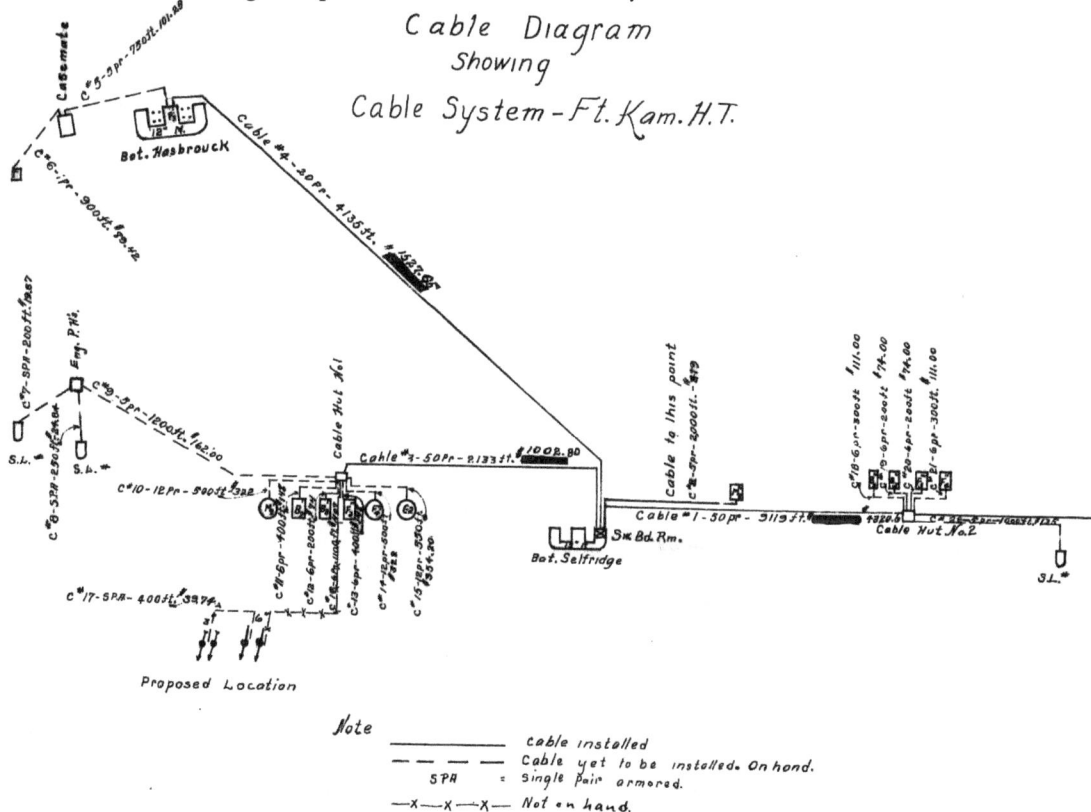

Map for the buried cable line connecting the stations at Fort Kamehameha and Ahua Point in 1913. *NARA.*

The Coast Artillery Corps used both systems, depending on the issues at each site. The Army developed and adopted a series of depression position finding and separate azimuth instruments. Even then the preference was for laying out at least an auxiliary horizontal baseline system for the most critical gun batteries. Early in the 20th century the chief of artillery concluded that horizontal base lines were necessary for all batteries and fire commands, but they should be equipped with DPFs for target identification and as auxiliary vertical fire control stations. However, with the long-range capabilities of the 1920s, the horizontal system was only marginally effective at the maximum range. The final answer was radar. By the end of World War II, it was effective far beyond both observer vision and gun range of even the most powerful batteries.

The other half of this range-finding system was plotting. Plotting turned the raw observation data collected by the observing stations and the tide or meteorological stations into accurate azimuth and range input for each gun using a plotting board with a number of other data correction instrument all operated by a plotting detail. The most important piece of equipment used in seacoast artillery range and position finding was this plotting board, with its mechanical solution to the trigonometric problem of locating a distant point—the target. Specially developed boards—a large usually semicircular board with movable mechanical arms, acted as a sort of computer to provide azimuth and range solutions by geometry. Over time three types of plotting boards were developed for use with the horizontal base system: The Whistler-Hearn plotting board, the Cloke plotting and relocating board, and the 110° plotting board. For each of these information would come in from observation stations on phone. The arm setters would move/point the various gun arms on the board and deduce preliminary data of range and azimuth for a particular gun battery. That was then modified with wind and tide information and transmitted to the gun. Depending on the gun size, 12 or more men could be involved in this detail in the room.

Two types of structures were needed. Observing stations (sometimes collectively called base-end stations, though not technically true if the station was only equipped with a DPF instrument) were usually just concrete box-like stations. In the Hawaiian defenses this was generally a 10 by 10-feet structure, sunk into the ground to reduce visibility from offshore. Entry was by either a side or rear door if partly above ground, or by a manhole and ladder if buried. Usually, they were permanent concrete structures, although mobile batteries like 155 mm guns sometimes just had a wooden or sheet metal hut and earth berm serving as a station. Each had

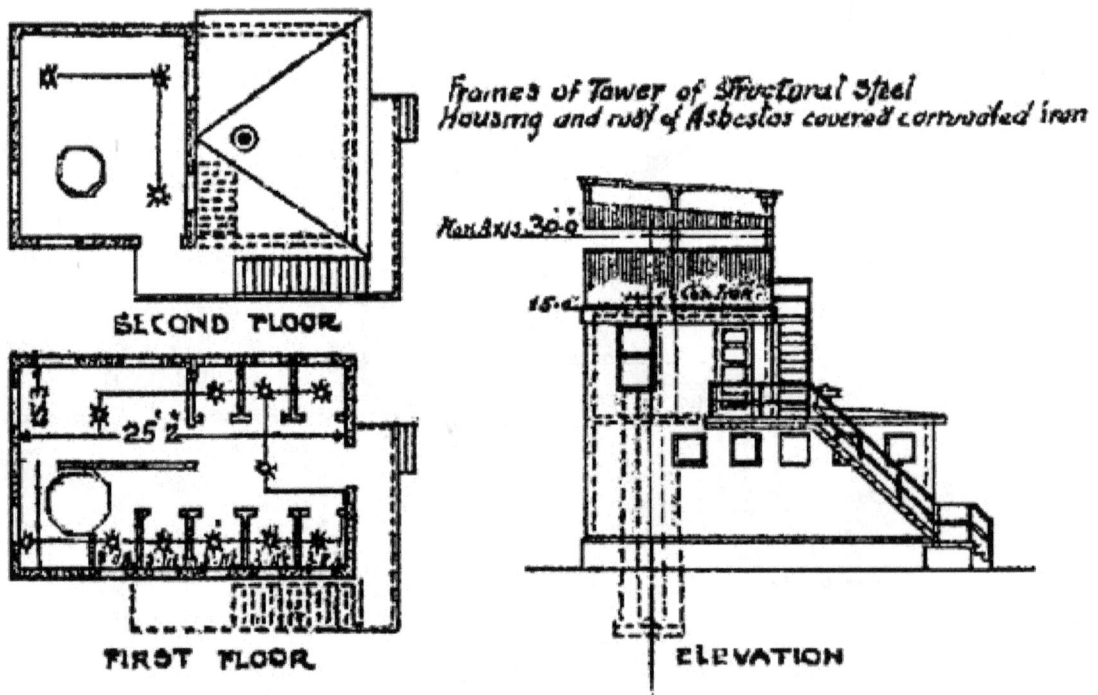

Plan for the "C" or command station of the Fort Kamehameha western tower sequence. *NARA.*

**Fire Control Stations
Ahua Point
December 28, 1919**

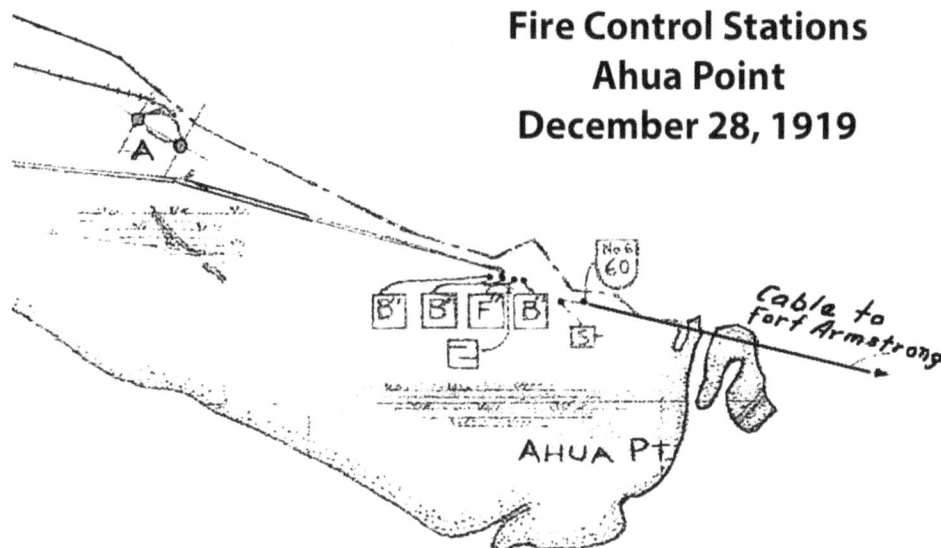

1919 Corps of Engineers map showing the various secondary station towers of the east group at Ahua Point. *NARA*.

an observation slit, usually protected with a steel shutter than could close, that had a good perspective on the required viewing field. Inside were stands to mount either DPF or Azimuth instruments and communication equipment depending on the station's requirements.

While stations were assigned to a specific battery, the information could be used by alternate battery assignments if necessary and preprogrammed into the plotting function. In some cases along low beaches, it was necessary to put stations atop towers. Generally, in Oahu, the towers were steel lattice structures with ladders or steps to the enclosed station and its instrument. In peacetime the instruments weren't kept in the station but stored in warehouses for security and protection from damage. The stations were equipped with telephones which were connected to the assigned battery plotting room. Besides lighting and an equipment shelf or locker there wasn't much else installed except a small electric generator provided either for primary or for back-up if commercial electrical power failed. Any crew comfort—latrines, beds, cooking stove, etc. were either in an adjacent tent encampment or a wooden dormitory. At least in Hawaii providing stoves for heating weren't necessary.

1922 photograph showing the edge of Battery Closson looking east towards the line of tower secondary stations at Ahua Point. *NARA*.

Line of battery secondary stations at Ahua Point sub-reservation of Fort Kamehameha. *NARA.*

Structures for seacoast fire control in Oahu generally date from one of three periods: the initial Taft fortifications of 1910-1916, the long-range interwar fortifications of 1922-1940, and the second world war fortifications of 1941-1945. Some of these structures were at or near the gun emplacement—the Battery Commander's Station, the plotting room, and, particularly with the earlier batteries, the primary observation station. As the range of guns increased the need to place stations at distant locations, sometimes on the opposite side of the island became necessary. Over the 40 years of the use of the Hawaiian harbor defenses, well over 100 base end or command observation stations were constructed. A brief description of numbers and locations for the three periods and a 1943 summary list of all locations provides some useful description of this important function.

The reservation at Queen Emma Point established in 1907 with its major fixed gun batteries was the fort with the heaviest allotment of armament, at least for the Taft Generation. The reservation was just above sea level and not well disposed for the use of vertical range finders, it was also not large enough to contain an adequate baseline for the longer range guns. That led to the purchase of the adjacent Ahua Point property for placement of secondary stations to make a practical horizontal baseline system.

The gun and mortar batteries at Fort Kamehameha employed baselines surveyed between a group of five primary stations located between Batteries Jackson and Hawkins on the far west side near the Pearl Harbor Channel, and a secondary group of stations at Ahua Point some 3,500-feet east of the primary group. The battle commander's station (later a Fort Commander or Groupment commander station) was located with the primary group. There were technically four equal and parallel base lines each with a separate set of paired primary and secondary stations for 12-inch Battery Selfridge, 6-inch Battery Jackson, and 12-inch mortar Battery Hasbrouck. The fourth line could be used as a spare, for a fire command, or even to separate the fire of the two pits of mortars at Battery Hasbrouck so that each pit could fire separately at different targets.

The four primary fire control stations were built in 1913-1914 in a growth of algaroba trees that partially concealed them. As described previously, they were arranged in a row, equidistant apart from each other. One tower had a double top, but the other three stations were identical to each other in size and construction consisting of structural steel towers 22-foot tall that were provided with a single observing level. Each station was equipped with a pedestal for a single observing instrument. These stations remained in service through most of World War II, although their functions changed in some cases as new batteries were assigned and others were deleted from the harbor defense plans. (2)

The Ahua Point secondary stations (B"), for Batteries Selfridge, Hasbrouck, Jackson and for Fire Commands Three and Four were the first of the installations at Ahua Point. They were built at the same time as their primary partners to the west on the fort. Each of these stations consisted of four identical single story observation rooms measuring ten feet, six inches square atop a 25-foot tall structural steel tower. The four stations were arranged in a line near the reservation's northeast boundary. Algaroba trees ranging from 15 to 20-feet in height to the southwest of these stations also partially screened them. After World War I, these stations were provided with an additional level. These four stations continued in service through August 1944, manned by range section personnel. (3)

The second major set of horizontal baselines were installed for the Defenses of Honolulu. At its western flank anchor (with the primaries located at the batteries of Randolph and Dudley, and the secondary station for Battery Harlow) was Fort DeRussy. The other (southern) end of the base lines had stations located atop Diamond Head crater of Fort Ruger for the secondary stations of the DeRussy gun batteries and the primary station for the Harlow mortars. The stations atop Leahi Peak were over 700-feet high and were high enough for the use of vertical triangulation with a DPF instrument. However, the army preferred the redundancy and better accuracy of a horizontal base for these emplacements. The construction of the stations at Leahi Peak are quite interesting and are described in more detail in Chapter 2.

There were two other, shorter baselines constructed. The Honolulu and Pearl Harbor mine defenses each had a separate baseline, and each had primary and secondary stations to provide target tracking for manually detonating the controlled mines. The two 3-inch batteries of the defenses had coincidence range finders at the BC at the batteries. When the Land Defense batteries were added to the harbor defenses, most of them did not have significant anti-ship roles. If called upon to fire at a landing in progress or a hostile army, they would function like field artillery for firing direction and control at the guns. The exceptions were at Fort Ruger. The 5-inch guns of Battery S.C. Mills could fire over waters to the east of Diamond Head. The battery was given a separate fire control station equipped with a self-contained coincidence range finder. The mortars of Battery

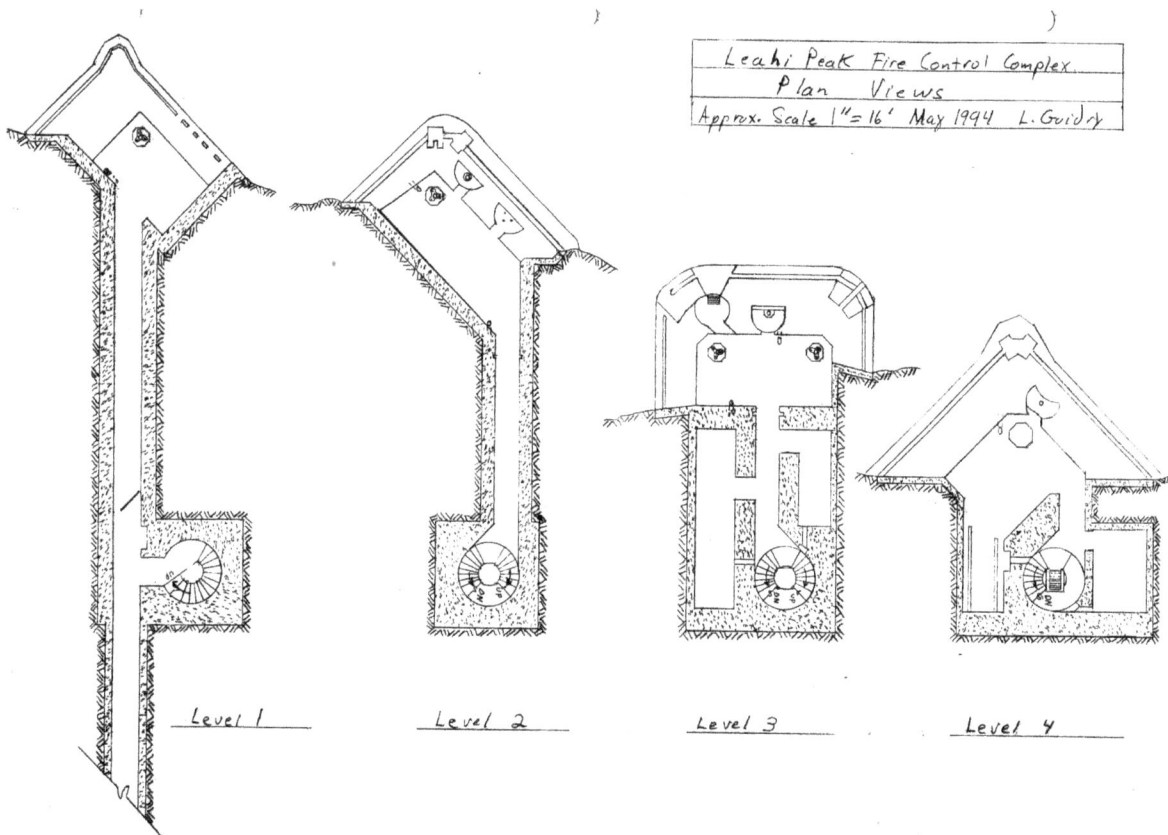

Leahi Peak station plan. *Guidry.*

Birkhimer had a dual-purpose role for land defense and seacoast defense. It was given a baseline from southern Diamond Head to Kaimuki Crater, and for land defense a station with a coincidence range finder on the northern rim of Diamond Head.

Finally, there were also a number of emergency fire control stations. These stations were meant to substitute for important locations that might be knocked out or otherwise rendered ineffective during combat. It was not like the elaborate and numerous emergency station network installed at Manila Bay, as there were only three emergency stations built. There was one each at Fort DeRussy (in the otherwise unused second crow's nest of Battery Randolph), one open station near but not at Leahi Peak, and one at Black Point.

Only three new stations were initially built for 12-inch Battery Closson. However, the very long-range 16-inch emplacements needed an entirely new systematic approach for the all-around fire from Battery Williston and then Battery Hatch. These guns could cover all but the northernmost point of Kahuku. To exploit their effectiveness virtually every angle and view of coastal Oahu needed to be placed under the scopes of the fire control details. An entire network of overlapping stations with their horizontal baselines and at heights to allow vertical rangefinding was needed, connected by protected communication cables. It took several years to complete, but this system formed the basis of the fire control network that continued to be used through the World War II years.

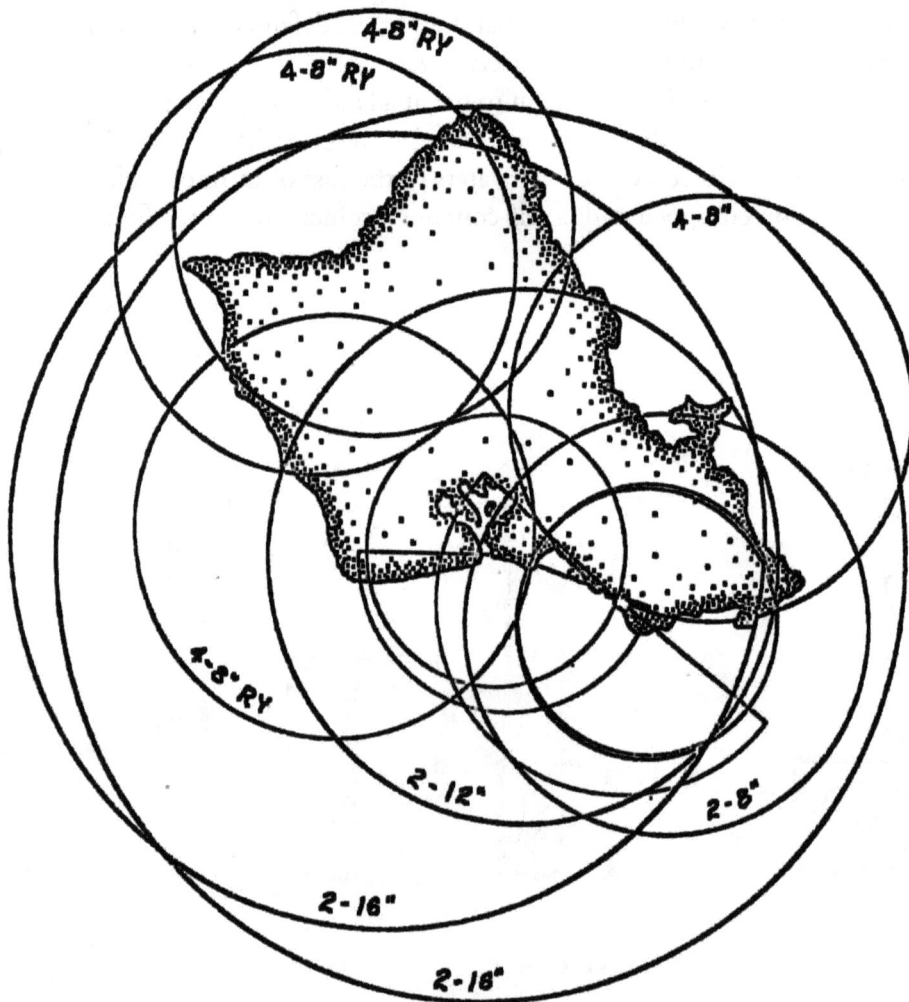

1936 Field of Fire map at the completion of the interwar long-range batteries. Note how coverage now extended completely around the island, necessitating a much expanded network of fire control stations. *NARA.*

Battery Hatch's plotting room received its target data from numerous fire control stations at various locations around the island. Initially the battery's fire control was set up to use data obtained from single stations equipped with DPF instruments. Several horizontal base lines were established by the late 1930s using pairs of stations providing simultaneous azimuth, and in some cases range data, to the battery plotting room where target location and directing data for the guns was developed. The M1907 DPF model had 15 different classes that were keyed to the height of the fire control station where it was installed. In the case of the type that served Battery Hatch, there were four variant or sub-classes for instruments set at heights from 280 to 1,140-feet above sea level. They could be used to obtain target data up to 20,000-yards out to sea.

Toward the end of the 1930s, the M1 model DPF became the newest iteration of this instrument and was slated for use in those stations that still required such instruments. The M1 DPF instruments had ten different classes that could be used at station elevations as low as 74-feet and as high as 1,395-feet. The minimum range of effectiveness was 20,000-yards and the maximum range that could be attained with an M1 DPF was 60,000-yards. The range and bearing data provided by the DPF were transmitted by Signal Corps telephone or cable to a plotting room at the gun battery. The manning details for these stations were usually comprised of two or three men per observing instrument.

For Battery Williston and Battery Hatch, in addition to the Battery Commander's stations and plotting rooms, 17 new observation stations were eventually commissioned. Some were completed in the early 1920s, but others weren't funded and constructed until right before the Second World War. Each station was referred to with an alphabetic letter, starting with site "A" in southwest Oahu and progressing counterclockwise around the island. There were some letter skips in the sequence. They were spread around the island of Oahu, virtually at every point except the northern Kahuku Point. It took a while to obtain purchase of the property. Several on the southern coast required towers, but most were on elevated peaks or ridges. As these were almost all at elevations over 400-feet, often a new road was necessary and no convenient camp available, aggravating construction time. At a few sites, cables were used to lift supplies to building sites. Communication cables had to be entrenched over many miles to reach the stations. While not sophisticated structures, all of this did take time.

One of the first fire control stations for Battery Williston was constructed in 1924. It was built on the western slope of Puu Pueo at Kaena Point, the island's western most point. Department engineers preferred stations with a height of about 400-500 feet. On the mountains of the western and eastern coasts there was a problem with visibility in low clouds if the station was placed too high. These early stations were just single room buildings, built of reinforced concrete for splinter-proof protection. The first stations of the alphabetic series were smaller when compared with stations built a decade later, measuring 8-feet wide and 13-feet deep. It was equipped with a single depression position finder (DPF) for determining both the range and azimuth of a target. Access to the station's interior was gained through a manhole in the rear of the station roof. When completed, it was designated Fire Control Station "S". (4)

It was found that some of the best sighting locations for these new stations were on very difficult sites to access. The mountains and ridges on both Oahu's ranges are very steep, an obstacle for construction and difficult for manning crew access and their regular supply. Most of the stations had at least rough roads for jeeps or trucks, but a few had to use mules or cable lines for supply. During the war years a manning detail maintained at the stations, using small wooden dormitories and basic garrison facilities. A few interesting descriptions survive of what life was like for these crews: (5)

> One old soldier who served at Fort Barrette's observation posts described the observers' daily routine. The first thing they did each morning was check in by telephone with the battery commander back at the fort. Any special orders for that day were received at that time. The rest of the day was spent scanning the post's sector and reading magazines. It could be boring duty and they were stuck up there until relieved. He said supplies were carried up with Army mules. Occasionally one of the sure-footed beasts would lose its footing and they were treated to the sight of an unfortunate mule tumbling-end-over-end down the mountainside.

When 16-inch battery of Battery Hatch at Fort Barrette was completed in 1934, it also received fire control stations. As this was during the depression, funds for their construction came from the allocations for seacoast fortifications and the economic recovery 'National Industrial Recovery Act' (NRA) appropriations. Most of these utilized pre-existing reservations used for Battery Williston with a new station adjacent or just near the older one. The property was already owned or leased by the army, an access road or trails were already built. It made practical sense to share the same manning detail (though it might have to be enhanced by a few men) and the small dormitory, mess hall, latrines, and other facilities. They were already served by or connected to the island's fire control communication cable.

Battery Hatch's initial stations were: Station "A" at Puu Manawahua; Station "B" at Puu Palailai, Station "S" on Puu Pueo at Kaena Point, Station "S' " at Kepuhi New on the Keaau Ridge and Station "U" at Puu O Hulu Kai were located at points west of the battery. Station "C" at Aliamanu Crater, Station "D" at Punchbowl Crater, Station "E" at Diamond Head Crater, the new Station "F" at Fort Weaver, Station "H" at Makapuu Point were located on Oahu's south shore east of Fort Barrette. Station "W" on Puuiki (sometimes spelled Puu Iki) on the northern slope of the Waianae Range, and Station "O" at Pupukea supplied data from targets off the north shore. On the island's northeast coast, Station "J" on Ulupau Head, Station "K" at Kahuku; Station "L" at Kaaawa, and at Station "M" on Kokololio Hill near Kaipapau also served Battery Hatch. All of these stations were equipped with M1907 DPFs. (6)

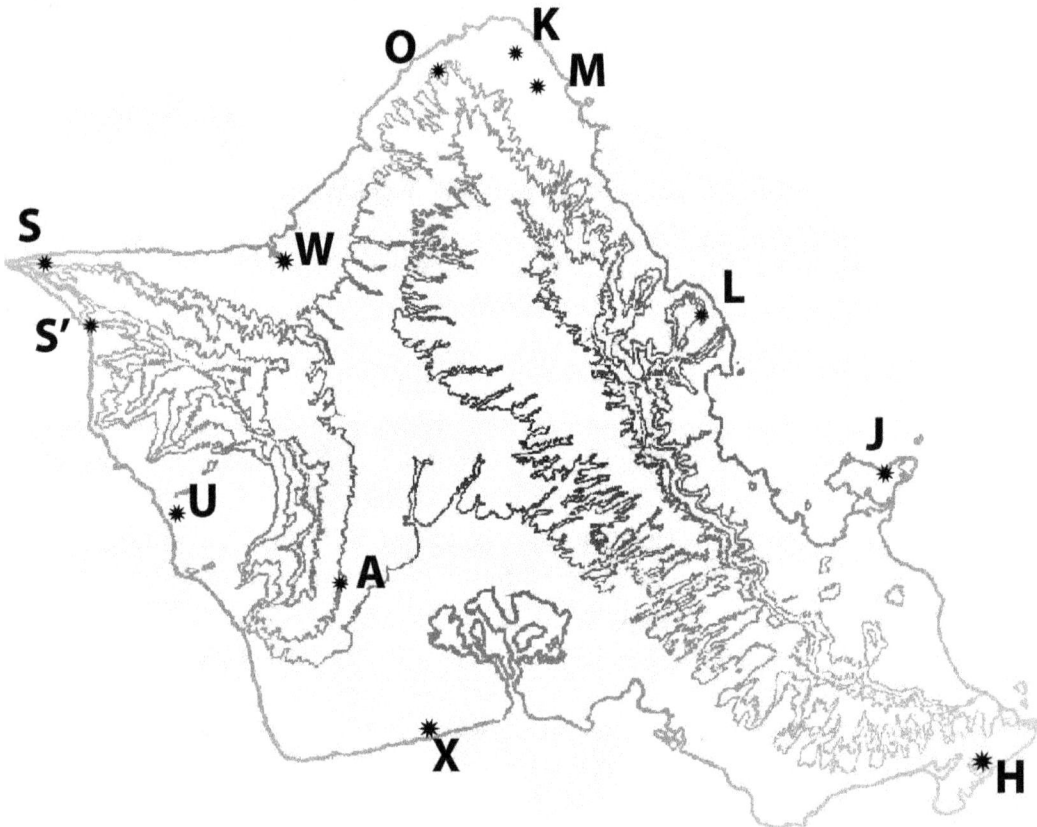

Map showing general locations for some of the new "Alphabetical" stations built to accommodate the needs of the new long-range batteries. *Williford Collection.*

Although some of the stations were completed in 1934 and transferred over to the 15th CA later that year, most of them were not equipped with their DPF instruments (for which there always seemed to be a chronic shortage) until sometime in 1936. Other stations, principally on Windward Oahu and the North Shore were not started until about 1940. The station at Kepuhi Nui had still not been equipped with its M1 type DPF in December 1940. When Battery Hatch was casemated in 1942 limiting its field of fire, fire control stations "S","S' ", "W", "O", "M", "L" and "J" were reassigned to other batteries. (7)

The only significant improvement in the design of these additions was in their size. They were larger with more room coming from greater dimensions. Most were 15 by 16-feet instead of the previous standard of 8 or 10 by 13-feet. At its full deployment, Battery Hatch had 12 stations assigned, along with the Battery Commander's Station at new Fort Barrette. Only one new reservation was acquired.

Details of many of the permanent stations for the long-range batteries on Oahu can be found in Appendix IX.


Fire control stations needed to have secure telephone connection to the command and plotting stations they supplied with observation information. The Army's Signal Corps provided the cable and communications equipment along with cable huts to service junctions and facilitate repairs along with independent fire control switchboards. The equipment and system used between the stations and forts for Pearl Harbor and Honolulu were not unusual for their times. However, when the outlying stations for the long-range guns were built the existing system was vastly expanded. With the completion of the last of the alphabetic-series stations at the end of the 1930s, the Army's telephonic communication network encircled the entire island of Oahu with an extensive cable system. The system was a military network, not reliant on the commercial Oahu telephone system. With it the military had "hard-wired" Oahu. (8)

Known as the "Command and Fire Control Cable System" it was an extensive network of armored, buried communications cable. Cable huts were located near the fire control stations for connecting the various communication phones and equipment. Usually these were given the same designation as the alphabetic title held by the paired station. Major telephone exchanges were provided with bombproof concrete structures, some with facilities for a manning detachment. In more remote areas trunk stations were generally given splinter-proof protection. The wire used was commercial-grade type, the phones and headsets standard types in production for the signal corps.

The Coast Artillery Corps weren't the only users of this communication system. The Command and Fire Control Cable System served all the military needs for secure communications. The cable connected all the army's forts and reservations and ammunition depots, headquarters centers, offices, and command posts. The airfields of the Army Air Corps, and the Navy station at Pearl Harbor were also connected. There were switchboards and even remote pedestal terminals to allow units in the field to conveniently plug into the telephone line without needing to deploy major lengths of wire.

Pre-war the heavy, permanent batteries and their observation stations were directly connected to the cable system. The wartime proliferation of temporary and emergency batteries added an immediate challenge to the communications network. Eventually all the 155 mm Panama Mount batteries, the 8-inch railway gun sidings, the temporary 4, 5, and 7-inch gun batteries were all connected to the network. As completed the new 8-inch ex-carrier turrets and the 14-inch gun batteries from USS *Arizona* and their new fire control stations were fully incorporated. The last leg to be connected was along the windward coast, connecting the new stations of the North Groupment near Kahuku to the Kaneohe Bay defenses at Fort Hase.

The assortment of mobile and temporary batteries built in 1941-1942 understandably did not receive sophisticated fire control systems. Guns in this category included the prepared 8-inch railway siding batteries, the 155 mm Panama Mount emplacements, and the ex-navy 3, 4, 5, and 7-inch emergency batteries. They did have rudimentary facilities to direct fire. Most of these emplacements had a Battery Commander's Station and an associated plotting room. These batteries were either equipped with a self-contained CRF instrument, a DPF instrument (if at sufficient height or on a tower), or an azimuth instrument and a single secondary station. Most of these stations were quickly constructed and provided only with marginal splinter-proof protection, thin concrete walls and roof with earth embankment or metal roof cover. Some of the these were just unprotected metal sheeting or even wooden walled structures.

However, a few of these evolved into more important sites with a full set of fire control stations. The 8-inch railway guns on concrete pads (Batteries Sylvester and Kahuku), the two 7-inch batteries (Hulu and Harbor)

Site "B" built in 1929 with a series of cascading stations near Puu Palialai for long-range batteries. *Williford Collection.*

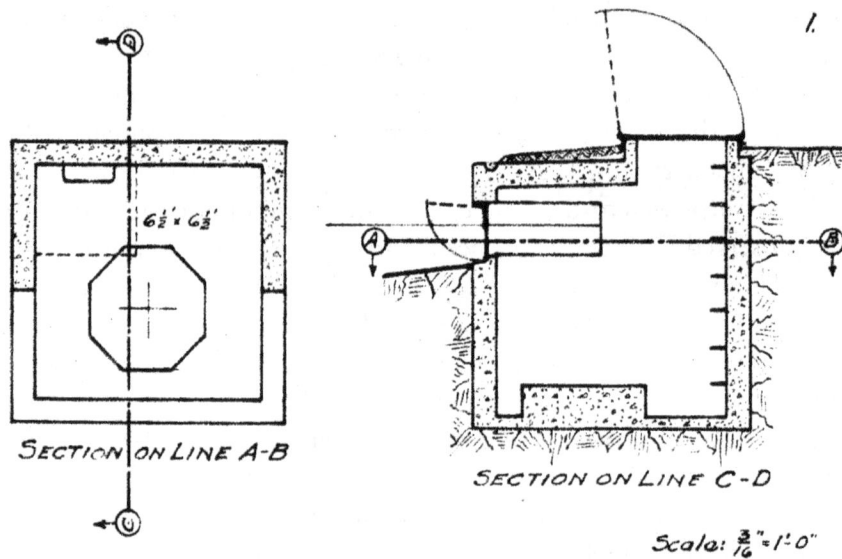

Site "C" Stations at Salt Lake. Originally built as a secondary for Battery Birkhimer, after World War I additional stations were added for Battery Closson and Williston. Note small dimensions for this first generation of dispersed stations. *NARA.*

are good examples. The new permanent 6 and 8-inch gun batteries completed for Kaneohe Bay and the Oahu modern program were to get the standard in-battery plotting room and a network of 3-4 observation stations. Some of these could be sited adjacent to the alphabetic stations from the previous 16-inch batteries, but others received new sites and were built to be fully permanent. Finally, the ex-navy turret batteries were to be fully equipped with fire control equipment and stations. All four of the 8-inch ex-aircraft carrier turret batteries and the two *Arizona* turrets received their own off-site complement of observation stations.

As mentioned above, the casemating of Battery Hatch at Fort Barrette and Battery Closson at Fort Kamehameha resulted in the elimination of a number of fire control stations that were no longer needed. As the modernized batteries would be restricted to just a 240° forward field of fire to the southern front of the defenses, the requirement for observation stations for the northern sector was obviated. Some of this area was still covered by non-converted Battery Williston, but several observation stations would be no longer needed for Hatch. These could be immediately converted to use for the new heavy batteries being built to cover the northern sector. Stations at alphabetic sites L, M, J, O, S, S', and W were reassigned.

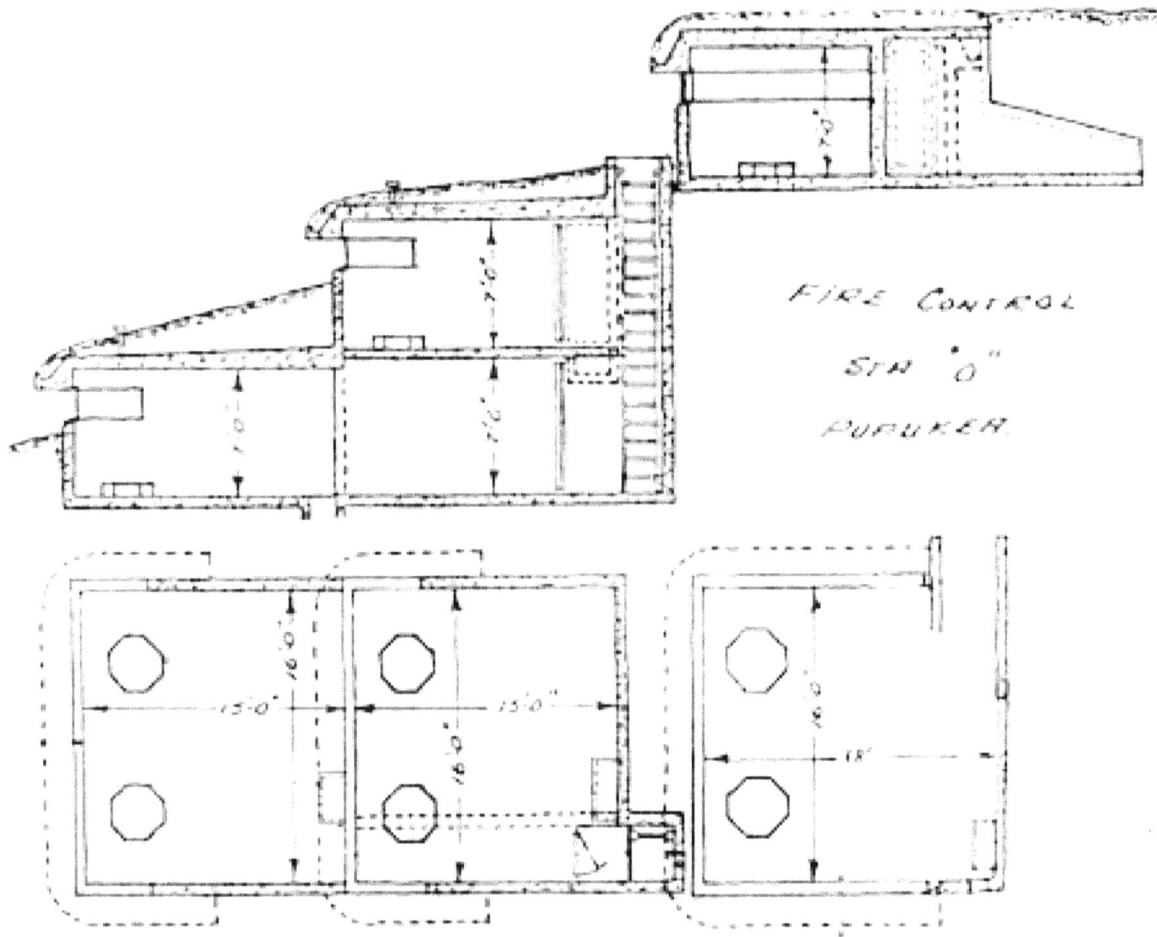

Station "O" near Pupukea. Initially for Battery Williston, in the late 1930s another station was added for Battery Hatch. *NARA*.

Station "U" on the Puu O Hulu ridge above the western Oahu coastline. This picture is during completion in 1934. *NARA*.

The total number of stations required in 1942-43 to provide observation to all the regular and temporary batteries on Oahu was impressive. As emplacements phased in and out the number varied, but a few examples hinting at the total are instructive. A letter of January 25, 1943, requested the need for 61 gasoline-driven, 2.5-kV AC-generator sets to be used at observation stations either not having access to commercial power, or requiring a back-up set in case commercial power was lost. (9)

One list survives of all the current or under-construction stations dated May 1, 1943. There are 134 assignments of the list, though some are housed adjacent or above or below each other. That list shows as follows: (10)

List of Oahu Base End Stations as of March 1, 1943:

| HSAC REFER. NUM. | SIZE | NAME AND POSITION | LOCATION |
|---|---|---|---|
| 1 | Double | School | Kamehameha School |
| 2 | Single | B. C. School 155 mm | Battery School 155 mm |
| 3 | Double | Station D Left | Punchbowl |
| 4 | Single | Station D Left Top | Punchbowl |
| 5 | Double | Station D Lower Middle | Punchbowl |
| 6 | Double | Station D Right Lower | Punchbowl |
| 6A | Double | Station D Right Top | Punchbowl |
| 7 | Single | B. C. Punchbowl | Punchbowl |
| 8A | Single | Roundtop | Roundtop |
| 8B | Double | Roundtop | Roundtop |
| 9 | Single | B. C. Randolph | Fort DeRussy |
| 10 | Single | B. C. Dudley | Fort DeRussy |
| 11 | Single | B-2 Harlow | Fort DeRussy |
| 12 | Double | West Birkhimer | Fort Ruger Diamond Head |
| 13A | Single | Station E Lowest Level | Fort Ruger Diamond Head |
| 13B | Single | Station E Second Level | Fort Ruger Diamond Head |
| 13C | Double | Station E Third Level | Fort Ruger Diamond Head |
| 13D | Single | Station E Top Level | Fort Ruger Diamond Head |
| 14 | Double | Leahi Group | Fort Ruger Diamond Head |
| 15 | Double | East Birkhimer | Fort Ruger Diamond Head |
| 16 | Double | B' Granger Adams | Fort Ruger Diamond Head |
| 17 | Double | Wilhelmina | Wilhelmina Rise |
| 18 | Double | Wailupe | Wailupe |
| 19 | Double | Station H' | Koko Head |
| 20 | Single | B.C. Koko Head 155 mm | Koko Head |
| 21A | Single | New Station H' Lower Level | Koko Head |
| 21B | Single | New Station H' Second Level | Koko Head |
| 22 | Single | B. C. Willy 155 mm | Wiliwilinui |
| 23A | Double | Station H Lower Level | Makapuu Head |
| 23B | Single | Station H Top Level | Makapuu Head |
| 24 | Double | New Station H Separate | Makapuu Head |
| 25 | Single | B. C. Wilridge 155 mm | Wiliwilinui |
| 26A | Single | Papaa Upper Level | Puu Papaa |

| 26B | Single | Papaa Lower Level | Puu Papaa |
|---|---|---|---|
| 27 | Double | Papaa | Puu Papaa |
| 29A | Double | Podmore North | Kaiwa Ridge |
| 29B | Single | Podmore South Lower Level | Kaiwa Ridge |
| 29C | Single | Podmore South Upper Level | Kaiwa Ridge |
| 30 | Single | Wailea North Upper | Wailea Point |
| 31 | Single | Wailea South Lower | Wailea Point |
| 32 | Single | Station J South | Ulupau Head |
| 33 | Double | Station J North | Ulupau Head |
| 34 | Double | Station J Middle | Ulupau Head |
| 34A | Single | B. C. Pennsylvania | Mokapu |
| 35 | Single | Heeia Upper | Heeia |
| 36 | Single | Heeia Lower | Heeia |
| 37 | Single | Heeia Middle | Heeia |
| 38 | Double | Heeia Separate | Heeia |
| 39 | Single | Heeia Lower | Heeia |
| 40 | Single | Heeia Upper | Heeia |
| 41 | Single | B-2 Loko 155 mm | Puu Kiolea |
| 42 | Single | B' Loko 155 mm | Lae O Kaoio |
| 43A | Single | B.C. BCN 302 Lower | Lae O Kaoio |
| 43B | Double | Kualoa 155 mm Upper | Lae O Kaoio |
| 44 | Double | Station L | Kaaawa |
| 45 | Double | Old Station M | Kokololio |
| 46 | Single | New Station M | Kokololio |
| 47 | Single | New Station M | Kokololio |
| 48 | Double | Monument | Monument 350 |
| 49 | Double | Old Station K Lower | Kahuku Military Reservation |
| 50 | Double | Old Station K Middle | Kahuku Military Reservation |
| 51 | Double | New Station K Upper | Kahuku Military Reservation |
| 53 | Double | Lena | Paumalu |
| 54 | Double | Station O Lower | Pupukea |
| 55 | Double | Station O 2nd Level | Pupukea |
| 56 | Double | Station O Top | Pupukea |
| 58 | Double | Station Y | Chain Gate |
| 59 | Single | Old Station Y | Chain Gate |
| 60 | Single | Ashley | Ashley Station |
| 62 | Single | B. C. Waimea 155 mm | Battery Position Waimea |
| 62A | Double | Reservoir | Waimea Reservation |
| 63 | Single | Station 63 Separate | Kawailoa |
| 64 | Double | Old Station T Right | Kawailoa |
| 65 | Double | Station T-3 Left | Kawailoa |
| 66 | Double | Station T | Kawailoa |
| 67 | Single | B. C. Pine 155 mm | Battery Position Pine |
| 68A | Single | B. C. Opaeula | Opaeula |
| 68B | Single | DPF Opaeula T | Opaeula |
| 69A | Single | B.C. Brodie | Battery Brodie |
| 69B | Single | DPF Brodie West of Battery | Battery Brodie |

| | | | |
|---|---|---|---|
| 70 | Double | Station W East Lower | Kamananui |
| 71 | Double | Station W West Upper | Kamananui |
| 72 | Double | Puuiki | Puuiki |
| 73 | Single | DPF Kaena | Battery Position Kaena |
| 74 | Double | Station S Lower | Kaena Point |
| 75 | Single | Station S Upper | Kaena Point |
| 76 | Double | Station S' Lower | Kepuhi |
| 77 | Single | Station S' Upper | Kepuhi |
| 78 | Double | Station U | Puu O Hulu |
| 79 | Single | Station U | Puu O Hulu |
| 80A | Single | Station U | Puu O Hulu |
| 80B | Double | Station U | Puu O Hulu |
| 80C | Single | Station | Puu O Hulu |
| 81 | Single | Kahe Middle | Kahe Point |
| 82 | Single | Kahe West | Kahe Point |
| 83 | Single | Kahe East | Kahe Point |
| 84 | Single | Station A Top | Manawahua |
| 85 | Single | Station A 3rd Level | Manawahua |
| 86 | Single | Station A 2nd Level | Manawahua |
| 87 | Single | Station A Lowest | Manawahua |
| 88 | Single | Station B 3rd Level | Palailai |
| 88A | Single | Station B Separate | Palailai |
| 89 | Double | Station B Top | Palailai |
| 90 | Double | Station B Middle | Palailai |
| 91 | Single | Station B 2nd Level | Palailai |
| 91A | Double | Fort Barrette Tower | Barbers Point |
| 91B | Double | Waimanalo Tower | Waimanalo |
| 92 | Double | Makakilo | Puu Makakilo |
| 93 | Double | Makakilo | Puu Makakilo |
| 94 | Double | Station X | Oneula |
| 94A | Double | Oneula Tower | Oneula |
| 95 | Double | Station F | Fort Weaver |
| 96A | Single | C1 Inclosure West Right | Fort Kamehameha |
| 96B | Single | C1 Inclosure Middle Right | Fort Kamehameha |
| 96C | Single | C1 Inclosure Middle | Fort Kamehameha |
| 96D | Single | C1 Inclosure Middle Left | Fort Kamehameha |
| 96E | Single | C1 Inclosure West Left | Fort Kamehameha |
| 97 | Double | Ahua Tower | Ahua Point |
| 98 | Single | Station G Tower | Ahua Point |
| 99 | Single | Station C Lower Single | Salt Lake |
| 100 | Double | Station C East Double | Salt Lake |
| 101 | Single | Station C Higher Single | Salt Lake |
| 102 | Double | Station C West Lower | Salt Lake |
| 103 | Double | Station C West Upper | Salt Lake |
| 104A | Double | B. C. Battery Hasbrouck | Fort Kamehameha |
| 104B | Double | Pearl Harbor Groupment C.P. | Fort Kamehameha |
| 105A | Single | C2 Inclosure West Right | Ahua Point |

| | | | |
|------|--------|------------------------|--------------|
| 105B | Single | C2 Inclosure Middle Right | Ahua Point |
| 105C | Single | C2 Inclosure Middle Left | Ahua Point |
| 105D | Single | C2 Inclosure East Left | Ahua Point |
| 200 | Single | Marconi | Marconi Pass |
| 201 | Single | Nuuanu | Nuuanu Pali |
| 202 | Single | Kalihi | Kalihi Pali |
| 203 | Single | Waiahole | Waiahole |

By 1940 the Army began deploying radar for use with seacoast defenses. A prototype navy fire control shipboard set was acquired and modified with lobe switching which enabled it to track a target. The set was tested, the modification approved, and a production order was placed for 20 sets. This first type became the SCR-296. It was utilized for fire control and operated with a 40-cm wavelength at a 700 MC frequency. The SCR-296A was the standard World War II fire control radar utilized by the U.S. Coast Artillery for engaging surface targets. It was authorized for issue to all modern batteries of 6-inch and larger size. The function of the radar was to provide the range and azimuth of the target vessel to the plotting room of the battery. The set was generally installed with a prefabricated metal lattice tower containing the antenna; an operating room containing transmitter/receiver, indicating panels, power panels, and communication devices; and two 25 kW gasoline generators. The operating room was in a prefabricated metal building and the generators in two other prefabricated metal buildings. In high-risk areas, the operating room could be located in bombproof structures such as the battery plotting room bunker, which was the case for the radar operating rooms for Battery Hatch, Battery French, and the two 14-inch turret batteries. In many locations the antenna tower could be directly on the roof of the operating room.

Mounted on top of the tower was an antenna on a platform. This antenna was used for both transmission and reception. It was a directional antenna formed of five curved ribs on which was mounted a metal screen reflector. The whole apparatus was approximately 6-feet long by 6-feet wide. This antenna was capable of being remotely controlled in azimuth and had an electronic data transmission device, which indicated its position on indicators located in the operating room. The antenna did not continuously rotate as in some modern radar but was directed at the target and in receiving functioned as a modern radio direction finder. The antenna was housed in a round wooden housing shaped much like a contemporary water tower. The radar system normally operated on commercial AC power. In isolated locations and as emergency back-up each set was provided with two portable gasoline powered generators.

Recent aerial photograph overlooking the stations on Puu O Hulu. *McGovern Collection.*

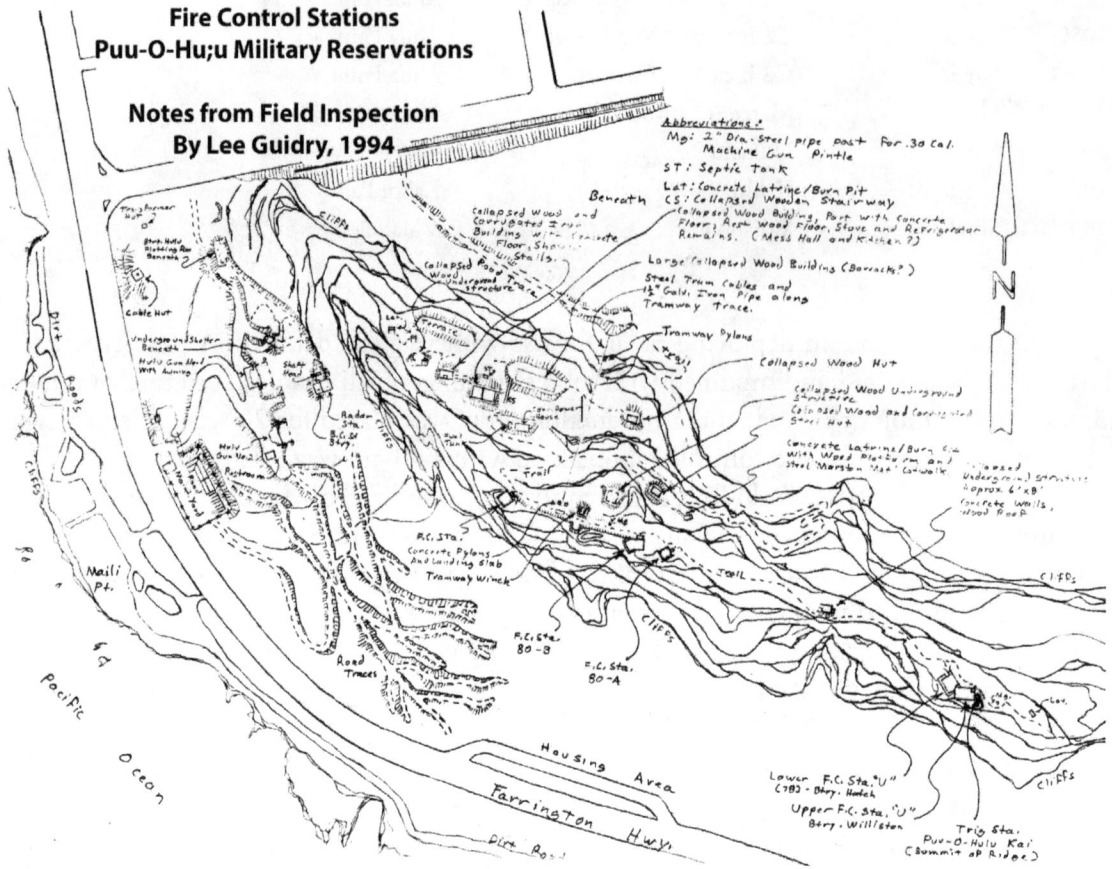

Field inspection notes from Lee Guidry of the little village of structures on the top of Puu O Hulu that progressively grew through the Second World War. *Guidry.*

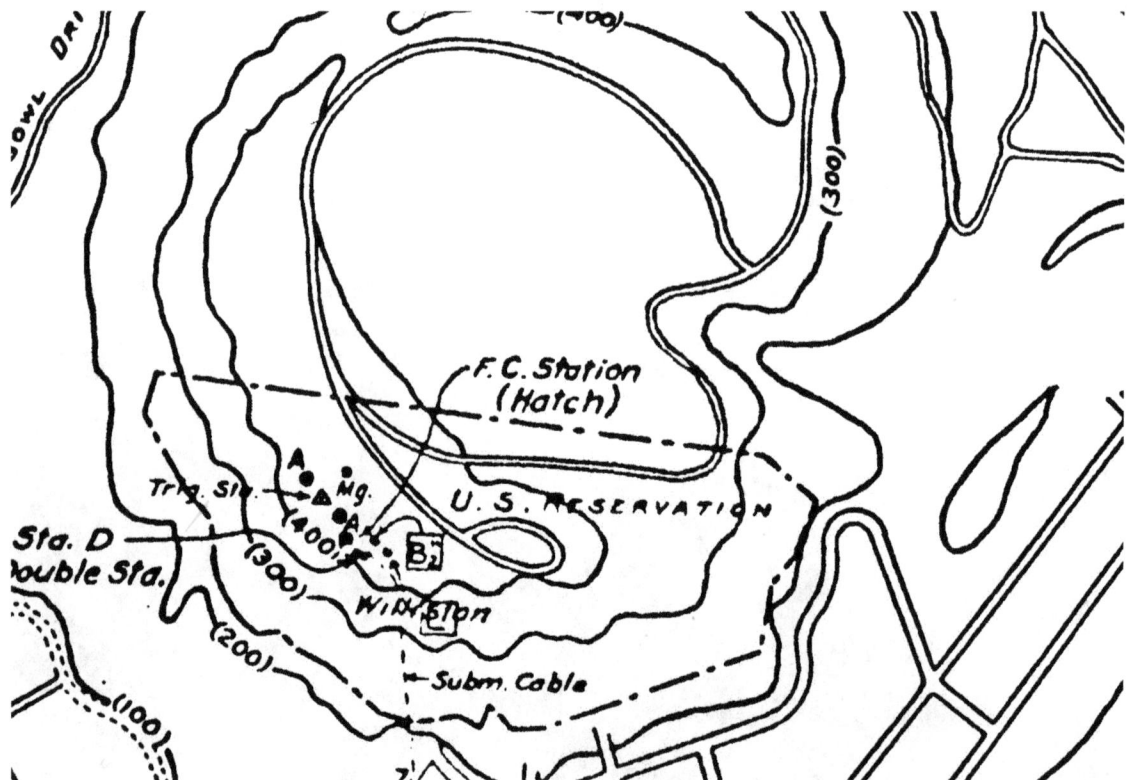

Map of the south rim of Punchbowl, showing the location of the fire control station "D" emplaced here prior to World War II. *NARA.*

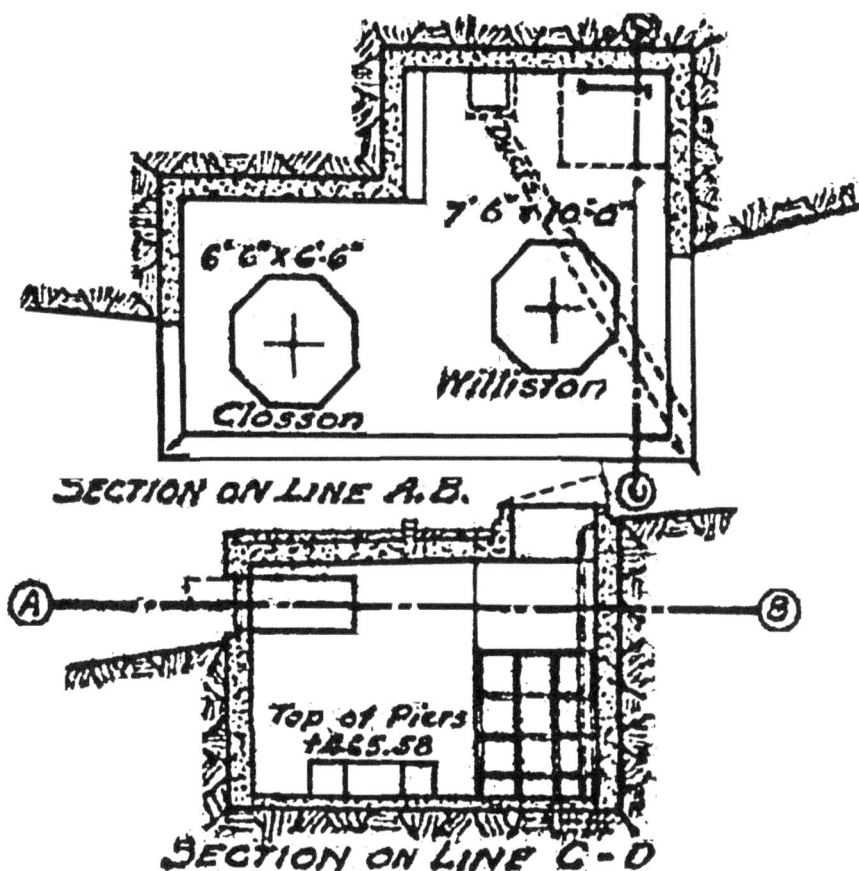

Some stations in the 1930s were built as doubles—fitted for two instrument stands to support two different functions or even two different batteries. Such was the case for station "D" at the Punchbowl. *NARA.*

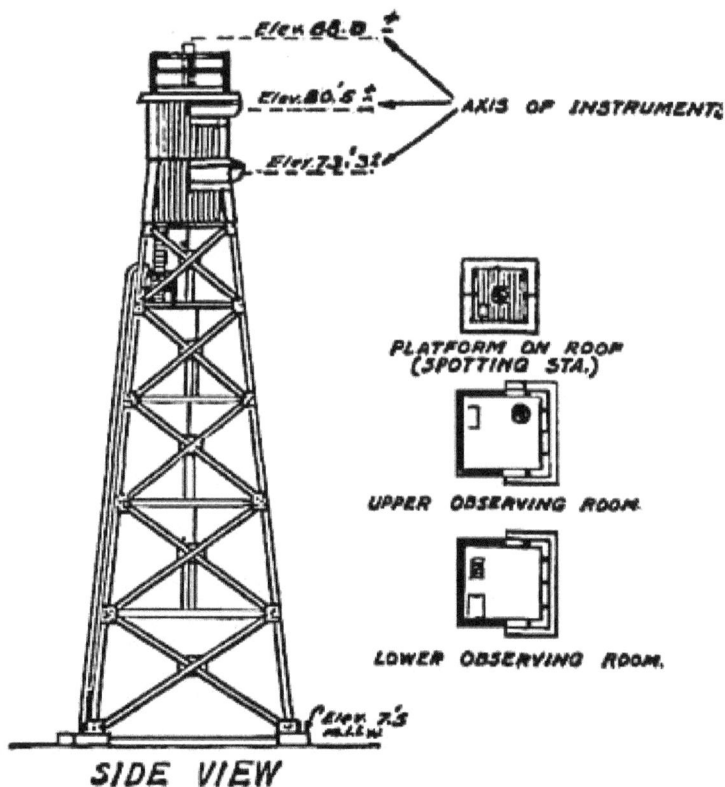

In the southwest of Oahu it was necessary to rely on steel towers to give the DPF instruments the necessary height to function properly. This was the new station "F" from 1932 at Fort Weaver. *NARA.*

## SECTION A-A

## PLAN

8"-0" x 12"-0"

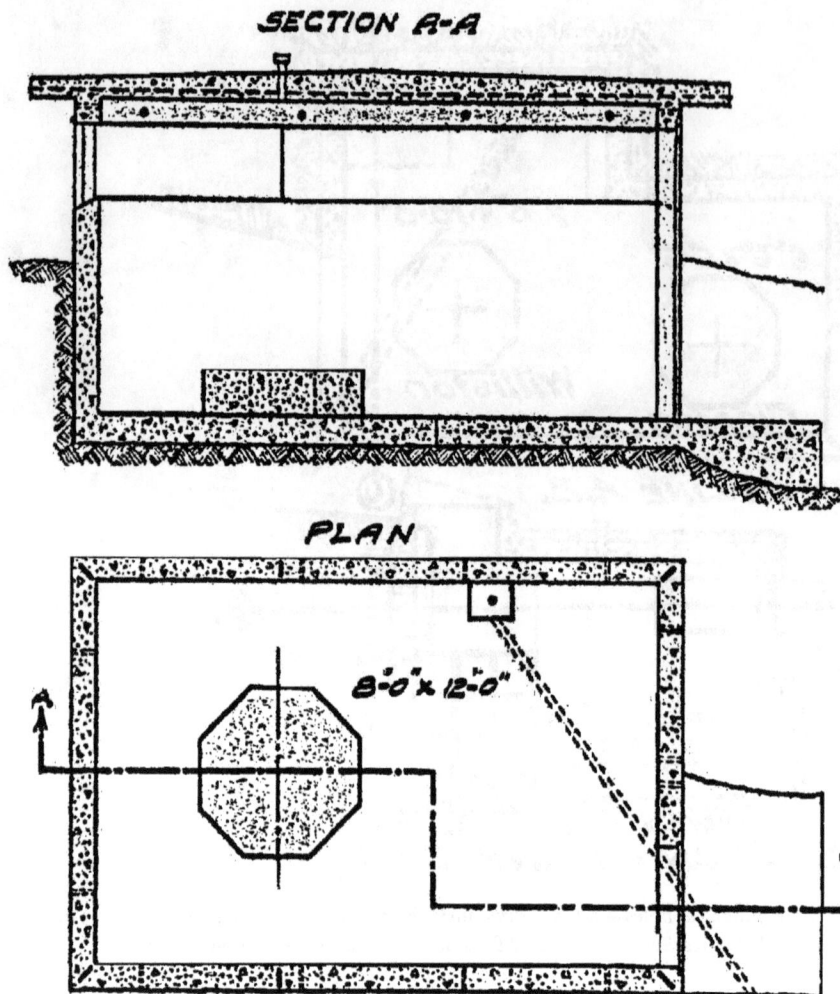

The original configuration of station "J" at Ulupau Head. It was rebuilt to accommodate a second instrument during the war. *NARA.*

The most important components of the radar were inside the operating room. This room contained the radar transmitter, receiver, power control panel, high voltage rectifier, and operating controls with a standard operating crew of 5 men: The NCO chief of the section, the range operator, the azimuth operator, the range reader, and the azimuth reader. The operating room required a minimum of nine cable pairs for telephonic communication and it was equipped with 4 telephones. As they became available, some sets were equipped with an electrical data transmission which fed continuous range and azimuth information to gun data M-1 or M-8 computers in the plotting room.

On August 4, 1942, the War Department allocated to the Hawaiian Department four SCR-582A surveillance radar sets and 12 SCR-296A fire control radars to provide monitoring of the waters around Oahu. Of the 12 SCR-296A radars, three were assigned to the Kaneohe harbor defense batteries, and four to the batteries on the North Shore and five to the Pearl Harbor Defenses. The SCR-296A radar sets were, however, not fully installed, and operational until January 1944.

Radar sets were installed for Battery Closson, Granger Adams, Williston, and Hatch. New 6 and 8-inch batteries for Kaneohe Bay (French, Cooper and DeMerritt) received operating rooms and radar at or near their BCS. In the case of 6-inch batteries No. 300, 301, and 304 at Punchbowl the operating room was worked into the design of the battery structure itself. The same would have been done with the other new batteries that were begun, and presumably the two additional batteries approved but never started. In addition, the more important temporary or expedient batteries of 1942 were provided with SCR-296A radar sets. This included the 7-inch ex-navy guns at Battery Hulu and Harbor. Also, the 8-inch railway M1918 Battery Sylvester at

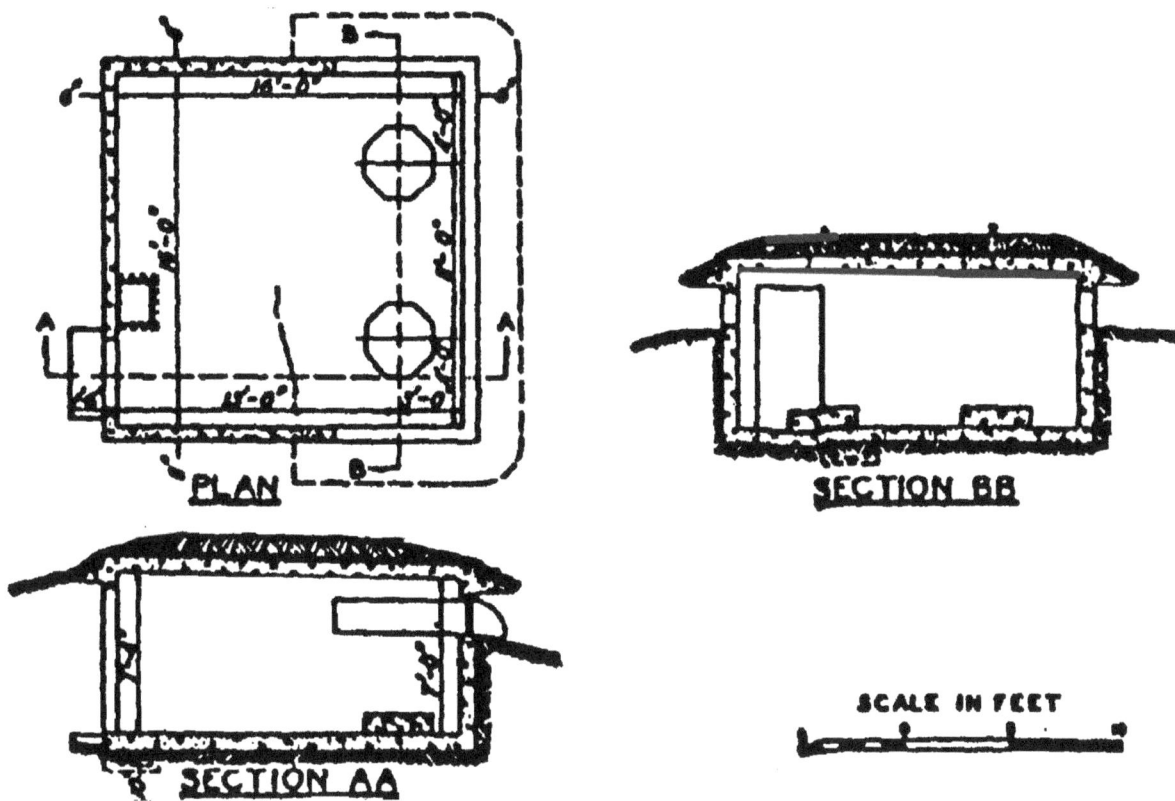

PLAN

SECTION BB

SECTION AA

SCALE IN FEET

Construction of additional stations continued up to and into the war. This was station "M" at Kokololio Gulch, completed in the summer of 1941 initially to serve 16-inch guns, however during the war new positions were added to support both 8-inch and 14-inch turret batteries. *NARA.*

Two stations, one a double and the other a single, were emplaced during the war at Podmore, on Kaiwa Ridge. They are still there today as shown in this recent aerial photograph. *Williford Collection.*

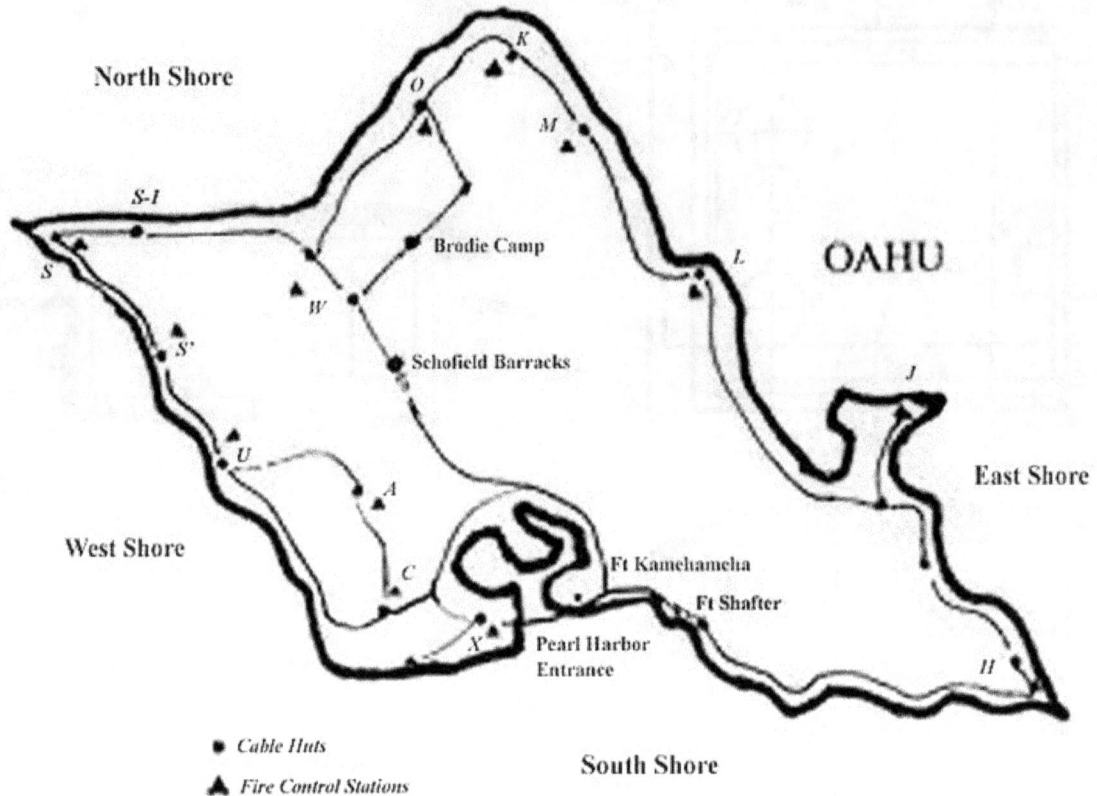

Map showing the major trunk connections of the Command and Fire Control Cable System that connected the various stations and virtually encircled the Island of Oahu. *Bennett Collection.*

Kaneohe Bay, Kahuku in northern Oahu, and Haleiwa at the north shore were all given radar fire control. The two 14-inch turret batteries had radar operating rooms built into the underground emplacements with radar tower directly above them. The four 8-inch turret batteries received separate standard operating rooms, generators, and towers with SCR-296A. Twenty radar operating sets, were installed or planned, depending on how far along the installation came on the uncompleted emplacements.

In the earliest Endicott period of U.S. coast defense use of searchlight to find or illuminate hostile targets was still in its experimental stage. By the time of the Taft Board Report in 1906, the importance and use of searchlights was of a high priority. The Board recommended the installation of 36-inch and 60-inch searchlights (indicating the diameter of the projecting mirror) and their electrical plants at all defended harbors. Within a few years it was realized that the 36-inch searchlight was not powerful enough for use and the 60-inch light became the only one suitable. Regular, but small appropriations were made over the next few years for the purchase and installation of the searchlights at various harbor defenses. Many of the searchlights installed during the period 1901-1917 were "fixed", that is located in a structure of some sort for concealment during the day, with their electrical generator. The Army also examined mobile searchlight models as well; the searchlights and electrical generators made to be carted by trucks and set up at a predetermined position. Over the years, the mobile searchlights became more reliable, durable, and rugged. By the late 1930s, the Coast Artillery switched to using mobile searchlights and replaced fixed searchlights where at all possible.

The U.S. Army searchlights used an electric arc generated by a current running through a carbon rod as the source of light. The actual light source was an incandescent ball of vapor formed in the high-intensity arc held in the crater of the positive carbon. A reflecting parabolic mirror was located behind the arc, which reflected the light forward in a parallel pattern, which produced a well-defined narrow beam of light. The basic design of the searchlight remained the same over the years, with a number of refinements in electrical equipment,

AAIS positions were also constructed at various locations around the island and tied into the cable system. Here is the one being built at Kaena Point in May 1942. *NARA.*

One of the earliest searchlight positions of the defenses was built at the Ahua Point reservation east of Fort Kamehameha. Here it is under construction, but showing the rotating circular supports for its lowering tower structure. *NARA.*

motorized parts, controls, etc. The earlier fixed searchlights used a turntable and trunnions on the drum, supported by arms, motorized to alter the drum's elevation and azimuth, electrical cabling, a controller for power, azimuth, and elevation (located some distance away from the searchlight itself), and of course the electrical generating plant. By the later part of the 1910s, the 60-inch light was in general use. Searchlights in a given area were generally grouped under one commander, either a gun command or a mine command, or (rarely in seacoast defense) as a separate searchlight command.

Searchlights were located with consideration to effectively search the approaches to the harbor area to locate and identify the enemy. The lights were positioned in advance of the outside line of the primary armament at a distance one-half of the effective range of the light. Lights were also placed to illuminate potential landing areas for raiding parties, to cover mine fields and channels leading to the inner harbors against torpedo boats and minesweepers. In general, they were located as close to the water's edge as possible to maximize their effective range of between 8,000 and 15,000-yards. In some situations, they were installed on elevated platforms or "disappearing" towers. Some lights were co-located so that one could be a backup in case the other malfunctioned or was knocked out by enemy action. Only one of these was used at a time to prevent confusion caused by crossing beams. The development of portable searchlights expanded the tactical positioning in that the searchlights did not have to be on the fixed government reservations but could be deployed temporarily anywhere they were required.

Fixed searchlights were provided with a shelter. The shelters functioned to protect the searchlight from the elements, to house the electrical generator, and to provide concealment for the light when it was not in use. In many places searchlights were mounted on small carts which were moved back and forth on a few feet of track from a recessed shelter to the operating position. Searchlights were also housed in fake water towers, or beach cottages and the like, which had faces that could be opened for use. Other searchlights on the tops of hills were lowered vertically into pits with a movable roof, or they rolled back down an inclined plane. The disappearing towers mentioned above also functioned as a method of concealment. This all became moot with the adaptation of the mobile light since the light could simply be moved from the exposed position when not in use.

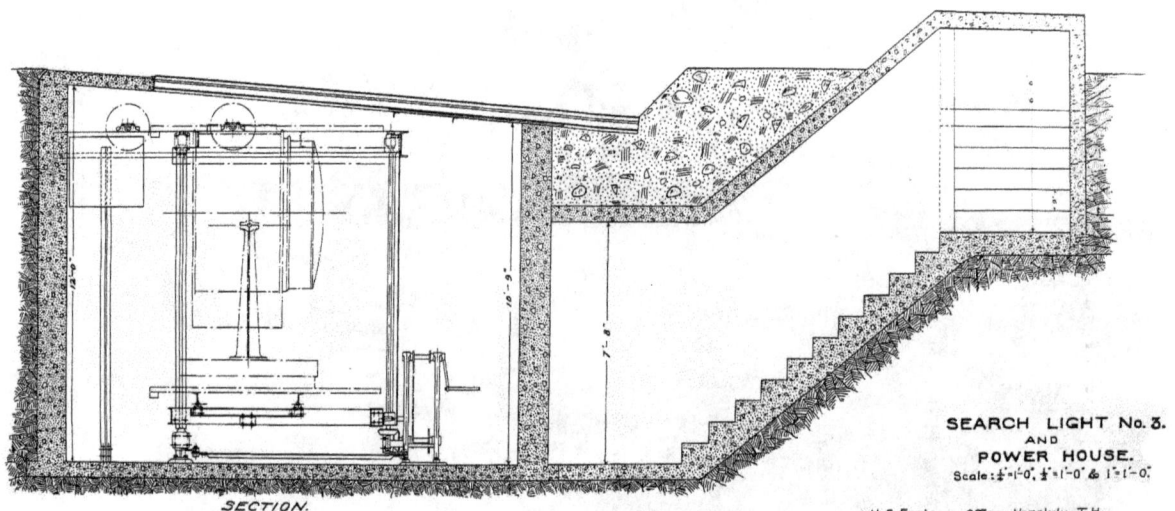

SEARCH LIGHT No. 3.
AND
POWER HOUSE.
Scale: ¼"=1'-0", ½"=1'-0" & 1"=1'-0".

U. S. Engineer Office, Honolulu T. H.
Feb. 10, 1913.

Respectfully submitted to the Chief of Engineers;
approval recommended.

W. A. Wooten
Major, Corps of Engineers.

SECTION.

Also used in this early generation were buried, or at least protected, concrete shelters for lights and their power generators. This is the plan for Searchlight No. 3 at Fort Ruger in 1913. *NARA.*

*Searchlight No.1 (New No.16), Diamond Head, Fort Ruger, T.H., operating position*

Also at the Diamond Head reservation of Fort Ruger was an emplacement for Searchlight No.1, which could be raised or lowered as needed. *NARA.*

*60" Searchlight, Sand Island, T.H., day position*

A 60-inch seacoast searchlight in its shelter at Sand Island. *NARA.*

Searchlights were designated by consecutive numbers facing seaward from right to left. Mobile searchlights were designated in the same manner usually in their own sequence; predetermined locations were numbered consecutively from right to left (although for some reason those on Oahu seemed to be numbered in the exact opposite way) facing seaward and searchlights assigned to that position were given that number. Early on (1910s) searchlights were assigned as either searching or illuminating lights and as either fort, fire, or mine command lights. Later this "permanent" assignment was abandoned, and lights were assigned based on tactical demands. Certain searchlights were assigned to search their respective areas. Once targets were spotted, that searchlight, or another, was designated to illuminate that target. Other lights would pick up the searching job, and once the target passed out of the range of one illuminating light, another would be assigned to it.

In 1914 the searchlight squad consisted of: a controller operator, a searchlight operator, an assistant, a watcher, an engineer, and a telephone operator. By 1937 this was simplified somewhat to a light commander (NCO), a controller operator, a light operator, and a power plant operator. Mobile searchlights may have been assigned more men for drivers and to assist in the set-up of the equipment. Though initially horse drawn, the development of the motorized vehicle, especially during the WW I years, brought the mobile searchlight to the forefront of tactical use. Initially the mobile motor unit consisted of two trucks, one mounting the 25 kW generator, the other mounting the searchlight. Later, the development of lighter searchlights and a rubber wheeled trailer resulted in the elimination of one of the trucks. Additional searchlights were developed for anti-aircraft work, and eventually the standardized units of either General Electric or Sperry manufacture of the M1941 & M1942 designs were used during WW II for both AA and HD defense. (11)

The initial batch of searchlight positions for the Taft Generation forts of both Pearl Harbor and Honolulu channels were constructed on the reservations of the first four forts. Most of these were types of 60-inch lights, but two were just 36-inch size. All the Pearl Harbor lights were at Fort Kamehameha or the Ahua Point reservation, while the Honolulu lights were either at Fort Armstrong or Ruger; Fort DeRussy was not assigned a searchlight.

A Summary List of Oahu Seacoast Searchlight Positions in included in Appendix X.

Some searchlights were even placed on towers, like No. 4 also at Ahua Point. Here it is shown in 1937. *NARA.*

# CHAPTER 15
## LAST PLANS AND SHUTDOWN

By mid-1944 it was clear the Allies were winning the war. In the Pacific the front was approaching the outer islands of Japan, and the offensive capabilities of the enemy were so reduced that the danger of significant attack on the Hawaiian Islands was remote. A draw down of troops assigned to the harbor defenses began. On May 27, 1944, the Harbor Defenses of Honolulu and Pearl Harbor on Oahu's shore were combined into a single harbor defense command and the 15th Coast Artillery was assigned to man those harbor defenses. The elements of the 16th Coast Artillery still posted in the Honolulu harbor defenses were transferred to Oahu's Windward side to man the Harbor Defenses of Kaneohe Bay and North Shore. In most cases these transfers did not require the movement of the troops as only the unit designations were transferred. The personnel of the 41st Coast Artillery Regiment were used to provide the personnel assets of the 16th Coast Artillery.

In August 1944 the garrison at Fort Kamehameha was composed of elements of the 53rd and 54th CA (HD) Battalions, the successor organizations of the 15th CA. In February 1945, these units were inactivated, and the garrison reduced to a handful of separate batteries of coast artillery. By the middle of 1946, the harbor defense post was garrisoned by a single coast artillery unit, Battery A, 2274th Hawaiian Seacoast Artillery Command (HSAC). Old Battery Selfridge and Jackson had been in a reduced maintenance status for some time late in the war. Soon after the Japanese surrender the guns and carriages were removed for scrap. With an important air base adjacent, the heavy, modern anti-aircraft guns continued in service postwar.

Towards the end of the war Fort DeRussy's garrison was substantially reduced. Battery Randolph and Dudley were no longer manned, and the AA defenses were the responsibility of the Battery A 750th AAA Battalion. The searchlights at Fort DeRussy were now manned by details from the 856th Coast Artillery Searchlight Battery (Separate). The former personnel of the 16th Coast Artillery batteries manning the guns still in service on Oahu's south shore became administratively attached "detachments" of the firing batteries of the 15th Coast Artillery but generally continued in the same locations manning the same armament they had manned as part of the 16th Coast Artillery. (1)

The personnel at Fort DeRussy provided caretaking duties for the 6-inch and 14-inch armament as well as performing beach defense duties and the usual post functions. Through the remainder of 1944 and into 1945 the size of the detachment would gradually decline as personnel were reassigned to other duties. Fort DeRussy became the Army's preeminent recreation center. The center also served the other services on Oahu. Dances at the Maluhia Dance Hall on the post drew thousands of service men from all over Oahu. The post also housed as one of its tenant units the Overseas Motion Picture Unit that provided movies to the far-flung reaches of the Central Pacific Area. On August 12, 1944, the out-of-service 14-inch guns of Battery Randolph, the 6-inch guns of Battery Dudley and their disappearing carriages were ordered salvaged.

With the dismantling and scrapping of Fort DeRussy seacoast guns, and the removal of the AA battery, in the summer of 1946, the fate of the post was up in the air. However, Fort DeRussy continued to function as an Army installation and in 1949 it underwent a renovation. The barracks, officers' quarters and other buildings built in the early 20th century that had occupied the portion of the post south of Kalia Road were demolished leaving many of the more recent structures built during the war between Kalia Road and Kalakaua Avenue in place. The post's close proximity to Honolulu and Waikiki made it an appropriate location for a station of the Army Military Police and the joint Hawaiian Armed Services Police (HASP). In 1957 alone 2,259,382 servicemen used the recreational spaces and facilities at the post. That same year, almost 2,000 civilian reservists of the Army, Navy, and Coast Guard trained at the Reserve Training Center on the post.

In the 1960s, efforts were made to demolish the fort's pair of seacoast gun batteries. Battery Dudley was removed in 1969 but at an extraordinary cost, and the destruction of the massive Battery Randolph was abandoned. Eventually the 14-inch gun battery was converted into The United States Army Museum of Hawaii, which was dedicated on July 4, 1976, and opened its doors to the public on December 7 of that bicentennial year.

At Fort Ruger in May 1944, the 41st CA and the 57th CA were inactivated and most of the personnel were reassigned to the stations atop the south rim of Diamond Head. The 15th CA was assigned the entire south shore of Oahu, and manned Fort Ruger's batteries, at least until it too was inactivated in mid-August 1944. Thereafter, the guns of Fort Ruger were manned by a succession of newly constituted coast artillery battalions and separate batteries. In 1945 the old Hawaiian Artillery Command was resurrected. The Hawaiian Seacoast Artillery Command was also reorganized and redesignated the 2274th Hawaiian Seacoast Artillery Command. HAC established its headquarters at Fort Ruger alongside that of the 2274th HSAC. Soon after, in late March 1945, the War Department authorized scrapping the mortars of Batteries Harlow and Birkhimer at a time to

The armament at Battery Randolph had been authorized for salvage in August 1944, though it took a couple of years before scrapping was actually accomplished. *Williford Collection.*

Seacoast mortars, awaiting shipping to continental scrapyards, lies haphazardly in a yard at Sand Island in 1948.
*Bennett Collection*

be selected by the commanding general, Pacific Ocean Area. That time was immediately at the conclusion of hostilities in 1945.

In July 1946, the harbor defense commands on Oahu were again reorganized and HQ and HQ Battery, 2274th HSAC, became harbor defense headquarters for all the harbor defenses on Oahu. Two batteries of the 2274th were assigned to Fort Kamehameha, one battery to Fort Hase on the Mokapu Peninsula, and the remaining batteries to Fort Ruger. Each line battery supported small caretaking detachments and battery outposts along Oahu's shoreline. (2)

In 1947, Fort Ruger was designated the South Sector HQ and command post. In January 1950, the fort became the headquarters post for the Hawaii National Guard. In December 1955, 595 of Fort Ruger's 675 acres were transferred to Hawaii for use by the guard. The army retained the Palm Circle area until September 1974, when it and the parade ground area were turned over to the state. Much of the cantonment area have given way to the campus of the University of Hawaii's Kapiolani Community College. A small reservation on the east side of Diamond Head was retained as headquarters of the Hawaii Army National Guard. Battery Birkhimer served for many years as a Civil Defense facility, and later Federal Emergency Management Administration location. (3)

Over the years, the barracks and most other post structures of Fort Ruger have been demolished. The commanding officer's quarters, the Tennis Courts AA battery, and the Panama mounts that once occupied the reservation's northern end are now part of the campus of the University of Hawaii's Kapiolani Community College. The Hawaii Air National Guard and other government agencies used Battery Harlow's magazines and storerooms for storage. The tunnels, magazines, and storerooms of Batteries Hulings and Dodge were similarly used before eventually being abandoned. The nearly completed but unarmed Battery 407 tunneled into Diamond Head's south rim during World War II was refurbished and used for many years as a command post for the Hawaii Air National Guard. Point Leahi's fire control complex and much of Diamond Head Crater are a historical site administered by the State of Hawaii.

The long-range batteries of Closson, Hatch, and Williston were the last of the seacoast artillery guns to be decommissioned. Battery B, 53rd CA (HD) Battalion continued to man Battery Closson's guns through early 1945. By April 1945 the battery was left in service but manned only by a small detachment to serve as a caretaking unit. Still, the end of the war, the reduction of military manpower and the lack of any apparent enemy led to the inevitable. The battery was disarmed, and the armament scrapped in 1948. The exact same story applied to the two 16-inch batteries. At Fort Weaver the 16-inch battery continued to be manned until early 1945, when the guns were placed in maintenance status and Battery A was transferred to Sand Island. Battery Williston's guns were removed and scrapped in 1948 and the post was eventually transferred to the Navy Department. A detachment of the 15th Coast Artillery continued to provide the manning detachment for Battery Hatch's 16-inch guns at Fort Barrette. Postwar Fort Barrette was turned over to a caretaking detachment and the troops transferred to Fort Kamehameha. Battery Hatch was disarmed and cut up for scrap in 1948.

All four of the 8-inch, ex-aircraft carrier gun batteries were reduced to caretaker status in 1945. Caretakers, under a variety of unit designations, continued to care for the guns and emplacements until authorization came in 1948 for the end of the seacoast command and scrapping of the guns and salvaging of any usable equipment at the sites. At Battery Kirkpatrick, soon after the Japanese Surrender in 1945, the Trustees of the Bishop Estate initiated legal proceedings to regain the tracts of land taken by the Army during the war for the battery. After a lengthy period of negotiation and legal proceedings, the property was finally declared excess to the needs of the Department of Defense on May 31, 1949, and a negotiated settlement with the Bishop Estate was made that called for the Army to remove the guns, turrets, and "other metal in connection therewith." The legal proceedings dragged on until a final judgment was obtained and the settlement filed in the Federal District Court in Honolulu, on March 3, 1951. On that date the Wiliwilinui Ridge Military Reservation ceased to exist. The other 8-inch turret sites also reverted to private ownership.

Battery Arizona at Kahe Point was not completed, and Battery Pennsylvania at Kaneohe Bay was able to carry out one proof firing. The engineering and ordnance personnel soon left, replaced by the usual caretaker unit as much concerned to protect against theft as anything. There were several official inspections and assessments of what it would take to finish these two batteries in 1946, but no action taken to accomplish the needs. The incomplete guns, turrets, and internal battery equipment was scrapped in 1948.

For a while postwar Battery Arizona continued a shadowy existence. It was never actually transferred to the coast artillery or provided with its manning detachment. Even though work was suspended, such a large investment had been made that attempts to complete it lingered on. An inspection in March 1946 identified numerous modifications needed in its water system and shell cage hoist arrangements, but there is no indication any more work was undertaken. By 1948, the battery had been abandoned and the turret and guns sold for scrap.

By early 1945, the seacoast defenses of the North Shore and Kaneohe Bay had been expanded far beyond what War Department planners had foreseen in 1941. The gun defenses were quite formidable, complementing those on the island's south shore. By the last year of the war, the powerful 8-inch and long-range 6-inch guns had largely supplanted the need for the older 155 mm guns and the obsolescent railway mounted armament that had been the mainstay of the defenses at the onset of the conflict. Despite the massive reorganizations and reductions throughout the Harbor Defenses of Kaneohe Bay and North Shore, the number of personnel at the end of 1944 was still considerable. Fort Hase provided overall administrative and tactical command and control of the harbor defense command with Headquarters and Headquarters Batteries for the 16th CA Group and the 55th CA (HD) Battalion. Also posted at Fort Hase was a consolidated battery (the 852nd) for all the region's searchlights.

On the North Shore the 143rd CA Group operated the various batteries of the North Shore from its command post in Station "T" at Kawailoa until January 1945. With the continuing reorganization the 56th CA

Battalion assumed charge of the four batteries of 8-inch guns in the sector and established its headquarters at Station "T." Battery A of the 56th was posted at Kahuku Plantation Golf Course where its four officers and 168 enlisted men manned Battery Kahuku's four M1888 8-inch guns on their shielded barbette carriages. Battery B of the 56th manned the 8-inch turret battery at Opaeula Gulch into February 1945. Battery C was at Brodie Camp Number 4. Just four officers and 166 enlisted men of Battery D manned Battery Haleiwa's 8-inch Railway guns. (4)

The process of shutting down the defenses began in September 1944, when authorization was received to take the 155 mm GPF batteries out of service. A battalion continued to operate the three 8-inch batteries of the Saratoga Group until the gun group was inactivated on October 4. On February 13, 1945, the 55th Battalion was inactivated at Fort Hase and the personnel assets of its headquarters battery were absorbed. The 609th Battery was inactivated at Kahuku on April 10, 1945, and most of its personnel reassigned to the 606th CA (HD) Battery (Separate). Battery B, of the 55th was redesignated as the 610th CA Battery on February 15, 1945. It continued to man Battery No. 405 until April 10, when the battery of 8-inch guns was inactivated. When Battery 405 was placed in caretaking status the pair of 8-inch guns had been an active element of the harbor defenses for less than a year. Likewise, Battery D continued to man Battery No. 302 and the 90 mm guns of the AMTB battery until February 15, 1945 when the unit was redesignated as the 612th CA Battery at Kualoa Point; then Battery No. 302 was placed in maintenance status. (5)

On March 27, 1945, Headquarters Middle Pacific Ocean Area advised the War Department that Battery Sylvester was no longer required and that it could be removed. At this point just the 90 mm AMTB batteries at Kualoa Point (with the 612th CA Battery) and Pyramid Rock (with the 880th AA Battery) were left actively manned. In August 1945 that too ended and the guns authorized for removal and placed into storage.

In March 1945, the Artillery Officer, Pacific Ocean Area, conducted a survey of obsolescent armament and a report was submitted to the War Department on March 27. Shortly afterward the War Department directed that, while no longer required in the Pearl Harbor defenses, the 8-inch railway guns of Battery Browns Camp were to be placed in maintenance status pending the completion of Battery Construction Number 409 at Kaena Point and Battery Arizona at Kahe Point. The guns were grouped in four batteries—Kahuku, Sylvester, Haleiwa, and Browns Camp, unmanned but maintained on site until the end of the war, when they were all authorized to be scrapped. By late 1944 the 155 mm M1917/18 guns were no longer being actively used, and neither were most of the small, and generally remote Panama mount batteries. The guns were not scrapped until after the war and the reservations mostly sold back to private ownership.

On June 6, 1945, a special committee of the 2274th HSAC completed a study on postwar Hawaiian seacoast artillery needs. The report provided several succinct recommendations: (6)

Uncompleted projects at Battery No. 408, 409, and 305 be discontinued. These sites were not turreted and thus vulnerable, and outside the major antiaircraft shield emplaced around the major army forts.

Long-range batteries Closson and Hatch should have their casemating protection completed.

Long-range Battery Williston should be relocated or preferably replaced with an ex-navy turret emplacement.

Concrete bunkers should be emplaced surrounding and protecting the ex-navy 8-inch turret batteries that were otherwise lacking even splinter-proofing with their light shields.

Batteries Arizona, Pennsylvania, 407, 303, and 304 should be carried to completion.

Fire control SCR-296A radar should be replaced with the improved An/FPG-1 sets.

The discussion around Battery Williston was interesting. While it was the longest-ranged battery of the defense, its location close to the beach on the west side of Pearl Harbor made it particularly vulnerable to attack. The guns were purposely not casemated during the war, and there were concerns about whether the sandy soil could support the weight of casemates. Relocation to another site on the forward slope of Aliamanu crater or replacing the battery was considered, but finally a total replacement was though better. In this early postwar period, the army was highly attracted to navy-type turret armament. It seemed to offer the best combination of the widest field of fire with heaviest protection. On March 25, 1946, the local command recommended that Battery Williston be replaced by a triple 16-inch navy turret at the Aliamanu location. They had been recently advised by the War Department that two such turrets fabricated for the incomplete USS *Illinois* might become available. However, such work and expense in a rapidly fading service was not possible and nothing came of this proposal. (7)

Work on the incomplete USS *Illinois* (BB-65) was suspended in July 1945, even before the formal end of the war. Her triple 16-inch gun turrets were briefly offered to the Hawaiian defenses, but by this time the end was nigh for this branch of service. *NARA*

In June of 1950 the Coast Artillery Corps ceased to exist, its remnants were combined with the Antiaircraft Artillery and Field Artillery into a single branch of Artillery. Following World War II, the various coast artillery gun and fire control positions that had been established over the previous three decades were gradually scrapped out and abandoned. In many cases the former battle positions reverted to the prewar landowners. In other instances, former coast artillery battery tracts and other reservations were transferred for the use of various other military and naval commands on the island.

Fort Kamehameha was incorporated into Hickam Air Force Base, which is now a part of Joint Base Pearl Harbor-Hickam.  Fort DeRussy is still an active army post—the location of U.S. Army Museum of Hawaii, the Hale Koa Hotel, an Armed Forces Recreation Center, and the Daniel K. Inouye Asia-Pacific Center for Security Studies. The Fort Ruger military reservation has been divided up—Hawaiian National Guard property, the Hawaii Emergency Management Agency, a college, and a public park. In 1949 Fort Weaver was transferred to the U.S. Navy, it is now the Iroquois Point Navy Housing unit. The gun platforms and most of the supporting magazine structures of Fort Weaver were destroyed or buried. Fort Barrette is now the Kapolei Regional Park postwar, though the concrete structures still exist.

On July 1, 1950, Fort Hase became one of the numerous reservations comprising the Army Post of Oahu, the headquarters of which were at Fort Shafter. In 1951, the Marine Corps determined that the old Kaneohe Air Station would be suitable as a combined air-ground team training site. The Marine station absorbed the Army's inactive Fort Hase and the old cantonment area was converted for use of the Marines. Subsequently the 4th Marines and additional aviation elements were assigned to Kaneohe. Today Kaneohe Marine Corps Air Station has been renamed U.S. Marine Corps Base, Hawaii, includes all of the area once occupied by Fort Hase and is home to the 1st Marine Expeditionary Brigade (1st MEB). (8)

A significant number of the old seacoast artillery structures remain, though most are on military reservations and not generally accessible to the public. Since the end of the Second World War, eight gun batteries have been destroyed or buried—Batteries Barri, Chandler, Tiernon, Dudley, S.C.Mills, Granger Adams, Williston, and the 8-inch turret site at Salt Lake. Almost all of the transient mobile, antiaircraft and emergency battery sites are gone, overgrown, or altered to the extent of not being recognizable. However, most of the fire control and many of the permanent searchlight stations remain on remote or high ridges that have helped prevent them from being destroyed. More than a few are on well-known hiking trails, often highlighted with the bright colors of graffiti.

At Joint Base Pearl Harbor-Hickam, Marine Corps Air Station Kaneohe, and the Hawaiian National Guard portion of the old Fort Ruger base many of the old coast artillery structures are still used for storage or offices. The largest of the original Taft-era structures, 14-inch battery Randolph at Fort DeRussy, houses the U.S. Army Museum of Hawaii. It has on display the two surviving guns of the harbor defenses, a pair of 7-inch ex-navy MkII guns, originally mounted at Battery Harbor on Sand Island. The two 4.7-inch gun barrels of Battery Dodge are on display outside of the Hawaiian National Guard armory in Wahiawa. All of the other guns and equipment of the defenses were either relocated elsewhere or scrapped in the years following the shutdown of the late 1940s.

Over a period of forty years, the Harbor Defenses of Oahu grew from the defenses of Honolulu and Pearl Harbor into four commands overseeing the defense of Oahu by the end of the Second World War. Arguably the ultimate armament, both seacoast and antiaircraft, was more substantial than any other American harbor. The seacoast defenses in Oahu were never tested by attack by an invading fleet. When the call to action came, on the morning of December 7, 1941, it came in a totally unanticipated manner. The antiaircraft defenses of both the Army and Navy, in the face of an overwhelming surprise air attack of a size and skill never foreseen, proved wanting.

Under the perceived threat of returning enemy forces, perhaps intent on landing and occupation, the Oahu seacoast defenses were substantially reinforced. A totally unique (at least for the United States) armament scheme based on available ex-naval turrets helped double the number of long and intermediate-range weapons. This anticipated threat never materialized into action. Strategically the Japanese did not have the strength nor an overriding imperative to attempt to take the Hawaiian Islands, regardless of the fixed defenses. By the end of the war fixed defenses like these were strategically obsolete and they would persist for just a few more years in a constant state of declining readiness. Still, they represent a fascinating period of military defensive technology and have left a definite historical footprint on the island of Oahu.

# APPENDIX I

## ORDNANCE DEPLOYED IN PERMANENT, COMPLETED HARBOR DEFENSES IN OAHU, 1907-1948

| FORT AND BATTERY | ORIGINAL ARMAMENT | REPLACEMENT ARMAMENT |
|---|---|---|
| | | |
| **FORT ARMSTRONG** | | |
| BATTERY TIERNON | 2 x 3-in M1903/M1903 ped<br>No. 96/ No. 96<br>No. 95/ No. 97 | |
| | | |
| **FORT KAMEHAMEHA** | | |
| BATTERY CLOSSON | 2 x 12-in M1895M1A4/ M1917 BCLR<br>No. 1 Watervliet/ No. 26<br>No. 30 Watervliet/ No. 27 | No. 30 Watervliet/ No. 26<br>No. 1 Bethlehem/ No. 27 |
| BATTERY SELFRIDGE | 2 x 12-in M1895M1/M1901 DC<br>No. 12 Bethlehem/ No. 18<br>No. 13 Bethlehem/ No. 19 | No. 8 Bethlehem/ No. 18<br>No. 11 Bethlehem/No. 19 |
| BATTERY HASBROUCK | 8 x 12-in mortars M1908/<br>M1908 mortar carriage<br>No. 16/ No. 16<br>No. 15/ No. 15<br>No. 14/ No. 14<br>No. 7/ No. 13<br>No. 20/ No. 5<br>No. 17/ No. 6<br>No. 19/ No. 7<br>No. 18/ No.8 | |
| BATTERY JACKSON | 2 x 6-in M1908/ M1905MII DC<br>No. 7/ No. 32<br>No. 13/ No. 33 | |
| BATTERY BARRI | 2 x 4.7-in Armstrong/ Pedestals<br>No. 10999/ No. 9018<br>No. 11000/ No. 9083 | |
| BATTERY CHANDLER | 2 x 3-in M1903/ M1903 Ped<br>No. 13/ No. 56<br>No. 93/ No. 55 | |
| BATTERY HAWKINS | 2 x 3-in M1903/ M1903 Ped<br>No. 97/ No.97<br>No. 98/ No.98 | |
| | | |

| FORT DERUSSY | | |
| --- | --- | --- |
| BATTERY RANDOLPH | 1 x 14-in M1907/M1907 DC<br>No. 1/ No. 5<br>1 x 14-in M1907M1/M1907 DC<br>No. 3/ No. 2 | 2 x 14-in M1907M1/M1907 DC<br>No. 1/ No. 5<br>No. 3/ No. 2 |
| BATTERY DUDLEY | 2 x 6-in M1908/M1905MI DC<br>No. 1/ No. 18<br>No. 2/ No. 19 | |

| FORT RUGER | | |
| --- | --- | --- |
| BATTERY HARLOW | 8 x 12-in mortars M1890M2/ M1896M1<br>mortar carriages<br>No. 180/ No. 307 M2<br>No. 178/ No. 308 M2<br>No. 179/ No. 310 M2<br>No. 177/ No. 309 M2<br>No. 124/ No. 261 M1<br>No. 175/ No. 243 M1<br>No. 152/ No. 262 M1<br>No. 181/ No. 289 M1 | |
| BATTERY BIRKHIMER | 4 x 12-in mortars M1890M1 on mortar<br>carriages M1896M1<br>No. 119/ No. 118<br>No. 98/ No. 107<br>No. 62/ No. 106<br>No. 92/ No. 103 | |
| BATTERY DODGE | 2 x 4.7-in Armstrong/ Pedestals<br>No. 11933/ No. 11021<br>No. 11009/ No. 11019 | |
| BATTERY HULINGS | 2 x 4.7-in Armstrong/ Pedestals<br>No. 11005/ No. 11016<br>No. 11001/ No. 11015 | |
| BATTERY S. C. MILLS | 2 x 5-in M1900/M1903 Ped<br>No. 20/ No. 20<br>No. 18/ No. 21 | |
| BATTERY GRANGER ADAMS | 2 x 8-in M1888MII/M1918 RY<br>No. 37 Watervliet/ No. 24<br>No. 45 Watervliet/ No. 25 | |

| FORD ISLAND MILITARY RESERVATION | | |
|---|---|---|
| BATTERY ADAIR | 2 x 6-in Armstrong/ Pedestals<br>No. 12123/ No. 11159<br>No. 12137/ No. 11163 | |
| BATTERY BOYD | 2 x 6-in Armstrong/ Pedestals<br>No. 12134/ No. 11158<br>No. 12138/ No. 11164 | |
| | | |
| FORT WEAVER | | |
| BATTERY WILLISTON | 2 x 16-in M1919MII/M1919 Barbette<br>No. 6/ No. 4<br>No. 7/ No. 3 | |
| | | |
| FORT BARRETTE | | |
| BATTERY HATCH | 2 x 16-in MkIIM1/M1919M1 Barbette<br>No. 63/ No. 8<br>No. 62/ No. 7 | |
| | | |
| FORT HASE KANEOHE BAY DEFENSES | | |
| BATTERY PENNSYLVANIA | 3 x 14-in NavyMkVIIIM4/MIIM4 turret mounts<br>Serial numbers not known | |
| BATTERY FRENCH (301) | 2 x 6-in M1903A2/M1 barbette<br>No. 2/ No. 90<br>No. 12/ No. 91 | |
| BATTERY COOPER (302) | 2 x 6-in M1903A2/M1 barbette<br>No. 7/ No. 92<br>No. 36/ No. 93 | |
| BATTERY DEMERRITT (405) | 2 x 8-in Navy MkVIM3A2/M1 barbette<br>No. 212/ No. 15<br>No. 218/ No. 16 | |
| | | |
| KAHE POINT MILITARY RESERVATION | | |
| BATTERY ARIZONA | 3 x 14-in Navy MkVIIIM4/MIIM4 turret mounts<br>Serial numbers not known | |
| | | |

| OPAEULA | | |
|---|---|---|
| BATTERY RIGGS | 4 x 8-in Navy MkIXM2<br>No. 496<br>No. 497<br>No. 498<br>No. 503 | |

| BRODIE CAMP | | |
|---|---|---|
| BATTERY RICKER | 4 x 8-in Navy MkIXM2<br>No. 495<br>No. 507<br>No. 508<br>No. 509 | |

| SALT LAKE | | |
|---|---|---|
| BATTERY BURGESS | 4 x 8-in Navy MkIXM2<br>No. 516<br>No. 510<br>No. 518<br>No. 517 | |

| WILIWILINUI  RIDGE | | |
|---|---|---|
| BATTERY KIRKPATRICK | 4 x 8-in Navy MkIXM2<br>No. 519<br>No. 520<br>No. 521<br>No. 522 | |

# APPENDIX II

## NAMING CITATIONS FOR THE BATTERIES OF THE HARBOR DEFENSES OF HONOLULU, PEARL HARBOR, KANEOHE BAY, AND NORTH SHORE GROUPMENT

### BATTERY ADAIR

One of the two 6-inch gun batteries built on the Ford Island Military Reservation. It was named on General Orders No. 13 of January 16, 1917, for 1st Lt. Henry R. Adair. Adair was born in Astoria in 1882, and graduated class of 1904 from West Point. His career was almost entirely with the 10th Cavalry Regiment. This unit participated with the 1916 Mexican Punitive Expedition against Pancho Villa. Adair was killed, along with Captain Charles Boyd, in action with Carrancistas troops at Carrizal, Mexico on June 11, 1916.

### BATTERY AVERY J. COOPER

A modern 6-inch battery built as Battery Construction No. 302 at Lae-O-Kaoio near the Kualoa Airfield on the northern flank of the Defenses of Kaneohe Bay. The Battery was named on War Department General Orders No. 96 of August 27, 1946. It was named for Colonel Avery J. Cooper, of the army's Coast Artillery Crops. Avery was born in 1880 in The Dalles, Oregon. He had a long service in the artillery, including a stint as a member of the War Department General Staff, followed by being posted to command of the Second Artillery District in 1938, He died on October 23, 1944, in Los Angeles at the age of 64.

### BATTERY BARRI

A battery for two 4.7-inch pedestal guns built as half of an emplacement at Bishop's Point section of Fort Kamehameha. The emplacement was named on General Orders No. 23 of April 27, 1915, for Captain Thomas O. Barri. Barri was born in 1821 in Norwich, Connecticut, He joined the pre-war 5th Massachusetts Militia in 1856. At the start of the Civil War, he was appointed as Captain in the Regular 11th US Infantry. He was shot on the second day of the Battle of Gettysburg, and died of wounds the next day, July 2, 1863.

### BATTERY BIRKHIMER

This was the mortar battery built during 1915-16 inside the Diamond Head crater of Fort Ruger. It was named on General Orders No. 15 of April 25, 1916, for Brigadier General William E. Birkhimer, U.S. Army. Birkhimer was born in 1848 and joined Iowa volunteer forces in 1864 for Civil War service as a private. After the war he entered West Point, graduating in the class of 1870. He excelled in mathematics and artillery, serving most of his assignments in that branch, though an interest in legal affairs led to several judge advocate postings. During the Philippine Insurrection he won the Congressional Medal of Honor for combat in 1899. He retired in 1906 and died on June 10, 1914.

### BATTERY BOYD

The second battery for two 6-inch guns erected on the Oahu Ford Island reservation. It was named on General Orders No. 13 of January 16, 1917, for Captain Charles T. Boyd. Boyd was born in Sperry, Iowa in 1870. He graduated with the West Point Class of 1896. He had assignments during the Philippine Insurrection and was an observer to the Russo-Japanese War in Manchuria in 1904, Assigned to the 10th Cavalry, the unit participated with the 1916 Mexican Punitive Expedition against Pancho Villa. Boyd was killed, along with Lieutenant Henry Adair, in action with Carrancistas troops at Carrizal, Mexico on June 11, 1916.

## BATTERY CARROLL G. RIGGS

Formerly known as Battery Opaeula, with General Orders No. 96 of August 29, 1946, this battery was named for Colonel Carroll G. Riggs, Coast Artillery Corps. It was one of the twin 8-inch turret sites created from the gun turrets removed from USS *Lexington* early in the war. Colonel Riggs was killed on December 18, 1943, along with 11 others in the crash of a B-24 bomber in Queensland, Australia while serving with the Federalized 197th CA (AA), New Hampshire National Guard.

## BATTERY CHANDLER

A battery for two 3-inch guns built as a section of a four-gun emplacement at Bishop's Point, Fort Kamehameha. The emplacement weas named on General Orders No. 23 of April 27, 1915, for 2nd Lieutenant Rex Chandler, Coast Artillery Corps. Chandler had enlisted at the Signal Corps Aviation School in San Diego, California. On his first flight, piloted by Captain Lewis Brereton, the Curtiss F floatplane crashed in San Diego Bay, and Chandler drowned under the wreckage, on April 4, 1918.

## BATTERY CLOSSON

A long-ranged 12-inch battery constructed at Fort Kamehameha in the early 1920s. Named by War Department General Orders No. 13, 1922. The battery was named in honor of Colonel Henry Whitney Closson, 4th U.S. Artillery. Closson had been born in 1832 in Whitingham, Vermont, and graduated West Point in the Class of 1854. He served in the Civil War and was recognized for meritorious service at the siege of Port Hudson in 1863. He served with the 1st and 4th Artillery and did a long posting with the Board of Ordnance and Fortification. The Colonel retired after forty years of service in 1896, and died on July 15, 1915, in Washington D. C.

## BATTERY DODGE

A battery for two 4.72-inch guns of the Oahu Land Defense Project at Fort Ruger. It was named on War Department General Orders Number 36 of June 9, 1915. It was named in honor of Major Theodore Ayrault Dodge. Dodge was born in 1842 in Pittsfield, Mass. At the start of the Civil War, he enlisted in the New York volunteers and lost his right leg at the Battle of Gettysburg. He commissioned as a lieutenant in the U.S. Army in 1866. He retired in 1870, Subsequently he was a renowned author, writer of the multi-volume History of the Art of War. He did in 1909 in Paris.

## BATTERY DUDLEY

Named by General Orders 59 War Department May 6, 1911, This battery was ordered named for Brigadier General Edgar Swarthout Dudley, USA who had died January 9, 1911. General Dudley had entered military service during the Civil War as a second lieutenant in the 1st New York Light Artillery Regiment. He was mustered out in November 1864 and entered the U.S. Military Academy from which he graduated in 1870 and was assigned to the 2nd U.S. Artillery Regiment as a second lieutenant. He was promoted to first lieutenant in June 1870 and to captain and acting quartermaster in December 1892. On May 9, 1898, he was appointed lieutenant colonel and judge advocate, U.S. Volunteers during the War with Spain. Following the war, he reverted to the rank of major and judge advocate in the Regular Army and was promoted to lieutenant colonel in the Regular Army. Subsequently he was promoted to colonel and upon his retirement he was promoted to brigadier general.

## BATTERY FORREST J. FRENCH

A modern 6-inch battery built on the north shore of the Mokapu Peninsula as part of the Kaneohe Bay defenses. It was constructed as Battery Construction No. 301. Naming occurred by War Department General Orders No. 96 of August 27, 1946, for Colonel Forrest J. French of the Coast Artillery Corps. French had been born in 1902 in Toledo Ohio and graduated in 1924 from the Annapolis Naval Academy. He transferred to the army, and spent a career in the Coast Artillery Corps. While service in the Southwest Pacific Theatre as commander of the 10th Antiaircraft Group, he died in a plane crash in New Guinea on March 8, 1944.

## BATTERY GEORGE W. RICKER

One of the twin 8-inch turret batteries emplaced on Oahu and named on General Orders No. 96 of August 27, 1946. It had previously been known as Battery Brodie. Lieutenant Colonel George W. Ricker was a career Coast Artillery Corps officer. He was born in 1892 in Newburyport, Massachusetts. He served as commanding officer of the enlisted division of the Coast Artillery School from August 1940 to September 1941. Colonel Ricker and Maj. Gen. Herbert A. Dargue, commander of the First Air Force, were killed in an airplane crash on December 12, 1941, in California.

## BATTERY GRANGER ADAMS

The 8-inch gun battery at Black's Point, Fort Ruger was named on War Department General Orders No. 10 of November 23, 1934. It was named for Brigadier General Granger Adams, an eminent artillery in the U.S. Army. Adams was born in Williamson, New York, he graduated West Point in the class of 1876. Granger Adams excelled in the science of artillery, became an instructor at West Point, and finished his career with assignments as President of the Field Artillery Board from June 1910. He retired in 1916, and died on March 27, 1928.

## BATTERY HARLOW

On January 23, 1909, with War Department General Orders No. 15, this mortar battery at the Diamond Head reservation of Fort Ruger was named Battery Harlow in honor of Maj. Frank Stowell Harlow, Artillery Corps. An 1879 West Point Military Academy graduate, Harlow had been briefly assigned to the 9th Inf. before being reassigned to the 1st Artillery, in which he served until the creation of the Artillery Corps in 1901. Promoted to Major in 1904, Harlow died on August 11, 1906.

## BATTERY HASBROUCK

The battery for 12-inch mortars built on the Fort Kamehameha reservation. It was named on General Orders No. 59 of May 6, 1911, for Brigadier General Henry Cornelius Hasbrouck, U.S. Army. Hasbrouck had been born in Newburgh, New York in 1839, and was graduated in the West Point Academy in 1861. He served in the 4th Artillery during the Civil War. Subsequently he had many assignments, including in Cuba in 1899. The general retired in 1903 at Fort Adams and died December 17, 1910.

## BATTERY HATCH

The two gun emplacements were named in General Orders No. 10 of the War Department dated November 23, 1934 in honor of Brigadier General Henry J. Hatch who as a captain had commanded the 143rd Company, CAC, when it manned Battery Hasbrouck at Fort Kamehameha. Hatch's company had been awarded the first Knox Trophy in 1913 for its winning target practice that year. Later in his career Hatch was a member of Coast Artillery and Ordnance Boards. General Hatch had died December 31, 1931.

## BATTERY HAWKINS

A battery for two 3-inch guns emplaced on the western side of the Fort Kamehameha Reservation. It was named on General Orders No. 72 of November 1913 for Brigadier General Hamilton S. Hawkins of both Civil War and Spanish American War service and former commandant at West Point. He retired on October 4, 1898. General Hawkins died March 27, 1910.

## BATTERY HULINGS

A battery for two 4.72-inch guns of the Oahu Land Defense Project at Fort Ruger. It was named on War Department General Orders No. 36 of June 8, 1915, to honor Colonel Thomas M. Hulings. Hulings was born in 1835 in Lewistown, Pennsylvania and was a lawyer, He enlisted in the Pennsylvania volunteers for the Civil War. He was killed in action with the 49th Pennsylvania Infantry on May 10, 1864, at Spotsylvania Courthouse, VA.

## BATTERY JACKSON

A battery for two 6-inch disappearing guns built on the main gun line of Fort Kamehameha. It was named on General Orders No. 72 of November 1913 for Brigadier General Richard Henry Jackson, U.S. Army. A native of Ireland who after nine years of service in the 4th U.S. Artillery Regiment as an enlisted man, was commissioned as an officer and rose through the ranks during the Civil War reaching the rank of lieutenant colonel in the Regular Army, Brigadier General of Volunteers, and Brevet Major General of Volunteers by 1865. Following the war, he reverted to the rank of major and in 1888 attained the permanent rank of lieutenant colonel of the 4th Artillery. General Jackson died November 28, 1892.

## BATTERY LEWIS S. KIRKPATRICK

War Department General Orders No. 96 of August 27, 1946, named the emplacement for two 8-inch naval turret mounts in honor of Lt. Col. Lewis Spencer Kirkpatrick, 59th CA (HD) Regiment, who commanded Fort Drum during the Japanese siege of the Manila Bay forts during the early months of World War II. After he was ordered to surrender Fort Drum, he was taken to Fort Mills on Corregidor Island, where he succumbed to pneumonia while a prisoner of war on April 27, 1943. Kirkpatrick was a 1924 graduate of West Point and a native of Oklahoma City, Oklahoma.

## BATTERY LOUIS R. BURGESS

On August 27, 1946, War Department General Orders No. 96 formally renamed Battery Salt Lake for Colonel Louis R. Burgess, Coast Artillery Corps. The colonel died on December 17, 1938. Burgess was born in Salem, Wisconsin, he was a West Point graduate class of 1892. He had many artillery and coast artillery assignments, including leading the 56th Regiment to France in 1917 and being a commander of the Harbor Defenses of Pearl Harbor and Fort Kamehameha in 1922-24.

## BATTERY RANDOLPH

On January 28, 1909, with General Orders No. 15 the War Department ordered that the 14-inch battery be named for Major Benjamin Harrison Randolph of the Artillery Corps who had died on October 14, 1907. Randolph had graduated from the U.S. Military Academy in 1870 and had been assigned to the 3rd U.S. Artillery as a second lieutenant. He was promoted to first lieutenant in 1879 and to captain in 1898. He was appointed to the rank of Major in the Artillery Corps in 1901.

## BATTERY ROBERT E. DEMERRITT

A modern battery of two 8-inch guns, during construction known as Battery Construction No. 405. It was located on the Puu Papaa Military Reservation, just to the south of the Kaneohe Bay peninsula and base. Naming was authorized by General Orders No. 96 of August 27, 1945, for Colonel Robert E, DeMerritt, USA, Coast Artillery Corps. DeMerritt was a career officer, serving mainly in antiaircraft units up to the Second World War. He died from non-combat injuries while in service at Fort Davis, North Carolina on July 25, 1942.

## BATTERY S.C. MILLS

The battery for two 5-inch guns at Black Point, Fort Ruger was named Battery S.C. Mills by General Orders No. 36 of June 9, 1915, honoring Colonel Stephen C. Mills. Mills was a West Point class of 1877 graduate, who served with distinction in the Indian Wars in New Mexico in 1880 and 1882. Much of his service was in the office of Army Inspector General. He was Chief of Staff of the Philippine Division in 1907. Mills died while in service at the age of 60 at Fort Ticonderoga on August 2, 1914.

## BATTERY SELFRIDGE

Named in War Department General Orders Number 15, 1909. Battery Selfridge was named in memory of First Lieutenant Thomas E. Selfridge, 1st U.S. Field Artillery Regiment. Selfridge had assisted Alexander Graham Bell for experiments in Nova Scotia before being detailed for other aeronautical work At Fort Myer Virginia in April 1908, he died while being a passenger on a flight of Orville Wright's on September 17, 1908.

## BATTERY TIERNON

The 3-inch gun battery at Fort Armstrong was named on General Orders No. 59 on May 6, 1911, for Brigadier General John Luke Tiernon. Tiernon was born in 1841 in Madison, Indiana and inducted into service as an artillery lieutenant in Missouri in 1862. He had a long service with the field artillery branch, including being commander of artillery in the Philippines from 1899-1900. He died in Buffalo, New York on March 30, 1910, after more than four decades of Army service.

## BATTERY WILLISTON

The battery for two 16-inch barbette guns erected at Fort Weaver, Oahu. It was named on War Department General Orders No. 13 of March 27, 1922. It was named for Brigadier General Edward B. Williston. Williston was born in Norwich, Vermont in 1837, graduating from Norwich University in 1856. In 1861 he was commissioned into the 2nd U.S. Artillery and served in several wartime actions. Was awarded the Medal of Honor in recognition of his actions at Trevilian Station as part of the U.S. Horse Artillery Brigade on June 12, 1864. At the start of the Spanish American War was appointed Brigadier General of Volunteers. He retired in 1900, and died in Portland, Oregon on April 24, 1920.

# APPENDIX III

## DESCRIPTIONS AND HISTORIES OF OAHU 3-INCH MODEL 1917 AA BATTERIES

**Fort Ruger:** One battery for four guns built July-August 1917 on the crater ridge above Battery Harlow, on the northern rim. Transferred to troops January 15, 1918 at a cost of $2236.02. Also known as AA Battery 5. Location apparently proved inadequate, three guns were mounted in 1920, but by 1925 were relocated to new positions in the post cantonment area north of the crater. Second battery for two guns built July-August 1917 near tennis courts of the post cantonment on the north side of the fort. Transferred to troops January 15, 1918 for $3562.45. Armed with guns relocated from the other post position on the crater ridge. Third battery for two guns at the Black Point reservation near Battery Granger Adams. Built in 1937 and used guns moved from the cantonment battery on the main Fort Ruger reservation. While in service also known as AA Battery 4. A third position was apparently added right before the war.

**Fort DeRussy:** Two-gun battery built July 1917 off Kalia and Saratoga Road (positions No. 1 and 2). Transferred to troops January 16, 1918 for $1277.02. Guns mounted in 1920 but moved by 1934 to new positions. Second two-gun battery built July 1917 off Saratoga Road towards the fort's eastern boundary, referred to as positions No. 3 and 4. Transferred to troops January 16, 1918 for $1277.02. Guns mounted in 1920. Expanded with two new blocks to form a 4-gun square by building new blocks and transferring the guns from old positions No. 1 and 2. New work done from September 16 to October 5, 1934. Transfer to troops November 19, 1934 for $937.89.

**Punchbowl:** Two-gun battery built in July 1917, inside the Punchbowl crater on the southwest rim. Transferred to troops in 1918 for $1142.80 for the blocks and $297.54 for the separate magazine. Not armed in 1925, probably was never armed.

**Sand Island:** Four-gun battery built in July 1917. Originally in two sections, emplacements No. 1 and 3 built first together. Transferred to troops January 25, 1918 for $1030.76 for blocks and $287,86 for magazine of corrugated iron and timber. Then added were emplacements No. 2 and 4 built August to September 1928, transferring troops on January 30, 1929. No armament reported in either site in 1925, but subsequently did receive all four guns. The AA Detachment of Battery F, 55th CA was assigned to provide the manning detachment for the battery in April 1941.

**Fort Shafter:** Four-gun emplacement built from July to September 1917. It was transferred to troops January 17, 1918 for $1957.86. It consisted of four gun blocks (separated into two 2-gun positions) and two frame magazines. Later guns from original No. 1 and 2 relocated to Fort Barrette. Then emplacements No. 1 and 2 expanded by the addition of two new gun blocks near, to form a four-gun square. Construction done in August to September 1934, with transfer to troops coming on November 19, 1934 for a cost of $1456.58.

**Fort Kamehameha:** One battery for four guns built from September to October 1917. One section of two guns built west of the main gunline near Battery Hasbrouck (No. 1 and 2) and one two-gun section to the east nearer Battery Selfridge (No. 3 and 4) at Fort Kamehameha. Transferred to troops February 4, 1918 for $1192.14. Guns No. 1 and 2 mounted 1920, but due to filling in the area, had been dismounted in 1928. A second, four-gun battery built from October to November 1928. It was located on the Ahua Point Reservation on the far eastern end of Fort Kamehameha. Transferred to troops on December 31, 1918 for $3094. Known while in service as AA Battery No. 12. Used the two guns dismounted from positions No. 1 and 2 of the gunline battery. A third, four-gun battery built from February 17 to June 12, 1937 as a square-shaped position west of Battery Closson. Transferred to troops on May 21, 1937 for $2548. Utilized the guns located at the Ahua Point location previously.

**Ford Island:** A four gun emplacement (blocks on the points of a 200-foot square) was begun in September 1917, completed the following month, and transferred to troops on February 4, 1918. Construction cost to completion was $1866. The battery was placed at the northern end of the island (still within the army's portion) not far from Battery Adair. Just two of the emplacements (No. 1 and No. 2) were reported as mounted in 1920. Apparently, they did not serve here long.

**Schofield Barracks:** Four-gun emplacement built in two separated two-gun sections. First two emplacements built from September to October 1917. Transferred to troops January 23, 1918 for a cost of $1503.73 for blocks, and $759.44 for magazines. Located north of main cantonment. Two final emplacements not originally built until 1927. Constructed in February 1927. Transferred to troops September 27, 1927 for $2042.14. Was armed with all four guns also in 1927.

**Fort Weaver**: New four-gun emplacement originally built in a square formation east of Battery Williston. Work was done in January 1927, transfer to troops made on September 24, 1927 for a cost of $1957.86. All four guns were reported mounted in June 1927.

**Fort Barrette**: New four-gun emplacement built from March 4, 1937, to May 10, 1937. It was configured as a square of four guns, north of Battery Hatch. Transfer to troops made May 21, 1937 for a cost of $1416.51. In service it was also known as AA Battery No. 22. It was armed with four guns that had been in storage for a while at Fort Shafter.

Early 1921 photograph of the two 3-inch Model 1917 guns on Ford Island. *Smith Collection*

# APPENDIX IV

## DESCRIPTIONS AND HISTORIES OF OAHU 8-INCH RAILWAY FIRING SITES

**Gilbert Firing Point.** Four spurs terminating in firing position. The Gilbert tract between Browns Camp and the Kapolei Military Reservation (Fort Barrette) was established in 1923, as the first of the firing positions for the M1890M1 12-inch railway mortars of the Hawaiian Railway Battalion. The position was used extensively as a firing point for the railway mortars during the 1920s and early 1930s; and after 1934, by the 8-inch guns of the 41st CA, until the Browns Camp Battery Position was placed in service in 1937. The position was deleted from the railway artillery positions early in World War II and the location was not apparently utilized as a defensive position, though listed as an emergency alternate position.

**Browns Camp Military Reservation.** Four spurs terminating in firing position. This reservation, one of the several tracts that made up the extensive prewar Honouliuli Military Reservation was selected for a railway artillery battery battle position in 1937, work completed by July 1937. Extensively used for practice firing by 8-inch railway guns of 41st CA. The site was also expanded to include other coastal batteries after the commencement of World War II. This reservation absorbed the adjacent Awanui Military Reservation during the early part of 1942. Assigned to the Hulu Group, the position was manned on December 8, 1941, by Battery A, 41st CA (Ry) Regiment with four 8-inch M1888MIIA1 guns on M1918 Barbette Carriages. Battery A appears to have manned the Browns Camp Railway Battery at least through December 1943. Fire control of the battery was provided by a pair of 50-foot demountable towers that served as a battery commander's station and as a CRF position at the battery site. Other base end stations were distributed along the Leeward shore of Oahu at Kahe Point, Puu O Hulu and other locations. The 8-inch guns were used for target practice on a regular basis through October 1944.

**Waianae,** midway between Nanakuli and Makaha on the southwest coast. Four spurs terminating in firing position. The Waianae position was built under the jurisdiction of the District Engineer, and was inspected as completed on December 5, 1941.

**Puuiki** near Mokuleia on the north shore. Four spurs terminating in firing position. Land purchased 1937. The Puuiki Battery position begun in 1939, was completed by September 11, and the position transferred to the HSCAB on November 17, 1939. Scheduled for occupation after the Pearl Harbor attack guns were deployed instead to Kawailoa, and Puuiki was not subsequently used as a battery position.

**Kahuku** at Oahu's northernmost point. Four spurs terminating in firing position. Land purchased 1937. Site not developed but replaced in function by relocating the Laie position further north.

**Fort Kamehameha.** Two firing positions, each with four spurs terminating in firing position (so all eight guns of a battalion could have access to a practice firing spot). Firing spurs generally completed by June 28, 1941, but many problems found with track and ground stability. Repair work undertaken, but still less than satisfactory. Despite the problems, one position of four spurs accepted and kept in service, but little used, after the start of the war.

**Kawailoa** on the North shore, initially developed as an alternate position. Four spurs terminating in firing position. Land purchased 1937. The 3ʳᵈ Engineer Regt began construction of the Kawailoa alternate position by on November 1, 1939, completed it December 29, and transferred to the HSCAB on February 8, 1940. Occupied by four guns on December 9, 1941. Known as Battery Haleiwa, the battery occupied this site until 1944.

**Maili** on Leeward coast, initially developed as an alternate position. Four spurs terminating in firing position.

**Laie** on Windward coast, initially developed as an alternate position. Four spurs terminating in firing position. Site moved north (to a position subsequently known as "New Laie"), replacing in function the Kahuku site. Site known as Kahuku Palms Golf Course. Problems trying to camouflage such a position on this coast led to a change to emplace the guns permanently as railway upper carriages on fixed gun blocks. Two guns served in January 1942 on temporary spurs. Starting in May 1942 new fixed gun blocks built and four guns mounted (including the two temporary mounts) as Battery Kahuku.

**Camp Ulupau** on crater at Mokapuu Point. Four spurs terminating in firing positions. Placed just east of the future Fort Hase cantonment area. Construction started after the beginning of the war, guns emplaced on new railbed by December 24, 1941, Manned in early January by the newly arrived 1st Bn, 52ⁿᵈ CA (Ry). Guns re-emplaced on fixed gun blocks to serve as Battery Sylvester in the Harbor Defenses of Kaneohe Bay.

The first contingent of eight 8-inch Railway Guns assembled on 10-10 Dock, Pearl Harbor ready for transfer to the Army in July 1934.
*Smith Collection*

# APPENDIX V

## DESCRIPTIONS AND HISTORIES OF OAHU 155 MM PANAMA MOUNT EMPLACEMENTS

**Ahua.** As part of Pearl Harbor defenses, scheduled construction of Panama mounts to replace 5-inch emergency battery, likely never constructed, no further documentation on its existence has been found.

**Aliamanu.** Four Panama mounts for the Pearl Harbor defenses. February 1942 authorized as new alternate Panama mount site, built later that spring. Reserve position for Fort Kamehameha guns, it is not known if ever actively armed or manned except perhaps in training exercises.

**Ashley Military Reservation.** Near Ashley Station in northern Oahu. For four Panama mounts. North Shore Groupment defenses. Started May 17, 1939, completed October 18. 1939. Transferred January 3, 1940 for $2957.00. Initially unmanned at the start of the war, but took four 155, of the 11ᵗʰ Field Artillery of the 24ᵗʰ Division when deployed right after the Pearl Harbor raid. Armed and manned by Battery F, 57ᵗʰ CA from January 1942 to November 1943. Then unarmed until arrival of Battery C, 47ᵗʰ CA with 155 mm M1 guns in February 1944. This unit left in May, and once again 155 mm GPF guns were emplaced from July to September 1944. After that the emplacement was deactivated to the end of the war.

**Awanui (Browns Camp).** Four Panama mounts of the Pearl Harbor defenses. Started October 15, 1940, completed November 27, 1940. Transferred July 15, 1941 for $5844.20. This reservation on the Oahu southwest shore at Awanui Gulch on the shoreline east of Browns Camp and west of Camp Malakole was established in 1940, as an alternate battery position for 155 mm guns. On December 7, 1941, two 155 mm field position guns manned by a detachment of the 1ˢᵗ Battalion, 55ᵗʰ CA. After 1942 the battery was increased to a four-gun battery on Panama mounts also known as the Browns Camp 155 mm Gun Battery. Battery Awanui covered the small boat harbor, a potential enemy landing point on the south shore. It is doubtful that the battery position was manned after April 1945.

**Barbers Point Military Reservation.** Four Panama mounts. Built in two 2-gun sections about 500-feet apart northeast of Barbers Point Light. Started April 13, 1937, completed June 30, 1937. Transferred August 16, 1937 for $5174.79. Alternate position used before the war by the 1ˢᵗ Battalion, 55ᵗʰ CA. While manned on December 7th, does not appear to have been regularly used for armament subsequently.

**Dillingham.** Two Panama mounts of the North Shore Groupment defenses. Battery authorized to replace a battery of ex-navy 4-inch guns emplaced and then removed in May 1943. Located at the Anahulu Flats Reservation to cover Waialua Bay. Panama mounts constructed late in 1942. Guns with the 3ʳᵈ Bn, 57ᵗʰ CA served here between May 1943 and January 1944. Later in 1944 the battery was supplied with 155 mm GPFs twice in order to train units stationed here for short periods. It does not appear to have been used after September 1944.

**East Battery.** East was established in the early months of 1942 a short distance from the site of the East Beach Battery, but further inland and closer to the southern slope of Ulupau's west rim, and more importantly on permanent Panama mounts. During this period, four bombproof concrete magazines with a capacity of 100 powder charges and projectiles were constructed and then covered with a thick earth covering to form mounds some 15-feet high by the 34ᵗʰ Engineers. The mounds were supported on their exterior by several layers of hardened cement bags. Reinforced concrete for four Panama mounts for the guns was then poured atop the mounds. Upon completion of this construction on March 3, 1942, the two 155 mm guns temporarily emplaced at Wailea Point were returned to the new battery position and the pair of guns at Battery East Beach was moved to their new emplacements atop the mounds. Battery East was retained in position through September 1944, at least.

**Ewa.** February 1942 authorized as a new alternate Panama mount site. Built a few hundred yards northeast of fire control station "A" in early 1937. Initially manned by Battery C, 55th CA Armed and manned by Battery D, 57th CA in June 1942, and Battery A, 48th CA in early 1944.

**Homestead.** Four Panama mounts for the Pearl Harbor defenses. Authorized to replace an emergency 5-inch battery. Work finished November 9, 1942. As an alternate position, was not regularly armed. Maintained by 804th CA (HD) Battery (Separate) until May 23, 1943. Thereafter used only intermittently for training, however records state 238 round were fired in 1944 from this location.

**Kaena.** Two Panama mounts of the North Shore Groupment defenses. Authorized to replace an emergency 4-inch battery being removed in mid-1943. New Panama mount emplacements completed by May 1943. Served with guns by 3rd Bn, 57th CA until removed on January 31, 1944. Positioned inactivated and unused until the end of the war.

**Kahana.** Four Panama mounts of the Kaneohe Bay defenses. To replace an emergency 5-inch battery. In May 1944 work was underway. After decommissioning the ex-navy 5-inch guns at this battery site in 1944, the location was assigned to several units as a location for 155 mm guns. At times these were the more modern 155 mm M1 guns, but in 1944 four older GPFs were emplaced on the Panama mounts recently constructed. These were manned by Battery B, 16th CA. By late summer that unit departed, and the battery site was abandoned as far as armament was concerned.

**Kahe.** Four Panama mounts of the Pearl Harbor defenses. February 1942 authorized as new alternate Panama mount site. Work started February 10, 1942. Armed and manned by Battery B, 55th CA from late spring 1942. Served until late 1943, and again with Battery B, 48th CA in the summer a few months in 1944.

**Kam.** Four Panama mounts in the Pearl Harbor defenses. Established at Fort Kamehameha during the Department maneuvers of May 1939, when a company of the 3rd Engineers was assigned to build four Panama mounts as a firing point for 155 mm guns. They were located on the shore between Batteries Hawkins and Jackson and covered the approaches to the Pearl Harbor channel. Magazine bunker revetments were built of corrugated iron, placed behind vertical pipes and backed up with a large amount of dirt. The battery position was transferred to manning detachment of the 55th CA on March 25, 1940. Not armed with guns until after the start of the war. Armed and manned by Battery A, 55th CA in June 1942. After a reorganization of units in mid-1942, they were taken over by the 803rd CA (HD) Battery (Separate). Manned until end of 1943.

**Kapoho.** Four Panama mounts of the Kaneohe Bay defenses. February 1942 authorized as new alternate Panama mount site for the batteries at Fort Hase. Located on Papaa Ridge, being built in the summer of 1942. Fire direction to the east and northeast. Outfitted with the usual magazines, plotting room BC station and coincidence rangefinder position. It is not clear if the circular Panama mounts were ever actually completed. No evidence of being armed or regularly garrisoned. Site abandoned by mid-1943.

**Kawailoa.** Four Panama mount in the. North Shore Groupment defenses. Located a few hundred yards north of Camp Kawailoa, on the "E" Reservation. Built in October 1939. In service with troops of 11th FA on December 7, 1941, on January 24, 1942 turned over. Armed and manned by Battery E, 57ᵗʰ CA. Several other units manned the emplacement continuously until August 1944. Frequently used for firing exercises and training by several 155 mm and even 8-inch turret batteries.

**Koko Head.** Four Panama mounts of the Honolulu defenses. The four concrete Panama mounts were completed on January 22, 1941, and transferred on July 22, 1941, at a cost of $3,351.49. Trunnion elevation (in battery) of the four guns was over 650-feet. The battery site also included 3,960 linear feet of five-strand barbed wire. At the start of World War II, Battery E, 55ᵗʰ CA (TD) Regiment, based at Fort Ruger, established the position as an alternate site for Battery Ruger and one gun was moved to the position. A second 155 mm GPF was towed to the position by mid-March and both guns were test fired before both guns were returned to Battery Ruger on March 19, 1942. There was no evidence that the Koko Head Battle Position was consistently manned after March 1942. A permanent concrete battery commander's station was constructed below the battery, some 400-yards southwest of the approximate center of the battery.

**Loko.** Four Panama mounts of the Kaneohe Bay defenses. This battery derived its name from nearby Ka Lae O Kaoio Point at the north end of the Kualoa Airstrip. It was not developed or armed until Battery A, 57ᵗʰ CA was assigned to the location on January 20, 1942. Panama mounts were built for four 155 mm guns early in 1942. Batteries of the 57ᵗʰ CA continued to man the position at Battery Loko and AMTB Battery Number 8 until May 1944. The 155 mm guns were removed from Battery Loko on May 9, 1944, and relocated at Battery Wili, the 155 mm gun position on Wiliwilinui Ridge when the battery was transferred from the HD of Kaneohe Bay and North Shore to the HD of Honolulu and Pearl Harbor.

**Mokuleia.** Four Panama mounts of the North Shore Groupment defenses. Located about two miles inland from Puuiki railway mount position on the Makalehe-Makua trail. February 1942 authorized as new alternate Panama mount site, probably completed in early summer 1942 by the 34ᵗʰ Engineer Regiment. Never armed or fully garrisoned.

**Nanakuli.** Four Panama mounts of the Pearl Harbor defenses. Authorized in September 1942 to replace 5-inch battery at this position. Panama mounts constructed and guns taken for Department war reserve for mounting. Manned by Battery E, 57ᵗʰ CA. Not actively manned, armament removed and battery use discontinued after about May 1943.

**Palailai.** Four Panama mounts of the Pearl Harbor defenses. February 1942 authorized as new alternate Panama mount site. Constructed in the spring of 1942 near the summit of Puu Palailai. Reservation used from the 1920s for a complex of fire control stations. Gun battery manned by Battery C, 15ᵗʰ CA from Fort Barrette in 1942-43, though generally unarmed as an alternate site.

**Papaa.** Four Panama mounts of the Kaneohe Bay Defenses. February 1942 authorized as new alternate Panama mount site. On the northeast slope of Puu Papaa, overlooking Kailua Bay. Like Kapaho Battery there is no evidence that the emplacement was ever built, at least with the positions and support rooms originally intended. Never armed, and apparently site abandoned in 1943-44 with the completion of other elements of the Kaneohe Bay defenses.

**Pine.** Four Panama mounts in the North Shore Groupment defenses. Located at an elevation of 500-feet in the pine forest at a ridge four miles inland from Waialua Bay, north of the Haleiwa-Opaeula Road. February 1942 authorized as new alternate Panama mount site, construction accomplished in the summer of 1942. Intended as a fallback position for the 57ᵗʰ CA. Not armed, garrison present only irregularly.

**Punchbowl.** Four Panama mounts of the Honolulu defenses. As early as 1929 plans were proposed to emplace four Panama mounts on the Punchbowl's south rim for 155 mm guns. In the late 1930s, two 360° type Panama mounts were emplaced on the eastern side of the reservation at an elevation of about 300-feet. Here a broad field of fire could be obtained to seaward as well as all around coverage of the Honolulu area. These emplacements were not provided with guns, however, until after World War II had begun. Two additional mounts were approved early in January 1941 and brought to completion and on July 22, 1941. The position was not occupied, however, until October 1942. On October 21, the 807th emplaced two guns on the Panama mounts, to the right of the original mounts. During the period the Punchbowl was manned by the 807th, the battery conducted its live fire practices at an Artillery Range at Schofield Barracks. During the summer of 1943, the four guns at Punchbowl were removed and prepared for movement to other islands in the Central Pacific Base Command Area. The Panama mounts at the battery site were modified for new 155 mm M1 guns. The first three of the new guns arrived on June 24, and the fourth on July 8. The battery received movement orders in November 1943 and moved from the Punchbowl with its M1A1 155 mm guns. Subsequently the Punchbowl Battery was temporarily unmanned, but four newly arrived M1918 GPF guns, part of a shipment of 25 received from the mainland in the summer of 1943, were emplaced on the Panama mounts. These guns remained in place through 1944.

**Pupukea.** Four Panama mounts of the North Shore Groupment defenses. February 1942 authorized as new alternate Panama mount site. Completed in late summer 1942, and manned initially by Battery F, 57th CA. Located about four miles in from Waimea Bay, north side of the Kamananui stream. It fired to the north. Not manned on a regular schedule, though used in 1944 for target practice with the groupment defenses.

**Pyramid.** This battery was established as a new position for the four 155 mm M1918M1 GPF guns on Panama mounts previously positioned in field emplacements at North Beach. The new position was located just to the south of Pyramid Rock by Battery C, 57th CA in January 1942. 34th Engineers prepared these new battery emplacements. The completed battery position consisted, in addition to the gun emplacements, of four powder magazines with a capacity of powder charges for 100 rounds, a plotting room, a battery commander's station and a coincidence range finder station. The battery served until mid-1944.

**Ranch.** Also referred to as **Kahuku**, these four Panama mounts were assigned to the North Shore Groupment defenses. At the Kolaepkahipa Ridge on the Kahuku Ranch property. Started March 11, 1941, completed April 16, 1941. Transferred July 23, 1941. Built in two pairs, one facing north, the other pair mostly east. Was not armed on December 7, 1941 due to shortage of personnel. Guns issued from department war reserve and crew of Battery B, 57th CA arrived in January 1942. The unit rearmed with 155 mm M1 guns and took them when the unit redeployed elsewhere in the Pacific in late 1943, the 155 mm GPF guns stayed for a while at Battery Ranch on their Panama mounts. Emplacement modified in mid-1944 for M1 155 mm guns of Battery B, 45th CA, but appears to have changed back again later. It remained armed, served by small caretaker detachment until April 1945.

**Round Top.** Four Panama mounts of the Honolulu Harbor defenses. On Roundtop Hill behind and above Fort DeRussy. February 1942 authorized as new alternate Panama mount site. Work started February 10, 1942. Fallback position of the 2nd Bn, 55th CA. Rarely, if ever, used actively armed, though garrisoned by a small detachment for maintenance purposes.

**Ruger.** Four Panama mounts. On the northwest corner of the cantonment area to the north of Diamond Head crater. Atop a rise covering the western approached to Fort Ruger through Kapiolani Park with the Honolulu defenses. Started February 8, 1937, completed May 10, 1937, Transferred September 2, 1937 for $7282.34. Unarmed at the start of the war, its guns having been temporarily moved to Kaneohe Bay. Issued 155 mm guns from the war reserve in January 1942, fired test rounds on February 6, 1942. Armed and manned by Battery E, 55th CA until February 1943 when it, and the four guns, relocated to Battery Wili. Then remained unarmed through the end of the war.

**Sand Island Military Reservation.** Four Panama mounts of the Honolulu defenses. Started February 15, 1937, completed April 16, 1937. Transferred September 12, 1937 for $6393.91. Besides gun positions, also had a small room for generator. Armed and manned by Battery F, 55th CA from November 27, 1941 to October 1942. Guns relocated to Punchbowl battery in December 1942. Emplacements remained unoccupied to the end of the war.

**School (Kalihi).** Four Panama mounts of the Honolulu Harbor defenses. On the campus of Kamehameha School in Aiewa Heights to the east of Fort Shafter. February 1942 authorized as new alternate Panama mount site, from April 1943 to replace a 4-inch emergency battery at Kalihi, then being disarmed. Does not appear to have been regularly armed or garrisoned.

**Waimea.** Three Panama mounts, of the 180° type, assigned to the North Shore Groupment defenses. February 1942 authorized as new alternate Panama mount site. Built in the summer of 1942 on a ridge of the north bank of Kaiwikoele Stream and Waimea Valley, about three miles inland from Waimea Bay. Constraints with the local topography and economy combined to change the three emplacements finished when work was suspended in October 1943 and Battery F, 57th CA departed. Work was not completed on placing the mount traversing rails or other buildings. Never armed, never completed.

**Weaver.** Four Panama mounts. On Keahi Point, between Battery Williston and the beach at Fort Weaver for the Pearl Harbor defenses. Started April 16, 1934, completed August 25, 1934. Transferred September 18, 1934 for $5912.04. The first Panama mount battery built on Oahu. Frequently used for practice firings, and continuously armed up to December 7, 1941; armed and manned by Battery C, 55th CA in June 1942. Thereafter was not occupied, though maintained as a possible alternate position.

**Willy.** Also known as just Will. Four Panama mounts of the Honolulu Harbor defenses at the Wiliwilinui Military Reservation. Authorized February 1942 as new alternate Panama mount site, not actually constructed until later that year. Site is about 3000-feet south of 8-inch turret Battery Wilridge at an elevation of 500-feet. In December 1942 fired first testing rounds with guns by Battery E 55th CA. Manned by several units through most of 1943 and 1944, when it was disarmed and reduced to caretaker status by the troops manning Battery Wilridge.

**X-Ray.** Also called **Oneula** 155mm Battery. Four Panama mounts of the Pearl Harbor defenses. An alternate position built in early 1942, near fire control station "X" on the beach at Oneula. Originally site had a single ex-navy 5-inch gun from Battery Ahua, but this was removed when the Panama mounts were constructed. Not known if ever armed with 155 mm guns or actively manned.

# APPENDIX VI

## DESCRIPTIONS AND HISTORIES OF OAHU FIXED 240 MM HOWITZERS SITES

**Kaaawa Battery.** The first of the new fixed emplacements for the 240 mm howitzer was built as a single emplacement at Kaaawa just east of Puu O Mahie and Kahana Bay. Construction was from March 3, 1927 to April 7, 1927. Transfer to troops made on September 27, 1927 for a cost of $2267.51.

**Laie Battery.** A pair of emplacements for 240 mm howitzers was constructed from March 23, 1931 to May 29, 1931. It was transferred to troops on January 22, 1932 for $5487.76. Site was abandoned for armament prior to the start of the Second World War.

**Makua Battery.** A battery consisting of three emplacements, was constructed on the Makua Military Reservation south of Kaena Point on Oahu's western shore. This battery was composed of three circular concrete mount emplacements built from July 17, 1930 to November 29, 1930. It was transferred to troops on January 22, 1932 for $9150,51. Two platforms were emplaced on one of the small tracts of the reservation, the remaining howitzer emplacement being located on separate nearby tract. These howitzers were not normally emplaced at the battery position, instead being stored at the Hawaiian Ordnance Depot at Aliamanu Crater. In accord with the howitzers' changed function in 1941 the emplacements at Makua were supplanted in 1939 and 1941 by newly prepared emplacements located further inland. The traverse racks of the Makua emplacements were removed and reused in the new emplacements at Batteries Kahili and Kolekole.

**Pupukea Battery.** Another set of emplacements was also constructed to cover less well-defended portions of the Oahu shoreline. This battery site of two emplacements was built at Pupukea on the north side of Waimea Valley on the island's North Shore. Construction was done from May 23, 1927 to June 30, 1927. It was transferred to troops on September 27, 1929 for a cost of $4216.74.

**Ulupau Battery.** On May 20, 1927 a battery site was completed for two howitzers at the Kuwaahoe Military Reservation in the shadow of Ulupau crater's west rim on the Mokapu Peninsula. Using materials supplied by the Hawaiian Department Quartermaster, two concrete gun blocks of the same type as that constructed at Kaaawa were built for two 240 mm M1918 howitzers to be mounted on M1918A1 pedestal mounts were constructed by troop labor between April 11, and May 20, 1927. It was transferred to troops on September 27, 1927 for $4515.75.

**Waimanalo Battery.** A third set of two emplacements was constructed on the Waimanalo Military Reservation in October and November 1929 these were reconstructed and completed by May 29, 1931 and transferred to the coast artillery on January 22, 1932 for $1501.73 (remodel cost) One of the sites had been Major Johnson's prototype emplacement of 1923, but was remodeled with improvements while the second emplacement was new. These emplacements were located to cover the waters of Waimanalo Bay on the island's northeastern shore.

The six later 240 mm Emplacements for Field Artillery usage:

**Anahulu Flats Battery.** A small reservation in the foothills of the Koolau Range east of Kawailoa was acquired in the months just before the Japanese attack in 1941. Anahulu Flats had long been an area of interest for the US Army on Oahu and had been considered as a location for a pair of 16-inch guns during the period between the two world wars. By the late 1930s the area was selected a field artillery site for a 240 mm howitzer battery. Construction was probably done in the first half of 1941. On July 16, 1944, the Area Artillery Officer recommended that the howitzers be taken out of service and their provisional manning detachments be disbanded, as they were no longer required for the defense of Oahu.

**Kalihi Battery.** In May 1939, a small military reservation encompassing about 0.20 of an acre at the Kalihi Pali above the Kalihi Valley was set aside by Executive Order on August 2, 1939. The tract, along with a right of way easement, was acquired from the Bishop Estate. The Kalihi Battery was begun by 3rd Engineers, on January 27, 1941, and completed February 24, transferring on June 26, 1941. When the field artillery had selected the site more than 1,000-feet above sea level, they had not taken into account the difficulty of providing ammunition, a new roadway was required. The battery remained incomplete, however, in terms of its plotting room, battery commander's and fire control stations, magazines, quarters, and other support facilities. Among the various temporary field works built on the Kalihi Pali was a single base end station that provided target data to batteries on both the north and south sides of the Koolau Range. When the U.S. entered World War II, the battery site was manned by detailed personnel from the 90th Field Artillery Battalion, 25th Infantry Division. The battery was finally placed in service in 1942 and remained active until mid-1944. Various other field fortifications were built on the reservation during World War II. In the latter months of 1948 the tract of land occupied by the howitzer battery at Kalihi Pali was returned to the Territory of Hawaii by Executive Order of the President.

**Kolekole Battery.** In December 1936, Major General Hugh A. Drum, commanding the Hawaiian Department recommended that a project for a new battery site at Kolekole on the Schofield Barracks Reservation near Kolekole Pass in the Waianae Range. The battery's access road down the steep western face of the Waianae Range was completed in 1939. Construction of the emplacement was undertaken on June 15, 1939. Construction proved difficult to access and the location was moved slightly. The battery was finally completed on March 30, 1941 and on April 1 the battery was transferred for $293.75. The Kolekole position was manned by elements of the 90th Field Artillery Battalion, 25th Division in December 1941. The howitzers remained assigned here until July 1944 when they were taken out of service and their manning detachments reassigned.

**Kunia Battery.** The 11th Field Artillery Brigade recommended that a fourth howitzer battery site be established at Kunia in addition to the three whose locations had been previously approved for construction. This battery could cover the entrance to Pearl Harbor, Barbers Point, and Pokai Bay areas. The Kunia reservation on the Waikele Plain south of Wheeler Field was established in 1940. Construction of the emplacements began March 24, 1940, proceeded well and was completed on May 20. It was transferred to troops on October 21, 1941 for $4689.39. It was one of three such battery positions manned by elements of the 90th Field Artillery Battalion, 25th Division in December 1941. The battery site at Kunia was used as a firing point for live fire target practice by the units manning the howitzers. The reservation was enlarged later in the war becoming a replacement camp during the last months of the war.

**Pupukea Battery.** This was the older 1927 position taken in and refurbished. The position was one of the few original howitzer positions retained for use by the Field Artillery when they took over the howitzers from the coast artillery in the Mid-1930s and was used as an alternate position for the howitzers during World War II.

**Quadrupod Battery.** In December 1936, Major General Hugh A. Drum, commanding the Hawaiian Department, recommended that a project for three new battery sites for the howitzers be authorized by the War Department. Each of the positions was to be occupied by two howitzers on all around fire mounts. The three sites were at Quadrupod, Kalihi and Kolekole on the Schofield Barracks Reservation. This one was on a small tract acquired near Paalaa Peak on the summit of Quadrupod, in the Koolau Range in 1941. Construction began on May 13, 1941 and was completed on June 28 1941, Transfer to troops was made on October 28, 1941. Here one of the new battery positions for two 240 mm howitzers was established at an elevation of some 1,650 feet. This battery served as an alternate location to the older emplacements at Pupukea built in 1927 This battery of two emplacements was initially manned by field artillery from the 24th Division in December 1941. In 1942, a provisional field artillery battery was assigned to the Quadrupod battery. By July 16, 1944, it was recommended that the howitzer battery position be closed and its manning detachment be disbanded.

After they were withdrawn from coast artillery use, many of the 240 mm howitzers were assigned to Oahu's field artillery units.
*Smith Collection*

# APPENDIX VII

## DESCRIPTIONS AND HISTORIES OF OAHU WAR EMERGENCY GUN BATTERY SITES

Descriptions and Histories of Oahu Emergency Navy 5-inch AA Gun Battery Sites

**Battery No.1, Hickam Field.** Located at the border of the AAF Hickam Field and the main Pearl Harbor gate. Constructed and initially manned by U.S. Navy crew from USS *California* (which also was the source of these guns), Site selected January 20, 1942, completed by February 9, 1942. Four circular blocks arranged in a diamond formation. Had ready ammunition lockers inside the revetments, separate main magazine. Capable also of surface coverage. Disarmed about the end of 1943.

**Battery No.2, West Loch.** Located west of West Loch, a site the navy at first described as Honouliuli. Used guns from USS *West Virginia*. Navy planned, built, and initially manned when first completed on February 9, 1942. Gun crew sheltered in adjacent tents. Subsequently turned over to the army. Disarmed about the end of 1943.

**Battery No.3, Puuloa.** Located on navy land adjacent to Fort Weaver on the east side of the Pearl Harbor entry. Used guns from USS *California*. Completed by the navy by May, 1942. May have been initially manned by personnel from USS *Utah*. Capable of surface coverage. Transferred to army, who referred to it as Ewa Beach Battery. Became an important training site for other crews with this type of armament. Disarmed about the end of 1943.

**Battery No.4, Ewa Mooring Mast.** Located adjacent to the old navy airship mooring mast at Ewa Airfield (which later became the Marine Corps Air Station in April 1942). Built on a coral sand clearing. Navy originally manned but transferred then to army operation. The only battery of this series to use navy 5-inch/38 dual purpose guns rather than 5-inch/25. Guns originated from damaged destroyers USS *Downes* and *Shaw*. Disarmed about the end of 1943.

**Battery No.5, Fort Kamehameha.** Located on the Ahua Point reservation on the far eastern side of Fort Kamehameha. Also begun by the navy but completed by army engineers before operational status. Built from May to June 1942. Consisted of gun blocks, two separate magazines, an air bottle storage room, machinegun pillbox, and battery commander's station. Manned by the 710th CA (AA) Battery (Separate) initially. Active until disarmed in mid-1944.

**Battery No.6, Waipio.** Located on navy land leased for sugar production on the eastern shore of the Waipio Peninsula. Construction started by the navy, finished by the army and their contractor Hawaiian Construction Co. Finished November 20, 1942. Manned by the 711th CA (AA) Battery (Separate) initially. Disarmed about the end of 1943.

**Battery No.7, Ford Island.** Located on the southern end of the island. Started by the navy, completed by and manned by the army. Probably finished in late 1942 but disarmed a year or a year and a half later in early 1944.

**Battery No.8, Aiea Heights.** Located on the Aiea Heights Military Reservation along with other coast defense elements. Detailed plans were completed on April 8, 1942. Construction work undertaken by Hawaiian

Contractors under army supervision. Guns mounted September 18, 1942, emplacement reported complete on November 20, 1942. Manned by the 710th CA (AA) Battery (Separate) initially Disarmed about the end of 1943.

**Battery No.9, Sand Island.** Located on the Sand Island Military Reservation on the western side of Honolulu Harbor. Final battery started by the army for ex-navy guns. Scheduled for completion on October 15, 1942, guns reported mounted September 18, 1942. Disarmed about the end of 1943.

Descriptions and Histories of Oahu War Emergency 3-inch Gun Battery Sites

**Battery Kii,** South Ulupau Head, Fort Hase, Defenses of Kaneohe Bay. Simple emplacement for two loaned ex-navy 3-inch Mk VI guns on Mk MVII pedestals erected for seacoast defense. Battery built in mid-1942 by the army's 34th Engineer Regiment. Manned by Battery D, 57th Coast Artillery. Served at least until mid-1944.

**Battery Puka,** North Ulupau Head, Fort Hase Defenses of Kaneohe Bay. Simple emplacement for three loaned ex-navy 3-inch Mk VI guns on Mk MVII pedestals erected for seacoast defense. Battery built in mid-1942 by the army's 34th Engineer Regiment. Manned by Battery C, 41st Coast Artillery. Served at least until mid-1944.

**Battery Wailea,** Wailea Point, Oahu. Separate reservation southeast of Kaneohe Bay, along the shore covering Waimanalo Bay. Utilized spare army 3-inch Model 1903 seacoast guns and pedestal mounts. The 3-inch guns were emplaced at the locations formerly occupied by the searchlights on the spine of Puu O Lanikai. The Number 1 gun emplacement was positioned in front of, and below the Number 2 gun emplacement. Both of the circular concrete gun pads were surrounded by a low rock masonry wall and each emplacement was provided with small service and fuse magazines. A battery commander's station of reinforced concrete construction was built on the crest of the ridge behind the upper emplacement. It was built by the 34th Engineer Regiment in the spring of 1942. Apparently hastily planned and built, it later was criticized and had to be rebuilt. Served until early 1945 when the armament was declared obsolete and removed.

Descriptions and Histories of Oahu War Emergency 4-inch Gun Battery Sites

**Battery Kaena.** In the spring of 1942, an emergency battery of two 4-inch naval guns was emplaced near Kaena on Oahu's north shore, west of the Army Air Force's Mokuleia Airfield. Both guns were mounted on specially designed mounts in hastily prepared field fortifications. The battery was constructed with troop labor supervised by Major Davis of the 34th Engineer Regiment. The emplacements were revetted with sand or stone-filled boxes and sandbags. Battery Kaena was some 2.5 miles east of the KPMR in an area that covered the approaches to the only sandy beaches east of the point suitable for amphibious landings. Battery G, 57th CA, initially manned Battery Kaena until relieved by HQ Battery, 3rd Bn, 57th CA, which manned the guns from September 1942 to May 1943. After withdrawal guns replaced with a new emplacement for 155 mm Panama mounts.

**Battery Dillingham.** One pair of 4-inch guns, Battery Dillingham, was emplaced on the Dillingham Ranch near the Mokuleia Beach railway artillery battle position. Each of the guns were provided with a demountable, wooden firing platform with hold down bolts that was constructed by the department shops at Beretania and Miller Streets in Honolulu. The fire control system employed was a CRF station and a plotting room, both of which were initially constructed during the summer of 1942 in an exposed position. After an inspection a concrete splinter proof combined BCS, fire control station for the CRF, and plotting room at the battery were constructed in September 1942. Commercial electric power was to be extended to the battery installation and an emergency power source provided in the form of a 5-kV generator. *During its initial assignment to the North Shore, Battery G, 57th CA manned the battery.* A pair of 155 mm GPF guns was received on May 21, 1943 as a replacement for the 4-inch navy guns recalled by that service. They were emplaced on the left and right flanks of the old 4-inch emplacements.

**Battery Dodge.** This old emplacement on the eastern rim of Diamond Head Crater at Fort Ruger received and mounted two navy 4-inch guns in April 1942. One gun each was placed in the casemates formerly mounting 4.72-inch guns in 1916 as a land defense battery. Work reported completed in June 1942. Manned in September 1942 by Battery E, 55th CA. In September 1942 the emplacements holding the guns needed to be modified. In order to obtain 19° gun barrel elevation, concrete needed to be removed from the overhanding protection for both casemates. Also for gun No. 2 a niche needed to be cut in the chamber side wall to accommodate the sight to allow maximum traverse. Guns removed with navy recall by October 1943.

**Battery Kalihi,** also known as **Battery School.** This emergency battery consisted of three 4-inch guns. Initially Mokuoeo Island in Honolulu Harbor was suggested as the battery site, but finally selected was a site near Kalihi Stream from which the guns could cover the waters of Keehe Lagoon. The request for the land required was submitted in May 1942. Construction of the battery at Kalihi Stream received final approval on May 26 however dredging operations required relocation to east at Kamehameha School on Kapalama Heights across the Kahili Valley from Fort Shafter. The battery was built by a detachment of the 34th Engineer Regiment. The guns were emplaced during the late summer of 1942. A manning detail for the guns was provided by a detachment of the 805th CA Battery, and later by a detachment of Battery F, 15th CA Regiment.

**Kaneohe Bay Battery.** Headquarters, HCAC, assigned Lieutenant Colonel Donald L. Dunlop as the first commander of the Harbor Defenses of Kaneohe Bay. Dunlop sought improvements in the defenses of the Kaneohe Bay. In concert with Commander Martin, USN, at the naval air station, he arranged to mount a pair of obsolescent 4-inch naval deck guns on pedestal mounts on Keanaiki Hill at the naval air station early in 1942. The guns were not part of the 25 guns loaned by the navy to the army, but independently obtained by Martin from navy stocks. It was initially manned by Marines from the station's detail. The battery consisted of two widely separated concrete gun blocks revetted with sandbags. After a detailed inspection, the District Engineer determined that the battery was not very well laid out and the pair of 4-inch naval guns was emplaced too far apart for efficient fire control and direction. Still, once emplaced, the battery succeeded the 155 mm guns of Battery Pyramid as Kaneohe Bay's harbor defense examination battery until the emplacement of 3-inch Army AA guns a few months later. The 4-inch guns were still manned by a 24-man detail of Marines through May 1, 1943.

Descriptions and Histories of Oahu War Emergency 5-inch Gun Battery Sites

**Battery Wailea.** Located on the southern shore of Wailea Point at the northeast end of Bellows Army Air Field. In the spring of 1942, two 5-inch Mark VI Mod 6 naval guns pedestal mounts were made available on a temporary basis and emplaced here. These guns were part of the eleven loaned by the Navy in January. A provisional manning detachment was formed with personnel from Battery D of the 16th Coast Artillery, Batteries C, 41st CA, and from Batteries C and D of the 57th CA and designated locally as Battery "W". In the meantime the Hawaiian Coast Artillery Command undertook action to prepare a quasi-permanent battery at Wailea Point during the spring of 1942. A detachment of the 34th Engineer Regiment assisted by coast artillery personnel was authorized to build a battery commander's station, plotting room, magazines, and gun blocks for two relocated 3-inch seacoast guns. By mid-summer 1942, the 5-inch naval guns were removed and taken to Kahana to become part of the armament at that battery emplacement.

**Battery Homestead.** A new site relocated during planning from a location known as Battery Makua. Equipped with three 5-inch/51 guns, it was sited on the west coast of Oahu. Although this battery had the highest priority of the first three, the exact location for the battery had not been determined at the end of January 1942. An initial proposed position at Makua was not found to be feasible because of its close proximity to the beach. A board of officers located a suitable site on the Ohikilolo Ranch on the Keaau Homesteads between Barking Sands and Makaha about 1,000-yards inland. The new reservation was just over 58 acres. The battery was constructed by Hawaiian Constructors, a consortium of four civilian contractors in the early spring of 1942. By April 4, 1942, the concrete had been poured for the battery's three emplacements, work was complete by April 20. However, the battery still exhibited numerous shortcomings. The guns had no bunker revetments, a splinter proof small arms magazine was still needed and the battery commander's station located on a ridge was totally exposed to view from seaward. Camouflage for the battery commander's station tower was provided and control of the dusty conditions at the station was undertaken. Although a requested small arms magazine was not approved, the men of the battery eventually constructed one using scrap materials. Major General Henry T. Burgin visited the battery about the middle of April 1942. Because there were few if any other military installations in the area at the time, he believed that the battery would receive major attention from enemy fire directed at the west shore. He believed the battery's guns were important enough to deserve casemating. Finally, the additional overhead protection to the three gun emplacements was approved by the Hawaiian Department, but the priority they were given was so low that the project was never implemented. Battery Homestead's manning detail of three officers and 75 enlisted men was initially composed of the 804th CA (HD) Battery (Separate). This unit had been created just in May 1942. at Fort Weaver. By May 22, 1942, in response to the urgings of the department commander, the 804th proof fired the three 5-inch guns. The three emplacements were completed in all respects by August 30, 1942 and the battery designated Battery Homestead on September 6, 1942. A year later, in October 1943 the 5-inch batteries were authorized for deletion, by early 1944 they were inactivated and the guns disposed.

**Battery Ahua.** Sited on the eastern Fort Kamehameha reservation at the Ahua sub-post. Originally intended for three ex-navy 5-inch/51 guns. Authorized by the Hawaiian Department on January 31, 1942. Work begun in February and by April 1942 the project was reported as about 60% complete. Late in 1942 recommendation by the Department Engineer, that the guns be casemated was disapproved by the Hawaiian Department. As built the guns had its three blocks, with projectile and powder magazines. On October 3, 1942, approval was made for improvement of the local defense positions, camouflage and dust control, suitable ventilation and illumination of the magazine interiors, along with installation of standby lighting and extension of commercial power to the battery site. A splinter proof and gas proof plotting room measuring 16-feet wide and 19-feet long, and the battery commander's station atop a 15-foot timber frame tower was enlarged and a similar tower that was used as a fire control tower were both provided with a measure of splinter proofing. Its manning detail was initially provided by a detachment from Battery B, 15th CA, Later the detachment was reorganized into the 803rd CA (HD) Battery (Separate). One of the naval guns was moved to Battery XRAY on Oneula Beach during 1943. An October 1943 survey of the seacoast artillery by the commanding general, recommended that the 5-inch naval batteries be eliminated from the Hawaiian Department Defense Project and approved by the War Department. It is believed that the emergency battery was inactivated at that time. Plans were developed to prepare four replacement emplacements for 155 mm guns on Panama mounts at the battery site; it is uncertain whether the plans were implemented.

**Battery Oneula.** An emergency battery for three 5-inch/51 ex-navy battery construction project at Oneula near the beach. As it was located near fire control station "X", it was also known as Battery XRAY. Authorization for the new batteries of this type were made by the Hawaiian Department on January 14, 1942. The Oneula Battery was sited just to the west of Fire Control Station "X" and about two miles west of Fort Weaver. When nearly half complete in early March 1942, this battery was armed with at least one 5-inch gun. By April, the battery position was about 60% complete. Consideration was given to the casemating of the 5-inch guns in May 1942, but this aspect of the project was not viewed favorably by the Hawaiian Department. As the battery neared completion, the 808th CA (HD) Battery (Separate) was activated on May 28, 1942 at Fort Kamehameha. Its cadre was provided by transferred personnel from Battery A, 15th CA and Battery E, 57th CA to that were added a large contingent of coast artillery replacements from the mainland. Detachments of the 808th Coast Artillery Battery were also assigned to man batteries at Nanakuli and Puu O Hulu in 1942 and 1943. Inadequate sandbag bunkers protected the gun emplacements and were eventually replaced in November 1942 by earth filled timber-walled bunkers. The fishnet camouflage over the guns was sagging because of poor supports and it was recommended that the nets be replaced with garnished chicken wire frames. The magazines had virtually no backfilling protection around and over them and garnished chicken wire was recommended over these structures as well. The wooden building that served as a temporary plotting room was too small and a splinter-proof and gas-proof plotting room was replaced. Other improvement to the battery site included a 70-man mess hall, and an emergency 5-kV electrical generator. Local defense of the battery was hampered by the existence of surface coral that made the digging of firing trenches and emplacements for the battery's four .30 caliber machine guns difficult. Most of these changes were performed with the supervision of the 34th Engineer Regiment. The battery was manned by the 808th CA Battery from May 29, 1942, until May 23, 1943. The emplacement was reduced to two guns, however, when one of the 5-inch guns was moved to the battery at Kahana Bay in September of 1942. A survey of October 1943 recommended that the 5-inch naval batteries be eliminated. It was manned through December 1943 and then inactivated.

**Battery Kahana.** An emplacement for three 5-inch/51 ex-navy guns constructed in the vicinity of Makalii Point overlooking Kahana Bay of the Oahu east coast. Originally these guns were to be a battery near Kalihi Stream near Honolulu. However, when the site for an 8-inch railway line at Laie was shifted north to Kahuku, coverage of Kahana Bay was lost. In mid-September 1942, the army was advised that three additional 5-inch naval guns (The two that had been emplaced at Wailea Point and a third from the south shore of the island.) would become available. In early October a proposal was made to emplace three 5-inch naval guns in the vicinity of Makalii Point overlooking Kahana Bay. On October 31, the order was given to begin the project and to take advantage of natural cover among trees and thick undergrowth. The fortification elements consisting of the gun emplacements, the magazines, plotting room, combination battery commander's station and DPF station were all sited on high ground inland on the north side of Kahana Bay. The month of November 1942 was use obtaining right of entry to the site. By month's end concrete work of the emplacements was progressing rapidly and the engineers would soon be ready to commence construction of the magazines and battery commander's station. Work continued on the battery into 1943. It was completed on May 8, the last of the emergency batteries composed of naval guns to be established on Oahu. It fired a service practice consisting of 12 rounds on February 23, 1943. Detailed personnel formed a provisional battery ultimately designated the 715th CA (HD) Battery (Separate). On October 21, 1942, the still incomplete 5-inch gun battery was officially named Battery Kahana. The 715th CA Battery provided the manning detachment for Battery Kahana until May 1943, when the battery was placed in maintenance status. An October 1943 survey recommended that the 5-inch naval batteries be eliminated from the department's defense project at a time to be selected by the Commanding General. This recommendation was subsequently approved by the War Department. By September 1, 1945, Battery Kahana's 5-inch guns had been removed.

**Battery Nanakuli.** With the commencement of hostilities in 1941, an emergency naval battery composed of 5-inch/51 caliber MkVII guns and one 3-inch naval gun, also on an open mount was established for training by 3rd Defense Battalion of the U.S. Marines at nearby Camp Andrews. These two 5-inch naval guns were not a part of the original eleven 5-inchers loaned to the Army early in 1942, but would come under the army's control in the spring of 1942. Temporary relief of the marine detachment by the coast artillery came in April 1942. This assignment to the Army was, at first, not considered to be a permanent arrangement and the Army deferred any extensive improvement of the battery's gun emplacements and bivouac area where the troops lived in tents. The battery site was finally taken over in the early summer of 1942 by a detachment from the 808th CA (HD) Battery (Separate). An inspection in July 1942 found the site in poor defensive condition. The sandbag bunkers around the guns were deteriorating, the cut and cover magazines were inadequately protected, the barbed wire entanglements were insufficient for close-in ground defense. While the Army proposed that the battery be relocated it was decided there would be no move and that the battery and camp's transfer to the Army would be a permanent arrangement. Finally, by September the 34th Engineer regiment had issued orders to carry out improvements to the site. The disintegrating sandbag gun bunkers were replaced and backfilled up to the bunker tops. The magazines were enlarged to provide cut and cover concrete splinter proofing and expanded to allow a capacity of 500 rounds for each 5-inch gun. A wood and revetted splinter proof plotting room with an overhead protection of seven feet was constructed. The entry was provided with offset entrances and an emergency exit. Pipe mounts were installed for the .30 caliber machine gun emplacements, and the barbed wire entanglements were extended to provide an adequate all-around ground defense. In addition to the 5-inch guns, four 155 mm GPF guns were taken from the war reserves at the ordnance depot and assigned to the 808th Battery. The 155s were emplaced on Panama mounts in September 1942. It is doubtful that the 5-inch guns were retained in an active status after late May 1943. The 155 mm guns were also removed when the 808th departed the location.

Descriptions and Histories of Oahu War Emergency 7-inch Gun Battery Sites

**Battery Hulu.** Sited at the foot of Puu O Hulu Kai. It was initially projected to be composed of three 7-inch/45 naval broadside guns. Ultimately built for just two guns. The emplacements were provided with low concrete walls around their fronts at the foot of the slopes of Puu O Hulu Kai. It consisted of two separate, reinforced concrete casemates with an open front embrasure. The casemates held magazines for powder and shell. Nearby were plotting room, reserve power, battery commander's station, radar operating room. Upon completion of the emplacements in August 1942, the battery was manned by a detachment from the 808th CA (HD) Battery (Separate) at Oneula Beach. The detachment consisted of two officers and about 42 enlisted men. On October 21, 1942, the battery's name of Battery Hulu was formally announced by the Headquarters of the HSAC. On May 24, 1943, the detachment at Battery Hulu rejoined the main body of the 808th CA Battery at Oneula where the 808th was redesignated Battery B, 41st CA. An October 1943 survey recommendation that the 7-inch guns be eliminated from the defense project was and was subsequently approved. While manned by Battery E of the 15th CA, during 1944, some seventy rounds were fired from the pair of 7-inch guns in various target practices. It is probable that Battery Hulu was placed in maintenance status after the departure of its last manning unit in April 1945.

**Battery Harbor.** Emplacement site was on Sand Island to cover the approaches to Honolulu Harbor. Intended for four guns, though initially only two emplaced pending renovation of two more. Each of the four emplacements had a three-foot high concrete revetment around it. Separate earth covered splinter-proof projectile and powder magazines were provided to the rear. A small fuse storage magazine, and an earth covered plotting room was also provided. A battery commander's station was established in the lower level of a two-story fire control station set atop a 50-foot high steel frame tower at the rear of the battery. The battery was completed about the end of September 1942, when it was manned by a detachment of Battery F, 55th CA Regiment, followed by the 807th CA (HD) Battery (Separate). An October 1943 survey recommended that the 7-inch naval batteries be eliminated from the Hawaiian Department Defense Project. It is believed that Battery Harbor was placed in a reduced manning status at the end of November 1944. During that year, however, four target service practices were fired expending a total of sixteen rounds, and one regular target practice in that thirty-two more rounds of 7-inch ammunition was fired. The battery was probably disarmed in early 1945. Two of 7-inch guns of the type emplaced at Battery Harbor and their pedestal mounts are also extant being preserved and displayed at the US Army Museum of Hawaii.

A navy 5-inch emergency emplacement during the war, while not on Oahu, it is typical of the simple type of emplacement used for these guns. *Williford Collection*

# APPENDIX VIII

## DESCRIPTIONS AND HISTORIES OF OAHU AMTB BATTERY EMPLACEMENTS

**Fort Weaver Anti-Motor Torpedo Boat Battery Number 1.** An Anti-Motor Torpedo Boat (AMTB) battery built at Fort Weaver, coving the entrance into Pearl Harbor. It was built in late 1943 and consisted of two fixed and two mobile mounts for 90 mm guns plus a section of 37 mm automatic guns. In the latter months of the war, the 37 mm armament was replaced by 40 mm automatic weapons. This battery was manned by Battery D, 15th CA from late 1943, until August 1944, then by Battery D, 54th CA and finally by the 881st AAA Battery. By the end of December 1945 the 878th was inactivated and the armament of the AMTB Battery placed in storage.

**Fort Kamehameha Anti-Motor Torpedo Boat Battery Number 2.** One of the standard AMTB batteries of the 1943 Program erected near the Pearl Harbor Channel on the western extreme of the Kamehameha reservation. It consisted of the usual two 90 mm on fixed M3 carriages and two on mobile M1 carriages, and a pair of 37 mm automatic weapons on M3 mobile mounts. It was initially manned by Battery E, 15th CA Regiment upon its completion in July 1943. When the 15th CA was inactivated on August 15, 1944, it is believed that Battery D, 54th CA (HD) Battalion assumed manning responsibilities. The battery was inactivated on November 21, 1946.

**Battery Ala Moana Anti-Motor Torpedo Boat Battery Number 3.** During World War II a site in this ocean side park in the Waikiki section of Honolulu was initially set up as a dummy antiaircraft gun battery "manned" by the 64th CA Regiment in the fall of 1942. In 1943, the site was used for an additional AMTB battery. It had four 90 mm M1 AA guns (two on fixed mounts, two mobile) and two automatic 37 mm guns. The 879th AAA Gun Battery (AMTB) manned the battery in 1945. It was continued in service through the end of the war and afterward by the 879th that was gradually reduced to a small caretaking unit. On December 31, 1945, the AMTB Battery was removed.

**Battery Sand Anti-Motor Torpedo Boat Battery Number 4.** On July 14, 1943, construction was begun on an Anti-Motor Torpedo Boat Battery at the Sand Island Military Reservation. Battery B, 16th CA and Engineer personnel built the battery and the fixed guns were emplaced by August 6. Work continued on into September and the first trial shots were fired on September 9. The barracks and other facilities for the battery personnel were generally complete by September 14, 1943. Battery F, 16th CA was transferred to Sand Island where it took over the AMTB Battery that it manned until May 30, 1944. The 878th Antiaircraft Artillery Battery then manned Battery Sand until April 1945. After the end of World War II the Battery Sand was inactivated, its armament placed in storage. The battery also served as an Antiaircraft Alert Battery for Honolulu Harbor.

**Fort Hase Anti-Motor Torpedo Battery Number 7.** At Kaneohe Bay's Fort Hase plans were developed by November 1942 to replace the 3-inch AA guns with an AMTB battery of 90 mm and 37 mm guns near Pyramid Rock. Construction of the emplacements were finally begun in July 1943. The 3-inch guns of the temporary AMTB battery functioned as the Examination Battery until the new armament was finally provided. This new armament, emplaced just to the south of Pyramid Rock, on the heights of Pali Kilo was designated AMTB Battery Number 7. The battery's section of two 37 mm automatic weapons was set up near the foot of the naval air station's runway to cover the Sampan Channel and the inner waters of the bay. The battery functioned as the Harbor Defense Examination Battery through the end of the war. Battery G 98th CA manned these guns until December 12, 1943. Early in 1944, Battery B, 41st CA (HD) Regiment, succeeded this detachment at the Examination Battery. The 880th manned the Examination Battery until early August 1945, when it was discontinued. Its guns were placed in storage at the end of the war, On October 24, 1942 plans were finalized for a battery to be established amidst the flats and the Molii fishponds at Kualoa Point. By late fall 1943, two 90 mm M1 Dual Purpose guns on mobile M3 Carriages and two 37 mm Automatic Weapons on M3 mobile carriages had been emplaced at Kualoa Point. The remaining pair of fixed 90 mm guns was positioned at Wailea Point near Bellows Field. The manning detachment for the AMTB battery was provided by Battery C of the 57th CA Regiment. On October 30, 1944 the primary mission of the AMTB batteries was reversed making air defense their principal mission with surface action against torpedo boats secondary. The 612th Battery actively manned the AMTB Battery until the end of the war in August 1945.

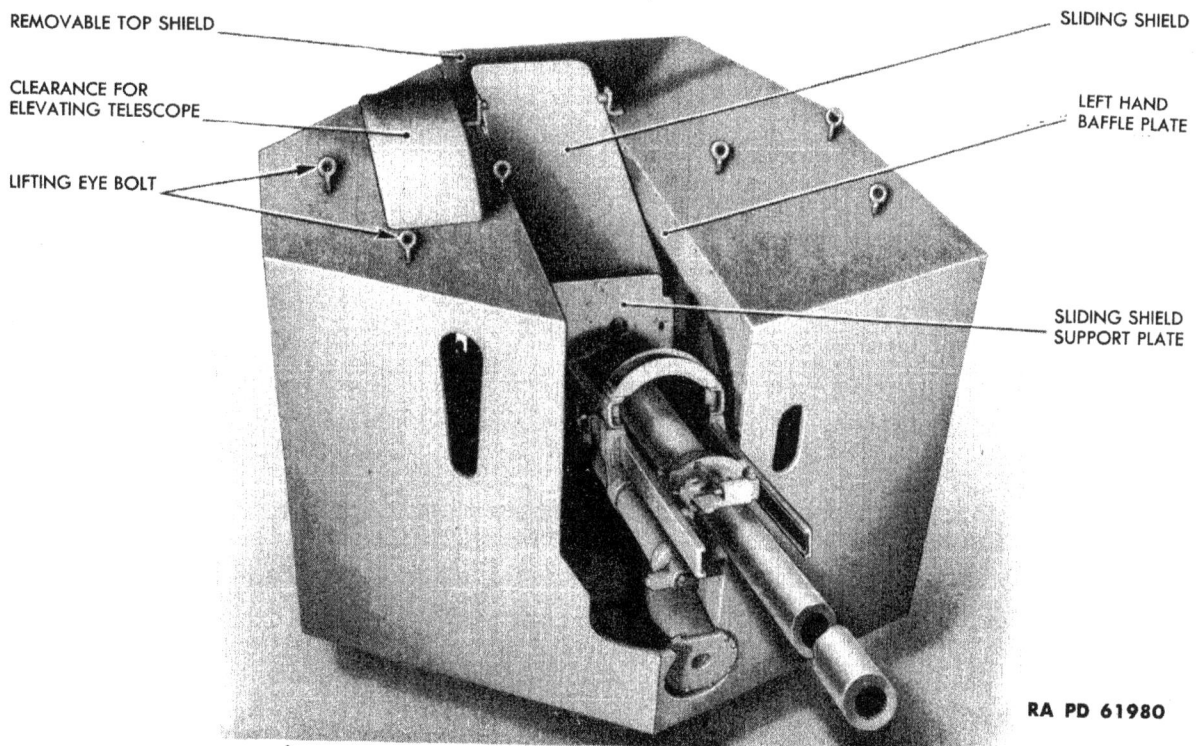

REMOVABLE TOP SHIELD

CLEARANCE FOR ELEVATING TELESCOPE

LIFTING EYE BOLT

SLIDING SHIELD

LEFT HAND BAFFLE PLATE

SLIDING SHIELD SUPPORT PLATE

RA PD 61980

*Figure 110. Shield for all-around protection (90-mm gun).*

The 90 mm anti-motor torpedo boat M3 mounting of the type deployed on Oahu. *Army Technical Manual.*

# APPENDIX IX

## DESCRIPTIONS AND HISTORIES OF OAHU LONG-RANGE BATTERY STATIONS, SUMMARY LIST

**Station A.** This station near the crest of Puu Manawahua was built in 1939. It was part of a collection of stations located at an elevation of 2,300-feet, just below the Puu's summit. Eventually it consisted of a set of three separate single fire control stations and a power generator building in addition to a group command post and several other structures consisting of dormitories and messing facilities, etc. Station A, serving a 16-inch battery was transferred to the custody of the commanding officer of the HD of Pearl Harbor on November 15, 1939. At that time it consisted of a combination command post for the Barrette Group (Batteries Hatch and Williston) and fire control station structures. The command post was the largest of the structures and was built on two levels, one of which served as the operations and plotting room, and the other as an observation room. A small fire control station for the Browns Camp battery equipped with a DPF instrument was physically connected to the observation station of the command post. Sometime in the early 1940s, a third level was added to the station. In 1941, the Antiaircraft Intelligence Service (AAIS) established an observation post (AAIS Station No. 12) on the puu's summit.

**Station B.** A grouping of fire control stations on Puu Palailai, Waianae Range, was built and transferred in 1929. Initially the fire control complex consisted of a Battle Command Station and secondary stations (B") for Batteries Williston and Closson in descending order down the slope. These appeared to be one single three-tiered building with its stations staggered in echelon down the puu's slope. In 1934, a fourth station was added to the lowermost portion of the structure to serve as base end station for Battery Hatch. Like the other fire control stations erected in the mid-1930s for Battery Hatch, this new station was somewhat larger than the original fircontrol stations. The most elevated and the largest of the structures was the Battle Command Station. This command post was later redesignated the command post for the Ewa Group composed of batteries emplaced along the shore of Oahu west of the Pearl Harbor entrance. This station measured 16-feet by 13-feet. In 1940, the complex was designated as Station "B".

**Station C.** Salt Lake Hill (north rim of Aliamanu Crater) was selected prior to World War I as the location for a secondary fire control station for Battery Harlow at Fort Ruger. In succeeding years additional fire control stations were built into the crest of the crater's north rim at elevations varying between 475 and 480-feet above sea level. In the early 1920s, a fire control and observation post was erected for Battery Closson, and a battle command observation post was constructed nearby. These two small stations were identical in design and measured only six and a half-feet side dimensions. In 1924, when Battery Williston was placed in service, the battle commander's station was redesignated as the "C" station for Battery Williston. Although this station was intended to be equipped with a DPF, its instrument was not supplied until the mid-1930s. With the completion of Battery Hatch at Fort Barrette in 1934, yet another fire control station was added to the complex. This structure was turned over to the coast artillery on July 2, 1934. It too was dug into the slope of the crater rim at an elevation of about 477-feet above sea level. It, however, reflected a vast improvement over the earlier stations it was more commodious having interior measurements of 16-feet in width and 15-feet in depth. Further, it was equipped to receive two M1907 DPF instruments.

**Station D**. This complex was ultimately composed of five fire control structures on Punchbowl's south rim. The first station built at the site was a tertiary station (B''') completed about 1911 for the seacoast baseline serving the mortars of Battery Harlow at Fort Ruger. This square station, dug into the slope of Punchbowl, measured about five-feet per side. When Battery Closson was completed in 1920, one of its fire control stations was constructed on the Punchbowl's rim. In 1924 a new double fire control station was built at the Punchbowl. This double station served as emergency stations (E) for Battery Closson, replacing the smaller Closson station (that was reassigned as a coast artillery observation post), and the newly completed Battery Williston at Fort Weaver. This new station of reinforced concrete was more than double the size of the original Closson station and was also dug into the crest of the rim at about 475-feet elevation about 50-feet south of Battery Harlow's B" station and was entered through a manhole in the left rear of the station roof. The combined Closson/Williston station was equipped with two M1907 DPFs and transferred to the coast artillery garrison on May 24, 1924, having cost some $2,570.89 to construct. A fourth station was built into the crater rim at an elevation of about 495-feet above sea-level in early 1934 to serve Battery Hatch. It too was a partially dug in concrete splinter-proof building measuring approximately 15-feet by 16-feet. The fire control complex was designated Station "D" and upon being equipped with its M1907 DPF instrument, was transferred to the troops on July 2, 1934. The station was manned on December 7, 1941 and remained in service through World War II. The final station erected on the south rim of Punchbowl crater was a single station that served as the battery commander's station for Battery Punchbowl. This structure was partially modified later in the war as the battery commander's station for Battery Construction Number 304.

**Station E**. This station, also on the south rim of Diamond Head, Fort Ruger was built in 1917, and transferred to the Fort Ruger garrison on March 9, 1917. The station that initially served as the primary station for Battery Birkhimer's west baseline was essentially of the same design and dimensions as the battery's east baseline primary station. The station was in service for this function for only a brief period during World War I and when the land defense system was de-emphasize following the war, the station was briefly abandoned. In the 1920s the station was assigned as Fire Control Station E as a base end station for Battery Williston at Fort Weaver.

**Station F**. This station was built at Fort Weaver on a site some 8-feet above low tide in a cluster of trees about 600-feet east of the Number 2 emplacement of Battery Williston, near the Pearl Harbor Channel in 1923. The station was transferred on September 21 of that year. It consisted of a single story 10-foot square observing room atop a 50-foot structural steel tower. The station walls and roof were provided with a covering of corrugated iron and asbestos siding. The station was to be equipped with a DPF instrument, but it is doubtful that it was ever supplied. The station was, however, used as the location for the controllers of fixed seacoast Searchlight Numbers 2 and 3, later redesignated as Searchlights 5 and 6. The station was the responsibility of the range section of Battery A, 15th CA until the station was abandoned in 1932. The "F" Station was replaced with a new tower located 300-feet to the northwest of Battery Williston's plotting room and 1,000-feet north of emplacement No.1. It was begun in 1932, and transferred to the garrison on December 2, of that year. The station consisted of a 70-foot tall rigid steel tower supporting an enclosure containing two observing levels and a spotting station on the roof of the upper story 89-feet above the ground. The station served as the primary observation post and BCS for Battery Williston. The station was equipped with both DPF and azimuth instruments.

**Station H.** At the island's eastern extremity of Makapuu Point, another coast artillery fire control station and observation post, Station "H" was completed by 1930. This station, the first of several that would eventually occupy the Makapuu Headland, was built above and to the rear of the Makapuu Head Lighthouse at an elevation of 660-feet near the highest point on the ridge of the point. Like many of the early stations constructed on Oahu in the 1920s to serve Battery Williston, it too was a small one-room station of reinforced concrete equipped with a pedestal to support a single DPF.

**Station J.** This station was built in 1931 on the opposite side of the island atop the west rim of Ulupau crater. This small concrete coast artillery observation post was designated Station "J." It was partially dug into the crest of the rim near Ulupau's highest elevation some 650-feet above sea level. The structure was transferred to the Coast Artillery on January 22, 1932, having cost $3,162.86 to construct. Station "J" was, slightly larger than others and consisted of a single observation room measuring about 16-feet in length and 15-feet in width. Entry to the observation post was gained at the building's south or rear end. The station was provided with a concrete pedestal for a DPF instrument and telephones for data transmission. While many of the fire control stations were equipped with their communication equipment, telephones were not yet in place when Station "J" was placed in the custody of the coast artillery. The station was not manned prior to World War II except during periodic exercises and maneuvers. In 1940, as part of a general program of improvements and modernization, Station "J" was rebuilt to accommodate two fire control instruments. In the small ravine directly to the rear of the station, a dormitory of frame construction was built for the use of the small fire control detachment. During this overall improvement of Station "J," a small reinforced concrete powerhouse was built for an electric power generator. The reconstruction of the station was completed by September 27, 1940, and the engineers transferred the improved station back to coast artillery custody on November 5, 1940.

**Station L.** A specific fire-control site located on a separate tract of land at an elevation of about 450-feet above sea level on the northern slope of Puu Manamana. Fire Control Station "L" was established in 1941 as a fire control station for Battery Williston's 16-inch guns. Initially, a single position for one DPF was constructed as a temporary station to the west of the proposed site for the permanent station. The permanent station was completed during the summer of 1941 and transferred to the coast artillery on September 3, 1941. Initially the station had been planned as a double fire control station for the two 16-inch gun batteries, Hatch and Williston, on the island's south shore. When the decision was made to casemate Battery Hatch, the station was assigned to Battery Williston alone. Station "L" was a single level splinter proof concrete structure measuring 18-feet square. Partially dug into a niche in the slope the station was provided with an earthen cover for camouflage. Entry to the structure was gained through a doorway in the left rear corner of the station. The quarters and other support facilities for the range detail were located at the foot of the slopes on the Kaaawa Military Reservation. In addition to serving Battery Williston, the station is known to have also served during World War II as the B6S6 base end station for Battery 405 and later was assigned as the B5S5 base end station for Battery Pennsylvania's 14-inch turret guns.

**Station M.** This reservation was established in 1941 as a Coast Artillery Fire Control and Observation station for the 16-inch batteries on the south shore. The site was located on the extreme right flank of the Northern Groupment above Kokololio Gulch between the towns of Laie and Hauula on Oahu's Windward side. There on the slopes of Kaipapau Hill at an elevation of about 500-feet above sea level, Fire Control Station "M" was built for two double fire control stations that were initially authorized for this location. The first of the two stations authorized was a simple one-story affair built of reinforced concrete and dug into the hillside some 518-feet above sea level. It was to be provided with both an M1 DPF and an azimuth instrument upon completion. These two instruments were still awaiting procurement when the Japanese attacked Oahu. The interior of the station consisted of a single room that measured 16-feet wide and 15-feet deep. A doorway in the right rear afforded access to the station. The isolated nature of this station necessitated building of a bunkhouse, or dormitory, of frame construction and a latrine by the engineer troops that built the station. When completed it was initially planned to serve as a base end station for Batteries Williston and Hatch. It was transferred on August 23, 1941. This station was also assigned as a base end Stations for the 8-inch railway battery position at Kahuku, and the secondary station for 155 mm Battery Ranch. By the spring of 1942, two more fire control structures, one of which was a two-tiered station, were approved for construction by the Hawaiian Department. They would serve as a base end station for the pair of naval turret batteries on the north shore (Batteries Ricker at Opaeula and Riggs at Brodie Camp Number 4) and for Battery 302 (Battery Cooper) at Lae O Kaoio. These two structures were built near the military crest of Kaipapau Hill at an elevation of 569-feet above sea level. Near the end of the war one of the stations was also assigned as the B6S6 base end station for Battery Pennsylvania.

**Station O.** Fire Control Station "O" was sited on a tract of land just under half an acre in area at the edge of the bluff at Pupukea, some 350-feet above the ocean. This site about 2,500-feet inland from the shore had been acquired by on June 29, 1927. When NIRA funds became available, the station was built. Upon completion the structure consisted of a two-tiered concrete building dug into the edge of the bluff. A one-level station served Battery Williston at Fort Weaver, while a second was for a potential command group. In the mid-30s a third station was added for Battery Hatch at Fort Barrette. The upper level was set back from the lower one and had a 15 foot by 16-foot room beneath it and to the rear of the lower observing level. This space was used as a dormitory for the fire control detachment that usually consisted of about six or eight men. Like the new station on Puu Pueo and other stations built in the mid-1930s to serve the 16-inch gun batteries, the interiors of the observation rooms measured 15 feet long and 16-feet wide. Each observing level was arranged to receive two M1907 DPF instruments. Access to the station interior was gained through a manhole and vertical shaft in the rear roof of the upper level. To aid in camouflaging the structure, a thin covering of earth was applied to the top of the structure's foot thick reinforced concrete roof.

**Station S.** This station was built on the western slope of Puu Pueo at Kaena Point. As completed, it was a one-room reinforced concrete structure dug into the western slope at an elevation of 565-feet. It measured only 8-feet wide and 13-feet deep. It had a pedestal base for a single depression position finder (DPF) instrument. Entry to the station's interior was gained through a manhole in the rear of the station roof. When completed, it was designated Fire Control Station "S" and was transferred on April 10, 1924. The station's DPF was not installed, however, until June 1936. The electricity to power the instrument lights was provided by an Edison wet cell battery. In 1934, a second fire control station was built at Station "S" to provide fire-control data to Battery Hatch at Fort Barrette. This new station built with NRA funds, was directly in front of, and several feet below, the original (Williston) Station "S" at an elevation of approximately 560-feet above sea level. This station, also built of reinforced concrete, reflected an improved design and was somewhat larger than the Williston station, measuring 16-feet wide and 15-feet deep. This splinter proof station was entered at ground level through a baffled entry at its right rear. It was equipped to receive two M1907 DPF instruments.

**Station S'.** One of the first fire control stations built for Battery Williston's 16-inch guns at Fort Weaver was located near the crest of Kepuhi Point. This single station, equipped with one DPF instrument was designated as Fire Control Station "S Prime" (S'). It was built in the latter part of 1923 and transferred to the Coast on April 10, 1924. The station was dug into the mountain slope at an elevation of about 1,160-feet and was entered through a manhole in its roof. The structure was small, measuring only 12 by 8-feet. The station's observation slots were provided with splinter proof steel shutters painted earthen tone-down paints to help camouflage the position. On November 30, 1934, a second fire control station was built about 100-feet above the first station. This new double station (two DPF instruments) was built to serve Battery Hatch's newly emplaced guns at Fort Barrette. It represented an improved style of station being a somewhat larger. The high elevation of these two stations required installation of a tramway that enabled the fire control details on the heights to move supplies up and down the steep slopes of the ridge. The tram was however, not generally used to move the personnel up and down the puu. The site also had temporary wooden theater of operations dormitories, mess halls, and other support facilities for the men stationed on the ridge during exercises, or in time of war. By November 1942, Battery Hatch was in the process of being casemated. This reduced that battery's field of fire and its station at Kepuhi was no longer required for the 16-inch guns. The station was then reassigned to the newly emplaced 8-inch naval turrets of Batteries Brodie and Opaeula of the North Shore Groupment. In 1944, a third fire control station was authorized for Kepuhi to serve Battery Homestead, and Battery Construction Number 409 at Kaena Point then under construction. This station was also to be built at an elevation of 1,260-feet. The new station was also projected for assignment to the 14-inch turret mounted guns of Battery Arizona as its B2S2 base end station.

**Station U.** The initial reservation site was acquired near the crest of Puu O Hulu Kai as the location for one of Battery Williston's fire control stations. This small eight-foot by 12-foot station, was equipped for a single M1907 DPF instrument. The structure was dug into the slope of the hill and was entered through a manhole in the right rear of its roof. It was transferred to troops on April 7, 1924. Two additional stations with the same dimensions as Battery Williston's were built at Station "U" later in the 1920s. One served Battery Closson's 12-inch guns at Fort Kamehameha. The third station was erected to serve the railway mortar positions on Oahu's Leeward shore. Each was of an identical design with the exception that the latter 1920s stations were entered through doorways in the rear of the structure rather than a manhole. Each of these stations were equipped with DPFs.

**Station W.** A new station built on the North Shore in 1934 just south of Haleiwa on the slopes of Puuiki at the north end of the Waianae Range at an elevation of 545-feet above sea level. Designated Fire Control Station "W" this reinforced-concrete structure was also a single observation room measuring 15-feet deep and 16-feet wide. It too was entered through a manhole in the rear of the station roof. It was also equipped with two M1907 DPF instruments. Station "W" provided target data for both Battery Williston and Battery Hatch. Both Stations "O" and "W" were transferred to the custody of the Commanding Officer of the Harbor Defenses of Pearl Harbor on November 30, 1934. During coast defense exercises, the range details of Batteries A and C of the 15th CA Regiment manned the stations.

**Station X.** This double station, about halfway between Oneula Beach and Ewa Beach, was located at the end of a road some two miles west of Battery Williston. Station "X" was built in 1924 as a fire control station for Battery Williston and consisted of a single-story house constructed of corrugated asbestos sheathing on a steel frame atop a 35-foot tall structural steel tower. Station "X" was located in a grove of algaroba trees that left only the upper six-feet of the tower visible above the foliage. The observing room was to be equipped with a DPF instrument. A year after its completion, however, the station still lacked its fire control equipment. This station after receiving its instrument may have also functioned during World War II as the battery commander's station for the 5-inch battery at Oneula.

# APPENDIX X

## DESCRIPTIONS AND HISTORIES OF OAHU SEARCHLIGHT SITES

### Fort Kamehameha Searchlights

The searchlights at Fort Kamehameha had a short service life, for many years the fort had no lights in service. The first position was intended initially for two fixed 36-inch searchlights to be installed during the 1910s. The lights were powered by a searchlight power plant located north of Battery Hawkins. These lights were removed following World War I but were replaced for a few years by two 60-inch General Electric searchlights that were temporally mounted on wooden towers. These lights were moved to Fort Weaver in 1922 and 1924 where they were permanently installed. No lights were reinstalled at the Fort Kamehameha location until 1941, when two 60-inch portable seacoast searchlights, Numbers 7 and 8, were positioned near Battery Hawkins and the Pearl Harbor Channel. These lights were mounted atop a pair of 50-foot towers that were procured in 1941. When the portable 60-inch lights were installed in 1941, they were powered by their own mobile power plants, and controlled from a nearby fire control tower. The lights were manned by a detail from Headquarters Battery, 15th CA Regiment until Battery G, 15th CA was activated in August 1942. Battery G manned the lights until August 1944, when the 15th was inactivated.

### Ahua Point Searchlights

The initial searchlight emplaced on Ahua Point was a 60-inch General Electric projector obtained in 1911. It was installed prior to August 13, 1913 on a Scherzer Rolling Lift Tower that when elevated to its operating position placed the searchlight 70-feet above sea level. The light could be controlled either electrically or manually. It was powered by the Ahua Point power plant located in a concrete powerhouse located about halfway between the secondary fire control towers and the tower for searchlight Number 6. This power plant was protected on three sides by an earthen revetment. The power plant consisted of a General Electric gasoline operated four-cycle engine that was directly connected to a 25 kW General Electric generator that was purchased in 1912. The cost of the power plant was $6,300 while the cost of the searchlight and its tower was $12,500. This light designated as Searchlight Number 6 was transferred to the garrison on August 13, 1914. Searchlight Number 6 was redesignated as Searchlight No. 1 for the Harbor Defenses of Pearl Harbor in 1924 and renumbered again during the period between the two world wars, as Searchlight No. 9.

In January 1933, a plan for a second 60-inch searchlight on a disappearing tower was projected for Ahua Point. The cost for a disappearing tower was nearly three times that of a rigid steel frame tower. Consequently the latter type tower was finally settled upon and an 85-foot tower was built on the point in 1937. The searchlight shelter atop the tower measured 15-feet by 12-feet. The light was powered by a 43-54 h.p. General Electric gasoline powered engine and a 25 kW generator that was housed in a splinter proof concrete powerhouse about 295-feet from the tower. The searchlight, its tower and power plant were transferred to the garrison on August 19, 1937. This light was designated SL 10. The searchlight's distant electric controller was located in the B" station for Battery Hasbrouck. Searchlight Numbers 9 and 10 remained in service through World War II. They were manned by a detail from Headquarters Battery, 15th Coast Artillery Regiment until Battery G, 15th CA was activated in August 1942. Battery G manned the lights until at least August 1944.

### Fort Armstrong Light Number 5

Searchlight Number 5 was a 36-inch seacoast searchlight mounted on a short track of railway on a small car. Its power was provided by a power plant located in the left flank traverse of 3-inch Battery Tiernon at the small Fort Armstrong reservation. This plant consisted of a General Electric 4-cycle gasoline engine that powered a 25 kW General Electric generator. The plant was transferred to the Fort Armstrong garrison on June 17, 1911. This power plant proved insufficient for the demand, and a reserve plant of the same type and capacity as the initial plant was purchased in 1913 and installed at a cost of $5,800 in a separate 12 by 16-foot concrete power plant building in the rear of Battery Tiernon in 1914. Its plant furnished electric service to Battery Tiernon, the mine casemate, and other torpedo structures, as well as to power Searchlight Number 5 in the event of accident to the main power plant. It was transferred on June 22, 1914. The mine casemate and powerplant and all traces of the searchlight were demolished in 1973.

### Fort Ruger Searchlights

Fort Ruger was an important position on the eastern flank of the defenses for searchlight installations even during the first generations of harbor defense. As early as 1910 installation of seacoast searchlights were planned for Fort Ruger. The first two of four searchlights at Fort Ruger were installed in 1911; one at Diamond Head and the other at Kupikipikio (Black) Point. A third light was emplaced at Diamond Head in 1913 and the fourth in 1920. After construction of Battery Granger Adams in 1934 at Kupikipikio Point, it was necessary to move the light there to the southeast slope of Diamond Head where it was eventually renumbered as Searchlight Number 16. Initially the searchlights at Diamond Head were manned by details from the coast artillery companies at Fort Ruger.

### Black Point Searchlight Number 1

The 60-inch seacoast searchlight for Kupikipikio Point arrived on Oahu in 1910 and was installed in a searchlight emplacement at the east end of Kupikipikio Point. The emplacement was a concrete house measuring 15-feet square and was dug into the point. The 60-inch General Electric Searchlight was set on an elevator platform with a lift of some ten-feet. The mechanism for lifting the searchlight to the operating position was accomplished by a device consisting of three screws connected by a triangular geared plate. The searchlight shelter was secured on top by a sliding metal roof. Power for the searchlight was supplied by a 25 kW 128-volt General Electric Generator powered by a 50-Horsepower General Electric 4-cylinder gasoline engine, that was installed in a separate concrete powerhouse dug into the slope of Kupikipikio Hill. The light and its power plant were transferred to the Fort Ruger garrison on January 4, 1911. In the mid-1930s Searchlight Number 1 was moved to the southeast slope of Diamond Head and redesignated as Searchlight Number 16 after Battery Granger Adams was built directly behind the searchlight emplacement. The light in its new position was retained in service through the end of World War II.

### Fort Ruger Searchlight Number 2

This 60-inch General Electric seacoast searchlight was installed in 1912 at the southern slope of Diamond Head at an elevation of about 90-feet above shoreline. It was on an elevator platform bringing it from a depressed pit. As elevated the fixed light was housed in a concrete shelter with a sliding steel roof. The light was transferred to the Fort Ruger Garrison on January 11, 1913. It was renumbered as Searchlight Number 14 in the years between the two world wars and remained in service through the end of World War II. Power for the searchlight and interior illumination of the powerhouse was supplied by a 25 kW 128-volt General Electric Generator powered by a 50-Horsepower General Electric 4-cylinder gasoline engine, which was installed in a separate concrete powerhouse located in an excavation into the side of a gulch between spurs of Diamond Head's south slope. The power plant was transferred to the garrison in December 1912. This searchlight site, although still intact, is located in a thick growth of underbrush and is difficult to access.

### Fort Ruger Searchlight Number 3

This seacoast searchlight was a fixed 60-inch General Electric device installed on a lift platform housed in a concrete shelter with a sliding steel roof. The shelter covered a pit containing the searchlight on an elevator platform dug into a knoll of a spur on the southwest slope of Diamond Head at an elevation of 123-feet above the shoreline near the Territorial Reservoir. When the searchlight was to be placed in service the roof was slid back and the light winched up to its operating position. Power for the searchlight was supplied by a 25 kW 128-volt General Electric Generator powered by a 50-Horsepower General Electric 4-cylinder gasoline engine, which was installed in a separate concrete powerhouse inside a tunnel excavated into the slope of a spur on the southwest slope of Diamond Head. The light and its power plant were transferred to the Fort Ruger garrison on January 4, 1911. The light was renumbered as Searchlight Number 13 in the years between the two world wars and remained in service through the end of World War II. This searchlight site shelter is still intact.

### Fort Ruger Searchlight Number 4

This seacoast searchlight emplaced on the western slope of Diamond Head was a fixed 60-inch General Electric searchlight installed on a movable railway tracked cart. The searchlight shelter was a 10-foot wide concrete tunnel 35-feet long with a crown of 13-feet that was located at an elevation of 257-feet above sea level. This light was placed in operating position by moving the truck-bearing searchlight down the tunnel to the entrance where it was placed in operation. The power generator for the light was also housed in the tunnel and consisted of a 25 kW 128 volt General Electric Generator powered by a 50-Horsepower General Electric 4-cylinder gasoline engine. Construction of the emplacement was completed in 1920, and transferred to the garrison on June 28, of that year. The light was renumbered as Searchlight Number 15 in the years between the two world wars and remained in service through the end of World War II. This searchlight site is still intact though overgrown.

New Fixed Searchlights for the Oahu South Coast in the 1930s and 40s

### Fort Weaver Searchlight Positions

These two General Electric 60-inch searchlights, initially numbered Nos. 7 and 8, were received in the coast artillery district in 1911, and first temporally mounted on wooden towers at Fort Kamehameha. The lights were "Sperryized" in 1920, and in 1922, one searchlight (Number 2) was moved to the Iroquois Point Military Reservation, which had just been renamed Fort Weaver. The light was positioned in a dense growth of algaroba trees about 3,000-feet west of Battery Williston's Number 1 emplacement near the west end of the fort adjacent to the Puuloa Naval Reservation. It was housed in an eight-foot wide and 10-foot deep shelter of structural steel and corrugated iron that was placed atop a 100-foot tall, Type C, M1919 structural steel tower. The powerhouse for the searchlight was built about 300-feet east of the tower and consisted of a 25 kW gasoline powerplant set.

The second searchlight (to become Number 3) was moved from Fort Kamehameha to Fort Weaver in 1924 and installed about 500-feet west of Searchlight Number 2. It too was housed in a shelter identical to that housing the other searchlight. Its power plant was housed about 300-feet north of the Number 3 searchlight tower. Both lights were controlled from Fire Control Station "F." These two searchlights were renumbered in the period just prior to World War II with lights number 2 and 3 becoming light numbers 5 and 6. During the early months of the war, the lights were manned by a detail from Headquarters Battery, 15th CA until Battery G the searchlight battery of the 15th was organized in August 1942. Battery G operated the lights until August 1944 when the 15th CA was inactivated. The 856th CA continued to man the lights at Fort Weaver until the end of the war.

### The Sand Island Position

On November 9, 1935 a seacoast searchlight (to be numbered Number 12) was received by the Honolulu harbor defense command for installation as a fixed light on Sand Island. The light was a 60-inch Model 10 MkV Sperry High Intensity (H.I.) Searchlight manufactured by the General Electric Company. The light was contained in a housing 40-feet above sea level atop a 26-foot high welded steel rigid tower that was completed on July 3, 1937. When placed in operation, the light on a movable cart was winched out of the shelter and moved along steel rails on the deck of the searchlight platform at the top of the tower. A gasoline engine of 25 kW of power although the light required only 18 kW to operate effectively powered the searchlight. While the light could be controlled at the tower, it was normally operated from a controller booth nearly 600-feet to the northwest.

A pair of 60-inch seacoast searchlights on fixed mounts atop two six legged rigid towers 20-feet, six-inches tall constructed of welded steel were erected on Sand Island in 1941. A corrugated galvanized iron shed measuring 12-feet square atop the tower housed the light. The lights were controlled from a 50-foot tall tower acquired in 1941 that was erected 597-feet to the northwest of the searchlight towers. The searchlights were manned by Battery D, 16th CA (HD) Regiment prior to World War II, and by a detail from Headquarters Battery, 16th CA during the early months of the war. In the early summer of 1943, Battery G, 55th CA was activated at Fort Ruger, to operate the searchlights in the HD Honolulu. The 856th CA (Searchlight) Battery (Separate) manned the lights at the end of the war.

### Kaneohe Bay Portable Searchlights

The new Project for the Defenses of Kaneohe Bay approved in mid-1941 recommended numerous 60-inch searchlights to sweep the waters of the bay and its approaches and illuminate targets for the new gun emplacements. By this time the service had switched to using just the newest generation of portable units, there were no more fixed searchlight positions built anew on Oahu. These particular searchlights at Kaneohe Bay were designated as the Ulupau Searchlight Group in 1941, and manned by a detachment of Battery D, 16th CA (HD) Regiment from Fort DeRussy on the south shore. The group consisted of five positions each provided with two portable 60-inch lights. One of the searchlights was an M1937 light while the remaining lights were M1941 searchlights, all manufactured by the Sperry Company. Two of the lights Numbers 21 and 22 (later renumbered as Numbers 25 and 26), were located at Makapuu Point.

A second pair Numbers 23 and 24 (later renumbered Numbers 27 and 28), were emplaced on concrete pads atop the ridge at Wailea Point and later moved to the towers near the shore of Waimanalo Bay. These portable searchlights each had 600,000,000 candle power and were positioned at elevations of 75 and 140-feet respectively, one above the other on the crest of the ridge. Two more searchlights, Numbers 25 and 26 (Later renumbered Numbers 29 and 30), were sited on a promontory of land known as Kii Point some 35 to 45-feet above the waters of Kailua Bay at the south end of Ulupau's east rim. These two lights, mounted atop 35-foot steel frame towers, were 70 to 80-feet above the waters of the bay. They later provided illumination for the guns of the Battery Kii, and reduced the uncovered zone northeast of Moku Manu, in addition to sweeping the waters of Kailua Bay.

Searchlights Numbers 27 and 28 (Later renumbered Numbers 31 and 32), were initially positioned on concrete platforms on the west slope of Keanaiki Hill south of Pyramid Rock 50-feet above sea level. Early in 1942, Searchlight Number 31 was moved to a new location on North Beach almost halfway between Pyramid Rock and the west slopes of Ulupau. A reserve searchlight was later set up near the small boat landing at the south end of the air station to cover the waters of the anchorage that lay off the naval air station.

### Lae O Kaoio Portable Searchlights

Across the bay at Ka Lae O Kaoio Point two more searchlights Numbers 29 and 30 (later renumbered Numbers 34 and 35) were emplaced near the foot of the cliffs above and behind the 155 mm guns of Battery Loko on the edge of the shore north of Kualoa Point. These lights had prepared concrete pads enclosed on front and sides by a low concrete wall. Nearby was a concrete shelter for the portable power generator that supplied the electricity for the searchlights. All of these portable lights could be utilized in an antiaircraft mode when necessary, thus supplementing the antiaircraft searchlights deployed in the Kaneohe area. Battery D of the 16th CA (HD) Regiment, posted at Fort DeRussy was a general service coast artillery battery serving on Oahu. On the brink of World War II it had been relieved of that duty and was assigned the task of providing a manning detachment for an observation post atop the west rim of Ulupau crater and responsibility for manning the Ulupau Searchlight Group's 12 seacoast searchlights along the northeast coast of Oahu from Makapuu Point to Punaluu just to the north of Kahana Bay.

West Coast Portable Searchlight Positions

### Makua Seacoast Searchlight Positions

Two 60-inch 600,000,000-candle power portable searchlights were installed early in World War II at Barking Sands on Oahu's southwestern (Leeward) shore at an elevation of 50-feet above sea level. They were atop two 35-foot tall "demountable" towers. The controller for the two lights was set up on a similar tower in a small emplacement near the searchlights. These searchlights provided target illumination for the 5-inch guns of Battery Homestead as well as the various coast artillery fire control and observation posts along the island's southwestern shoreline. The manning details for the lights were initially provided by a detail from Headquarters Battery, 1st Battalion, 41st CA, until August 1942 when Battery G, the searchlight battery of the 15th CA was activated.

### Kepuhi Point Searchlight Positions

Kepuhi Point, near Makaha, was the location for portable searchlight Numbers 52 and 53. Both of these 60-inch searchlights and their DEC were emplaced on small concrete platform type emplacements on the slopes of Keaau Ridge at an elevation of about 100-feet above sea level below the fire control stations. The manning details for the lights were provided by headquarters elements of Headquarters Battery, 1st Battalion, 41st CA until Battery G, 15th CA was activated in August, 1942. Battery G manned the lights until August 1944. the 856th CA (Searchlight) Battery (Separate) finally the lights until the end of the war.

### Puu Maelileli Searchlight Positions

Two portable 60-inch Searchlights: Numbers 55 and 56 were emplaced at Puu Maelileli. The two lights and their controllers were located on concrete pads on the slopes of the puu at an elevation of about 100-feet and were sited to provide illumination for the batteries emplaced along the island's Leeward Coast. The manning details for the lights were provided by headquarters elements of Headquarters Battery, 1st Battalion, 41st CA until Battery G, 15th CA was activated in August, 1942. Battery G manned the lights until August 1944, when the 15th CA was inactivated. The 856th CA (Searchlight) Battery (Separate) that continued to man the lights until the end of the war.

### Puu O Hulu Searchlight Positions

This installation at Puu O Hulu Kai, consisted of two 60-inch portable seacoast searchlights, Numbers 57 and 58. These lights were set up on concrete emplacements on the northwest slopes of the puu at an elevation of about 100-feet above sea level. The manning details for the lights were provided by headquarters elements of Headquarters Battery, 1st Battalion, 41st CA until Battery G, 15th CA was activated in August 1942. Battery G manned the lights until August 1944. when the 15th CA was inactivated. The 856th CA (Searchlight) Battery (Separate) continued to man the lights until the end of the war.

### Limoloa Gulch Searchlight Positions

Situated near the shore at Limoloa Gulch, were two portable 60-inch Searchlights and their director, all mounted atop 35-foot portable towers. As the site was only 20-feet above sea level, the towers barely gave the lights the minimum height to render them reasonably effective. The manning details for the lights were provided by headquarters elements of Headquarters Battery, 1st Battalion, 41st CA until Battery G, 15th CA was activated in August, 1942. Battery G manned the lights until August 1944. The lights were manned and served until the end of the war.

### Barbers Point Searchlight Positions

Just to the northwest of Barbers Point two portable 60-inch Seacoast Searchlights, Numbers 61 and 62 were located. Because of the low elevation of the site, five-feet above sea level, two 50-foot searchlight towers were procured; on which a light was mounted. Another tower of the same height was also procured and erected to house the DEC. During World War II, these two searchlights served as both AA and seacoast lights and were used regularly by the Hawaiian Antiaircraft Artillery Command and the Marine Corps Aviation and AA batteries at Ewa Marine Corps Airfield.

A second set of searchlights was positioned to the east of Barbers Point. This position was equipped with a pair of 60-inch Portable Seacoast Searchlights, Number 1 and 2. Again, because of the five-foot elevation of the site above sea level, two 50-foot searchlight towers were procured on which the two lights were to be mounted, and a third provided for the director. After their erection in World War II, these searchlights were frequently used to search for aircraft that had crashed in the waters off Pearl Harbor in addition to their seacoast defense mission. Seacoast searchlight drills were held on a weekly basis and the Barbers Point lights were frequent participants in seacoast artillery night target practices. The manning detachment for the position was initially provided by headquarters elements of Headquarters Battery, 1st Battalion, 55th CA until Battery G, 15th CA was activated in August, 1942. The personnel of Battery G were reassigned to the 856th CA (Searchlight) Battery (Separate) that continued to man the lights until the end of the war.

### Ewa Beach Searchlight Positions

This position on Ewa Beach consisted of two portable 60-inch Searchlights (Searchlight Numbers 3 and 4). Because of the low elevation of the site two 50-foot searchlight towers were procured early in WWII. The two lights were mounted atop these towers. A controller tower of the same height was also procured and erected. The two lights were exercised on a regular basis either in weekly searchlight drills, night practices by the coastal batteries at forts Weaver and Kamehameha and occasionally to search for downed aircraft in the waters off Fort Kamehameha and Pearl Harbor. The manning details for the lights were provided by headquarters elements of Headquarters Battery, 1st Battalion, 55th CA until Battery G, 15th CA was activated in August 1942. Personnel of Battery G were reassigned to the 856th CA (Searchlight) Battery (Separate) that continued to man the lights until the end of the war.

### North Shore Portable Searchlight Positions

Immediately after the Pearl Harbor attack, the three officers and 108 enlisted men of Battery G, 57th CA, the regimental searchlight battery, moved out to the North Shore and established its base camp near the Dillingham Ranch at Kawaihapai. It soon deployed detachments to the light positions at Pupukea, Ashley, Mokuleia, along the North Shore, and at Laniloa Point near Laie and Punaluu Point on Windward Oahu. At about the same time the installation of additional lights was begun at Kaena Point, at the west end of Camp Kawaihapai, and on the Paumalu Heights near Waialee on the North Shore, as well as Kahuku Point, and a second position between Kahuku and Makahoa Points, and on the island's windward coast. In order to operate effectively the searchlights needed to be a minimum of 60-feet above the water so that the spray and haze of wave action did not interfere with the searchlight beam. Typically only one searchlight of each pair of lights was used at a time. As the lights of each pair were usually separated by several hundred yards, alternating use of the lights, would defeat an enemy attempt to precisely locate the lights at night. During daylight hours both the searchlights and the towers were camouflaged and kept from the view of enemy observers. Battery G, 57th CA, provided the manning detachments at 12 searchlight sites along Oahu's northern shoreline.

Each of the sites were arranged to receive two portable 60-inch searchlights, their distant electric controllers and power plants. Although these searchlight locations had been selected prior to World War II, the searchlights were still in the process of being placed in service when hostilities began. The 24 searchlights, their portable powerplant equipment, and the distant electric controllers provided to the North Shore Groupment were a mixture of the M1929 MVI Duplex, M1934, M937, M1939, 1940 and 1941 Sperry types that had been deposited in the Hawaiian Department's War Reserves prior to the war. In most cases the lights were in good condition but many of the old M1929 Duplex truck mounted powerplants were considered poor.

In the locations where the searchlight positions were at, or near sea level, the searchlights were mounted atop demountable 50-foot steel frame towers. Typically, these towers were erected at elevations that varied between ten and 60-feet above sea level. The towers, some of which were 35-feet tall, enabled the searchlights to operate at an acceptable elevation. The site for Searchlight Numbers 36 and 37, at Makalii Point near Kahana Bay, was only five-feet above sea level. Consequently, two 50-foot towers were ordered for that site. Another pair of 50-foot towers was required for Searchlights 50 and 51 at the Mokuleia site that was barely ten-feet above the water. In all but two of the searchlight locations, 50-foot searchlight and controller towers were erected at North Shore locations in 1942.

These exceptions were Searchlights 52 and 53 at Kaena Point, and Searchlight Numbers 47 and 48 at Pupukea. At these two locations, the elevation of these searchlight sites was high enough that towers were not required for the searchlights or their controllers. These three sites had concrete slab platforms with low retaining walls around the front and sides built for the searchlights, while a splinter proof concrete booth was constructed to house the lights' electric controllers. The site at Pupukea was typical of the three installations not mounted atop towers. A two-room reinforced concrete structure containing a controller booth for its two M1940 and M1941 Sperry portable searchlights and a small bunk room for the use of the searchlight detachment, was dug into the edge of the cliff line 140-feet above Waimea Bay and consisted of two rooms. The controller booth was similar in appearance to the observation rooms of the fire control stations along the shoreline. It was equipped, however, with an M1940 Distant Electric Controller for the two searchlights instead of a DPF, or azimuth instrument. Typically the manning detachment for a pair of 60-inch searchlights was composed of about eight enlisted men from Battery G. At each of the searchlight installations, a small concrete splinter proof shelter was provided to house an M1940 wheeled 31.25KW, 125 amperes, 1500 RPM portable differential compound generator set that powered the two searchlights. Later in the war Searchlights 40 and 41 were placed on the high ground in the rear of the town of Kahuku.

# APPENDIX XI

## 1934 DETAIL MAPS OF THE HARBOR DEFENSES OF PEARL HARBOR AND HONOLULU HAWAII

Another valuable set of source documents for major tactical structures (gun batteries, mine buildings, switchboards, power plants, fire control structures, searchlights, etc.) are the "confidential blueprint" series of maps generated by the U.S. Army Corps of Engineers from various dates after construction. These maps were updated periodically through out the periods of construction. This appendix consists of a set of maps dated 1934 that show the layout of the forts approaching the war years.

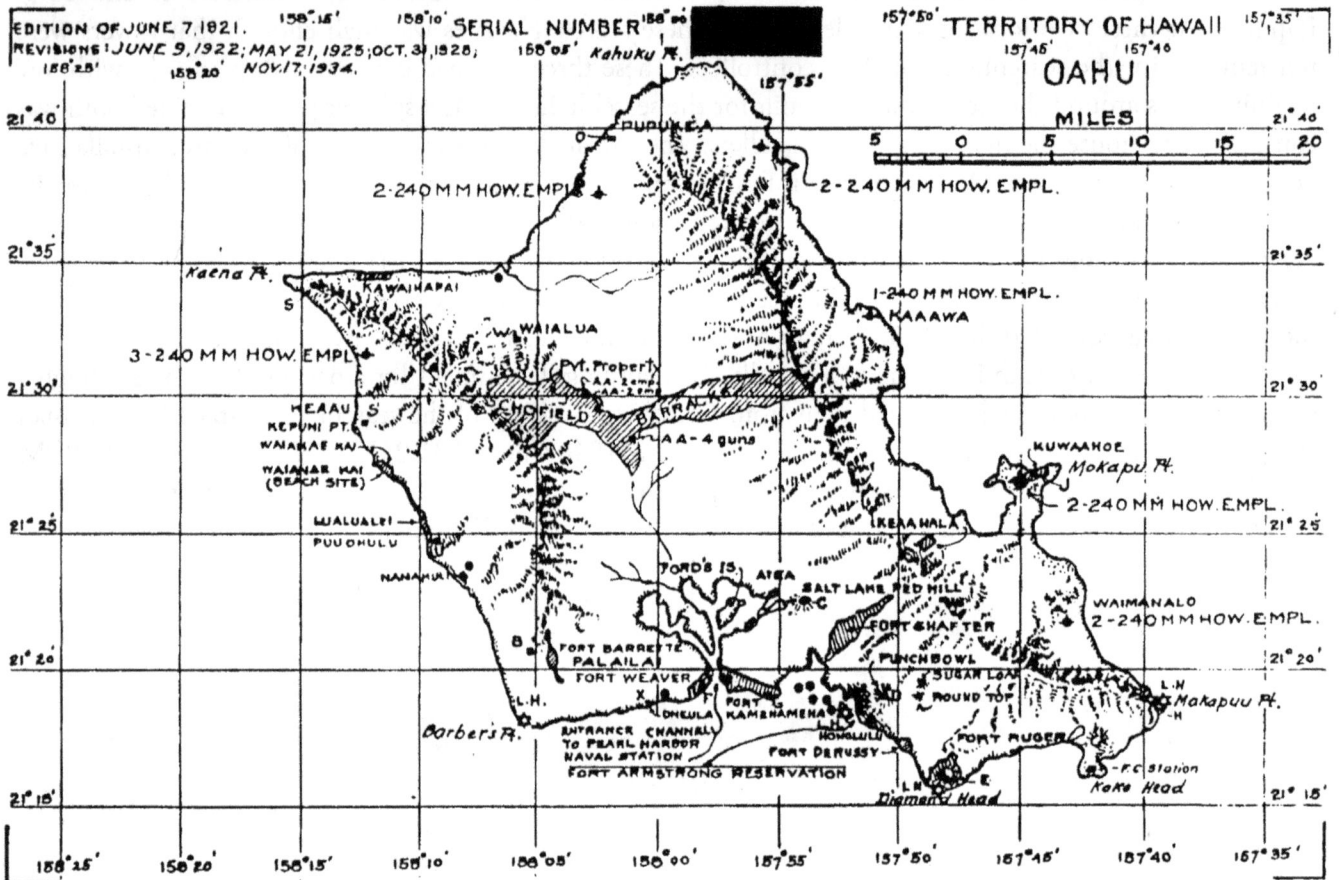

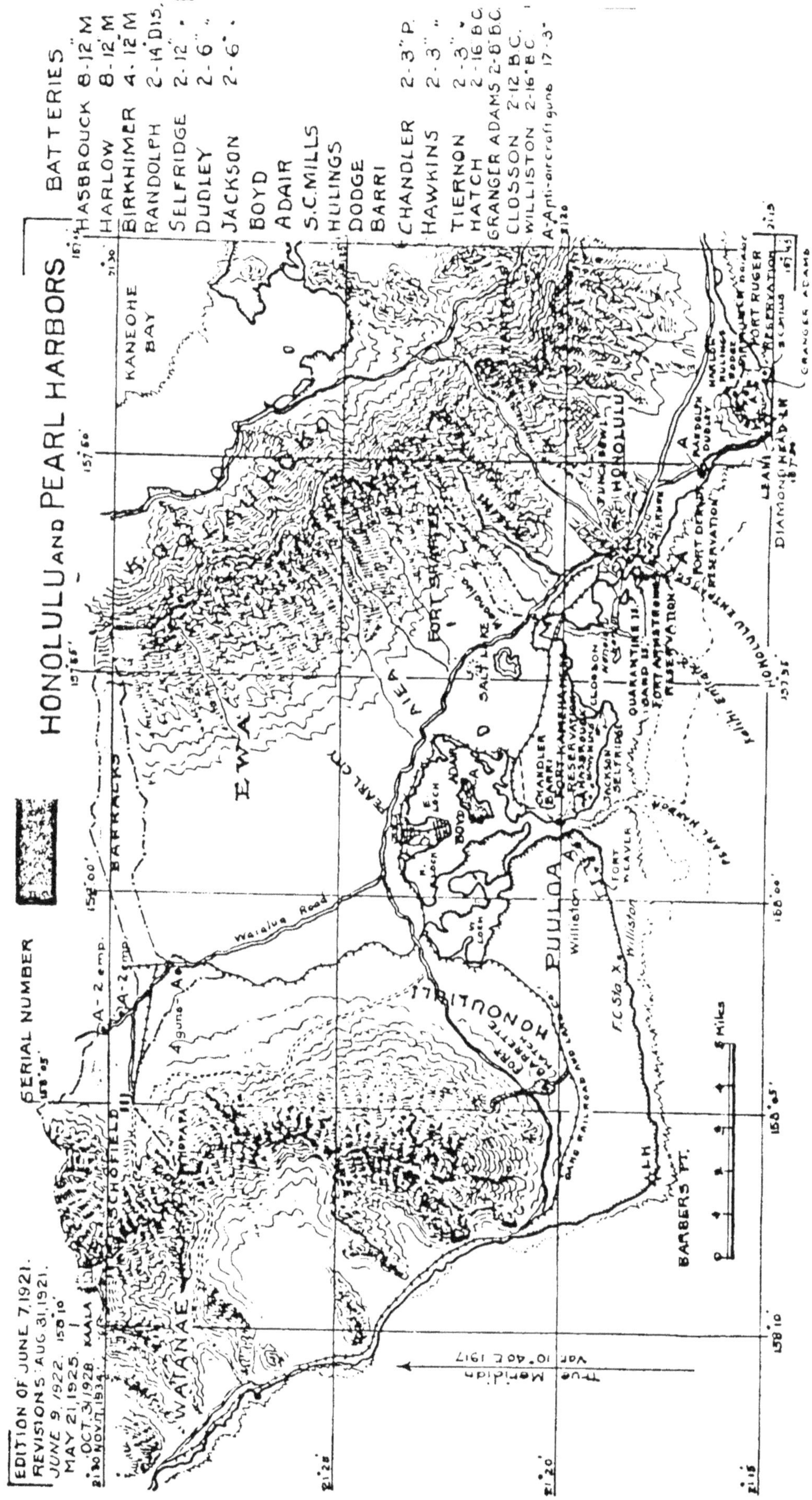

HONOLULU and PEARL HARBORS

EDITION OF JUNE 7, 1921.
REVISIONS AUG. 31, 1921.
MAY 21, 1925.
JUNE 9, 1922. OCT. 31, 1928. NOV. 1, 1934.

SERIAL NUMBER

| BATTERIES | |
| --- | --- |
| HASBROUCK | 8 - 12" M |
| HARLOW | 8 - 12" M |
| BIRKHIMER | 4 - 12" M |
| RANDOLPH | 2 - 14" D15. |
| SELFRIDGE | 2 - 12" |
| DUDLEY | 2 - 6" |
| JACKSON | 2 - 6" |
| BOYD | |
| ADAIR | |
| S.C. MILLS | |
| HULINGS | |
| DODGE | |
| BARRI | |
| CHANDLER | 2 - 3" P. |
| HAWKINS | 2 - 3" |
| TIERNON | 2 - 3" |
| HATCH | 2 - 16" B.C. |
| GRANGER ADAMS | 2 - 6" B.C. |
| CLOSSON | 2 - 12" B.C. |
| WILLISTON | 2 - 16" B.C. |
| A - Anti-aircraft guns | 17 - 3" |

WAIANAE

SCHOFIELD BARRACKS

KANEOHE BAY

EWA

PEARL HARBOR

HONOLULU

FORT SHAFTER

AIEA

FORT RUGER

DIAMOND HEAD

PUULOA

HONOULIULI

Waialua Road

BARBERS PT.

Miles

True Meridian
Var 10° 40 E 1917

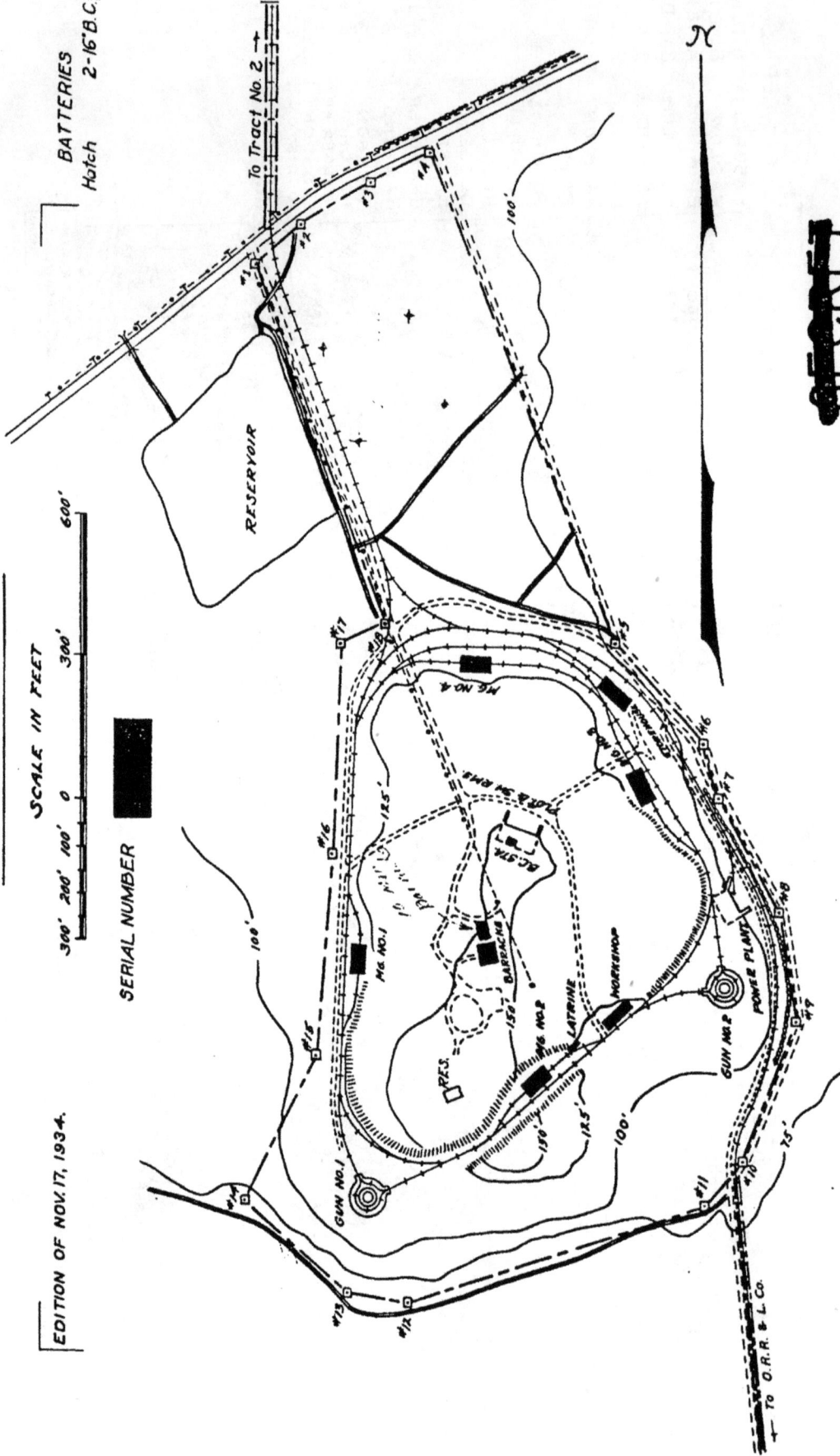

PEARL HARBOR, T.H.
FT. BARRETTE
TRACT NO. 1

SCALE IN FEET

600'

300'

300' 200' 100' 0

SERIAL NUMBER

EDITION OF NOV. 17, 1934.

BATTERIES
Hatch   2-16 B.C.

SECRET

To Tract No. 2

RESERVOIR

GUN NO. 1

GUN NO. 2

MG NO. 1

MG NO. 2

MG NO. 4

BARRACKS

BC. STA.

LATRINE

WORKSHOP

POWER PLANT

P.F.S.

To O.R.R. & L. Co.

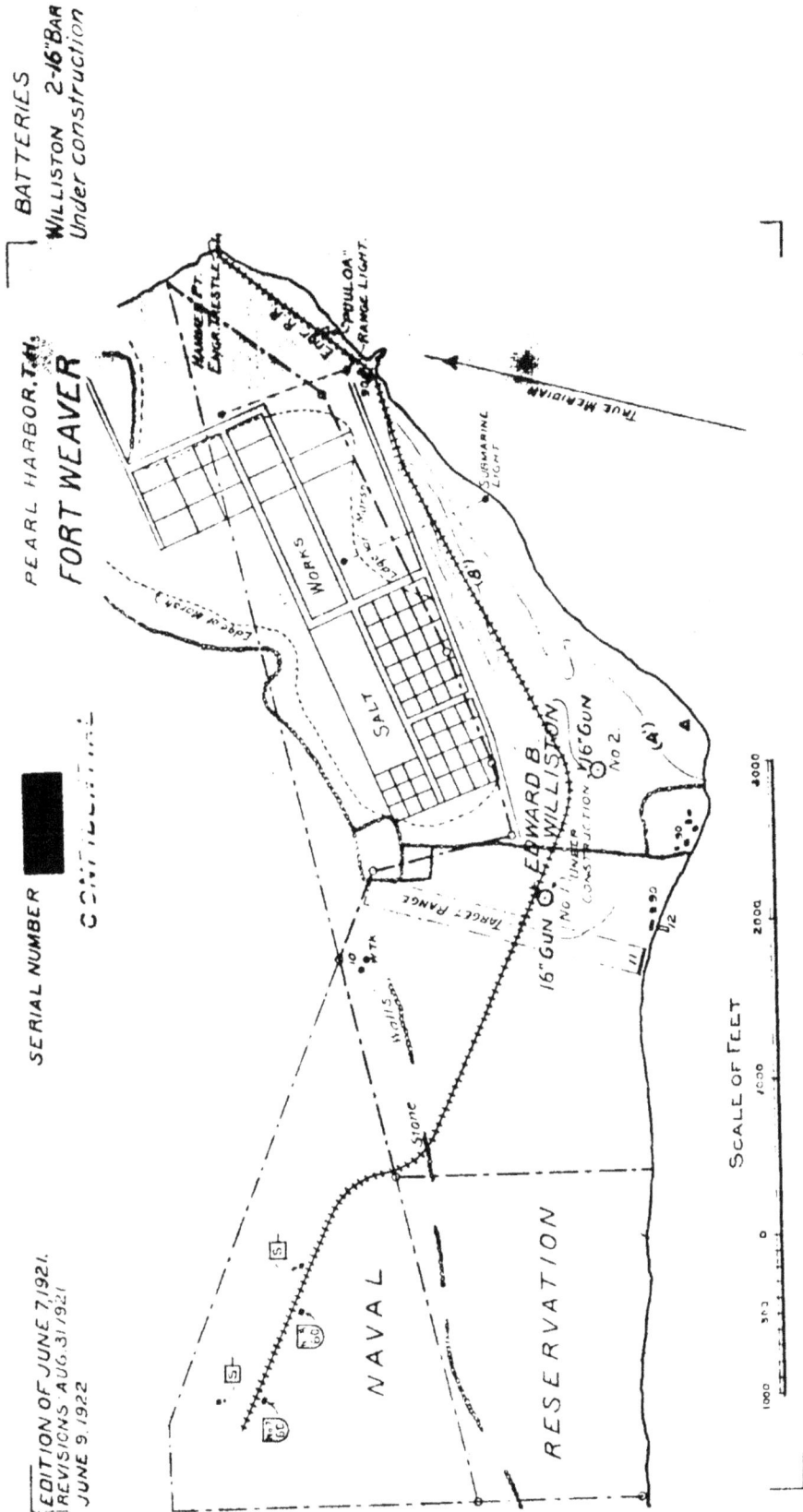

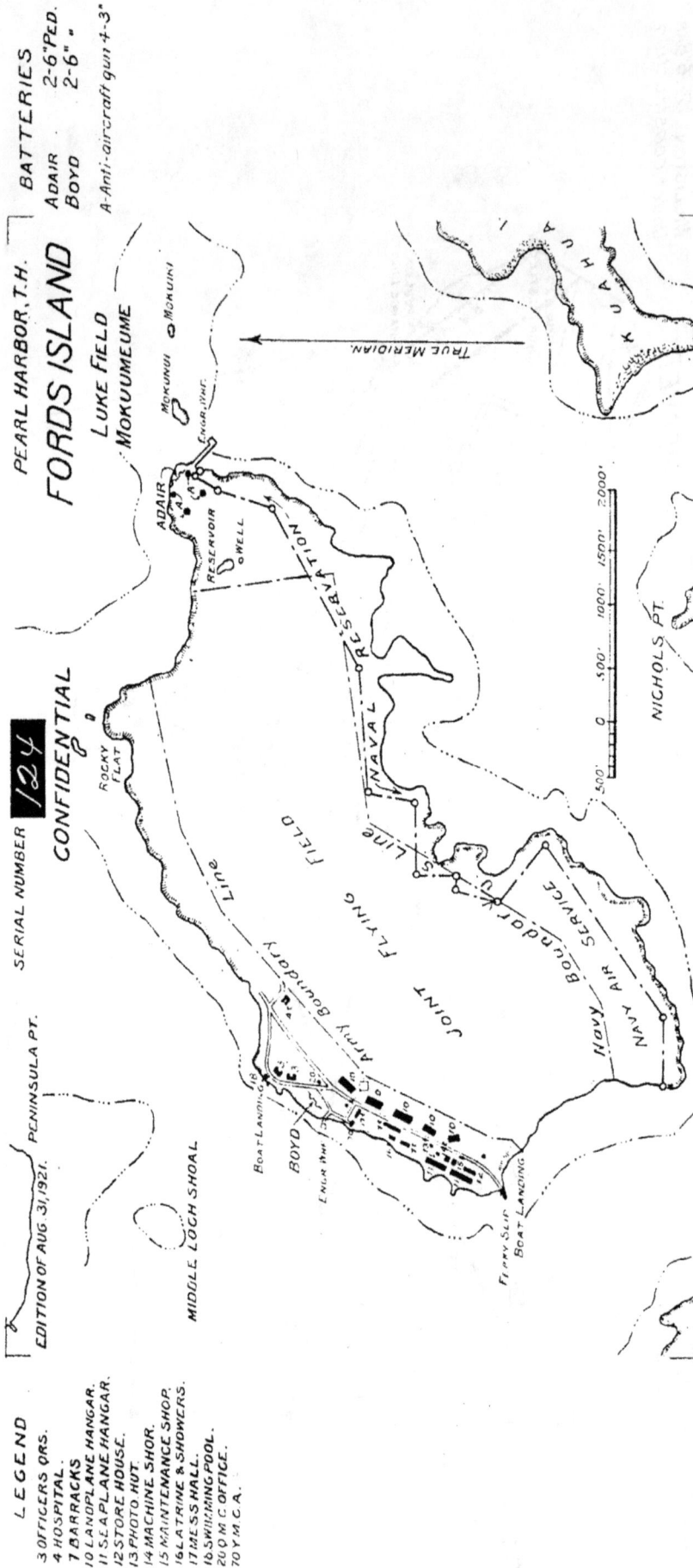

BATTERIES
ADAIR    2-6"PED.
BOYD     2-6"  "
A-Anti-aircraft gun 4-3"

PEARL HARBOR, T.H.

FORDS ISLAND

LUKE FIELD
MOKUUMEUME

MOKUNUI  MOKUIKI

TRUE MERIDIAN

KUAHUA

SERIAL NUMBER 124
CONFIDENTIAL

ADAIR
RESERVOIR
WELL

NAVAL RESERVATION LINE

ROCKY FLAT

NICHOLS PT.

0    500'   1000'   1500'   2000'

JOINT FLYING FIELD

JOINT

Line

Army Boundary

NAVY BOUNDARY

NAVY AIR SERVICE

PENINSULA PT.

EDITION OF AUG 31, 1921.

MIDDLE LOCH SHOAL

BOAT LANDING

BOYD

ENGR WHF.

FERRY SLIP
BOAT LANDING

LEGEND
3 OFFICERS QRS.
4 HOSPITAL.
7 BARRACKS.
10 LANDPLANE HANGAR.
11 S.E.A PLANE HANGAR.
12 STORE HOUSE.
13 PHOTO HUT.
14 MACHINE SHOP.
15 MAINTENANCE SHOP.
16 LATRINE & SHOWERS.
17 MESS HALL.
16 SWIMMING POOL.
20 O M C OFFICE.
70 Y.M.C.A.

PEARL HARBOR, T.H.
BISHOP POINT

CONFIDENTIAL.

SERIAL NUMBER 124

BATTERIES.
BARRI........2-4.7" PED.
CHANDLER..2-3" "

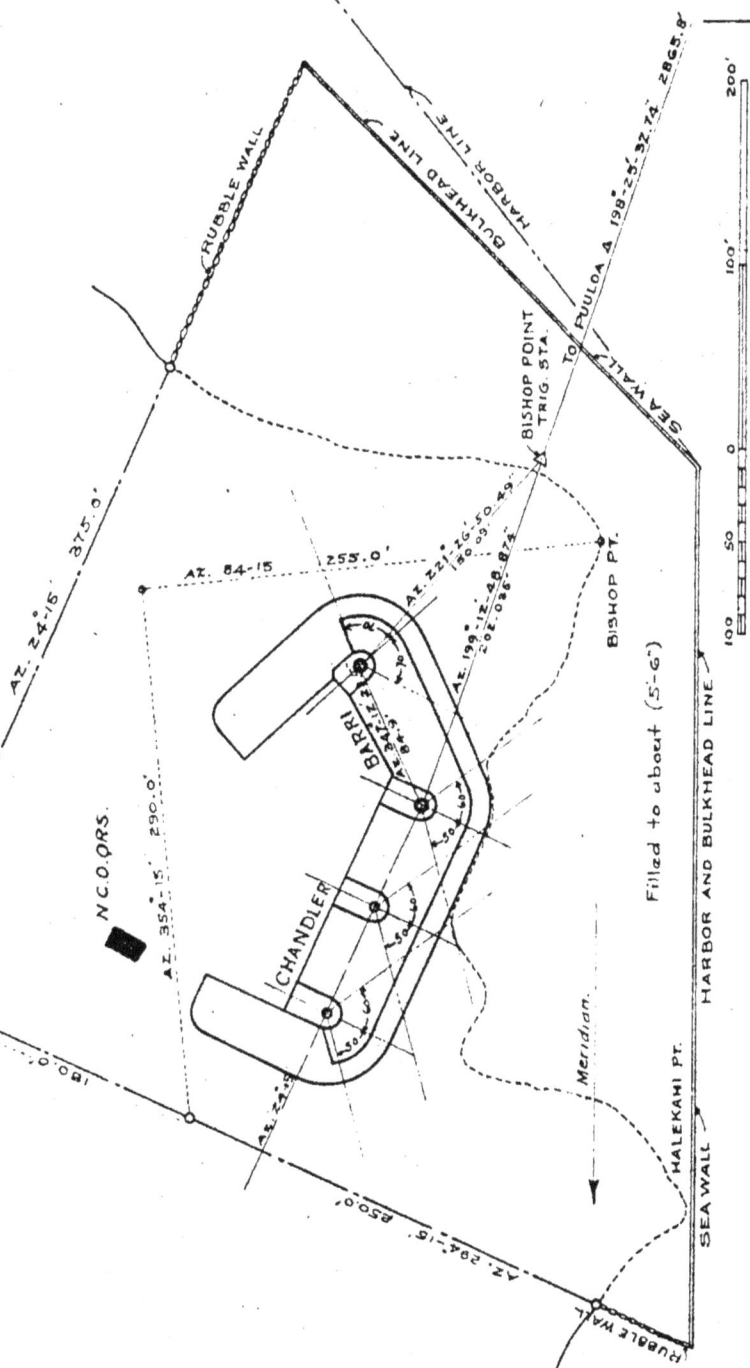

EDITION OF DEC. 26, 1919.
REVISIONS: JUNE 7, 1921.
AUG. 31, 1921.

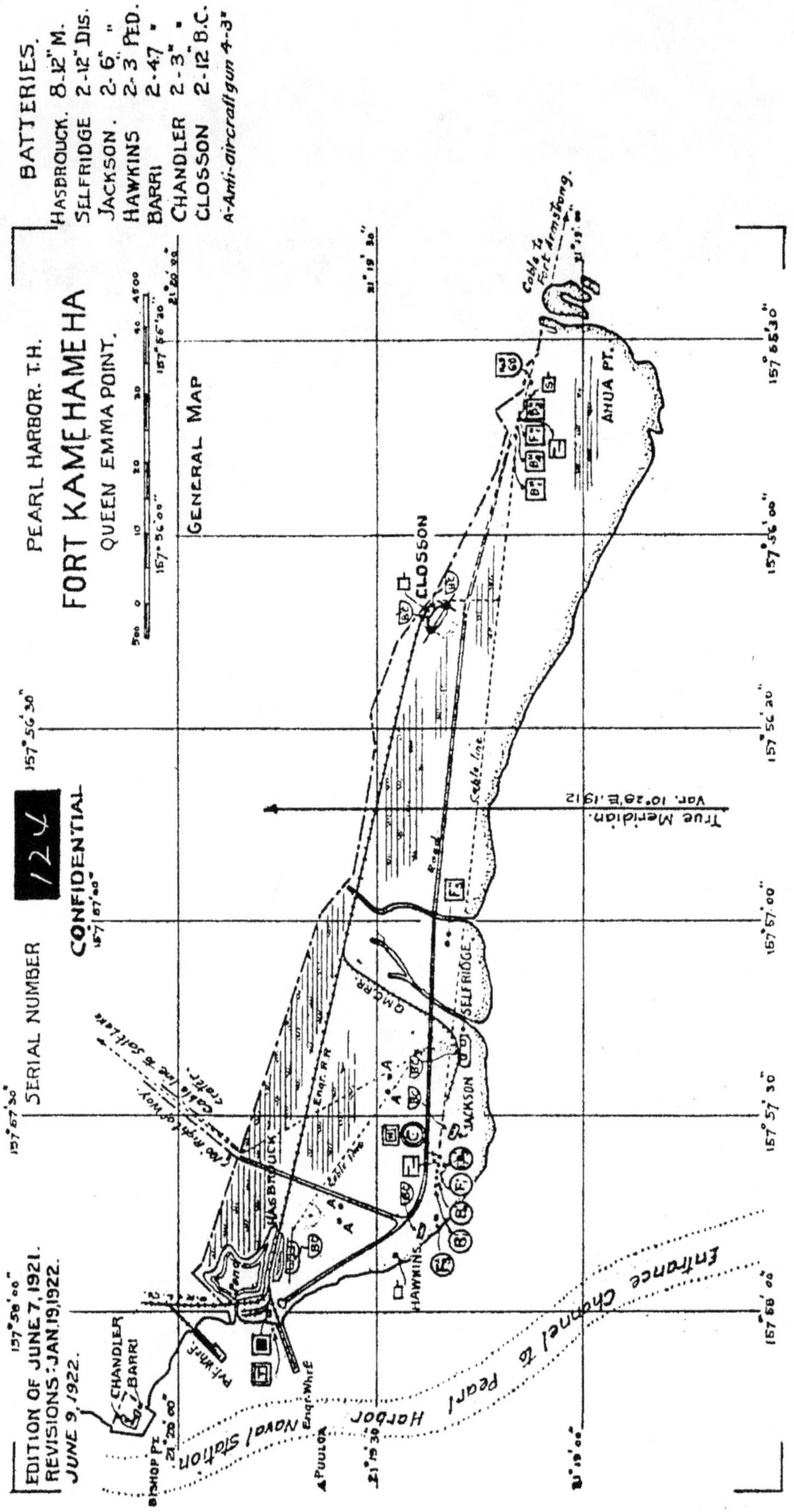
SERIAL NUMBER
124
CONFIDENTIAL
PEARL HARBOR. T.H.
FORT KAMEHAMEHA
QUEEN EMMA POINT.
GENERAL MAP
BATTERIES.
HASBROUCK. 8-12" M.
SELFRIDGE 2-12" DIS.
JACKSON. 2-6" "
HAWKINS 2-3" PED.
BARRI 2-4.7 "
CHANDLER 2-3" "
CLOSSON 2-12 B.C.
A-Anti-aircraft gun 4-3"
EDITION OF JUNE 7, 1921.
REVISIONS: JAN.19,1922.
JUNE 9, 1922.
CHANDLER
BARRI
BISHOP PT.
PUULOA
Engr. Wharf
HAWKINS
HASBROUCK
JACKSON
SELFRIDGE
O.M.C.R.R.
CLOSSON
AHUA PT.
Cable line
Road
Cable to Fort Armstrong
True Meridian.
Var. 10°28'E. 1912
Entrance Channel to Pearl Harbor Naval Station
157°58'00"
157°57'30"
157°57'00"
157°56'30"
157°56'00"
157°55'30"
21°19'30"
21°19'00"

SERIAL NUMBER 124

PEARL HARBOR, T.H.

CONFIDENTIAL FORT KAMEHAMEHA D-1.

EDITION OF JUNE 7, 1921.
REVISIONS: JAN. 19, 1922.

BATTERIES
HASBROUCK... 8-12"M.
HAWKINS........2-3"PED.
A-Anti-aircraft gun 2-3"

Scale of Feet.

LEGEND
1 ADMIN BUILDING.
2 COMM. OFFICERS QRS.
3 OFFICERS QRS.
4 HOSPITAL.
5.
6 N.C.OFFICERS QRS.
7 BARRACKS.
10 BATH HOUSE.
11 SWIMMING POOL.
12 TENNIS COURT.
13 CARPENTER SHOP.
14 GARAGE.
15 HAND BALL COURT.
16 MINE PLANTERS ST. HO.
109 TRANSFORMER STA. AND
    TEL. EXCHANGE.
110 FIRE APPARATUS HOUSE.
111 SEWER EJECTOR PUMP HO.
112 GRAND STAND.
116 COURT MARTIAL AND
    POST OFFICE.
117 REEL HOUSE.
41 ENGR. STORE HOUSE.
70 Y.M.C.A.

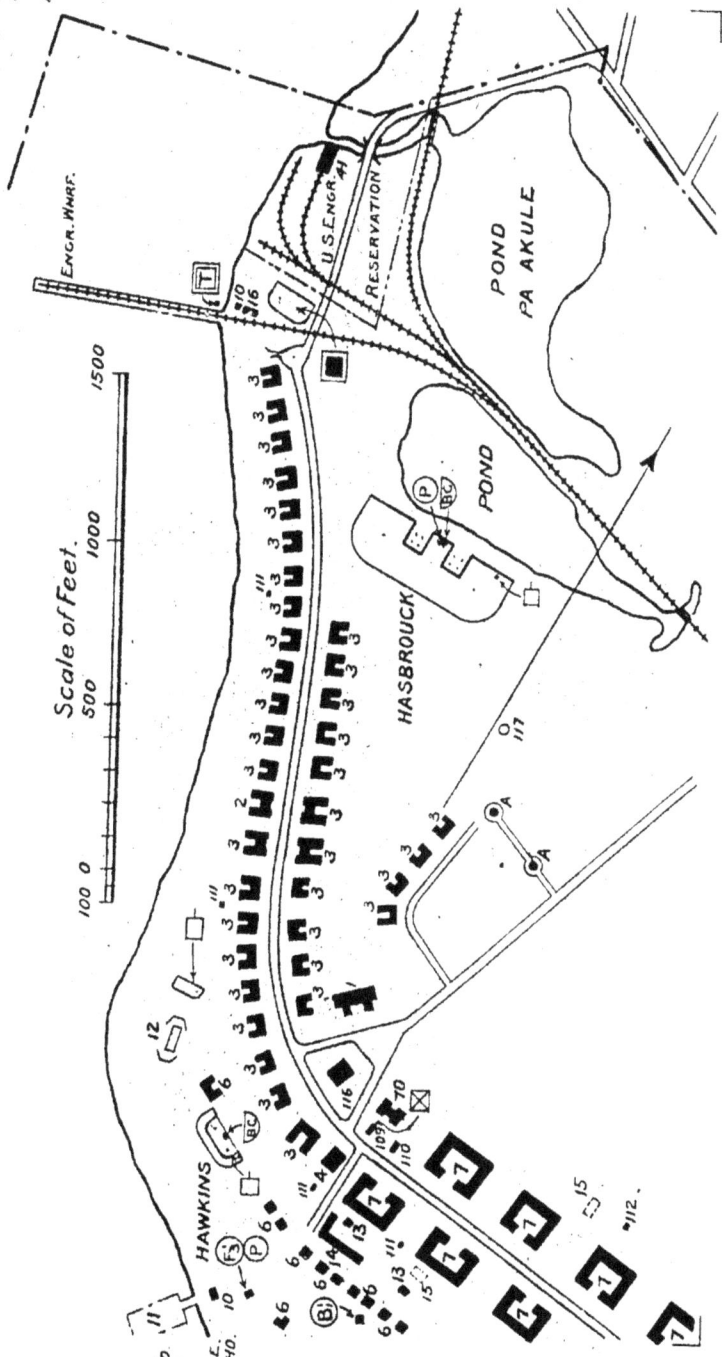

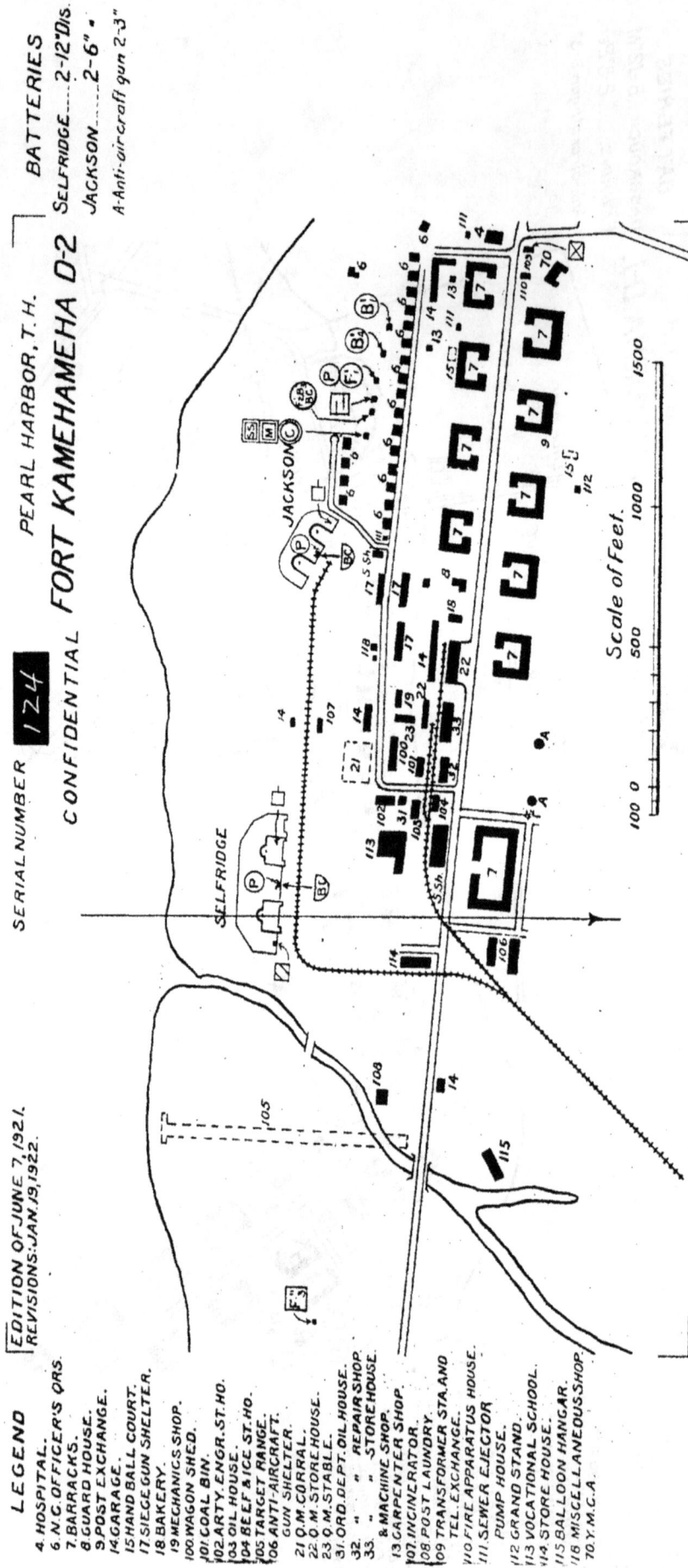

CONFIDENTIAL **FORT KAMEHAMEHA D-2**

*PEARL HARBOR, T.H.*

SERIAL NUMBER **124**

EDITION OF JUNE 7, 1921.
REVISIONS: JAN. 19, 1922.

BATTERIES

SELFRIDGE ____ 2-12"Dis.
JACKSON ____ 2-6" *
A-Anti-aircraft gun 2-3"

Scale of Feet.

LEGEND

4. HOSPITAL.
6. N.C. OFFICER'S QRS.
7. BARRACKS.
8. GUARD HOUSE.
9. POST EXCHANGE.
14. GARAGE.
15. HAND BALL COURT.
17. SIEGE GUN SHELTER.
18. BAKERY.
19. MECHANICS SHOP.
100. WAGON SHED.
101. COAL BIN.
102. ARTY. ENGR. ST. HO.
03. OIL HOUSE.
04. BEEF & ICE ST. HO.
105. TARGET RANGE.
106. ANTI-AIRCRAFT.
    GUN SHELTER.
21. O.M. CORRAL.
22. O.M. STORE HOUSE.
23. O.M. STABLE.
31. ORD. DEPT. OIL HOUSE.
32. "   "   REPAIR SHOP.
33. "   "   STORE HOUSE.
& MACHINE SHOP.
13. CARPENTER SHOP.
07. INCINERATOR.
08. POST LAUNDRY.
09. TRANSFORMER STA. AND
    TEL. EXCHANGE.
110. FIRE APPARATUS HOUSE.
111. SEWER EJECTOR
    PUMP HOUSE.
112. GRAND STAND.
113. VOCATIONAL SCHOOL.
114. STORE HOUSE.
115. BALLOON HANGAR.
116. MISCELLANEOUS SHOP.
70. Y.M.C.A.

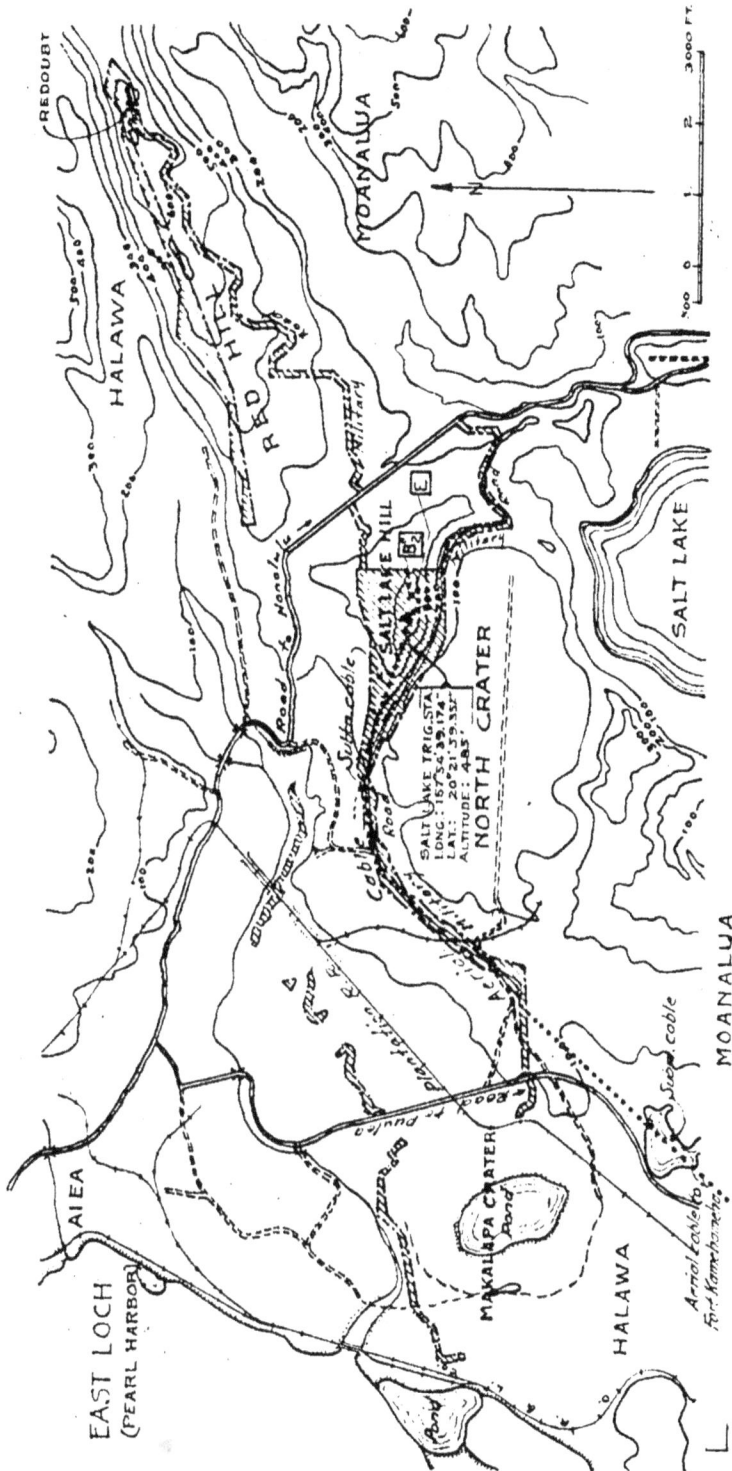

LEGEND
EDITION OF DEC 26, 1919.
REVISIONS: JUNE 7, 1921.
U.S. Military Reservations. AUG. 31, 1921.

SERIAL NUMBER

CONFIDENTIAL

ISLAND OF OAHU, T.H.
MAKALAPA CRATER, SALT LAKE,
RED HILL.

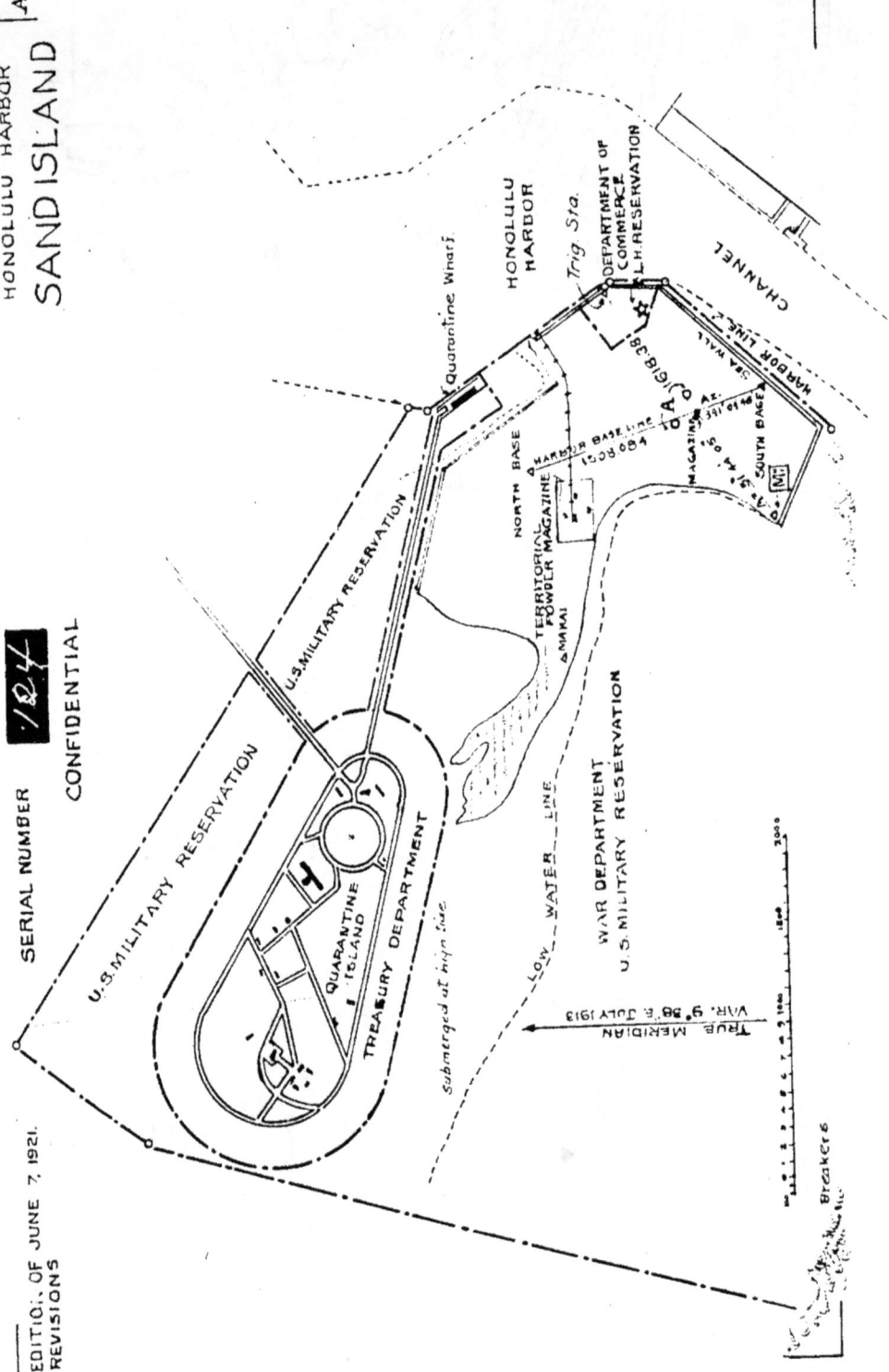

HONOLULU HARBOR

SAND ISLAND

A ···· EMP FOR 2 - 3"
ANTI AIR CRAFT GUNS

Quarantine Wharf

Honolulu Harbor

Trig Sta.

DEPARTMENT OF
COMMERCE
L.H. RESERVATION

CHANNEL

HARBOR LINE

SEA WALL

SOUTH BASE

HARBOR BASELINE TO A. (C.) 1818.084

NORTH BASE

MAGAZINE

TERRITORIAL
POWDER MAGAZINE
MAKAI

SERIAL NUMBER 124

CONFIDENTIAL

U.S. MILITARY RESERVATION

U.S. MILITARY RESERVATION

QUARANTINE
ISLAND

TREASURY DEPARTMENT

Submerged at high line

LOW WATER LINE

WAR DEPARTMENT
U.S. MILITARY RESERVATION

TRUE MERIDIAN
VAR. 9° 38' E. JULY 1913

EDITION OF JUNE 7, 1921.
REVISIONS

Breakers

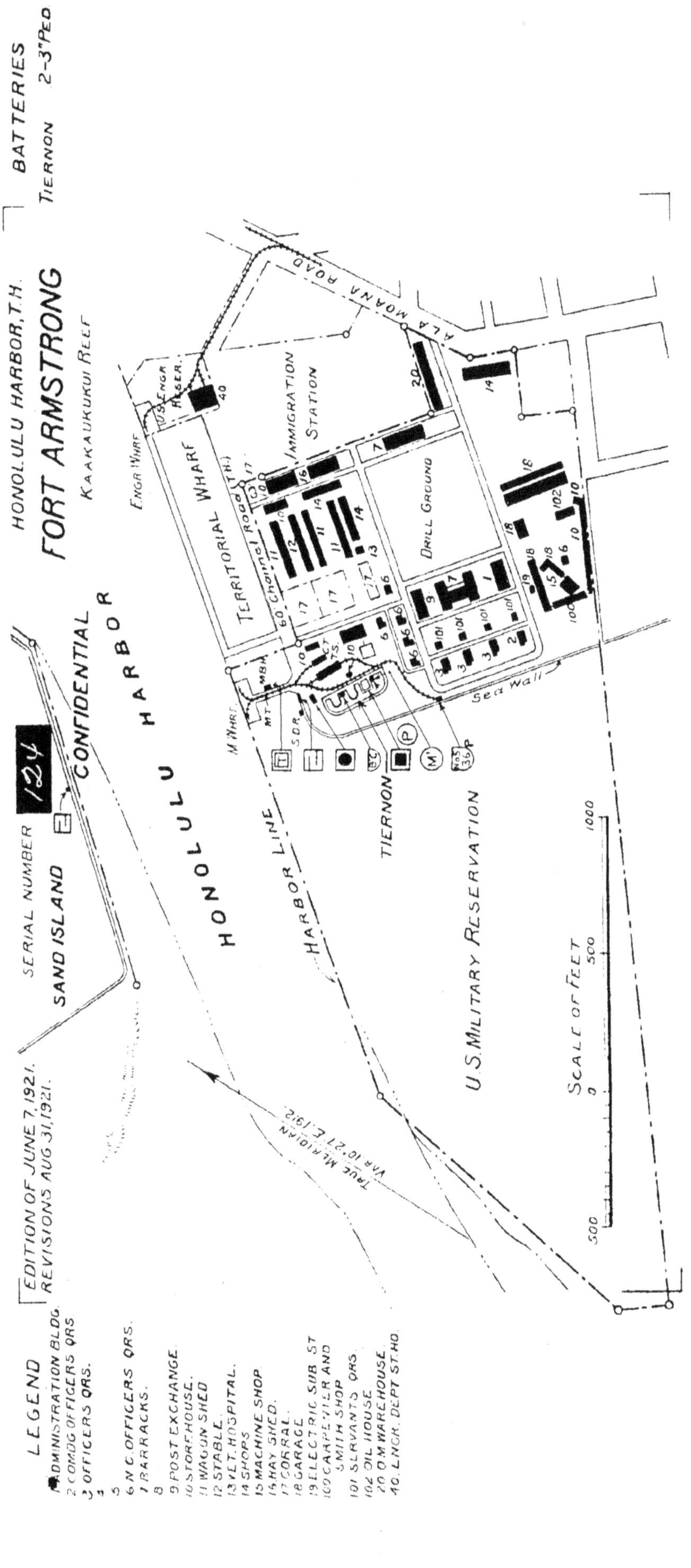

BATTERIES
TIERNON    2-3"PED

HONOLULU HARBOR, T.H.
FORT ARMSTRONG
KAAKAUKUKUI REEF

SERIAL NUMBER 124
SAND ISLAND
CONFIDENTIAL

HONOLULU HARBOR

EDITION OF JUNE 7, 1921.
REVISIONS AUG 31, 1921.

HONOLULU LINE

HARBOR LINE

TIERNON

U.S. MILITARY RESERVATION

TRUE MERIDIAN
VAR 10.27 E. 1912.

SCALE OF FEET
500    0    500    1000

LEGEND
1 ADMINISTRATION BLDG.
2 COMDG OFFICERS QRS
3 OFFICERS QRS.
4
5
6 N C OFFICERS QRS.
7 BARRACKS.
8
9 POST EXCHANGE
10 STOREHOUSE.
11 WAGON SHED
12 STABLE.
13 VET. HOSPITAL.
14 SHOPS
15 MACHINE SHOP
16 HAY SHED.
17 CORRAL.
18 GARAGE
19 ELECTRIC SUB ST
100 CARPENTER AND SMITH SHOP
101 SERVANTS QRS
102 OIL HOUSE
20 Q M WAREHOUSE
40 L ENGR DEPT ST. HO.

ENGR WHARF
U.S. ENGR. RESER.
TERRITORIAL WHARF
IMMIGRATION STATION
DRILL GROUND
SEA WALL
ALA MOANA ROAD

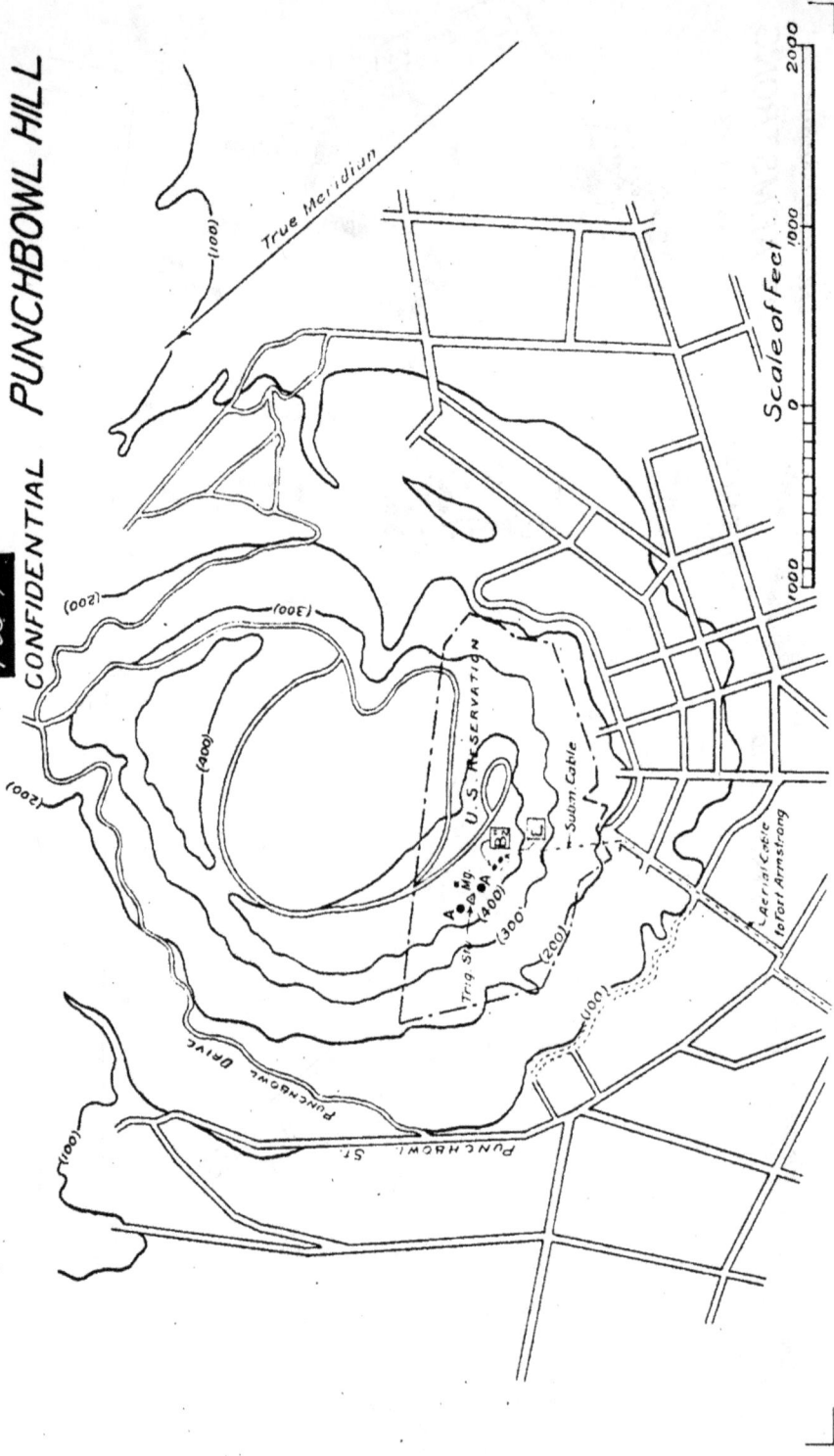

ISLAND OF OAHU, T.H.

CONFIDENTIAL **PUNCHBOWL HILL**

SERIAL NUMBER 124

EDITION OF AUG. 31, 1921.

A - Platform for
Anti-aircraft gun - 2-3"

Scale of Feet

True Meridian

U.S. RESERVATION

Subm. Cable

Aerial Cable
to Fort Armstrong

PUNCHBOWL DRIVE

PUNCHBOWL ST.

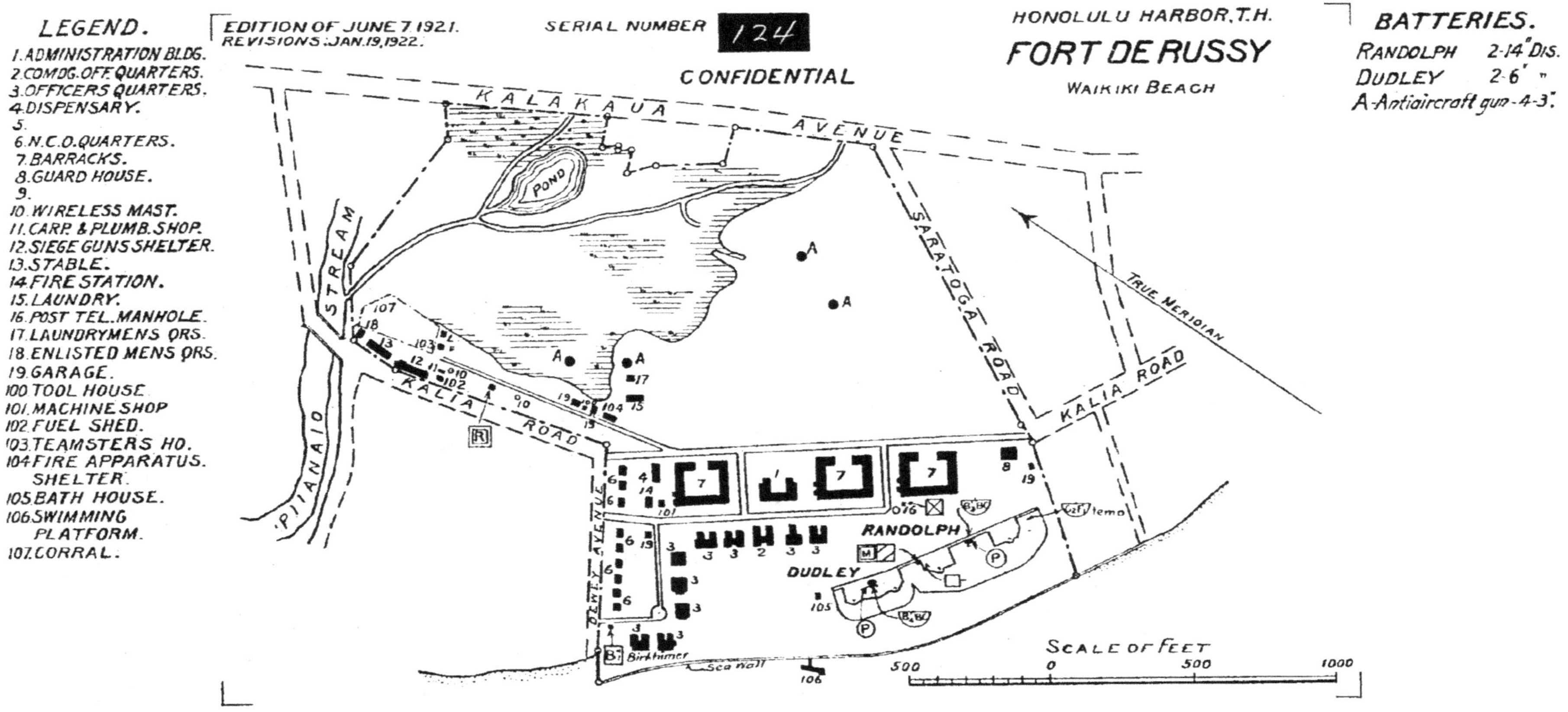

LEGEND.
1. ADMINISTRATION BLDG.
2. COMDG. OFF. QUARTERS.
3. OFFICERS QUARTERS.
4. DISPENSARY.
5.
6. N.C.O. QUARTERS.
7. BARRACKS.
8. GUARD HOUSE.
9.
10. WIRELESS MAST.
11. CARP. & PLUMB. SHOP.
12. SIEGE GUNS SHELTER.
13. STABLE.
14. FIRE STATION.
15. LAUNDRY.
16. POST TEL. MANHOLE.
17. LAUNDRYMENS QRS.
18. ENLISTED MENS QRS.
19. GARAGE.
100. TOOL HOUSE.
101. MACHINE SHOP
102. FUEL SHED.
103. TEAMSTERS HO.
104. FIRE APPARATUS. SHELTER.
105. BATH HOUSE.
106. SWIMMING PLATFORM.
107. CORRAL.
EDITION OF JUNE 7 1921.
REVISIONS: JAN. 19, 1922.
SERIAL NUMBER 124
CONFIDENTIAL
HONOLULU HARBOR, T.H.
FORT DE RUSSY
WAIKIKI BEACH
BATTERIES.
RANDOLPH 2-14" DIS.
DUDLEY 2-6" "
A-Antiaircraft gun-4-3".
KALAKAUA AVENUE
SARATOGA ROAD
KALIA ROAD
DEWEY AVENUE
STREAM
PIIANAIO
POND
TRUE MERIDIAN
RANDOLPH
DUDLEY
Birkhimer
Sea wall
SCALE OF FEET
500 0 500 1000

BATTERIES

| | | |
|---|---|---|
| HARLOW | 8-12" M. | |
| BIRKHIMER | 4-12" M. | |
| S.C. MILLS | 2-5" Ped. | |
| HULINGS | 2-4.7" | |
| DODGE | 2-4.7" | |
| A-Anti-aircraft guns. | 6-3" | |
| B. | 12-6 pdrs. | |

HONOLULU HARBOR T.H.

FORT RUGER

DIAMOND HEAD AND WAIALAE

GENERAL MAP

SERIAL NUMBER 124

CONFIDENTIAL

EDITION OF JUNE 7, 1921.
REVISIONS: JAN. 19, 1922.

LEGEND

1. ADMINISTRATION BUILDING.
2. COMDG. OFFICER'S QUARTERS
3. OFFICERS QUARTERS.
4. DISPENSARY.
5. HOSPITAL STWD'S. QRS.
6. N.C. OFFICERS QRS.
7. BARRACKS.
8. GUARDHOUSE.
9. POST EXCHANGE.
10. FIRE STATION.
11. STOREHOUSE.
12. MACHINE SHOP.
13. POST LAUNDRY.
14. POST SUPPLY OFFICE.
15. STABLES & WAGON SHED.
16. SIEGE GUNS SHELTERS.
17. OIL HOUSE.
18. HYDROLITIC TANK.
19. BALLOON HANGAR.
100. SERVANTS QRS.
101. SEPTIC TANK.
102. BARBERS SHOP.
103. CLASS ROOM.
31. ORDN. & ENGR. ST. HD.
70. SERVICE CLUB.

HONOLULU HARBOR, T.H.

FORT RUGER D-1

SCALE OF FEET

100 0 500 1000

SERIAL NUMBER 124

CONFIDENTIAL

EDITION OF JUNE 7, 1921.
REVISIONS: JAN. 19, 1922.

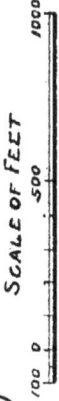

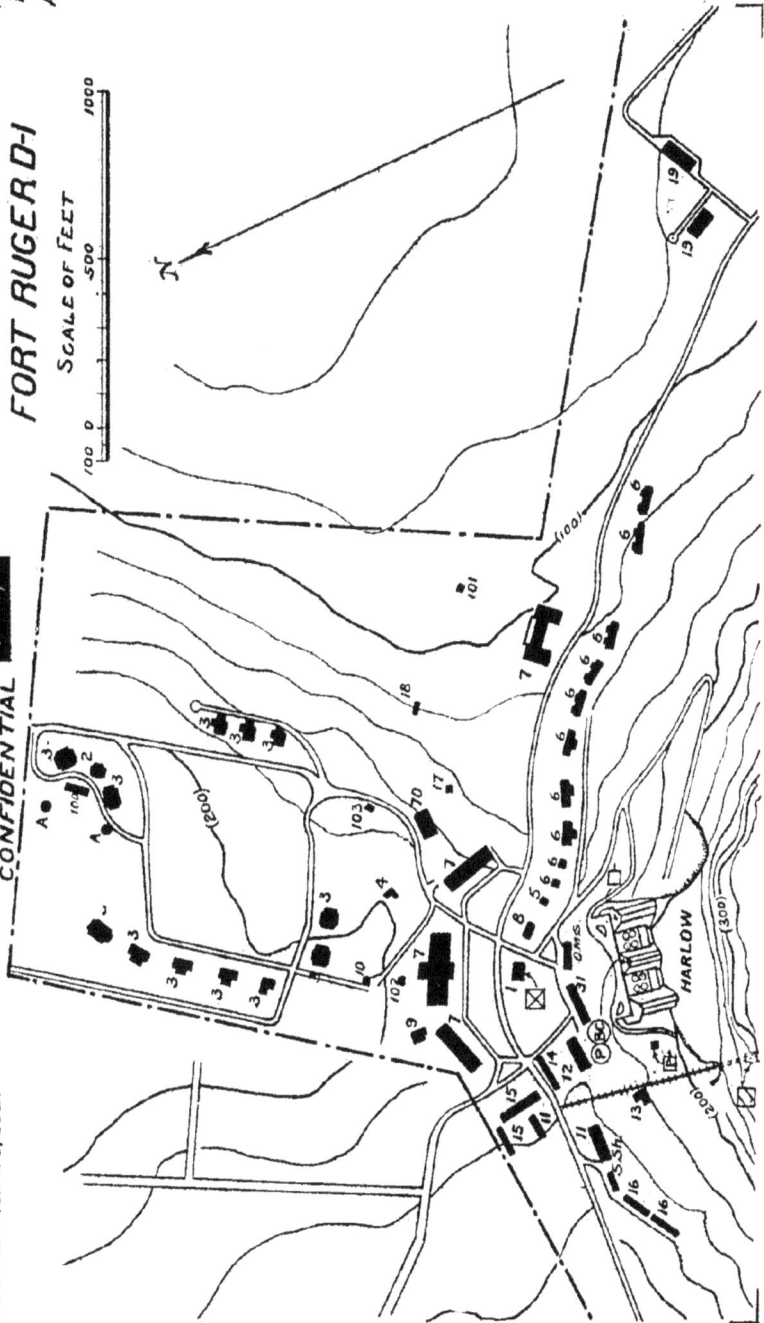

BATTERIES

HARLOW........8-12"M.
A-Anti-aircraft gun 2-3"

LEGEND

1 ADMINISTRATION BLDG.
2 COMDG. OFFICERS QRS.
3 OFFICERS QRS.
4 DISPENSARY
5 HOSPITAL STWD'S QRS.
6 N.C. OFFICERS QRS.
7 BARRACKS.
8 GUARD HOUSE.
9 POST EXCHANGE.
10 FIRE STATION.
11 STORE HOUSE.
12 MACHINE SHOP.
13 POST LAUNDRY.
14 POST SUPPLY OFFICE
15 STABLES & WAGON SHED
16 SIEGE GUN SHELTER.
17 OIL HOUSE.
18 HYDROLITIC TANK.
19 BALLOON HANGAR
100 SERVANTS QRS
101 SEPTIC TANK.
102 BARBERS SHOP.
103 CLASS ROOM.
31 ORDN. & ENGR. ST. HO.
70 SERVICE CLUB.

HARLOW

# NOTES TO SOURCES

## Archival Sources
### National Archives I, Washington D.C.

For the harbor defenses in Oahu, the records for engineer, ordnance, signal corps, and the Adjutant General are split by date between Archives I and Archives II at College Park, Maryland. Generally records dating during and before the First World War are at Archives I, later records are at Archives II. However, all the records for the Coast Artillery and its commands are housed entirely at Archives II. The major holdings consulted for this study at National Archives I were:

RG-77 Records of the Chief of Engineers

> Entry 103, General Correspondence, 1894-1923.

> Files arranged by topics and enumerated with an assigned index number. Contains copies of "Reports of Completed Batteries" These were summary reports completed each year listing all coast defense tactical structures built under Corps of Engineer supervision with date of transfer, cost, details of armament and equipment. Consulted were versions for 1910, 1912 and 1915 all in file No. 79705.

> Entry 220, Coast Defense Fortification File 1898-1920. Contains summary sheets of individual engineer works by harbor defense.

> Entry 225, Correspondence, Blueprints, and Reports Relating to Defense.

RG-92 Records of the Quartermaster General

> Entry 89, Correspondence of the Office of the Quartermaster General, 1890-1914

RG-94 Records of the Adjutant General

> Entry 25, Adjutant General Document File, 1890-1917

RG-156 Records of the Chief of Ordnance

> Entry 28, Correspondence of the Chief of Ordnance, 1894-1913

> Entry 29, Correspondence of the Chief of Ordnance, 1910-1915

### National Archives II, College Park, MD

National Archives II in College Park Maryland holds the major army textual records for the technical and administrative services (engineers, ordnance, quartermaster, transportation, adjutant general) for the interwar and Second World War period. In additional all the records for the Coast Artillery Corps are held here. For this work the most important documents are the Record Group 494, the historical correspondence holdings for the old Hawaiian Department. Also, the College Park facility holds all cartographic, still, and motion picture archives, regardless of date. Major holdings consulted were:

RG-77 Records of the Chief of Engineers

Entry 393, Historical Records of Buildings at Active Army Posts. 1905-1942. Contains by post a large number of sheets for individual post structures and buildings, mostly garrison structures rather than batteries or emplacements. Often known as the "Quartermaster Building File", these pages with photos are often the only surviving record of individual barracks, quarters, storehouses, etc.

Entry 1006, Correspondence of the Chief of Engineers, 1915-1942, Harbor Defense Files.

Entry 1007, Correspondence of the Chief of Engineers, 1915-1942, Geographical Files, Harbor Defenses (classified). Contents arranged by the War Department decimal system. File No. 600.914 contains copies of "Reports or Completed Works" or simply "RCWs". These successors to the Reports of Completed Batteries date from around 1920-1945. They are forms filled out for each tactical structure (gun batteries, searchlights, power plants, switchboards, cable huts, mining structures, etc.). Usually they are accompanied by a one-page drawing, but not photographs.

Entry 1009, Correspondence of the Chief of Engineers, Geographical Files (unclassified)

RG-156 Records of the Chief of Ordnance

Entry 36, Correspondence of the Chief of Ordnance, 1915-1941

RG-165 Records of War Department General and Special Staffs

Entry 281, War Plans Division correspondence

Entry 518, Record of Proceeding of the Board to Revise the Report of the Endicott Board 1905-06.

RG-177 Records of the Chief of Arms, Office of the Chief of Coast Artillery

Entry 4, General Correspondence 1907-1918

Entry 8, General Correspondence 1918-1942

Entry 19, Proceedings of the Coast Artillery Board, 1905-1919

RG-407 Records of the Adjutant General

Entry 360, General Correspondence 1918-1945

Entry 365, Plans for the Defense of the United States and Possessions, 1920-1948

RG-494 Records of U.S. Army Forces in the Middle Pacific (World War II) 1922-47, included in these records are information about the administrative history of the command: the records of the Hawaiian Department 1922-45, the Military Government of Hawaii 1941-46, and the Central Pacific Base Command 1935-47.

These records were formerly held in Record Group 338.

The Military Personnel Records Section, National Personnel Records Center, St. Louis, MO.

This facility is the repository for organizational and personnel records of the US Armed Forces. The Organizational Records Unit contains microfilm files of the Morning Reports as well as Historical Data Sheets and Station Lists for Regular Army units. This latter material is also housed at The U.S. Army Center of Military History, Fort Leslie J. McNair.

### United States Army Military History Institute, Carlisle Barracks, PA.

Generally this facility does not contain agency or department type official records, but rather personal collections of records, diaries, photographs of army personnel and authors. It also has a good library of military publications (including technical and field manuals) as well as general military history books and manuscripts.

## PUBLISHED WORKS

### Books

Adelman, William C. *History of the United States Army in Hawaii 1849-1939.* Schofield Barracks, HI: 1939.

Allen, Gwenfread. *Hawaii's War Years.* Honolulu, HI: University of Hawaii Press, 1950.

Adjutant General's Office. *Army List and Directory of Officers of the Army of the United States.* Washington D.C.: GPO editions of 1909, 1910, 1915, 1920 and 1921.

Annual Report of the Chief of Coast Artillery. Washington D.C.: War Department and GPO, editions for 1908-1927.

Arakaki, Leatrice R. and John R. Kuborn. 7 December 1941 *The Air Force Story.* Hickam AFB, HI: Pacific Air Force Office of History, 1991.

Berhow, Mark A., ed. *American Seacoast Defenses, A Reference Guide,* Second Edition. McLean, VA: The CDSG Press, 2004.

Bogart, Charles H. *Controlled Mines: A History of their use by the United States.* Bennington, VT: Weapons and Warfare Press, 1985.

Braisted, Williams Reynolds. *The United States Navy in the Pacific, 1909-1922.* Austin, TX: University of Texas Press, 1971.

Call, Lewis W. *United States Military Reservations, National Cemeteries and Military Parks Title Jurisdiction, Etc.* Washington DC: GPO, 1907.

Campbell, John. *Naval Weapons of World War Two.* Annapolis, MD: Naval Institute Press, 1985.

Clay, Lieutenant Colonel (Ret.) Steven C. *U.S. Army Order of Battle 1919-1941,* Vol. I, the Arms: Major Commands Infantry. Fort Leavenworth, KS: Combat Studies Institute Press US Army Combined Arms Center, 2010.

_____. *U.S. Army Order of Battle 1919-1941,* Vol. II, the Arms: Cavalry, Field Artillery and Coast Artillery. Fort Leavenworth, KS: Combat Studies Institute Press US Army Combined Arms Center, 2010.

_____. *U.S. Army Order of Battle 1919-1941*, Vol. III, the Services: Air Services, Engineers and Special Troops Organizations. Fort Leavenworth, KS: Combat Studies Institute Press US Army Combined Arms Center, 2010.

Coletta, Paolo E., ed. *United States Navy and Marine Corps Bases, Domestic.* Westport, CT: Greenwood Press, 1985.

Conklin, Arthur B. *Historical Sketch of the Defense of Oahu by the United States, From Annexation of the Hawaiian Islands July 1898 to July 1918.* Honolulu, HI: 1913.

Construction of Barracks, Islands of Oahu and Panama Canal Zone, November 6, 1913. House of Representatives Doc. No. 276, 63rd Cong., 1st Sess. Washington D.C.: GPO, 1913.

Craven, Wesley Frank, and James Lea Cate, eds. *The Army Air Forces in World War II.* vol. 1, Plans and Early Operations: January 1939 to August 1942. Chicago: University of Chicago Press, 1948.

Cressman, Robert J. and J. Michael Wenger. *Steady Nerves and Stout Hearts, The Enterprise (CV6) Air Group and Pearl Harbor 7 December 1941.* Missoula, MT: Pictorial Histories Publishing Company, 1990.

_____. *Infamous Day: Marines at Pearl Harbor. Marines in World War II* Commemorative Series. Washington DC: History and Museum Division, US Marine Corps, GPO, 1992.

Cullum, George Washington. *Biographical Register of the Officers and Graduates of the U.S. Military Academy at West Point, NY.* Saginaw, MI: Seemann and Peters, 1910.

Devaney, Dennis M. Kaneohe, *A History of Change, 1778-1950.* Honolulu, HI: Dept. of Anthropology, Bernice P. Bishop Museum, 1976.

Dorrance, William H. *Fort Kamehameha: The Story of the Harbor Defenses of Pearl Harbor.* Shippensburg, PA: White Mane Publishing, 1993.

Dupuy R. Ernest and Trevor N. Dupuy. *Encyclopedia of Military History.* New York: Harper and Row, 1970.

Fry, John. *USS Saratoga CV-3: an illustrated history of the legendary aircraft carrier, 1927-1946.* Atglen, PA: Schiffer Publishing Ltd., 1996.

*National Survey of Historic Sites and Buildings, Hawaii History 1778-1910.* Washington DC: United States Department of the Interior, National Park Service, 1962.

*Hawaiian Antiaircraft Artillery Command,* Scrapbook, Fort Shafter, HI: 1946.

Heitman, Francis B. *Historical Register and Dictionary of the United States Army from its Organization September 29, 1787 to March 2, 1903.* Washington D.C.: GPO 1903.

Hough, Lieutenant Colonel Frank O., Major Verle E. Ludwig, and Henry I. Shaw Jr. *History of U.S. Marine Corps Operations in World War II.* vol. 1, Pearl Harbor to Guadalcanal. Washington, D.C.: Historical Branch, G-3 Division, Headquarters, U.S. Marine Corps. 1958.

Jones, James *From Here to Eternity.* New York: Charles Scribner's Sons, 1951.

Lewis, Emanuel Raymond. *Seacoast Fortifications of the United States: An Introductory History.* Washington DC: Smithsonian Institution Press, 1970.

Linn, Brian McAllister. *Guardians of Empire; The U.S. Army and the Pacific 1902-1940.* Chapel Hill, NC: The University of North Carolina Press, 1997.

Madsen, Daniel. *Resurrection, Salvaging the Battle Fleet at Pearl Harbor.* Annapolis, MD: Naval Institute Press, 2003.

Meeken, S.R. *A History of Fort Shafter, 1898-1974.* Fort Shafter, HI: 1974.

Miller, Edward S. *War Plan Orange: The U.S. Strategy to Defeat Japan, 1897– 1945.* Annapolis, MD: Naval Institute Press, 1991.

Morton, Louis. *Strategy and Command: The First Two Years.* United States Army in World War II. Washington, D.C.: Office of the Chief of Military History, Department of the Army, 1962.

*Pearl Harbor Attack: Hearings before the Joint Committee on the Investigation of the Pearl Harbor Attack.* Washington, D.C.: Government Printing Office, 1946.

Prange, Gordon W. *At Dawn We Slept: The Untold Story of Pearl Harbor.* New York: Penguin Books, 1982.

_____. *Dec.7 1941, The Day the Japanese Attacked Pearl Harbor.* New York: Warner Books, 1988.

Report of the Board of Review of the War Department to the Secretary of War (November 26, 1915) on the Coast Defenses of the United States, the Panama Canal, and the Insular Possessions. House Document No. 49, 64th Congress, 1st Session. Washington, D.C.: GPO, 1916.

Report of the National Coast Defense Board appointed by the President of the United States by Executive Order January 31, 1905. Washington D.C.: GPO, 1906.

Report of the Need of Additional Naval Bases to Defend the Coasts of The United States, Its Territories, and Possessions. House Document No. 65, 76th Congress, 1st Session. Washington DC: GPO, 1940.

Slackman, Michael. *Target: Pearl Harbor.* Honolulu: University of Hawaii Press, 1990.

Sullivan, Charles J. (Comp.). *Army Posts and Towns; the Baedeker of the Army.* Los Angeles: Haynes Corp, 1943.

Smith, Carl. *Pearl Harbor.* Oxford, UK: Osprey Publishing Ltd., 2001.

Thompson, Erwin N. *Pacific Ocean Engineers: History of the U.S. Corps of Engineers in the Pacific.* Honolulu, HI: 1981.

U.S. Army War College, *Order of Battle of the United States Land Forces in the World War (1917-1919),* Zone of the Interior. Vol. III, Washington D.C.: GPO, 1949.

U.S. Bureau of Yards and Docks. *Building the Navy's bases in World War II: History of the Bureau of Yards and Docks and the Civil Engineer Corps, 1940-1946.* Washington, DC: GPO, 1947.

War Department. *Seacoast Artillery Weapons.* (War Department Technical Manual TM 4-210). Washington DC: GPO 1944.

War Department. *Standard Artillery and Fire Control Materiel.* (War Department Technical Manual TM 9-2300). Washington D.C.: GPO 1944.

Williford, Glen M. *Racing the Sunrise; Reinforcing America's Pacific Outposts 1941-42.* Annapolis, MD: Naval Institute Press, 2010.

_____. *Pacific Rampart, A History of Corregidor and the Harbor Defenses of Manila and Subic Bays*. McLean, VA: Redoubt Press, 2020.

Williford, Glen M. and Thomas D. Batha. *American Breechloading Mobile Artillery 1875-1953*. Atglen, PA: Schiffer Publishing Ltd., 2016.

_____. and T. McGovern. *Defenses of Pearl Harbor and Oahu 1907-50*. Oxford, UK: Osprey Publishing Ltd., 2003.

Winslow, Anne Goodwin. *Fort Derussy Days, Letters of a Malihini Army Wife 1908-1911*. Honolulu: The Folk Press, 1988.

Winslow, Colonel Eben Eveleth. *Notes on Seacoast Fortification Construction*. Occasional Papers No. 61 of the Engineer School United States Army. Washington D.C.: GPO, 1920.

Articles

Addington, Larry H. "The U.S. Coast Artillery and the Problem of Artillery Organization, 1907-1954". *Military Affairs* Vol 40, No.1, Feb. 1975

"Army to Use Thousands of Armored Pillboxes". *The Honolulu Advertiser,* June 21, 1941.

"Army's Newest Guns Are Tested in Hawaii". *New York Times,* April 25, 1936.

Barnes, Lt. Col. H.C. "The Mission of the Coast Artillery Corps". *Coast Artillery Journal* Vol. 57, No. 6 (December 1922), pp. 479-485.

_____."A Regimental Organization for the Coast Artillery Corps". *Coast Artillery Journal* Vol. 60 (April 1924) pp.293-299.

Bennett, John D., "Antiaircraft Gun Emplacements on Oahu, T.H. 1941-1945". *Coast Defense Journal* Vol. 31, No. 1 (Feb, 2017) pp. 4-19.

_____. "Ashley Military Reservation (155 mm Gun Battery)". *Coast Defense Journal* Vol. 23, No. 3 (Aug. 2009), pp. 55-64.

_____. "Battery Arizona and the Kahe Point Military Reservation". *Coast Defense Journal* Vol. 19, No. 1 (Feb. 2005), pp.61-80.

_____. "Battery Construction No. 302: The Sentinel of Kaneohe Bay". *Coast Defense Journal* Vol. 17, No. 1 (Feb. 2003), pp. 49-63.

_____. Battery Granger Adams: Oahu's Original 8-inch Fixed Battery". *Coast Defense Journal* Vol. 25, No. 4 (Nov. 2011), pp. 17-31.

_____. "Camouflaging Seacoast and Antiaircraft Batteries on Oahu, T.H. 1941-45," *Coast Defense Journal* Vol. 21, No. 3 (Aug. 2007), pp. 24-42.

_____. "Camp Malakole, T.H.". *Coast Defense Journal* Vol. 17, No. 3 (Aug 2003), pp. 49-61.

_____. "Defending the Hawaii North Shore's Access to Pearl Harbor". *Coast Defense Journal* Vol. 30, No. 2 (May 2016), pp. 4-31.

_____. "Fort Barrette and the 16-Inch Guns of the Kapolei Military Reservation". *Coast Defense Journal* Vol. 18, No. 3 (Aug. 2004), pp. 55-72.

_____. "Kaena Point Military Reservation". *Coast Defense Journal* Vol. 19, No. 2 (May 2005), pp. 74-103.

_____. "Kahuku's Defenses". *Coast Defense Journal* Vol. 26, No. 4 (Nov. 2012), pp. 18-51.

_____. "Koko Head Fire Control and Searchlight Station and Military Reservation, Maunalua, Oahu, T.H.". *Coast Defense Journal* Vol. 23, No. 2 (May 2009), pp. 52-78.

_____. "Makapuu Point Military Reservation". *Coast Defense Journal* Vol. 22, No. 3 (Aug. 2008), pp. 4-25.

_____. "Oahu's Command and Fire Control Cable System". *Coast Defense Journal* Vol. 16, No. 4 (Nov. 2002), pp. 44-55.

_____. "Oahu's Solitary 8-inch 1940 Project Battery Construction No. 405 (Robert E. De Merritt)". *Coast Defense Journal* Vol. 24, No. 1 (Feb. 2010), pp. 33-60.

_____. "Oahu's World War Two 5-Inch Naval Antiaircraft Shore Batteries". *Coast Defense Journal* Vol.21, No. 1 (Feb. 2007), pp. 31-67.

_____. "Oahu's 8-inch Naval Turret Batteries 1942-1949". *Coast Defense Journal* Vol. 22, No. 1 (Feb. 2008), pp. 4-55.

_____. "Oneula Military Reservation". *Coast Defense Journal* Vol.26, No. 3 (Aug. 2012), pp. 34-50.

_____. "Puu-o-Hulu Military Reservation 1923-1945". *Coast Defense Journal* Vol. 20, No. 3 (Aug. 2006), pp. 50-78.

_____. "Recollections of Fort Kamehameha, c. 1924-25". *Coast Defense Journal* Vol. 19, No. 4 (Nov. 2005), pp. 41-50.

_____. "Sand Island's Military Past 1916-1945". *Coast Defense Journal* Vol. 16, No. 3 (Aug. 2002), pp. 66-83.

_____. "The 5-Inch Emergency Gun Batteries on Oahu, T.H.," *Coast Defense Journal* Vol. 17, No. 1 (Feb. 2003), pp. 88-99.

_____. "World War II Machine-Gun Pillboxes, Coast Defense & Other Stations in the Hawaiian Islands". *Coast Defense Journal* Vol. 22, No. 2 (May 2008), pp. 4-32.

_____. "World War Two Memories of the Coast Artillery on Oahu, John H. Varney as told to John D. Bennett". *Coast Defense Journal* Vol. 16, No. 4 (Nov. 2002), pp. 69-84.

_____. "1945 Study of Seacoast Battery Requirements, Hawaiian Islands". *Coast Defense Journal* Vol. 20, No. 1 (Feb 2006), Vol. 20, No. 1 (Feb. 2006), pp. 34-47.

Bennett, John D., and Bolling W. Smith. "Oahu's First 16-inch Battery: Edward B. Williston". *Coast Defense Journal* Vol. 26, No. 1 (Feb. 2011), pp. 85-101.

"Big guns may be on the way", *Hawaiian Gazette,* 1911.

Bogart, Charles. "Antiaircraft Gunner at Fort Barrette". *Coast Defense Journal,* Vol. 18, No. 2, (May 2004), p. 101.

Brice, C.S. "Searchlights of the 64th Artillery in the Hawaiian Department". *Coast Artillery Journal*, Vol. 61, No. 1 (July 1924), pp. 69-72.

Christian, Francis L. "Harbor Defenses of Honolulu". *Coast Artillery Journal* Vol. 72, No. 6 (June 1930), pp. 483-484.

Cloke, Col. H.E. "A Trip to Hawaii". *Coast Artillery Journal* Vol. 72, No. 6 (June 1930), pp, 475-479,

Colby, Elbridge. "Pearl Harbor 1873". *American Historical Review* Vol. 30 No. 3 (Apr. 1925), pp. 560-565.

Colladay, E.B. "The 240mm Howitzer". *Coast Artillery Journal* Vol. 61 (July, 1924), pp 59-61.

Dorrance, William H.  "Beach Defenses of Hawaii (1924-1942)". *Coast Defense Study Group Journal*, Vol. 8, Number 4 (November, 1994), pp. 51-58.

_____. "Delivering Ordnance to and on Oahu 1909-1915". *Coast Defense Study Group Journal*, Vol. 7, No. 4 (Nov., 1993), p. 11.

_____. "Early Fort Ruger, 1906-1915". *Coast Defense Study Group Journal*, Vol. 11, No. 4 (Nov., 1994), pp. 4-14.

_____. "Fort DeRussy". *Coast Defense Study Group Journal*, Vol. 12, No. 4 (Feb., 1998), pp. 16-29.

_____. "Fort Armstrong". *Coast Defense Study Group Journal*, Vol. 11, No. 3 (Aug., 1997), pp. 9-19.

_____.  "The 4-Inch Guns that Protected Hawaii". *Coast Defense Study Group Journal*, Vol. 9, No. 4 (Nov, 1994), pp. 55-59.

"Fort DeRussy Expands as Army Men Recreation Center". *Honolulu Star Bulletin*, August 22, 1942, p. 1.

"Fort Ruger Buildings Assuming Shape Modeled for a New England Climate". *Hawaiian Gazette*, September 22, 1911, p. 2.

"Fortifications for Entire Island". *Hawaiian Gazette*, June 13. 1911.

Gage, P.S. "Army Swimming in Hawaii". *Coast Artillery Journal* Vol. 73, No. 5 (Nov. 1930), pp. 434-443.

Gaines, William C. "A Military History of Diamond Head and Fort Ruger, Part I". *Coast Defense Journal*, Vol. 19, No. 2 (May 2005), pp. 4-41; Part II, *Coast Defense Journal*, Vol. 19, No. 3 (August 2005), pp. 4-46.

_____. "Antiaircraft Defense of Oahu, 1916-1945". *Coast Defense Journal*, Vol. 15, No. 2 (May 2001), pp. 22-67.

_____. "Railway Artillery on Oahu". *Coast Defense Journal*, Vol. 16, No. 3 (Aug. 2002), pp.22-58.

_____. "Sand Island Military Reservation". *Coast Defense Study Group Journal*, Vol. 8, Number 3 (Aug. 1994), pp. 49-53.

_____. "The Fifteenth Coast Artillery Regiment 1924-1944". *Coast Defense Study Group Journal*, Vol. 8, Number 2, (May, 1994), pp. 39-48.

_____. "The Sixteenth Coast Artillery (Harbor Defense)". *Coast Defense Study Group Journal*, Vol. 13, No. 4 (Nov. 1999), pp. 45-59.

_____. "Wiliwilinui Ridge Military Reservation". *Coast Defense Study Group Journal*, Vol. 8, No. 2 (May, 1994), pp. 58-66.

_____."155-mm Gun Employment and Emplacements on Oahu, T.H. 1921-1945". *Coast Defense Study Group Journal*, Vol. 13, No. 2 (May 1999), pp. 58-87.

_____. "240-mm Howitzers on Oahu 1922-1944". *Coast Defense Study Group Journal*, Vol. 16, No. 4 (Nov., 2002), pp. 4-28.

_____. "41st Coast Artillery 1918-1944". *Coast Defense Study Group Journal*, Vol. 9 No. 2, (May, 1995), pp. 61-70.

"Harbor Defenses of Honolulu". *Coast Artillery Journal*, Vol. 74, No. 6 (May-June 1931), pp. 308-309.

"Hawaiian AA Firing Point". *Coast Artillery Journal*, Vol. 83, No. 3 (May-June 1940), p. 250.

"HSCAB Newsletter". *Coast Artillery Journal*, Vol. 77, No. 5 (Sep.-Oct. 1935), pp. 169-170.

Jewell, Henry C. "History of the Corps of Engineers to 1915". *Military Engineer* Vol. 14 (1922): 306.

Johnson, W. A. "Modification of 240mm Howitzer Mount to Secure All Around Fire". *Coast Artillery Journal*, Vol. 60, No. 1 (Jan., 1924), pp. 31-37.

Kirchner, D. P. and E. R. Lewis. "Oahu Turrets". *Military Engineer*, Vol. 59 (Nov.-Dec., 1967), pp. 430-433.

Lawry, Nelson H. "Another Five Degrees: WWI Alterations to the Disappearing Carriage". *Coast Defense Study Group Journal* Vol. 10 No. 2 (May 1996) pp. 4-12.

Lawry, Nelson H and Glen M. Williford. "The Coast Artillery 14-inch Gun 1907-1918". *Coast Defense Study Group Journal* Vol. 12 No. 3 (Aug. 1998) pp. 4-31.

Lewis, E. R. and D. P. Kirchner. "Oahu Turrets". *Warship International*, 1992, No. 3. pp. 273-301.

Lovell, John R. "The Hawaiian Separate Coast Artillery Brigade Newsletter". *Coast Artillery Journal* Vol. 77, No. 6 (Sep.-Oct. 1934), pp. 367-371.

_____. "The Hawaiian Separate Coast Artillery Brigade Newsletter". *Coast Artillery Journal* Vol. 78, No. 6 (May-June 1935), pp. 226-228.

_____. "The Hawaiian Separate Coast Artillery Brigade Newsletter". *Coast Artillery Journal* Vol. 78, No. 5 (September October 1935), pp. 400-402.

Lovell, John R. and Robert N. See. "The Hawaiian Separate Coast Artillery Brigade Newsletter". *Coast Artillery Journal* Vol. 79, No. 2 (March-April 1936), pp. 143-144.

Lovell, John R. and William F. LaFrenz. "The Hawaiian Separate Coast Artillery Brigade Newsletter". *Coast Artillery Journal* Vol. 77, No. 1 (July-Aug. 1934), pp. 294-297.

Mann, Walter M. "The 30th Infantry in the Hawaiian Maneuvers". *Coast Artillery Journal*, Vol. 75, No. 2 (March-April, 1932), pp. 106-113.

"New Forts will Duplicate Ft. Shafter". *Hawaiian Gazette*, 1910.

"Six Hits and Four Close Shots". *Hawaiian Gazette*, September 22, 1911.

Taylor, Charles J. "Submarine Mine Wharf at Fort Armstrong, T.H.". *Professional Memoirs Corps of Engineers, U.S. Army*, Vol. 8 No. 42, Nov-Dec. 1916 pp. 746-753.

Starr, R.E. "Fort Shafter". *Coast Artillery Journal* Vol. 72, No. 6 (June 1930), pp. 486-488.

"The Big Review". *Coast Artillery Journal* Vol. 75, No. 6 (Nov.-Dec. 1932), pp. 474-475.

Waters, K.L. "The Army Mine Planter Service". *Warship International* Vol. 22 No. 4 (1985): 400-411.

Weber, Milan G., "The Hawaiian Separate Coast Artillery Brigade Newsletter". *Coast Artillery Journal* Vol. 82, No. 3 (May-June 1939), pp. 264-266.

_____. "The Hawaiian Separate Coast Artillery Brigade Newsletter". *Coast Artillery Journal* Vol. 82, No. 4 (July-Aug. 1939), pp. 367-368.

Williford, Glen. "They Did Make Mistakes, Engineering Errors in the Taft Era". *Coast Defense Journal* Vol. 24 No. 2 (May 2010) pp. 35-49.

## Websites:

Addison, Cordell, "Oral History", https://presidentlincoln.illinois.gov/oral-history/collections/addison-cordell/interview-detail/

Department of Commerce, Bureau of the Census, Thirteenth Census for the United States Taken in the Year 1910, "Statistics for Hawaii" https://www2.census.gov/library/publications/decennial/1910/abstract/supplement-hi.pdf

Hartwell, Joe, "Oral History of Hope F, Buck Wilmer, Battery A 55th Coast Artillery, Hawaii, 1938-41", https://freepages.rootsweb.com/~cacunithistories/military/55th_Arty_Hawaii.html

Housman, Staff Sgt. Crystal, California State Military History Program, "Cal Guard Soldiers were First Killed at Pearl Harbor", https://grizzly.shorthandstories.com/pearl-harbor-80th-anniversary/index.html

Krug, Kurt Anthony, Michigan Today, "This line of bullets missed me by 15 feet", https://michigantoday.umich.edu/2021/12/07/41/this-line-of-bullets-missed-me-by-15-feet/

National Archives Educator Resources, "The 1897 Petition Against the Annexation of Hawaii",

https://www.archives.gov/education/lessons/hawaii-petition#:~:text=House%20Joint%20Resolution%20259%2C%2055th,of%20the%20Territory%20of%20Hawaii.

National Park Service, Pearl Harbor National Memorial, "U.S. Army Casualties" https://www.nps.gov/perl/learn/historyculture/us-army.htm

Office of the Historian, Foreign Service Institute, U.S. Department of State, A Guide to the United States' "History of Recognition, Diplomatic, and Consular Relations, by Country, since 1776: Hawaii", https://history.state.gov/countries/hawaii

Payette, Pete, "American Forts, Hawai'i", https://www.northamericanforts.com/West/hi.html

Rosenfeld, Alan, Densho, "Sand Island (detention facility)", https://encyclopedia.densho.org/Sand_Island_(detention_facility)/

Yarborough, Tom, Warfare History Network, "How Many Japanese Planes Were Shot Down During Pearl Harbor?" https://warfarehistorynetwork,com/2018/12/31/how-many-japanese-planes-were-shot-down-during -pearl-harbor/

## Manuscripts

"A Report on construction of Batteries Arizona and Pennsylvania". Records of U.S. Army Forces in the Middle Pacific 1942-1946, RG 494, Entry 118, Box 441, Archives II, NARA, College Park, MD.

Barr, Earl L. "Diary". Records of the Adjutant General, RG 407, Archives II, NARA, College Park, MD.

Gaines, William C. "Guarding Oahu's Back Door: The History of the Harbor Defenses of Kaneohe Bay and North Shore Groupment 1914-1946". Unpublished Manuscript, Champaign, IL, 2005.

_____. "Installations and Organizations of the Coast Artillery Corps, U.S. Army on the Island of Oahu, Territory of Hawaii 1905-1946". Unpublished Manuscript, Champaign, IL, 2006.

_____. "The North Shore Coast Artillery Groupment on the Island of Oahu, Territory of Hawaii, 1922-1946: A Historic Resource Study". Unpublished Manuscript., Champaign, IL 1993.

Green, Maj. Thomas H. "Martial Law in Hawaii". TJAGS Unpublished Manuscript in the Judge Adjutant General's Office, Washington DC.

"History of Development of 16" Gun Materiel Used for Seacoast Defens". U.S. Army Ordnance Department, Research and Development Service, RG-156, Archives II, College Park, MD.

"History of the 2274th Hawaiian Seacoast Artillery Command, 1945". RG-407, Archives II, College Park. MD.

Larson, Harold. "Water Transportation for the United States Army 1939-1942". Office of the Chief of Transportation, Army Service Forces, 1944.

Rupp, Paul B. "History of the Harbor Defenses of Honolulu". Honolulu: Ft. Ruger Chaplain's Office, 1931, Archives II, College Park, MD.

U.S. Army, Middle Pacific. "Historical Review Corps of Engineers United States Army, Pacific Ocean Area" Three volume report of Hawaiian Corps of Engineer projects. Copy at NARA Archives II, RG 494, Entry 125, Boxes 1156-66. Also known as the "Helmboldt Report" by apparently being primarily written by Col. Henry E. Helmboldt.

U. S. Navy, 14th Naval District. "Administrative History of the Fourteenth Naval District, and the Hawaiian Sea Frontier - World War II". Unpublished Manuscript at the U.S. Navy Historical Center, Washington D.C.

Williford, Glen M. "American AA, Seacoast, and Railway Artillery". Unpublished book manuscript in the author's personal collection.

53rd AAA Brigade. "AA Defense of Oahu". Hawaiian Antiaircraft Artillery Command (HAAC) Fort Shafter, Hawaii. Unpublished Manuscript in author's collection.

# ENDNOTES

## CHAPTER 1

(1) Pete Payette, "Hawai'i", https://www.northamericanforts.com/West/hi.html.

(2) William C. Gaines, *Installations and Organizations of the Coast Artillery Corps, U.S. Army on the Island of Oahu, Territory of Hawaii 1905-1946*, (Champaign, IL: Scarp Associates, 2007) 206. Hereafter: Gaines, *Redbook*.

(3) "A Guide to the United States' History of Recognition, Diplomatic, and Consular Relations, by Country, since 1776: Hawaii" https://history.state.gov/countries/hawaii.

(4) "Pearl Harbor," *The American Historical Review*, Vol. 130, No. 3 (April 1925) 560-565.

(5) "The 1897 Petition Against the Annexation of Hawaii," https://www.archives.gov/education/lessons/hawaii-petition#:~:text=House%20Joint%20Resolution%20259%2C%2055th,of%20the%20Territory%20of%20Hawaii.

(6) William C. Adelman, *History of the United States Army in Hawaii 1849-1939* (Schofield Barracks, HI: 1939) 2. Hereafter Adelman, *History of the Army in Hawaii*.

(7) Ibid. 7.

(8) Williams Reynolds Braisted, *The United States Navy in the Pacific, 1909-1922* (Austin, TX: University of Texas Press, 1971, 36-40.

(9) Paolo E. Coletta, ed., *United States Navy and Marine Corps Bases, Domestic* (Westport, CT: Greenwood Press, 1985, 439. Hereafter: Coletta, *United States Navy Bases, Domestic*.

(10) *Report of the National Coast Defense Board appointed by the President of the United States by Executive Order January 31, 1905* (Washington D.C.: GPO, 1906).

(11) Brian McAllister Linn, *Guardians of Empire; The U.S. Army and the Pacific 1902-1940* (Chapel Hill, NC: The University of North Carolina Press, 1997), 196. Hereafter: Linn, *Guardians of Empire*.

(12) "Defense of the Naval Base – Oahu", NARA, Archives II, RG 395, Entry 6051, January 12, 1931.

(13) "Hawaiian Defense Project Revision of 1940", NARA, Archives II, RG 407, AG No. 78. Hereafter: "Hawaiian Project of 1940".

## CHAPTER 2

(1) William H, Dorrance, *Fort Kamehameha: The Story of the Harbor Defenses of Pearl Harbor* (Shippensburg, PA: White Mane Publishing, 1993), 6. Hereafter Dorrance, *Fort Kamehameha*.

(2) NARA, Archives I, RG 77 Entry 103 File 50800-131, Letter of September 27, 1910.

(3) NARA, Archives I, RG 77 Entry 103 File 62246-55, September 7, 1907.

(4) NARA, Archives I, RG 77 Entry 103 File 62246-70, December 30, 1907.

(5) NARA, Archives II, RG 77 Entry 1007 Geographical Files, Report of Completed Works, Battery Selfridge corrected to March 31, 1919. Hereafter: NARA RCW with individual structure and date.

(6) Adelman, *History of the Army in Hawaii*, 6.

(7) Winslow *Fort Derussy Days* 123.

(8) NARA RCW Battery Hasbrouck, corrected to March 31, 1919.

(9)    Dorrance, *Fort Kamehameha.*17, *Historic Context Study of Historic Family Housing Hawaii*, (Honolulu, HI: Mason Architects Inc, 2003) 2-4.

(10)    NARA, Archives I, RG 77 Entry 103 File 62246-209, November 23, 1912.

(11)    NARA, Archives I, RG 77 Entry 103 File 62246-225, April 12, 1913.

(12)    NARA, Archives I, RG 77 Entry 103 File 68054-1, May 27, 1908.

(13)    NARA, Archives I, RG 77 Entry 103 File 79705, Report of Completed Batteries Torpedo structures, December 31, 1915. Hereafter: NARA RCB with date edition.

(14)    NARA RCW Fire Control Station M'/2, corrected to March 31, 1919.

(15)    NARA RCB, Fire control in the Insular Possessions for Fire Control Stations in Honolulu and Pearl Harbors, Hawaii, December 31, 1915.

(16)    "Hawaii Coast Artillery District, Fire Control Stations, Coast Defenses of Oahu", Prepared by the Artillery Engineer, Fort Ruger, October 27, 1915 and NARA, Archives II, RG-77 Entry 1008 Decimal 675, File 43653, March 1, 1943, Subject: Fire Control Base End Stations 1 March 1943. Hereafter: NARA "Base End Stations March 1943".

(17)    Dorrance, *Fort Kamehameha*, 23-24.

(18)    Ibid., 56.

(19)    NARA RCB, December 31, 1915.

(20)    NARA RCB, Searchlight Number 6, Ahua Point, Fort Kamehameha, December 31, 1915.

(21)    Adelman, *History of the Army in Hawaii* 2, 9, 35, 60.

(22)    Charles J. Taylor "Submarine Mine Wharf at Fort Armstrong, T.H." *Professional Memoirs, Corps of Engineers*, U.S. Army, Vol. 8 No. 42, Nov-Dec. 1916, 746-753.

(23)    NARA, Archives I, RG 77 Entry 103 File 70893-1, April 13, 1909, NARA RCW Battery Tiernon, corrected to March 31, 1919.

(24)    NARA RCW, Fort Armstrong, Mining Casemate and Torpedo Structures, corrected to November 12, 1921. Fort Armstrong, Fire Control Station M'1, corrected to March 31, 191, Searchlight Number 5, Mining Casemate and Torpedo Structures, corrected to November 12, 1921.

(25)    NARA, Archives I, RG-77 Entry 103 File 68054-69, October 17, 1914.

(26)    NARA, Archives I. RG-77 Entry 103 File 68054-13, April 29, 1910.

(27)    NARA RCW, Fort Armstrong, Mining Casemate and Torpedo Structures, Corrected to November 12, 1921.

(28)    NARA RCB, December 31, 1915.

(29)    Maj. Arthur S. Conklin, *Historical Sketch of the Defense of Oahu by the United States*, (Honolulu, HI: Chief of Staff, Hawaiian Department, 1913) 2-3. Hereafter: Conklin, *Historical Sketch of the Defense of Oahu*. Adelman, *History of the Army in Hawaii*, 1.

(30)    NARA, Archives I, RG 77, Entry 103, File 29531-45 April 29, 1901, Proposed Project for the Defense of Honolulu.

(31)    William H. Dorrance "Heavy Artillery at Waikiki", *Coast Defense Study Group Journal*, Vol. 13, No. 4 (November 1999), 5-6. Hereafter: Dorrance, "Heavy Artillery at Waikiki".

(32) George Washington Cullum, *Biographical Register of the Officers and Graduates of the U.S. Military Academy at West Point, NY.* (Saginaw, MI: Seemann and Peters, 1910), Vol. IV, 473-474.

(33) NARA, Archives II, RG 92 Quartermaster General's Office, Plans, Fort DeRussy.

(34) NARA, Archives I, RG 77 Entry 103 File 62246-103 of November 18, 1908.

(35) Nelson H. Lawry and Glen M. Williford "The Coast Artillery 14-inch Gun 1907-1918". *Coast Defense Study Group Journal* Vol. 12 No. 3 (Aug. 1998) 15.

(36) NARA, Archives I, RG 77 Entry 103 File 70893-1 March 27, 1909.

(37) NARA RCB Fire Control Station and Meteorological Station Combined, December 31, 1915.

(38) "New Forts will Duplicate Ft. Shafter," *Hawaiian Gazette*, 1910, "Fortifications for Entire Island," *Hawaiian Gazette*, June 13. 1911.

(39) "Big guns may be on the way," *Hawaiian Gazette*, 1911.

(40) Dorrance, "Heavy Artillery at Waikiki" 10.

(41) "Historical Record of Buildings" NARA, Archives II, RG 77, Entry 393. Paul B. Rupp "History of the Harbor Defenses of Honolulu" (Honolulu, HI: Ft. Ruger Chaplain's Office, 1931) 8, Hereafter: Rupp, "History of the Harbor Defenses of Honolulu". Francis L. Christian, "Harbor Defenses of Honolulu," *Coast Artillery Journal*, Vol. 72, No. 6 (June 1930), 483-484. Hereafter: Christian, "Harbor Defenses of Honolulu".

(42) Ibid. P. S. Gage "Army Swimming in Hawaii," *Coast Artillery Journal*, Vol. 73, No. 5 (November 1930) 434-443.

(43) Quartermaster General's Office, Plans, Fort DeRussy, Records of the Chief of Engineers, RG 77, Archives II, NARA. Adelman, *History of the Army in Hawaii* 24.

(44) Conklin, *Historical Sketch of the Defense of Oahu* 2-3.

(45) Ibid. Adelman, *History of the Army in Hawaii* 1849-1939 1.

(46) NARA, Archives I, RG 77 Entry 103 File 62246-55, October 5, 1907.

(47) NARA, Archives I, RG 156, Entry 28, File 32204-69, July 30. 1910.

(48) Winslow *Fort Derussy Days*. 59.

(49) NARA, Archives I, RG 77 Entry 103 File 50800-281 December 5, 1912.

(50) "History of the Hawaiian Separate Coast Artillery Brigade, 1925-1943," NARA, Archives II, RG 407. Hereafter: "History HSAC Brigade". Adelman, *History of the Army in Hawaii* 7-8.

(51) Rupp, "History of the Harbor Defenses of Honolulu, 4-5.

(52) "Construction of Barracks, Islands of Oahu and Panama Canal Zone," November 6, 1913, House of Representatives Doc. No. 276, 63rd Cong., 1st Sess. Rupp, "History of the Harbor Defenses of Honolulu, 4-5. "Fort Ruger Buildings Assuming Shape Modeled for a New England Climate," *Hawaiian Gazette*, September 22, 1911, 2.

(53) "Six Hits and Four Close Shots," *Hawaiian Gazette*, September 22, 1911, 2.

(54) NARA, Archives I, RG 77 Entry 103 File 47839-35, July 28, 1909.

(55) NARA, Archives I, RG 77 Entry 103 File 68258-365, July 3, 1916.

(56)   Glen Williford "They Did Make Mistakes", *Coast Defense Journal* Vol. 24 No. 2 (May 2010) 44-48.

(57)   NARA, Archives I, RG 77 Entry 103 File 68258-365, July 3, 1916.

(58)   NARA RCW, Battery Birkhimer, corrected to July 7, 1933.

(59)   NARA, Archives II, RG-177, Entry 219 Coast Defense Inspection, December 29, 1919, Fort Ruger.

(60)   Rupp, "History of the Harbor Defenses of Honolulu 5-6. Adelman, *History of the Army in Hawaii* 24.

(61)   NARA RCW, Fire Control Switchboard Room, corrected to August 12, 1925.

## CHAPTER 3

(1)    Thirteenth Census of the United States "Statistics for Hawaii". https://www2.census.gov/library/publications/decennial/1910/abstract/supplement-hi.pdf.

(2)    Conklin, *Historical Sketch of the Defense of Oahu* 8-30.

(3)    NARA, Archives I, RG 77 Entry 103 File 62246, September 6, 1912.

(4)    NARA, Archives I, RG 77 Entry 103 File 68258-273, June 1, 1915, Funds for Land Defense.

(5)    NARA RCB, December 31, 1915.

(6)    NARA, Archives II, RG 177 Entry 4 File 6318-4, June 9, 1913.

(7)    NARA, Archives I, RG 77 Entry 103 File 68258-265 June 1, 1915.

(8)    NARA, Archives I, RG 77 Entry 103 File 93957-12, November 17, 1914.

(9)    NARA, Archives I, RG 77 Entry 103 File 93957-12, 6th Indorsement, May 28, 1915.

(10)   NARA, Archives I, RG 77 Entry 103 File 93957-94, January 7, 1916.

(11)   NARA RCW, Battery Adair, corrected to March 31, 1919.

(12)   NARA RCW, Battery Boyd, corrected to March 9, 1925.

(13)   NARA, Archives I, RG 77 Entry 103 File 68258-343 January 3, 1916.

(14)   NARA, Archives I, RG77 Entry 220 Engineer Notebook.

(15)   NARA RCW, Red Hill Redoubt, corrected to March 31, 1919.

(16)   NARA, Archives II, RG 77 Entry 1008 File 661 November 24, 1931, Subject: Inspection of Red Hill Redoubt.

(17)   NARA RCW, Battery Dodge, corrected to December 31, 1919.

(18)   NARA RCW, Battery Hulings, corrected to December 31, 1919.

(19)   NARA, Archives I, RG77 Entry 220 Engineer Notebook.

(20)   Ibid.

(21)   NARA, Archives I, RG 77 Entry 103 File 68256-465, July 23, 1917.

(22)   NARA RCW, Battery S. C. Mills, corrected to March 31, 1919 and March 10, 1925.

(23)   NARA, Archives II, RG 407 "Antecedents and Coats of Arms of Coast Artillery Organizations in the Hawaiian Coast Artillery District," May 9, 1922, Unit Jacket, 2nd Co., CAC.

(24)   NARA, Archives I, RG 77 Entry 103, File 68258-273, July 29, 1915.

## CHAPTER 4

(1)   Dorrance, *Fort Kamehameha,* 49-50.

(2)   Larry H. Addington, "The U.S. Coast Artillery and the Problem of Artillery Organization 1907-1954", *Military Affairs*, Vol 40, No.1 (February 1975), 2.

(3)   NARA, Archives I, RG 77 Entry 103 File 99735-674 and endorsements September 18, 1918.

(4)    NARA, Archives I, RG 77 Entry 103 File 99735-189 February 24, 1917, Abstract of Proceedings of a Board of Officers in Regard to a Project for Anti-Aircraft Guns in the Hawaiian Department.

(5)    John D. Bennett, "Antiaircraft Gun Emplacements on Oahu 1941-45" *Coast Defense Journal,* Vol. 20 No. 1 (February 2017) 4. Hereafter: Bennett, "Antiaircraft".

(6)   NARA, Archives II, RG 156 Entry 36A, File 400.356, February 19, 1920 and May 11, 1920 Subject: 3" Anti-Aircraft Mounts.

(7)   C.S. Brice "Searchlights of the 64th Artillery in the Hawaiian Department". *Coast Artillery Journal*, Vol. 61, No. 1 (July 1924), pp. 69-72.

(8)   Ibid.

(9)   R. E. Starr "Fort Shafter," *Coast Artillery Journal*, Vol. 72, No. 6 (June 1930) 486-488.

(10)   "64th CA (AA) Regiment, Regimental History" (Fort Shafter, HI: 1933).

(11)   Ibid.

(12)   NARA, Archives II, RG 407 Entry 360, War Diary, Hawaiian Coast Artillery Command, 7 to 31 December 1941. Hereafter: "War Diary, HCAC".

(13)   "The Big Review," *Coast Artillery Journal*, Vol. 75, No. 6 (November-December 1932), 474-75.

(14)    "HSCAB Newsletter," *Coast Artillery Journal*, Vol. 77, No. 5 (September-October 1935, 169-70.

(15)   Fulton Quintus Cincinnatus Gardner, "Random Recollections," [Unpublished memoir in Gaines' collection].

(16)   "Hawaiian AA Firing Point," *Coast Artillery Journal*, Vol. 83, No. 3 (May-June 1940), 250-51.

(17)   "HSCAB Newsletter," *Coast Artillery Journal*, Vol. 84, No. 1 (January-February 1941), 85-86.

(18)   Gaines, "Antiaircraft Defense of Oahu" 65.

## CHAPTER 5

(1)   *Report of the Board of Review of the War Department to the Secretary of War (November 26, 1915) on the Coast Defenses of the United States, the Panama Canal, and the Insular Possessions.* House Document No. 49, 64ᵗʰ Congress, 1ˢᵗ Session. (Washington, D.C.: GPO, 1916).

(2)   Eben Eveleth Winslow, Col. *Notes on Seacoast Fortification Construction.* Occasional Papers No. 61 of the Engineer School United States Army. (Washington D.C.: GPO, 1920) 151-53.

(3)   NARA RCW. Battery Closson, corrected to May 4, 1920.  And form corrected to March 1, 1938.

(4)   Lewis W. Call *United States Military Reservations, National Cemeteries and Military Parks Title Jurisdiction, Etc.* (Washington DC: GPO, 1907) 403-04.

(5)    War Department. *Standard Artillery and Fire Control Materiel.* (War Department Technical Manual TM 9-2300) (Washington D.C.: GPO 1944).

(6)    NARA, Archives II, RG 156 U.S. War Department Ordnance Department, Research and Development Section Monograph Series: The Record of Army Ordnance Research and Development, "History of Development of 16-inch Gun Material used for Seacoast Defenses", Page A.

(7)    NARA RCW, Plant for Battery Edward B Williston, corrected to August 12, 1925, Battery Edward B. Williston, corrected to August 12, 1925.  Plotting Room for Battery Williston and Switchboard Room, corrected to August 12, 1925, Plotting Room for Battery Williston and Switchboard Room, corrected to March 1, 1938.

(8)    Adelman, *History of the Army in Hawaii* 55.

(9)    War Department. *Standard Artillery and Fire Control Materiel.* (War Department Technical Manual TM 9-2300) (Washington D.C.: GPO 1944).

(10)   Williford, Glen M. "American AA, Seacoast, and Railway Artillery". Unpublished book manuscript.

(11)   Dorrance, *Fort Kamehameha,* 84-91.

(12)   NARA, RCW Battery Hatch, corrected to November 25, 1935.

(13)   John D. Bennett, "Fort Barrette and the 16-inch Guns of the Kapolei Reservation," *Coast Defense Journal,* Vol. 18, 55-72.

(14)   NARA, RCW Battery Hatch, corrected to November 25, 1935.

(15)   Ibid.

(16)    War Department. *Standard Artillery and Fire Control Materiel.* (War Department Technical Manual TM 9-2300) (Washington D.C.: GPO 1944).

(17)   NARA RCW, Battery Granger Adams, corrected to May 11, 1935.

## CHAPTER 6

(1)    NARA Archives I, RG 77 Entry 103, File 62246, September 6, 1912, Report of a Board of Officers on the Defense of Oahu (Macomb Board).

(2)    Conklin, *Historical Sketch of the Defense of Oahu.*

(3)    H.C. Barnes Lt. Col., "The Mission of the Coast Artillery Corps," *Coast Artillery Journal* Vol. 57, No. 6 (December 1922) 475-81.

(4)    NARA, Archives II, RG 156 Entry 712 Ordnance Record Cards, 1917-1946.

(5)    NARA, Archive II, RG 494 Entry UD-UP 118, "Firing Positions for Hawaiian Railway Battalion".

(6)    NARA, RCW Battery NARA, November 28, 1931, Fire Control Switchboard Room, Maili.

(7)    NARA, Archives II, RG 494 Entry UD-UP 118, January 15, 1942, "Use of 12" Railway Mortars as Dummy Artillery" and "Removal of 12-inch Railway Mortars."

(8)    NARA, Archives II, RG 494 Entry UD-UP 118, January 4, 1942, "Preparation of Dummy Gun Positions," and December 27, 1941, "Location of Guns, Dummy Guns.

(9)    NARA, Archives II, RG 494 Entry UD-UP 118, June 14, 1933 Memorandum to G-4 Hawaiian Department.

(10)   William C. Gaines, "41st Coast Artillery 1918-1944" *Coast Defense Study Group Journal,* Vol. 9 No. 2, (May, 1995).

(11)  NARA, Archives II, RG 494 Entry UD-UP 118, July 16, 1933, "Report on Selection of Firing Positions for 8-inch Railway Guns."

(12)  NARA RCW, General Plan, Kawailoa 8-inch Railway Firing Position, corrected to July 1, 1940.

(13)  NARA, Archives II, RG 494, Entry UD-UP 118, "Project for 8-inch Railway Gun Positions, Island of Oahu, T.H.," 6th ind., February 23, 1935.

(14)  Ibid.

(15)  NARA, Archives II, RG 494, Entry UD-UP 118, "Acceptance Inspection of Waianae 8" Railway Position".

(16)  Glen M. Williford and Thomas D. Batha. *American Breechloading Mobile Artillery 1875-1953* (Atglen, PA: Schiffer Publishing Ltd., 2016). Hereafter: Williford, *American Breechloading Artillery.*120-21.

(17)  War Department, *Seacoast Artillery Weapons.* (War Department Technical Manual TM 4-210) (Washington DC: GPO 1944) 135-44. Hereafter: *Seacoast Artillery Weapons.*

(18)  NARA, Archives II, RG 494 Entry UD-UP 118, File 660.2, "Recommendations for 100% Efficiency in defense of North Sector, Fortifying the Pupukea Plateau."

(19)  "NARA 1942 Hawaiian Seacoast Artillery Project".

(20)  William C. Gaines "155-mm Gun Employment and Emplacements on Oahu, T.H. 1921-1945", *Coast Defense Study Group Journal*, Vol. 13, No. 2 (May 1999), pp. 58-87.

(21)  Williford, *American Breechloading Artillery* 122-23.

(22)  Glen M. Williford, *Pacific Rampart, A History of Corregidor and the Harbor Defenses of Manila and Subic Bays.* (McLean, VA: Redoubt Press, 2020) 187-90.

(23)  W.A. Johnson, "Modification of 240mm Howitzer Mount to Secure All Around Fire," *Coast Artillery Journal*, Vol. 60, No. 1 (January, 1924), 31-37.

(24)  NARA, Archives II, RG 494 Entry UD-UP 118, File 662/3 Subject: Construction. Design and Modification of Mounts and Emplacements for 240mm Howitzers, 1923-1924.

(25)  NARA, RCW Emplacements for two 240mm Howitzers corrected to October 4, 1927 and February 20, 1932.

(26)  William C. Gaines "240 Howitzers on Oahu 1922-1944, *Coast Defense Journal*, Vol. 16, No. 4 (November, 2002) 4-28.

(27)  E.B. Colladay "The 240mm Howitzer" *Coast Artillery Journal* Vol. 61 (July), 1924, 59-61.

(28)  William C. Gaines, "The Oahu Howitzers" *Fort MacArthur Alert*, Vol. 6, No. 2. (Spring, 1994), 13-15.

## CHAPTER 7

(1)  NARA, Archives II, RG 165, Entry 280, File 3511.

(2)  *PH Investigation* Part 30 2602-03, December 22, 1941 Subject: Army Personnel of Hawaiian Department.

(3)  *PH Investigation* Part 22 319, November 30, 1941 Subject: Armel Personnel of Hawaiian Department by Station.

(4)  Joe Hartwell "Oral History of Hope F, Buck Wilmer, Battery A 55th Coast Artillery, Hawaii, 1938-41", https://freepages.rootsweb.com/~cacunithistories/military/55th_Arty_Hawaii.html

(5)    NARA, Archives II, RG 494 Hawaiian Department. Coast Artillery Projects January 24, 1942 and Notes on Fortification Projects Pertaining to Hawaiian Seacoast Artillery Command, July 30 1942. Hereafter: "NARA 1942 Hawaiian Seacoast Artillery Project".

(6)    Gaines, "Fifteenth Coast Artillery Regiment". 39-48.

(7)    "Names of Battery Positions, 1942".

(8)    Gaines, "Fifteenth Coast Artillery Regiment".

(9)    John R. Lovell, "The Hawaiian Separate Coast Artillery Brigade Newsletter," Coast Artillery Journal Vol. 77, No. 6 (September-October 1934) 367-371.

(10)    John R. Lovell and William F. LaFrenz, "The Hawaiian Separate Coast Artillery Brigade Newsletter," Coast Artillery Journal Vol. 77, No. 1 (July-August 1934) 294-297.

(11)    Christian, "Harbor Defenses of Honolulu" 483-484.

(12)    Adelman, History of the Army in Hawaii, 42, 50-51.

(13)    John R. Lovell and William F. LaFrenz, "The Hawaiian Separate Coast Artillery Brigade Newsletter," Coast Artillery Journal Vol. 77, No. 1 (July-August 1934) 294-297. John R. Lovell, "The Hawaiian Separate Coast Artillery Brigade Newsletter," Coast Artillery Journal Vol. 77, No. 5 (September-October 1934), 367-71.

(14)    John R. Lovell, "The Hawaiian Separate Coast Artillery Brigade Newsletter," Coast Artillery Journal Vol. 78, No. 6 (May-June 1935) 226-28.

(15)    John R. Lovell, "The Hawaiian Separate Coast Artillery Brigade Newsletter," Coast Artillery Journal Vol. 78, No. 5 (September October 1935) 400-02.

(16)    John R. Lovell and Robert N. See, "The Hawaiian Separate Coast Artillery Brigade Newsletter," Coast Artillery Journal Vol. 79, No. 2 (March-April 1936) 143-44.

(17)    Adelman, History of the Army in Hawaii, 51.

(18)    Milan G. Weber, "The Hawaiian Separate Coast Artillery Brigade Newsletter," Coast Artillery Journal Vol. 82, No. 3 (May-June 1939) 264-266 and Vol. 82, No. 4 (July-August 1939) 367-368.

(19)    "History HSAC Brigade". Gaines, "Fifteenth Coast Artillery Regiment", 35-44. William C. Gaines "The Sixteenth Coast Artillery (Harbor Defense)," Coast Defense Study Group Journal, Vol. 13, No. 4 (November 1999) 45-59.

(20)    NARA RCB, December 31, 1910. Erwin N. Thompson, Pacific Ocean Engineers: History of the U.S. Corps of Engineers in the Pacific (Honolulu, HI: 1981) 39-40. Hereafter: Thompson Pacific Ocean Engineers.

(21)    NARA RCW, H.D. Command Post, corrected to April 10, 1934.

(22)    Rupp, "History of the Harbor Defenses of Honolulu" 308-09. John R. Lovell, "The Hawaiian Separate Coast Artillery Brigade News Letter," Coast Artillery Journal, Vol. 74, No. 6 (Jan-Feb 1934) 46-48.

(23)    John R. Lovell "The Hawaiian Separate Coast Artillery Brigade News Letter," Coast Artillery Journal, Vol. 78, No. 5 (Sept-Oct 1935) 400-402. Adelman, History of the Army in Hawaii 57.

(24)    NARA RCW, Plotting Room for Battery Williston and Switchboard Room, Corrected to March 1, 1938.

(25)    "Army's Newest Guns Are Tested In Hawaii," New York Times, April 25, 1936.

(26)    William C. Gaines, "Fort Barrette, Oahu, Hawaii," Coast Defense Study Group Journal, Vol. 8, No. 4, (November, 1994) 46-50.

(27) John D. Bennett, "Recollections of Fort Kamehameha" Coast Artillery Journal, Vol. 19, No. 4.

(28) Adelman, History of the Army in Hawaii. 36.

(29) Christian, "Harbor Defenses of Honolulu".

(30) NARA RCW, Aliamanu Ammunition Storage, corrected to April 1, 1939.

## CHAPTER 8

(1) "War Diary, HCAC".

(2) PH Investigation Part 31, 3138-39.

(3) Ibid.

(4) Ibid.

(5) Crystal Housman, Staff Sgt., California State Military History Program, "Cal Guard Soldiers were First Killed at Pearl Harbor", https://grizzly.shorthandstories.com/pearl-harbor-80th-anniversary/index.html.

(6) PH Investigation Part 22, 271-281 Subject: Testimony of Lt. Saltzman.

(7) Dorrance, Fort Kamehameha, 114.

(8) Carl Smith Pearl Harbor (Oxford, UK: Osprey Publishing Ltd., 2001) Hereafter: Smith Pearl Harbor.52. Michael Slackman Target: Pearl Harbor. (Honolulu: University of Hawaii Press, 1990) 134. Hereafter: Slackman, Target: Pearl Harbor.

(9) Kurt Anthony Krug, Michigan Today, "This line of bullets missed me by 15 feet", https://michigantoday. umich.edu/2021/12/07/41/this-line-of-bullets-missed-me-by-15-feet/.

(10) Slackman, Target: Pearl Harbor,144.

(11) Gordon Prange, Dec.7 1941 The Day the Japanese Attacked Pearl Harbor (New York: Warner Books, 1988.) 215-16.

(12) PH Investigation Part 22, 324 Subject: Action and Disposition of 53rd CA Brigade (Antiaircraft) on 7 December 1941.

(13) "Comments on: How Many Japanese Planes Were Shot Down During Pearl Harbor?" at https://warfarehistory-network,com/2018/12/31/how-many-japanese-planes-were-shot-down-during -pearl-harbor/.

(14) Robert J. Cressman and J. Michael Wenger Steady Nerves and Stout Hearts, The Enterprise (CV6) Air Group and Pearl Harbor 7 December 1941 (Missoula, MT: Pictorial Histories Publishing Company, 1990) 37-38. Hereafter: Cressman and Wenger, Steady Nerves.

(15) Smith Pearl Harbor 61-62.

(16) Charles Bogart, "Antiaircraft Gunner at Fort Barrette", Coast Defense Journal, Vol.18 No. 2, (May 2004), 101-04. Hereafter: Bogart, "Antiaircraft Gunner at Fort Barrette".

(17) PH Investigation Part 22, 324 Subject: Action and Disposition of 53rd CA Brigade (Antiaircraft) on 7 December 1941.

(18) "Morning Reports", 55th CA (TD) Regiment.

(19) S.R. Meekin, History of Fort Shafter, (Fort Shafter, 1974) 43.

(20)  PH Investigation Part 22, 324 Subject: Action and Disposition of 53rd CA Brigade (Antiaircraft) on 7 December 1941.

(21)  Cressman and Wenger, Steady Nerves, 55-57.

(22)  PH Investigation Part 22, 325 Subject: Army Casualties, 7 December 1941, Hawaiian Department. National Park Service, Pearl Harbor National Memorial, "U.S. Army Casualties" https://www.nps.gov/perl/learn/history-culture/us-army.htm.

(23)  Arakaki, The Air Force Story, 138.

(24)  "War Diary, HCAC".

(25)  Cressman and Wenger, Steady Nerves, 62.

(26)  NARA, Archives II, RG 165  Entry 281 File 3444-14, December 20, 1941.

(27)  Thompson Pacific Ocean Engineers 108.

(28)  NARA, Archives II, RG 407 Entry 234A File G-4/33822, December 12, 1941.

(29)  NARA, Archives II, RG 165 Entry 281 File 3444-14, December 8. 1941.

(30)  Glen M. Williford, Racing the Sunrise; Reinforcing America's Pacific Outposts 1941-42. (Annapolis, MD: Naval Institute Press, 2010) 282-303.

(31)  "Morning Reports", Headquarters and Headquarters Battery, 2nd Battalion, 57th Coast Artillery Regiment, December 1-17, 1941.

(32)  William C Gaines, "Guarding Oahu's Back Door: The History of the Harbor Defenses of Kaneohe Bay and North Shore Groupment 1914-1946" (Unpublished Manuscript, Champaign, IL, 2005. Hereafter: Gaines "Guarding Oahu's Back Door".

(33)  "Morning Reports", HHB, 2nd Battalion, 57th Coast Artillery Regiment, December 11-24, 1941. "Morning Reports", Battery C, 57th Coast Artillery Regiment, December 11-24, 1941. "Morning Reports", Battery D, 57th Coast Artillery Regiment December 11-24, 1941.

(34)  "War Diary, HCAC".

(35)  Ibid.

(36)  "Morning Reports", Battery D. 55th Coast Artillery Regiment, December 7, 1941.

(37)  "War Diary, HCAC".

(38)  Ibid.

(39)  NARA, Archives II, RG 494 Entry UD-UP 118 February 8, 1942, Subject: Emergency Projects being carried on under supervision of US Engineer Office.

(40)  Alan Rosenfeld, Densho, "Sand Island (detention facility)", https://encyclopedia.densho.org/Sand_Island_(detention_facility)/

(41)  "Morning Reports", 16th Coast Artillery Regiment.

(42)  NARA, Archives II, RG 407, Entry 234, File 370.5 Movement Copper.

CHAPTER 9

(1)   NARA, Archives II, RG 19, Entry A1 1206C, January 29, 1942, Subject: "Salvaged Ordnance Material".

(2)   NARA, Archives II, RG 494 Entry UD-UP 198, March 16, 1942, Subject: "Proposed Location of 5" Naval Anti-Aircraft Guns".

(3)   NARA, Archives II, RG 494, Entry UD-UP 198 December 6, 1943, Subject: Return to Storage 5".25 caliber AA Battery".

(4)   NARA, Archives II, RG 494 Entry UD-UP 198, March 16, 1942, Subject: "Proposed Location of 5" Naval Anti-Aircraft Guns", John D. Bennett, "The 5-Inch Emergency Gun Batteries on Oahu, T.H.," Coast Defense Journal Vol. 17, No. 1 (Feb. 2003), 88-99. Hereafter: Bennett, "5-inch Emergency Gun Batteries".

(5)   Bennett, "5-inch Emergency Gun Batteries".

(6)   NARA, Archives II, RG 494, Periodic Reports, HSAC, Jan.-May 1943, "Morning Reports", Battery G, 98th Coast Artillery Regiment, January 1, 1942-July 31, 1943 Letter Charles W. Tucker to William Gaines January 31, 1992.

(7)   NARA, Archives II RG 494 Entry UD-UP 118, April 10, 1942. "Installation of 4-inch Navy Guns,", and April 20, 1942 "4 in Navy Pedestal Mounts."

(8)   Gaines "Guarding Oahu's Back Door" 118.

(9)   NARA, Archives II, RG 494 Entry UD-UP 198, November 14, 1944, Obsolescent Seacoast Artillery Batteries Authorized for Removal by War Department [1944]. Hereafter: NARA, "Obsolescent Seacoast Artillery".

(10)  NARA, "Obsolescent Seacoast Artillery". NARA, Archives II, RG 494, Entry UD-UP 118, September 6, 1942, "Information concerning Geographical Names of Batteries, Armament, Manning Personnel and Coordinates of Batteries in H.S.A. Command".

(11)  Bennett, "5-inch Emergency Gun Batteries", 81-99.

(12)  "Names of Battery Positions, 1942".

(13)  Ibid.

(14)  NARA, "Obsolescent Seacoast Artillery".

(15)  U.S. Department of the Army. Puu-O-Hulu Military Reservation, Real Estate Files, Corps of Engineers, Fort Shafter, HI.

(16)  "Hawaiian Project of 1940" Ordnance Annex Revision of 1940. NARA "Base End Stations March 1943".

(17)  NARA, Regional Archives San Bruno, Ca., U.S. Engineer Honolulu, T. H. RG 77 Box 24, June 25, 1934 "Final Report Concrete Machine Gun Emplacements, Oahu, T.H.

(18)  John D. Bennett "Sand Island's Military Past," Coast Defense Journal, Vol. 16, No. 3 (August 2002) 66-82.

(19)  Walter M. Mann, "The 30th Infantry in the Hawaiian Maneuvers," Coast Artillery Journal, Vol. 75, No. 2 (March-April, 1932), 106-13.

(20)  Ibid.

(21)  NARA, Archives II, RG 494, Entry UD-UP 118, February 8, 1942 "Emergency Projects Being Carried out Under Supervision of U.S. Engineer Department". 8. Peter T. Young, "Ammo Tunnels", https://imagesofoldhawaii.com/ammo-tunnels/.

## CHAPTER 10

(1)    John Campbell *Naval Weapons of World War Two* (Annapolis, MD: Naval Institute Press, 1985) 127-28. Hereafter: Campbell, *Naval Weapons.*

(2)    John Fry *USS Saratoga CV-3: an illustrated history of the legendary aircraft carrier, 1927-1946* (Atglen, PA: Schiffer Publishing Ltd., 1996) 111.

(3)    E. R. Lewis and D. P. Kirchner. "Oahu Turrets" *Warship International,* 1992, No. 3, 281, Hereafter: Lewis and Kirchner "Oahu Turrets", "NARA 1942 Hawaiian Seacoast Artillery Project"

(4)    NARA, Archives II RG 494 Entry UD-UP 198, File 472.3 January 30, 1942 Letter.

(5)    NARA, Archives II RG 494 Entry UD-UP 198, File 472.3 February 15, 1942 Letter.

(6)    NARA, Archives II RG 407 File 472, April 9. 1942 copy of BuShips letter C-L-9-1 "USS Lexington – Advance Planning Report of Alterations to be Accomplished at Forthcoming Availability".

(7)    Lewis and Kirchner "Oahu Turrets".

(8)    *Historical Review Engineers Pacific Ocean Area* 323-26.

(9)    John D. Bennett "World War Two Memories of the Coast Artillery on Oahu, John H. Varney as told to John D. Bennett" *Coast Defense Journal* Vol. 16, No. 4 (Nov. 2002) 74-75.

(10)    NARA, Archives II, RG 494 Engineers General Correspondence, March 30-April 5, 1942 "Location Navy 8-inch Turret Guns", and July 30, 1942 "Notes on Fortification Projects Pertaining to Hawaiian Seacoast Artillery Command" and October 20-31, 1942 "Additional Engineer Work Required at Battery Salt Lake".

(11)    *Historical Review Engineers Pacific Ocean Area* 323-326. "NARA 1942 Hawaiian Seacoast Artillery Project" Batteries, Armament, Manning Personnel and Coordinates of Batteries in H.S.A. Command, September 6, 1942 Subject: "Names of Battery Positions, 1942".

(12)    NARA, Archives II, RG 494 UD-UP 118 "Correspondence Pertaining to Proof Firing of the 8"/55 Caliber Twin Mount Navy Guns at Opaeula, August 10. 1942" and "Names of Battery Positions, 1942" and "Correspondence Pertaining to Location of Navy Eight-Inch Turret Guns Hawaiian Seacoast Artillery Command", July 30, 1942.

(13)    NARA, Archives II, RG 494 UD-UP 118 "Correspondence Regarding Location Navy 8-inch Turret Guns", March 30-April 5, 1942.

(14)    *Historical Review Engineers Pacific Ocean Area* 323-26.

(15)    NARA, Archives II, RG 494 UD-UP 118 "Layout of Battery Kirkpatrick and Vicinity, 13 January 1947".

(16)    Campbell, *Naval Weapons,* 127.

(17)    Lewis and Kirchner "Oahu Turrets", 282.

(18)    *Historical Review Engineers Pacific Ocean Area* 326.

(19)    Lewis and Kirchner "Oahu Turrets", 282-83.

(20)    *Historical Review Engineers Pacific Ocean Area* 327.

(21)    NARA, Archives II RG165 Entry 257 File 471.45, April 4, 1943.

(22)    *Historical Review Engineers Pacific Ocean Area* 328-29.

(23)  *Historical Review Engineers Pacific Ocean Area* 329-30.

(24)  *Historical Review Engineers Pacific Ocean Area* 328-29.

(25)  "Project for Kaneohe Bay, March 1942", Annex B, Fire Control Installations.

(26)  NARA, Archives II Cartographic Section, Battery Pennsylvania, 14" Naval Guns, Mokapu Point, Oahu, T.H. General Layout and Vicinity Map, U.S. Engineers Office, Honolulu, T.H., May, 1944.

(27)  *Historical Review Engineers Pacific Ocean Area* 328-329. NARA, Archives II, RG 494 Entry UD-UP 118, December 22, 1942, Subject: "Report of Conference, 14" Naval Gun Turrets".

(28)  NARA, Archives II, RG 494 Entry UD-UP 118, July 16, 1942, Subject: "Selection of Sites for Navy Guns".

(29)  NARA, Archives II Cartographic Section, Battery Arizona, 14" Naval Guns, Mokapu Point, Oahu, T.H. General Layout and Vicinity Map, U.S. Engineers Office, Honolulu, T.H., May, 1944.

(30) NARA, Archives II, RG 494 Entry UD-UP 118, December 6, 1944 Subject: "Status of Batteries Arizona and Pennsylvania.

(31)  NARA, Archives II, RG 494 Entry UD-UP 118, June 21, 1945, Subject: "Batteries Arizona and Pennsylvania, Removal of Construction Service Priority".

## CHAPTER 11

(1)  Emanuel Raymond Lewis, *Seacoast Fortifications of the United States: An Introductory History* (Washington DC: Smithsonian Institution Press, 1970).

(2)  *Seacoast Artillery Weapons* 104-113. War Department. *Standard Artillery and Fire Control Materiel.* (War Department Technical Manual TM 9-2300) (Washington D.C.: GPO 1944), 138-39.

(3)  *Historical Review Engineers Pacific Ocean Area* 315.

(4)  Ibid. 310-14.

(5)  *Seacoast Artillery Weapons* 104-113. War Department. *Standard Artillery and Fire Control Materiel.* (War Department Technical Manual TM 9-2300) (Washington D.C.: GPO 1944), 138-39.

(6)  R. T. Ward, "Discussion: Military Survey of Oahu," Professional Memoirs, Corps of Engineers, U.S. Army, (Washington D.C.:1914). Conklin, Historical Sketch of the Defense of Oahu 13-16. (14-9) NARA, Archives II, RG 494 Entry UD-UP 118, May 16, 1944, Subject: "Seacoast Artillery Gun Position Construction Program".

(7)  Dennis M. Devaney, Marion Kelly, Polly Jae Lee, and Lee S. Motteler, *Kaneohe: A History of Change* (Honolulu, HI: Bess Pr. Inc, 1982) 113-15.

(8)  *Report of the Need of Additional Naval Bases to Defend the Coasts of The United States, Its Territories, and Possessions.* House of Representatives Document No. 65, 76th Congress, 1st Session. 25.

(9)  *Building the Navy's Bases in World War II: The History of the Bureau of Yards and Docks and the Civil Engineer Corps 1940-1946*, Vol II (Washington, DC: GPO, 1947) 138.

(10)  Coletta, *United States Navy Bases, Domestic,* 250-53.

(11)  NARA, Archives II, RG 407 History of the Hawaiian Seacoast Artillery Command, March 17, 1943.

(12)  NARA, Archives II, RG 494 Entry UD-UP 118, April 8, 1941, Subject: "Defense of Naval Air Station, Kaneohe Bay, Oahu, T.H.".

(13)    NARA, Archives II, RG 494 Entry UD-UP 118, April 14, 1941, Subject: "Protection of seacoast defense batteries." *PH Investigation* Part 24, 1869-72.

(14)    NARA, Archives II, RG 407 July 22, 1941 "War Garrison for Initial War Operations, Hawaiian Department".

(15)    "Project for Kaneohe Bay, March 1942" Annex B, Fire Control Installations.

(16)    NARA, Archives II, RG 494, Entry UD-UP 118, June 27, 1942, "Construction of Harbor Defenses of Kaneohe Bay".

(17)    *Historical Review Engineers Pacific Ocean Area* 312-14.

(18)    *Ibid.* 314-16.

(19)    *Ibid.* 314-16.

(20)    "Project for Kaneohe Bay, March 1942" Land Required for Fortifications of the Harbor Defenses of Kaneohe Bay.

(21)    NARA, Archives II, RG 494 Entry UD-UP 118, June 27, 1942 Construction of Harbor Defenses of Kaneohe Bay.

(22)    *Historical Review Engineers Pacific Ocean Area* 318-19.

(23)    "Project for Kaneohe Bay, March 1942", Annex B, Fire Control Installations.

(24)    "History of the Area Artillery Officer", 12.

(25)    "Morning Reports", Battery B, 41st Coast Artillery, May 29, 1944. Battery C, 41st Coast Artillery Regiment, December 1, 1943 - May 19, 1944.

(26)    *Historical Review Engineers Pacific Ocean Area* 318-20.

(27)    NARA, Archives II, RG 494 Entry UD-UP 118, May 31, 1945, Subject: "Study of Seacoast Battery Requirements Hawaiian Islands".

(28)    "Morning Reports", 57th CA (TD) Regiment, 1940-1944. "Morning Reports", 852nd CA (SL) Battery (Separate) 1944-1945.

(29)    NARA, Archives II, RG 494 UD-UP 118 File 43101, "Modernization of Seacoast Artillery Oahu, T.H.".

(30)    Historical Review Engineers Pacific Ocean Area 318.

(31)    War Department. Standard Artillery and Fire Control Materiel. (War Department Technical Manual TM 9-2300) (Washington D.C.: GPO 1944) 128-29.

(32)    "AA Defense of Oahu," unpublished report 53rd AAA Brigade, Hawaiian Antiaircraft Artillery Command (HAAC) (Fort Shafter, Hawaii), 12.

## CHAPTER 12

(1)    "NARA 1942 Hawaiian Seacoast Artillery Project".

(2)    Historical Review, Corps of Engineers United States Army, Pacific Ocean Area, Vol. I, 306-310. Hereafter: Historical Review Engineers Pacific Ocean Area.

(3)    William C. Gaines "The Fifteenth Coast Artillery Regiment 1924-1944", Coast Defense Study Group Journal, Vol. 8, No. 2 (May, 1994), 39-48. Hereafter: Gaines, "Fifteenth Coast Artillery Regiment".

(4)   Interview of Cordell Addison held at the Abraham Lincoln Presidential Library and Museum, https://presiden-tlincoln.illinois.gov/oral-history/collections/addison-cordell/interview-detail/.

(5)   NARA, Archives II, RG 494 Entry UD-UP 118 File 380/31, October 30, 1943, Subject: Deficiencies in Sea-coast Armament.

(6)   "Fort DeRussy Expands as Army Men Recreation Center," Honolulu Star Bulletin, August 22, 1942.

(7)   William C. Gaines, "Antiaircraft Defense of Oahu, 1916-1945," Coast Defense Journal, Vol. 15, No. 2 (May 2001) 22-67. Hereafter: Gaines, "Antiaircraft Defense of Oahu".

(8)   NARA, Archives II, RG 494 "History of the Area Artillery Officer", Headquarters, Army Forces, Middle Pacif-ic, 25 January 1942, to 15 October 1945, Fort Shafter, Honolulu, T.H.  54-55. Hereafter: "History of the Area Artillery Officer".

(9)   Organizational Records Unit, Military Personnel Records Section, National Personnel Records Center, St. Lou-is, MO. Hereafter: "Morning Reports" with unit designation. "Morning Reports", 55th CA (TD) Regiment, 1940-1944.

(10)   NARA, Archives II, RG 156 U.S. War Department Ordnance Department, Research and Development Section Monograph Series: The Record of Army Ordnance Research  and Development, "History of Development of 16-inch Gun Material used for Seacoast Defenses".

(11)   Historical Review Engineers Pacific Ocean Area 306-07.

(12)   PH Investigation Part 24, 1869, Subject: Protection of Seacoast Defense Batteries. Historical Review Engineers Pacific Ocean Area 307-08.

(13)   Historical Review Engineers Pacific Ocean Area 306-308.

(14)   NARA, Archives II, RG 494, Entry UD-UP 118, February 8, 1942 "Emergency Projects Being Carried out Under Supervision of U.S. Engineer Department". 8. Peter T. Young, "Ammo Tunnels", https://imagesofoldha-waii.com/ammo-tunnels/.

(15)   NARA, Archives II, RG 165 Entry 281, File WPD 789-4.

(16)   NARA, Archives II, RG 494, Entry UD-UP 118, February 8, 1942 "Emergency Projects Being Carried out Under Supervision of U.S. Engineer Department". 8. Peter T. Young, "The Hole" https://imagesofoldhawaii.com/the-hole/.

(17)   Thomas H. Green, Maj. Gen., "Martial Law in Hawaii", TJAGS unpublished manuscript. Hereafter: Green, "Martial Law in Hawaii".(18) U. S. Navy, 14th Naval District, Administrative History of the Fourteenth Naval District, and the Hawaiian Sea Frontier - World War II (Washington D.C.: U.S. Navy Historical Center).

(19)   Historical Review Engineers Pacific Ocean Area, Vol. II 551-54.

(20)   Gaines, Redbook. 176.

(21)   John D. Bennett, "World War II Machine-Gun Pillboxes, and Coast Defense and Other Stations in the Hawai-ian Islands", Coast Defense Journal, Vol. 22 No. 2 (May 2008).

(22)   "Army to Use Thousands of Armored Pillboxes," The Honolulu Advertiser, June 21, 1941.

(23)   Green, "Martial Law in Hawaii".

(24)   ibid.

CHAPTER 13

(1)     NARA, Archives II, RG 494 Entry UD-UP 118, "Provision of three (3) Panama Mount 155mm Batteries and one (1) 8-inch railway gun battery position for the Defense of Kaneohe Bay, Oahu", September 18, 1941.

(2)     NARA, Archives II, RG 494 Entry UD-UP 118, "Location of Batteries on KNAS".

(3)     Andrew W. Clement, "Seacoast Artillery Radar", Coast Artillery Journal, Vol. 91, No. 3 (May-June, 1948), 9. Danny R. Malone, "The 296A Radar," Coast Defense Study Group Journal Vol. 5, No. 2 (May, 1991) 18-25.

(4)     "Morning Reports", Headquarters Battery, 98th Coast Artillery Regiment, December 7, 1941. "Morning Reports", Battery G, 98th Coast Artillery Regiment, December 1-7, 1941.   Gordon Prange, At Dawn We Slept: The Untold Story of Pearl Harbor, (New York: 1981) 519.

(5)     "Project for Kaneohe Bay, March 1942", Annex B, Fire Control Installations.

(6)      Site Survey Report, Harbor Defense Command Post-Radar Operating Room Tunnels, Kaneohe Marine Corps Air Station, May 28, 1993. SCARP William C. Gaines.

(7)     "History of the Area Artillery Office", 12.  Danny R. Malone, "Seacoast Artillery Radar 1938-46," Coast Defense Study Group Journal, Vol. 3, No. 4 (November, 1989), 5-7.  Andrew W. Clement, "Seacoast Artillery Radar", Coast Artillery Journal Vol. 91, No. 3 (May-June, 1948), 9.

(8)     NARA, Archives II, RG 494 Entry UD-UP 118, November 8, 1941, "New Construction Base  Kaneohe, Oahu, T.H.".

(9)     NARA, Archives II, RG 494 Entry UD-UP 118, November 10, 1941, "New Construction Ulupau Military Camp".

(10)    NARA, Archives II, RG 494 Entry UD-UP 118, December 13, 1941, "Status of Projects at Ulupau".

(11)     "Project for Kaneohe Bay, March 1942", AMTB Defense, Kaneohe Bay, Plan of S/L Illumination Arc 10,000 Yd Radius, Chart Number 4. Overlay, Harbor Defenses of Kaneohe Bay, November 9, 1942.

(12)    NARA, Archives II, RG 494 Entry UD-UP 118, Fortification File 662/5, Engineer Files, "New 16-inch Guns Batteries: Site Selection and Land Acquisition for North Shore Battery, Oahu, 1936".

(13)    "Hawaiian Project of 1940".

(14)    "History HSAC Brigade".

(15)    NARA, RCW Battery Kahuku.

(16)    "Morning Reports", Headquarters and Headquarters Battery and 1st and 3rd Battalions, 57th Coast Artillery Regiment, January, 1942.

(17)    NARA, Archives II, RG 494, Entry UD-UP 118, "Construction of North Shore Railroad Connection" February 19, 1941, "Railway Now Joins Waialua, Wahiawa," The Honolulu Advertiser, December 24, 1941, 4.

(18)    NARA, Archives II, RG 494 Entry UD-UP 118, "Railway Equipment for the 8-inch Railway Artillery Units - Hawaiian Department" and inds. April 30, 1942 and, "Rolling Stock - Railway Artillery" September 23, 1942.

(19)    NARA, Archives II, RG 494 Entry UD-UP 118, "Installation of 8" Railway Guns at Kahuku and Ulupau," February 3, 1942.

(20)    NARA, Archives II, RG 494 Entry UD-UP 118, "Transfer of Completed Work, Vicinity of Kahuku, Oahu, T.H.," February 11, 1943.

(21)  NARA, Archives II, RG 494, Periodic Reports, HSAC, Jan.-May 1943.

(22)  "Morning Reports", 55th Coast Artillery (Harbor Defense) Battalion 1944-1945, "Morning Reports", 56th Coast Artillery (Harbor Defense) Battalion 1944-1945, "Morning Reports", 606th Coast Artillery (Harbor Defense) Battery (Separate) 1945-1946.

(23)  "Morning Reports", 809th Coast Artillery (Harbor Defense) Battery (Separate) May, 1942 – May 1943; 810th Coast Artillery (Harbor Defense) Battery (Separate) May, 1942-May, 1943, "Morning Reports", Battery B, 41st Coast Artillery (Railway) Regiment January 1942-May 1942; "History of the Area Artillery Officer", 11.

(24)  NARA, Archives II, RG 494 Entry UD-UP 118, September 6, 1941.

(25)  "Morning Reports", Battery E, 54th Coast Artillery (Harbor Defense) Battalion. NARA. Archives II, RG 407 Record of Events, January 1, 1943 to January 7, 1945

(26)  "History of the 2274th HSAC".

(27)  Pearl Harbor Attack: Hearings before the Joint Committee on the Investigation of the Pearl Harbor Attack. (Washington, D.C.: Government Printing Office, 1946) Part 22, 170-71, Subject: Testimony of Major General Henry Burgin before the Roberts Commission. Hereafter: PH Investigation, part and subject.

(28)  Ibid., Part 28, 1356, 1362. Part 22, 164-42. Part 28, 1355-87.

(29)  Ibid., Part 28, 1355-87 Testimony of Major General Henry Burgin before the Army Pearl Harbor Board. Pt. 22, 159-63 Testimony of Major General Maxwell Murray before the Proceedings of the Roberts Commission.

(30)  Shelby L. Stanton, Order of Battle U.S. Army World War II, (Novato, CA: Presidio Press, 1994) 472.

(31)  Gaines, "Antiaircraft Defense of Oahu" 66.

(32)  "History of the Area Artillery Officer" 15. Hawaiian Antiaircraft Artillery Command, Scrapbook, (Fort Shafter,HI: 1946).

(33)  Bennett, "Antiaircraft" 16-17.

(34)  "NARA 1942 Hawaiian Seacoast Artillery Project". NARA, Archives II, RG 494 Entry UD-UP 118, December 30, 1941, Subject: "Urgent War Department Construction Engineers General Correspondence".

(35)  NARA, Archives II, RG 494 Entry UD-UP 118, September 6, 1941.

(36)  Historical Review Engineers Pacific Ocean Area Vol. II 554-56.

## CHAPTER 14

(1)  Mark A. Berhow, ed. American Seacoast Defenses, A Reference Guide, Second Edition (McLean, VA: The CDSG Press, 2004) 265-79.

(2)   NARA RCW, Fort Kamehameha, Fire Control Station F"/3, corrected to March 31, 1919 and Fort Kamehameha, Fire Control Station B'/3 Pertaining to Battery Jackson, corrected to March 31, 1919 and Fort Kamehameha, Fire Control Station F'/4, corrected to March 31, 1919 and Fort Kamehameha, Fire Control Station B'/4 Pertaining to Battery Selfridge, corrected to March 31, 1919.

(3)  NARA RCW, Fort Kamehameha, Fire Control Station F""/3, Base End Station, Ahua Point, East Station, corrected to March 31, 1919 and Fort Kamehameha, Fire Control Station F"/4, Base End Station, Ahua Point, East Station, Next to F""/3, corrected to March 31, 1919 and Fort Kamehameha, Fire Control Station B"/3 Pertaining to Base End Station Battery Jackson, at Ahua Point, corrected to March 31, 1919 and Fort Kamehameha, Fire Control Station B"/4 Pertaining to Battery Selfridge, corrected to March 31, 1919.

(4)    NARA RCW, Fire Control Station "S," Kaena Point, Oahu, T.H. corrected to February 20, 1932.

(5)    Dorrance, Fort Kamehameha, 88.

(6)    "Hawaiian Project of 1940".

(7)    NARA RCW Fire Control Station "S", Kaena corrected to December 5, 1934 and  Fire Control Station "O", Pupukea, for Batteries Hatch and Williston, Forts Barrette and Weaver, T.H., corrected to December 5, 1934 and  Fire Control Station "B," Puu Palailai for Battery Hatch, Fort Barrette, T.H., corrected to July 14, 1934 and  Fire Control Station "C," Salt Lake for Battery Hatch, Fort Barrette, T.H., corrected to July 14, 1934 and Fire Control Station "S'," Kepuhi, for Battery Hatch, Fort Barrette, T.H., corrected to December 5, 1934 and Fire Control Station "U," Puu O-Hulu For Battery Hatch Fort Barrette, T.H., corrected to July, 14, 1934 and  Fort Barrette, T.H. Plotting and Switchboard Rooms for Battery Hatch, corrected to November 25, 1935 and Battery Hatch, corrected to November 25, 1935 and  Fort Barrette, Power for Battery Hatch, corrected to November 25, 1935.

(8)    John D. Bennett, "Oahu's Command and Fire Control Cable System", Coast Defense Journal Vol. 16, No. 4, (November 2002), 42-54.

(9)    NARA, Archives II, RG-77 Entry 1008 Decimal 675, January 25, 1943 Subject: Power for Harbor Defense End Stations.

(10)    NARA, Archives II, RG-77 Entry 1008 Decimal 675, File 43653, March 1, 1943, Subject: Fire Control Base End Stations 1 March 1943. Hereafter: NARA "Base End Stations March 1943".

(11)    Mark A. Berhow, ed. American Seacoast Defenses, A Reference Guide, Second Edition. (McLean, VA: The CDSG Press, 2004) 377-92.

## CHAPTER 15

(1)    "Morning Reports", 16th CA (HD) Regiment 1942-1943. "Morning Reports", 15th CA (HD) Regiment 1942-1943 "Morning Reports", 750th AAA Bn. 1943-1944.

(2)    NARA. Archives II, RG 407 "History of the 2274th Hawaiian Seacoast Artillery Command 1945". Hereafter: "History of the 2274th HSAC".

(3)    "History of the Area Artillery Officer" 54.

(4)    "Morning Reports", Battery A, 56th Coast Artillery Battalion, 1944. "Morning Reports", Battery E, 54th Coast Artillery Battalion, 1944. "History of the 2274th HSAC".

(5)    "History of the 2274th HSAC", "Morning Reports", 56th Coast Artillery (Harbor Defense) Battalion. "Morning Reports", 606th Coast Artillery (Harbor Defense) Battery. "Morning Reports", 610th Coast Artillery (Harbor Defense) Battery.

(6)    NARA, Archives II, RG 494 OCCAS 660.2/63a, June 6, 1945, Subject: "Postwar Seacoast Artillery Study Hawaiian Group".

(7)    NARA, Archives II, RG 494, Entry UD-UP 118, March 25, 1946, Subject: "Status Battery Williston".

(8)    Historic American Buildings Survey, "U.S. Naval Air Station Kaneohe, Oahu, Administration and Operations Building" http://lcweb2.loc.gov/master/pnp/habshaer/hi/hi1000/hi1012/data/hi1012data.pdf.

# Index

*Ship name*s are *italicized*

## A

Adair, Henry R. 354

Ahua Point 12, 14, 27, 30-1, 61, 63, 69, 74, 80, 103-7, 135, 155, 188, 205-6, 276, 284, 319-23, 326, 332-3, 339, 342, 359, 363, 367, 371, 374, 385

Aiea Heights 74, 187, 199, 203, 371

*Akagi* 178-9

Alexander Young Hotel 156

Alexander, Barton S. 3, 5

Aliamanu 69, 79, 102, 141, 156, 176-7, 187-8, 214, 222-3, 241, 279, 284-7, 316, 326, 348, 363, 368, 380

Aliamanu Alternate Command Center 177, 286

Aliamanu Army-Navy Joint Operating Center 286

Aliamanu Ordnance Storage Facility 156, 176-7, 188

AMTB (anti motor torpedo boat batteries) 272-3, 276, 278, 347, 371, 365, 378-9

Anahulu Flats 148, 225, 304, 307, 363, 369,

Annexation (1898) 3-5, 11

Antiaircraft Artillery Intelligence Center 288

Antiaircraft defenses 10, 88-108, 124, 1349, 149-50, 154-6, 159-60, 164, 166, 178-9, 183, 185-93, 197-8, 200, 202, 210, 213, 250, 268, 272, 274, 276, 278, 287-8, 293, 299, 311-2, 314, 316, 347-9, 359-60, 378

Antiaircraft Project (1916) 88-91

*Arizona* 198, 228-30, 234, 242, 327

Armstrong, Samuel Chapman 31

Army Aviation 155, 248, 284, 349, 355, 390

Artillery District of Honolulu 31, 41, 59

Artillery District of Pearl Harbor 31

Ashley Military Reservation 140, 305, 311, 331, 363, 390

Awanui 269, 361, 363

## B

Baker, Newton D. 100

Balloons 30, 66, 88

Balloons, Barrage 176, 276

Barbers Point 56, 101, 103, 140, 183, 188, 332, 363, 369, 390

Barking Sands 374, 389

Barrette, John D. 114-5

Base end station list 330-33 (see also fire control)

Batteries, Gun

Battery Adair 75-9, 352, 354

Battery Ahua 198, 205-6, 276, 363, 371, 374

Battery AMTB No. 1 378

Battery AMTB No. 2 378

Battery AMTB No. 3 378

Battery AMTB No. 4 378

Battery AMTB No. 7 378

Battery AMTB No. 8 365

Battery Arizona 209, 228-30, 234, 240-2, 284, 346-7, 352

Battery Barri 71-4, 275-6, 349-50, 354

Battery Birkhimer 63-5, 80-3, 86, 124, 187, 324, 328, 344-5, 351, 354

Battery Boyd 75-7, 352, 354

Battery Burgess 223, 353, 357

Battery Chandler 71-7, 203, 276, 349-50, 355

Battery Closson 91, 93, 105-7, 124, 135, 150, 152-4, 177, 183, 197, 248, 274-6, 279-80, 309, 321, 324, 328, 346-7, 350, 355, 359, 380-1, 384

Battery Construction No. 301 (see Battery French)

Battery Construction No. 302 (see Battery Cooper)

Battery Construction No. 303 209, 246, 267-8, 303, 347

Battery Construction No. 304 246, 269-70, 336, 347, 381

Battery Construction No. 305 246, 270, 347

Battery Construction No. 405 (see Battery DeMerritt)

Battery Construction No. 407, 264, 265-6, 345, 347

Battery Construction No. 408 246, 265, 347

Battery Construction No. 409 246, 266-8, 347, 384

Battery Cooper (No. 302) 246, 251, 254, 258-60, 264, 331, 336, 347, 352, 354, 383

Battery DeMerritt (No. 405) 45-6, 251, 259, 260-5, 267, 336, 347, 352, 358, 382

Battery Dillingham 141, 204-5, 311, 363, 372

Battery Dodge 80-6, 204, 345, 351, 355, 373

Battery Dudley 21, 44-8, 58, 156-7, 164, 186, 270, 277-8, 323, 330, 343, 349, 351, 355

Battery East 363-4

Battery East Beach 364

Battery Ewa 198-201, 364, 371

Battery French (No. 301) 246, 251-7, 264-5, 336, 352, 356

Battery Granger Adams 118-24, 164, 166, 169, 279, 293, 330, 336, 349, 351, 356, 359, 386

Battery Harbor 208-10, 317, 349, 377

Battery Harlow 51-62, 64, 67, 166, 177, 323, 330, 344-5, 351, 356, 380-1

Battery Hasbrouck 18-24, 27-8, 30, 152, 155, 275, 316, 322-3, 332, 350, 356, 359, 385

Battery Hatch 117-20, 135, 169, 177, 197, 209, 241, 279-83, 304, 309, 324-9, 333, 336, 346-7, 352, 356, 360, 380-4

Battery Hawkins 21, 23-7, 34, 74, 152-4, 194, 275-6, 322, 350, 357, 364, 385

Battery Homestead 141, 206-7, 209, 301, 364, 374, 389

Battery Hulings 80-3, 86, 345, 351, 357

Battery Hulu 208-9, 241, 268, 327, 336, 376

Battery Jackson 21-3, 26-7, 30, 152, 154, 275-6, 319, 322-3, 350, 357

Battery Kaena 141, 204-5, 315-6, 364, 372

Battery Kahana 414, 205, 301, 364, 374-5

Battery Kaneohe 203-4, 373

Battery Kirkpatrick 227-8, 243, 246, 253, 257

Battery Loko 141, 301, 331, 365

Battery Nanakuli 141, 205, 209, 365, 375-6

Battery Oneula 205-6, 367, 375, 384

# G

# H

# I

# J

The Coast Defense Study Group, Inc. (CDSG) is a tax-exempt corporation dedicated to study of seacoast fortifications. CDSG's purpose is to promote and encourage the study of coastal defenses, primarily but not exclusively those of the United States of America. The study of coast defenses and fortifications in- cludes their history, architecture, technology, strategic and tactical employment and evolution. The primary goals of the CDSG are the following:

- Educational study of coast defenses
- Technical research and documentation of coast defenses
- Preservation of coast defense sites, equipment and records for current and future generations
- Accurate coast defense site interpretations
- Assistance to groups interested in preservation and interpretation of coast defense sites
- Charitable activities which promote the goals of the CDSG

Membership is open to any person or organization interested in the study or history of the coast defenses and fortifications. Membership in the CDSG will allow you to attend the annual conference, special tours and receive the CDSG's quarterly journal and newsletter. For more information on the CDSG, please visit the CDSG website at cdsg.org or contact us at 24624 W. 96th Street, Lenexa, KS 66227-7285 USA, Attn: Quentin Schillare, Membership.

The CDSG Fund supports the efforts of the Coast Defense Study Group by raising funds for preservation and interpretation of American seacoast defenses. The CDSG Fund is seeking donations for projects sup- porting its goals. Donations are tax-deductible for federal tax purposes as the CDSG is a 501(c)(3) orga- nization, and 100% of your gift will go to project grants. Major contributions are acknowledged annually. The Fund is always seeking proposals for the monetary support of preservation and interpretation projects at former coast defense sites and museums. A one-page proposal briefly describing the site, the organization doing the work, and the proposed work or outcome should be sent to the address below. Successful proposals are usu- ally distinct projects rather than general requests for donations. Upon conclusion of a project a short report suitable for publication in the CDSG Newsletter is requested. The trustees shall review requests and pass their recommendation onto the CDSG Board of Directors for final approval. Send donations and grant requests to: CDSG Fund c/o Terry McGovern 1700 Oak Lane McLean, VA 22101- 3326 USA or use your credit card via PayPal on the cdsg.org website.

## The CDSG ePress
### The CDSG Press

CDSG Books and CDSD Gear ($ domestic / $ International) prices include domestic/international postage $US currency only (cash, check, money order or credit card via PayPal at the cdsg.org store), allow 6-8 weeks for delivery.

- *Notes on Seacoast Fortification Construction* by Col. Eben E. Winslow, 1920, 428 pp. 1994 reprint HC with drawings $45/$60
- *Seacoast Artillery Weapons Technical Manual* (TM) 9-210 by U.S. War Dept. 1944, 202 pp. 1995 reprint PB $25/$35
- *The Service of Coast Artillery* by F. Hines & F. Ward, 1910, 736 pp. 1997 reprint HC $40/$60
- *Permanent Fortifications & Sea-Coast Defences* by U.S. Congress, 1862, 544 pp. 1998 reprint HC
- $30/$45
- *American Coast Artillery Matériel* Ordnance Dept. Doc#2042 by U.S. War Dept., 1922, 528 pp., 2001 reprint HC $45/$65
- *American Seacoast Defenses: A Reference Guide* (3rd Edition) by Mark A. Berhow, (2015) 732 pp. HC $45/$80
- The Endicott & Taft Board Reports, reprint of original reports of 1886 and 1905 by U.S. Congress, 525 pp. 2007 reprint HC $45/$80
- *Artillerists and Engineers: The Beginnings of US Fortifications 1794-1815* by Col. Wade, U.S. Army. PB, 226 pp. $25/$40
- *WWII Harbor Defenses of San Diego,* 2021 by H.R. Everett, available from Amazon, $40 list price, ($50 if ordered from CDSG, domestic shipping only).
- CDSG Logo Hats each $20.00 domestic and $25.00 foreign. CDSG Logo Patches each $ 4.00 domestic & foreign.
- CDSG T-Shirts (XXXL, XXL, XL, L; Red, Khaki, Navy, Black) $18.00 Domestic and $26.00 Foreign.

Send order to: CDSG Press Attn: Terry McGovern 1700 Oak Lane, McLean, VA 22101-3326
Or order via our online store at cdsg.org.

### The CDSG Digital Library

The CDSG has digitized an extensive set of historic manuals, reports, records and documents on the harbor defenses of the United States Army.

- The CDSG provides back issues of the CDSG Publications (from 1985) in electronic format.
- The CDSG Documents covers a range of historical material related to seacoast defenses -- most are from the National Archives. Included are the annual reports of the chief of coast artillery and chief of engineers; several board proceedings and reports; army directories; text books; tables of organi- zation and equipment; WWII command histories; drill, field, training manuals and regulations; ordnance department documents; ordnance tables and compilations; and the ordnance gun and carriage cards.
- CDSG Documents related to specific harbor defenses. These PDF documents form the basis of the conference and special tour handouts that have been held at these locations. They include RCBs/RCWs; maps; annexes to defense projects; CD engineer notebooks; quartermaster building records; and aerial photos taken by the signal corps 1920-40. Please consult cdsg.org for more details.

Information on the CDSG ePress items can be obtained from Mark Berhow at berhowma@cdsg.org or by post to PO Box 6124, Peoria, IL 61601 USA, or at cdsg.org.

## McGovern Publishing Presents:

McGovern Publishing is comprised of two divisions: Redoubt Press (military titles) and Three Sisters Press (Rebecca, Rachel, and Alana) offering a range of subjects. McGovern Publishing is interested in new titles, especially those dealing with fortifications, please contact Terry McGovern at 703/538-5403 or at tcmcgovern@att.net if you have a title that you are seeking to have published.

Visit our website at www.mcgovernpublising.com, or post to 1700 Oak Lane, McLean, Virginia 22101-3326 USA

**Redoubt Press**
A Division of McGovern Publishing

THREE
SISTERSPRESS
A Division of McGovern Publishing

### *The Chesapeake Bay at War!*
The Coastal Defenses of Chesapeake Bay During World War II
by Terrance McGovern

The defense of America's seacoast has been one of key concerns since the earliest years of the Republic. American coastal defenses steadily evolved through the age of muzzle loading cannon, ever larger breech loading weapons, and finally to the culmination of large, long range guns capable of targeting the largest and most heavily armed warships of their age. By the end of World War II, the United States had some of the strongest defenses in the world. Given the importance of the U.S. naval bases around Norfolk, Virginia and the shipyards of Hampton Roads, the seacoast defenses protecting Chesapeake Bay contained the largest collection of firepower in the continental United States as they reached their apex during World War II.

This book tells the story of preparing the coastal defenses of the Chesapeake Bay for the coming of World War II and their operations from 1941 to 1945. Over a hundred rare black & white U.S. Army photographs and plans help document our nation's extensive efforts to defend against naval attacks and raids from Nazi Germany. A collection of over 50 recent color aerial photographs are also included allowing the reader to survey the surviving elements of these mighty defenses. A product of extensive research this book brings together rare images and the little-known military history of the Chesapeake Bay for the first time. Three Sisters Press (Rebecca, Rachel, and Alana) is pleased to offer this 70-page softbound book for $30 plus $5 for domestic shipping or $10 for foreign shipping.

### *American Defenses of the Panama Canal*
By Terrance McGovern.

The end of 20th Century has brought to a close the American involvement in the Panama Canal, a tremendous technological achievement that was started in the early days of the century. The Panama Canal took over 10 years to complete and involved the expenditure of millions of dollars and thousands of worker's lives. The importance of the Panama Canal to commerce and naval forces resulted in fortifications that match its size and cost. While the history of the Panama Canal has been recorded in numerous books and articles, the history of its defenses has not. This book hopes to remedy this situation by exploring the thirty-five years of fortification construction and use in the Panama Canal Zone.

The first part of this book discusses how the various fortification boards decided, with the help of Congress, on what defenses were needed and how evolving threats resulted in changes to these defenses. The second part of the book describes in detail the construction, service life, and current status of each major battery. Assisting this narrative are many current and period photographs, along with engineering drawings and site maps detailing the construction of these impressive fortifications. All serious students of seacoast defenses and the Panama Canal should have a copy.

This title was originally published in U.K based Fortress Study Group annual journal, *FORT: The International Journal of Fortification and Military Architecture* (Volume 26: 1998). This 116-page book is softbound using the same high- quality paper and printing process as *FORT.* As the most detailed and illustrated article to ever be produced on these defenses, the book has 45 maps and drawings, 43 historical B&W photographs, and 40 color photographs (taken in 1993 and 1999 by the author). This book is offered at a price of $50 plus $5 for domestic shipping or $10 for foreign shipping.

### *The Concrete Battleship*
Fort Drum, El Fraile Island, Manila Bay
by Francis J. Allen

Fort Drum on El Fraile Island in the Philippines is unique in the development of United States coastal fortifications. Fort Drum is part of a chain of forts built across the entrance of Manila Bay to defend the Bay from naval attack. The construction of Fort Drum began in 1909 by reducing tiny El Fraile Island to the low water mark. Over the next ten years a multi-deck concrete island was built to mount two twin 14-inch guns in superimposed Army designed armored turrets. The completed work rises 40 feet above sea level, it is 350 feet long and 144 across at its widest point. The exterior walls are up to 28 feet thick and the top deck attains a thickness of 20 feet of re-enforced concrete. The interior of the fort held a large engine room, powder and shell magazines, a mining casemate, storerooms and tankage, a accommodations for 300 personnel. The design of the fort followed a naval pattern with turrets, a cage mast, and secondary armament in side casemates. Due to these characteristics, Fort Drum became known as the "Concrete Battleship."

When completed in 1918, Fort Drum was the most powerful defense work in Manila Bay, but the advances in military technology during World War I already began to make the fort obsolete. The post-World War I reduction in military spending, the re- strictions of the Washington Naval Treaty of 1922, and economic depression of the 1930s resulted in Fort Drum being quickly reduced to caretaker sta- tus until the coming of World War II. Fort Drum became an important weapon during the Japanese siege of Corregidor and the other island forts during 1942 but only play a minor role during the American retaking of these islands in 1945. The battles of World War II would transform Fort Drum from an American-manned, fully operating fort to a burned-out hulk inhabited by lifeless Japanese sailors.

This revised and enlarged 64-page softbound volume tells the story of The Concrete Battleship in words, diagrams, and photographs from its inception to the present day. Redoubt Press is pleased to offer this book for $30 plus $5 for domestic shipping or $10 for foreign shipping.

### *The Delaware Bay at War!*
### The Coastal Defenses of Delaware Bay During World War II
by Terrance McGovern

The defense of America's seacoast has been one of the key concerns since the earliest years of the Republic. American coast defense steadily evolved through the age of muzzle loading cannon, ever larger breech loading weapons, and finally to the culmination in large, long range guns capable of targeting the largest and most heavily armed warships of their age. By the end of World War II, the United States had some of the strongest defenses in the world. Given the importance of the military- industrial complex along the banks of the Delaware River, including the large Philadelphia Naval Shipyard, the seacoast defenses protecting Delaware Bay had declined dramatically since the turn of  the century resulting in a whole program of modern coast artillery batteries and other defenses to be constructed starting in the late 1930's and reaching their reached their apex during the middle of World War II.

This book tells the story of preparing the coastal defenses of the Delaware Bay for the coming of World War Two and their operations from 1941 to 1945. Over a hundred rare black & white U.S. Army photographs and plans help document our nation's extensive efforts to defend against naval at- tacks and raids from Nazi Germany. A collection of recent color aerial photography is also included allowing the reader to survey the surviving elements of these generally unknown defenses. A product of extensive research this book brings together rare images and the little-known military history of the Delaware Bay for the first time. Three Sisters Press (Rebecca, Rachel, and Alana) is pleased to offer this 65-page softbound book for $30 plus $5 for domestic shipping or $10 for foreign shipping.

### *A Legacy in Brick and Stone – 2nd Edition*
### American Coastal Defense Forts of the Third System
### 1816-1867
By John R. Weaver II

The definitive history of the American Third System of Fortifications that defended our coastline for more than half of century, these architectural wonders were built from 1816 through 1867 from Maine through the Florida Keys to New Orleans, with two forts in San Francisco Bay. Most of these 42 masonry forts still stand guard along our shores, and open to the public. *A Legacy in Brick and Stone* provides the background of these famous Civil War forts – why they were built where they are, who built them, and how they functioned – as well as descriptions of each of the forts.

This revised and expanded edition has grown to 340 pages with 400 new photographs and drawings. John Weaver II, a nationally known expert on masonry coastal fortifications, has invested over 30 years of research into *A Legacy in Brick and Stone* to produce the only a full treatment of the magnificent American Third System.

 The book begins with a study of the history of the Coastal Fortifications Board, which developed and implemented this massive defense project. It then details the art of fortification of that period and describes the particular architecture components that were key to their design. A description of the development of the system over its 50-year life is followed by an analysis of how well certain forts held up under attack during the American Civil War. Approximately two-thirds of this volume is dedicated to a fort-by-fort description of the system. The overall defense scheme for each harbor is discussed, then each fort in that harbor is analyzed. The author uses unique photographs and drawings to answer many of the questions about of these forts that today's visitors ask. The book also documents the current status of these historic forts, including information about how to visit these forts today.

Redoubt Press is pleased to offer a deluxe hardcover edition with color illustrations $64.95 plus a $5 fee for domestic shipping and $10 for foreign shipping. Paperback edition with black and white ilustrations $39.95 plus a $5 fee for domestic shipping and $10 for foreign shipping.

## Seacoast Cannon Coloring Book
### By Brian B. Chin

The big seacoast guns thunder once again. From the decorated brass cannon of Spanish New World, to the powerful iron guns in the American Civil War, to the complex disappearing guns of the turn of the century, to the huge guns of steel and armor in World War II, this detailed picture book shows how coast artillery once protected American shores from enemy attack. This second edition brings to life the coast defenses that today remain as empty forts and batteries with our national, state, and local parks. The author, Brian B. Chin, is a well-known author and artist that has produce several works on military fortifications. This coloring book allow a new generation to learn the history of American seacoast defenses while having fun coloring in this 52-page softbound book.  It is hoped that book will help generate interest in preserving and interpreting these historic sites for future generations.

Three Sister Press (Rebecca, Rachel, and Alana) are pleased to offer this coloring book for $10 plus $5 for domestic shipping or $10 for foreign shipping.

## Pacific Rampart: A History of Corregidor and the Harbor Defenses of Manila and Subic Bays
### By Glen M. Williford

This is the definitive history of the American-built forts and harbor defenses of Manila and Subic Bay in the Philippines. This a heavily illustrated work tells the history of the fortified islands (the most famous being the island of Corregidor) from the fortifications built by the Spanish and the Americans to what remains today. Compiling research from the National Archives and many primary sources this work describes in great detail the defensive plans as well as the fortifications built between 1904 and 1940. The book describes the day by day of the fighting early in the Second World War that led to the surrender of these defenses, as well as the combat engagements in early 1945 when they were retaken. Consequently, it is simultaneously a "unit" history (the Coast Defense units stationed in the islands), a weapon / technical history (the artillery in the fixed gun batteries) and a combat history (the taking and then retaking of the fortress in World War II). The text is supported with thorough referenced endnotes, a bibliography, and six appendices. It contains over 340 illustration—black and white photographs, maps, and plat diagrams for many of the fort structures. The author has invested over 30 years of research into this work, making it an important addition to the body of knowledge on these historical defenses and to the story of Corregidor.

ISBN 978-1-7323916-3-5  Hardcover edition is $49.95 plus postage & handling for single orders. This book is also available for purchase from Amazon.